# Solid Waste Technology & Management

# Solid Waste Technology
# & Management

## VOLUME 2

**Edited by**

**THOMAS H. CHRISTENSEN**

Department of Environmental Engineering, Technical
University of Denmark, Lyngby, Denmark

WILEY

A John Wiley and Sons, Ltd., Publication

This edition first published 2011
2011 © Blackwell Publishing Ltd

*Registered office*
John Wiley & Sons Ltd, The Atrium, Southern Gate, Chichester, West Sussex, PO19 8SQ, United Kingdom

For details of our global editorial offices, for customer services and for information about how to apply for permission to reuse the copyright material in this book please see our website at www.wiley.com.

*Library of Congress Cataloging-in-Publication Data*

Christensen, Thomas H.
  Solid waste technology and management / Thomas H. Christensen.
    p. cm.
  Includes bibliographical references and index.
  ISBN 978-1-4051-7517-3 (cloth : alk. paper)   1. Refuse and refuse disposal.   I. Title.
  TD791.C44 2010
  628′.744–dc22

                                                                2010007989

A catalogue record for this book is available from the British Library.

ISBN : 9781405175173

Typeset in 10/12pt Times by Aptara Inc., New Delhi, India
Printed in Malaysia by Ho Printing (M) Sdn Bhd

# Contents

# Preface

*Solid Waste Technology & Management* is an international reference book on solid waste. The book holds 11 chapters written by 78 experts from around the world.

The need for a new book on solid waste with a broad coverage of all aspects has long been recognized by many professors and professionals. However, it is impossible for a single person to be an expert in all field of solid waste and if such a person existed, he would probably not have the time to write a comprehensive book of 1000 pages. Out of this schism, the idea emerged to involve many authors with a range of expertise and making a thorough edit of the contributions that emphasize the features of the book. This book has been developed over a 4-year period by the joint effort of 78 international solid waste experts. Members of IWWG, International Waste Working Group (www.iwwg.nu) and ISWA, International Solid Waste Association (www.iswa.org) as well as many other experts have contributed their expertise with the aim of supporting education and exchanging information on solid waste technology and management.

Great effort has been made by the authors in providing the draft chapters and into homogenizing the chapters in terms of terminology, approach and style. The remaining diversity in style and unavoidable repetitions still found in the book are hopefully many times compensated by the level of expert knowledge presented in the chapters.

This book would not have been possible without the dedicated contributions by the many authors (see List of Contributors), the continued secretarial work by Dr. Thomas Astrup, Thilde Fruergaard, Grete Hansen and Marianne Bigum as well as the graphical support by Ms. Birthe Brejl. These contributions are gratefully acknowledged.

Special thanks is given to the R98-foundation, Copenhagen for its generous gift to DTU in support of the book.

Copenhagen, January 2010
Thomas H. Christensen
Technical University of Denmark

**Referring to the book:**

If you refer to the book in general, a proper way of providing the reference would be:

Christensen, T. H. (eds.) (2010): *Solid Waste Technology & Management*, John Wiley & Sons, Ltd, Chichester (ISBN: 978-1-405-17517-3).

If you refer to more specifi information presented in a specifi chapter, full credit should be given to the authors of the specifi chapter by referring to, for example:

Hauschild. M. & Barlaz, M.A. (2010): LCA in waste management: Introduction to principle and method. In Christensen, T. H. (Eds.), *Solid Waste Technology & Management, Chapter 3.1*. John Wiley & Sons, Ltd, Chichester (ISBN: 978-1-405-17517-3).

# List of Contributors

**Andersen, Lizzi**  COWI, Denmark

**Angelidaki, Irini**  Technical University of Denmark, Denmark

**Arm, Maria**  Swedish Geotechnical Institute, Sweden

**Astrup, Thomas**  Technical University of Denmark, Denmark

**Barlaz, Morton A.**  North Carolina State University, USA

**Batarseh, Eyad S.**  University of Central Florida, Orlando, USA

**Batstone, Damien John**  University of Queensland, Australia

**Beaven, Richard**  University of Southampton, UK

**Berge, Nicole D.**  University of Central Florida, Orlando, USA

**Bertoldi, Marco de**  University of Udine, Italy

**Bigum, Marianne**  Technical University of Denmark, Denmark

**Bilitewski, Bernd**  Technical University of Dresden, Germany

**Birgisdottir, Harpa**  Technical University of Denmark, Denmark

**Bisbjerg, Peder**  EP&T Consultants Sdn. Bhd., Kuala Lumpur, Malaysia

**Björklund, Anna**  Royal Institute of Technology, Sweden

**Boldrin, Alessio**  Technical University of Denmark, Denmark

**Brogaard, Line**  Technical University of Denmark, Denmark

**Carlsbæk, Morten**  Solum Gruppen, Denmark

**Christensen, Thomas H.**  Technical University of Denmark, Denmark

**Christiansen, Ole Vennicke**  Danwaste Consult A/S, Copenhagen, Denmark

**Comans, Rob**  ECN, The Netherlands

**Cossu, Raffaello**  University of Padua, Padua, Italy

**Dalager, Søren**  Rambøll, Denmark

**Damgaard, Anders**  Technical University of Denmark, Denmark

**Diaz, Luis F.**  CalRecovery, California, USA

**Ecke, Holger**  Luleå Technical University, Sweden

**Ehrig, Hans-Jürgen**  University of Wuppertal, Wuppertal, Germany

**Eighmy, T. Taylor**  University of New Hampshire, USA

**Finnveden, Göran**  Royal Institute of Technology, Sweden

**Fruergaard, Thilde**  Technical University of Denmark, Denmark

**Hauschild, Michael**  Technical University of Denmark, Denmark

**Hjelmar, Ole**  DHI - Water, Environment & Health, Hørsholm, Denmark

**Holm, Peter E.**  University of Copenhagen, Denmark

**Hulgaard, Tore**  Rambøll, Denmark

**Jambeck, Jenna**  University of Georgia, Athens, USA

**Jansen, Jes la Cour**  Lund University, Sweden

**Jensen, Lars Stoumann**  University of Copenhagen, Denmark

| | |
|---|---|
| **Johnson, Annette** | EAWAG, Switzerland |
| **Karakashev, Dimitar** | Technical University of Denmark, Denmark |
| **Kjeldsen, Peter** | Technical University of Denmark, Denmark |
| **Knox, Keith** | Knox Associates, Nottingham, UK |
| **Körner, Ina** | Hamburg University of Technology, Hamburg, Germany |
| **Krogmann, Uta** | Rutgers University, New Jersey, USA |
| **Lagerkvist, Anders** | Luleå Technical University, Sweden |
| **Lauridsen, Jørn** | COWI, Svendborg, Denmark |
| **Leithoff, Hans** | Johann Heinrich von Thünen-Institut, Hamburg, Germany |
| **Lenz, Volker** | German Biomass Research Center, Leipzig, Germany |
| **Manfredi, Simone** | Technical University of Denmark, Denmark |
| **Matsufuji, Yasushi** | Fukuoka University, Japan |
| **McLaughlin, Michael J.** | CSIRO Land and Water/University of Adelaide, Australia |
| **Merrild, Hanna** | Technical University of Denmark, Denmark |
| **Møller, Jacob** | Technical University of Denmark, Denmark |
| **Nielsen, Joan Maj** | COWI, Lyngby, Denmark |
| **Nilsson, Per** | R98, Copenhagen, Denmark |
| **Oros, Christiane** | Technical University of Denmark, Denmark |
| **Pagh, Peter** | University of Copenhagen, Denmark |
| **Raga, Roberto** | University of Padua, Padua, Italy |
| **Rechberger, Helmut** | Vienna University of Technology, Austria |
| **Reimann, Dieter O.** | Bamberg, Germany |
| **Reinhart, Debra R.** | University of Central Florida, Orlando, USA |
| **Robinson, Howard** | Enviros Consulting, Shrewsbury, UK |
| **Roth, Liselott** | Royal Institute of Technology, Sweden |
| **Rotter, Susanne** | Technical University of Berlin, Germany |
| **Rowe, R. Kerry** | Queen's University, Kingston, Ontario, Canada |
| **Salhofer, Stefan** | University of Natural Resources and Applied Life Sciences, Austria |
| **Scharff, Heijo** | NV Afvalzorg Holding, Assendelft, The Netherlands |
| **Scheutz, Charlotte** | Technical University of Denmark, Denmark |
| **Simion, Federico** | Technical University of Denmark, Denmark |
| **Stegmann, Rainer** | Technical University of Hamburg-Harburg, Germany |
| **Stentiford, Edward** | University of Leeds, UK |
| **Tonini, Davide** | Technical University of Denmark, Denmark |
| **Unger, Nicole** | University of Natural Resources and Applied Life Sciences, Austria |
| **van der Sloot, Hans A.** | ECN, Petten, the Netherlands |
| **VanGulck, Jamie F.** | Arktis Solutions Inc., Yellowknife, Northwest Territories, Canada |
| **Vehlow, Jürgen** | Forschungcentrum Karlsruhe, Germany |
| **Wallace, Robert B.** | Solid Waste Engineering Consultant, Laguna Niguel, California, USA |
| **Wejdling, Henrik** | Danish Waste Management Association, (DAKOFA), Copenhagen, Denmark |
| **Willumsen, Hans** | LFG Consult, Denmark |

# 9

# Biological Treatment

# 9.1

# Composting: Process

**Edward Stentiford**

*University of Leeds, UK*

**Marco de Bertoldi**

*University of Udine, Italy*

Composting is the aerobic degradation of solid organic matter. In nature the process evolves spontaneously in plant litter decomposition and in animal residues and manure transformation. This chapter presents the basic aspects of the composting process with a view to utilization in technologies for composting of solid organic waste. This chapter addresses microbial activities in aerobically degrading waste, the temperature profile of composting waste, the factors affecting composting rates and the fate of pathogens during composting. The various composting technologies available are presented in Chapter 9.2 and mass balances and compost quality are presented in Chapter 9.3.

## 9.1.1 Definition of Composting and Compost

Compost is the useful product of the composting process.

### 9.1.1.1 Composting

Composting is a microbial aerobic transformation and stabilization of heterogeneous organic matters in aerobic conditions and in solid state. The process is esoergonic and energy is released. A part of this energy (about 50–60%) is utilized by microorganisms to synthesize ATP, the other is lost as heat. This heat can generate a temperature increase in the mass. The first phase of the composting process is mesophilic and starts the aerobic decomposition of easily degradable organic matter; this rapid decay of material releases a great quantity of energy in form of heat, which enhances the mass temperature and the degradation rates of the organic waste. Within a few days this gives rise to the thermophilic phase. Without control, the temperature can easily reach and exceed 70 °C. The main positive effect of operating at such high temperature is the reduction of pathogenic agents present in the waste (see later). In controlled composting processes

*Solid Waste Technology & Management*   Edited by Thomas Christensen
© 2011 Blackwell Publishing Ltd

this phase is limited in terms of temperature and exposure time (degrees and days) to obtain a balance between high stabilization rates and good sanitization, often to satisfy local legislation regarding sanitization conditions. The third phase, maturation includes not only the mineralization of slowly degradable molecules, but also the humification of lignocellulosic compounds. This phase can last some weeks, according to the composition of the starting material. During the microbial transformation intermediate metabolites are produced which can make the composting material phytotoxic. This phytotoxicity is completely overtaken at the end of the process; thereafter the final product becomes beneficial to plant growth. The composting process ideally should be stopped when the phytotoxicity is over. If the process goes on too long, there is an excessive loss in organic matter reducing the beneficial impacts of the final product. The composting process leads to the production of carbon dioxide, water, minerals and biologically stabilized organic matter. The latter, including part of the water and minerals, is called compost.

From a microbiological point of view, composting is a discontinuous process (batch) resulting from a sequential development of different microbial communities. The microorganisms involved in the composting process are normally present in the starting material. Only when the starting material is deficient in microorganisms should an inoculum be needed.

### 9.1.1.2   Compost

Compost is the result of a composting process and the preferred definition is: 'Compost is the stabilized and sanitized product of composting which is beneficial to plant growth'. It has undergone an initial rapid stage of decomposition and is in the process of humification. To be rightfully called compost, the organic matter must be biologically stabilized or 'cured' and should become a humus-like product. It should be degraded into fine particles, having lost its original identity. It must be a stable product which can be stored without further treatment and can be applied to land without damage to standing crops. If stabilization has not been achieved, phytotoxins are produced in the soil until completion of the decomposition stage (Zucconi *et al.*, 1981a, b).

In a technical context it may be useful to distinguish between 'stabilized compost' and 'mature compost':

- Stabilized compost is the condition of the material after it has passed through the fist stage of rapid biooxidation.
- Mature compost refers to compost that has been stored for an extended time allowing for substantial humification to take place.

Many adjectives have been used for compost; some of them are correct such as aerobic, solid state, biologically stabilized and sanitized. Some others are in conflict with the same definition of compost such as anaerobic, fresh and liquid state.

## 9.1.2   Microbial Biomass and Succession

Bacteria, actinomycetes and fungi are the main microbial biomass responsible for the degradation of the organic waste. Box 9.1.1 presents their main characteristics. Higher organisms, for example compost worms (e.g. *Eisenia foetida*) may be numerous in very mature compost or in dedicated vermiculture.

---

**Box 9.1.1   The Microbiology of Composting.**

**Bacteria**

Bacteria are small single-cell organisms proliferating by cell division. When easily degradable material is available this results in high bacterial growth rates. Bacteria in aerobic environments consist of versatile consortia capable of degrading many waste materials. These consortia are relative robust and will thrive under a variety of environmental conditions. Bacteria of different types can degrade waste under anaerobic as well as aerobic conditions. Bacteria

---

are often categorized according to their optimal temperature range as psycrophilic (5–15 °C), mesophilic (25–35 °C) and thermophilic (55–60 °C), but they may be active in a much wider temperature range (Finstein and Morris, 1975). A variety of bacterial genera have been identified in compost including thermophilic bacteria (e.g. *Bacillus, Clostridium, Thermus, Hydrogenobacter, Microbispora, Streptomyces*; Insam and de Bertoldi, 2007), nitrogen-fixing bacteria (*Azotobacter, Azomonas, Enterobacte, Klebsiella, Bacillus, Clostridium*; de Bertoldi *et al.*, 1983) as well as ammonia- and nitrite-oxidizing bacteria (*Nitrosomonas, Nitrosospira, Nitrosococcus, Nitrobacter, Nitrospira, Nitrococcus, Pseudomonas*; Insam and de Bertoldi, 2007).

## Actinomycetes

Actinomycetes constitute a large group of filamentous (thread-shaped) organisms. Actinomycetes grow slower than most bacteria but are enzymatically better equipped to degrade more complex substrates. High concentrations of actinomycetes may be seen in compost piles as blue-green or light green fluffy and dusty formations (Golueke, 1977). Genera as *Thermomonospora* and *Micropolyspora* have been identified in compost (Finstein and Morris, 1975). Actinomycetes do not thrive at low oxygen concentrations and at high temperatures. They are active in degrading hemicellulose, cellulose and lignin.

## Fungi

Fungi are more complex organisms than actinomycetes. Fungi are also sensitive to low oxygen concentrations and high temperatures, nonetheless several types of thermophilic fungi have been identified in compost (e.g. *Mucor pusillus, Penicillium duponti, Chaetomium thermophilum, Scytalidium thermophilum, Talaromyces thermophilus, Sporotrichum thermophile, Aspergillus fumigatu, Thermoascus aurantiacus, Humicola insolens, Micelia sterilia*; de Bertoldi *et al.*, 1983). Numerous mesophilic fungi have been identified in compost, for example Ascomycotina (e.g. *Chaetomium* sp., *Dasyscipha* sp., *Emericella nidulans, Mollisia* sp., *Thermoascus aurantiacus*), Basidiomycotina (e.g. *Armillaria mellea, Clitopilus insitus, Lentinus lepideus, Polyporus versicolo*) and Deuteromycotina (e.g. *Alternaria tenuis, Aspergillus amstelodami, Cephaliophora tropica, Cephalosporium, Cladosporium herbarum, Geotrichum candidum, Penicillium, Sporotrichum termophile*). Fungi are very important in the degradation of hemicellulose, cellulose and lignin (de Bertoldi *et al.*, 1983).

Mesophilic bacteria, actinomycetes and fungi start their activity in the first phase of composting until high temperatures inhibit their metabolism. Most of them die or drastically reduce their activity in the thermophilic phase. Only a few sporigenous bacteria and nonsporigenous thermophilic bacteria (such as *Bacillus, Clostridium, Thermus*) show metabolic activity above 70 °C (Finstein and Morris, 1975; de Bertoldi *et al.*, 1983).

During the thermophilic phase only organisms adapted to high temperatures can survive and metabolize organic matter. Previously flourishing mesophiles die off and are eventually degraded by the succeeding thermophiles. The high temperature is the direct consequence of heat produced during composting. Owing to the lower activity of the fewer thermophilic microorganisms and to the decreasing degradability of the organic waste remaining, the rate of release of heat reduces and the process slowly enters in a new mesophilic phase (second).

Mesophilic microorganisms start to recolonize the substrate, either starting from surviving spores and microorganisms or from microorganisms colonizing from the outside. This phase is characterized by an increasing number of organisms able to degrade the long polymers: Lignin, cellulose, pectins and hemicellulose. The prevalent microflora consists of fungi and actinomycetes. Bacteria also are present but in a reduced number. In the later phases of composting the number of cellulolytic bacteria decreases, while the number of cellulolytic fungi (eumycetes) increases. The fungi benefit from the decrease in temperature, pH and moisture content and from the higher oxygen concentrations caused by lower water content and lower degradability of the organic waste. The same environmental factors positively affect the presence and diffusion of actinomycetes. With the degradation of lignin, an enzymatic aerobic transformation restricted to a limited microbial group namely the higher fungi (basidiomycetes), humification of organic matter begins together with the production of aromatic compounds (e.g. geosmin). These processes gradually lead to the final biological stabilization of the product.

## 9.1.3 Degradation of Carbon

The prime objective of microbial activity is to ensure the continued viability of the microbial community. The composting process in biological terms is not very efficient and the result of this is that, as new biomass is formed, there is a large release of heat under aerobic conditions, i.e. the process is exothermic. Equation (9.1.1) shows in general terms what happens to carbon in the composting waste material:

$$[C] + O_2 + \text{microbial activity} \rightarrow \text{new biomass} + CO_2 + H_2O + \text{Heat} \tag{9.1.1}$$

Box 9.1.2 presents the stoichiometry of degrading organic waste.

The majority of composting systems use a mixture of waste materials some of which decay rapidly and some such as lignin, which have a much slower rate of biodegradation. Within this mixture we have a whole series of materials which are losing mass (and releasing heat) at their own individual rates. Therefore the composting mass is a complex system with each material having its own reaction rate and each of these reaction rates being temperature dependent. In full-scale systems where the whole of the composting material is not at one temperature the system itself becomes very complex.

The composting process is driven by the rate at which mass is lost and by determining how much of the original mass is remaining at a particular time we can access the amount of composting which has taken place. The biological degradation of organic material generally follows a first order relationship (Haug, 1993). On this basis we can represent the mass remaining at any one time by Equation (9.1.2) when different fractions of the mass are degrading at different rates:

$$M_t = M_0(f_0 + f_1 e^{-k_1 t} + f_2 e^{-k_2 t} \ldots + f_n e^{-k_n t}) \tag{9.1.2}$$

where $M_t$ = mass remaining at time $t$, $M_0$ = initial mass, $f_n$ = fraction of the mass with a reaction rate of $k_n$ and $f_0$ is the nonvolatile (ash content) fraction of the mass.

In the case of composting with a wide range of materials it has proved sufficient in most cases to use only two rate constants. One of these represents the high rate of degradation of the more putrescible fraction (fast rate) and the other the more slowly degrading materials such as cellulose (slow rate). In addition, one fraction of the mass is considered not to degrade. Fast and slow degradation rates which have been recorded for different wastes are shown in Table 9.1.1. These degradation rates are overall rates including the break down of the biomass synthesized during the microbial degradation process.

---

**Box 9.1.2   The Stoichiometry of Composting.**

The degradation of organic waste, neglecting the minor contributions of degradation of N and S to the mass loss, can be approached stoichiometrically by (after Tchobanoglous *et al.*, 1993):

$$C_a H_b O_c + 0.5(ny + 2s + 0.5b - 0.5nx - c) O_2 \rightarrow n C_w H_x O_y + s CO_2 + (0.5b - 0.5nx) H_2O$$

where $s = a - nw$. The term $C_a H_b O_c$ represents the initial mole composition of the waste and $C_w H_x O_y$ the mole composition of the stabilized waste.

Tchobanoglous *et al.* (1993) suggest that the initial composition of organic waste is $C_{30}H_{50}O_{25}$ and the final composition is $C_{12}H_{20}O_{10}$. Assuming that 40 % by weight of the initial dry organic waste ends up as dry stabilized compost, $n$ equals 1.0. This yields the reaction:

$$C_{30}H_{50}O_{25} + 18 O_2 \rightarrow C_{12}H_{20}O_{10} + 18 CO_2 + 15 H_2O$$

This suggests that 1 kg dry organic waste measured as volatile solids requires 0.71 kg of oxygen producing 0.40 kg of stabilized organic compost measured as volatile solids, 0.98 kg of carbon dioxide and 0.33 kg of water.

**Table 9.1.1** *First order rate constants (per day, base e) and waste degradabilities (%) determined by long-term respirometry (temperature range 20–25 °C; after Haug, 1993). Reprinted with permission from Practical Handbook of Compost Engineering by R. T. Haug, CRC Press, 9780873713733 © (1993) Taylor and Francis.*

| Waste materials | Test duration (days) | $K_t$ (per day) | | Degradability (%) | |
|---|---|---|---|---|---|
| | | Fast | Slow | Fast | Slow |
| Raw sludge | 60 | 0.15 | 0.05 | 19 | 81 |
| Raw sludge | 242 | 0.015 | 0.004 | 40 | 60 |
| Pulmill sludge | 200 | — | 0.0095 | 0 | 100 |
| Sawdust, pine | 90 | 0.15 | 0.02 | 71 | 29 |
| Sawdust, hardwood | 368 | — | 0.0081 | 0 | 100 |

Equation (9.1.2) shows that the longer the composting process runs the more is degraded. The total amount of heat released is of the order of 19 MJ/kg dry matter degraded. Evaporation of water takes 2.4 MJ/kg, suggesting that the composting process supplies plenty of energy for evaporation of the moisture in the waste material.

Box 9.1.3 shows an example of calculating degradation of organic waste during composting. The example illustrates that the major mass loss is loss of water during composting.

---

**Box 9.1.3   Fate of Mass During Composting.**

Considering 1 t of material having a moisture content of 60 % (600 kg) and a dry solids content of 40 % (400 kg). By using Equation (9.1.2) with only two rate constants we end up with the relationship shown below. This gives the mass of dry solids remaining at time $t$ which consists of nonvolatile and volatile solids.

$$M_t = 400\,(0.3 + 0.35e^{-k_f t} + 0.35e^{-k_s t})$$

where

$k_f$ = fast rate of breakdown
$k_s$ = slow rate of breakdown

The volatile solids include both a 'fast' and a 'slow' degrading fraction. Using a fast rate constant of 0.15 day$^{-1}$ and a slow rate constant of 0.01 day$^{-1}$, after 21 days the mass of dry solids remaining ($M_{21}$) can be calculated.

$$M_{21} = 400\left(0.3 + 0.35e^{-0.15\times21} + 0.35e^{-0.01\times21}\right)$$
$$= 400\,(0.3 + 0.015 + 0.284)$$
$$= 239.6\,\text{kg}$$

This example represents the output of a typical composting plant.

The figure below represents an initial mass of 400 kg of dry solids of waste which includes 30 % nonvolatile solids (ash) and 70 % volatile solids.

The composting process has reduced the volatile solids by:

$$\% \text{ loss in volatile solids} = \frac{0.7 - (0.015 + 0.284)}{0.7} \times 100$$
$$= 57.3\,\%$$

It is interesting to note that whilst in the example there has been a 160.4 kg loss of volatile solids at the same time there has been 439.6 kg of water lost. This is the normal occurrence in a composting plant with by far the largest mass loss being the water.

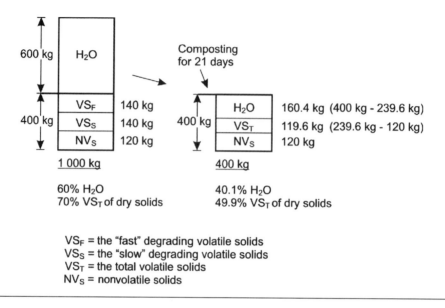

$VS_F$ = the "fast" degrading volatile solids
$VS_S$ = the "slow" degrading volatile solids
$VS_T$ = the total volatile solids
$NV_S$ = nonvolatile solids

If we consider the wide range of materials which are included in composting operations it would be extremely complex to produce a model which predicts overall rates of breakdown. To simplify considerations it is often convenient to consider materials which degrade rapidly and those which break down more slowly as was illustrated earlier. In slightly more general terms we can group organic materials depending on the rate at which they degrade. Table 9.1.2 shows a system of categorizing organic materials used by Bardos and Lopez-Real (1989) which gives an indication of what we can expect from materials routinely found in waste streams.

In practice when composting mixtures of materials, tests are required which reflect the degradation of the group as a whole and this often is referred to as the degree of stabilization. Box 9.1.4 presents a respirometer test that often is used for determination of stability or degradability of a waste by measuring its oxygen consumption over time.

***Table 9.1.2*** *The relative microbiological degradation rates of a range of organic materials encountered in composting (after Bardos and Lopez-Real, 1989). Reprinted with permission from Final report to the Directorate General Environment, Costs for municipal waste management in the EU by D. Hogg © (2002) European Commission.*

| Organic material | Degradation rate |
| --- | --- |
| Sugars Starches, glycogen, pectin, fatty acids and glycerol | Readily degradable |
| Lipids, fats and phospholids | |
| Amino acids | |
| Nucleic acids | |
| Protein | |
| Hemicellulose and cellulose | Slower to degrade (more relevant during maturation) |
| Chitin | |
| Low molecular weight aromatics and aliphatics | |
| Lignocellulose | Usually resistant |
| Lignin | |

**Box 9.1.4  Respiration Test Used in Measuring Degradability/Stability of Organic Waste. Reprinted with permission from Journal of Environmental Quality, 33, 1866–1876 by F. Adani, R. Confalonieri and F. Tambone, © (2004) Taylor and Francis.**

Respiration tests determine the uptake of oxygen into an organic waste sample or compost sample and express how active the microbial degradation is. The test is often expressed as a rate, e.g. $mgO_2/kgVS*h$, or as cumulative uptake over a number of days, e.g. $mgO_2/kgVS$ during 5 days. The respiration tests are defined as static or dynamic, depending on the absence (static) or presence (dynamic) of continuous aeration of the biomass (Adani *et al.*, 2003). The advantage of the dynamic methods is that oxygen is never depleted and can disperse and diffuse much better into the biomass, avoiding underestimation of the respiration activity as may occur in the static methods.

The respiration test can be performed in many different ways. In general, the shredded waste or compost is sieved and screened (to remove large inorganic contaminants) and a known amount is fed to a tight flask. Moisture is adjusted to a fixed content (e.g. 50–60 % w/w) or better at 75–80 % of the water holding capacity, and the temperature is controlled at 30–40 °C. Air is added to the reactor and the consumption of air is measured in many different ways, as describe below. A typical result of a respiration test is shown in the figure below (adapted from Adani *et al.*, 2004).

- In a closed flask the oxygen concentration is measured by an oxygen probe (Iannotti *et al.*, 1993) and the drop in concentration converted to an oxygen consumption.
- $CO_2$ being produced as the reaction product from the oxygen consumption is trapped in a closed flask in an aqueous solution of alkali (NaOH) or a soda lime trap. The amount of $CO_2$ is measured gravimetrically, titrimetrically or as a pressure drop. The quantified $CO_2$ is by a stochiometric calculation transferred to an oxygen consumption. The pressure drop approach is used in the commercial Oxytop.
- $CO_2$ is trapped as it is produced and the drop in pressure is compensated by addition of oxygen. The amount of oxygen added represents the oxygen consumption (Barberis and Nappi, 1996; Lasaridi and Stentiford, 1996). This approach is used in the commercial Sapromat, where oxygen is produced electrolytically.
- A constant flow of air or oxygen is supplied to the flask and concentrations in the inflow as well as in the outflow are measured. The difference represents the oxygen consumption (Adani *et al.*, 2001, 2003, 2004).

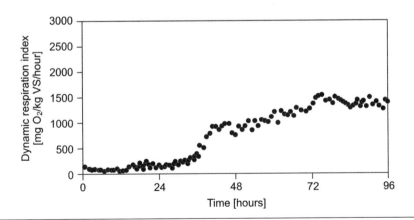

## 9.1.4  Degradation of Nitrogen and Sulfur

While carbon degradation is important for the overall mass loss during composting, nitrogen and sulfur degradation is important for the nutrient content of the compost and release of odors during the composting process.

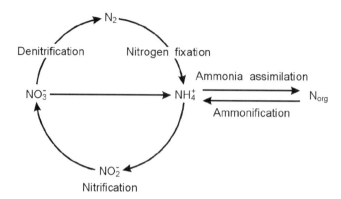

***Figure 9.1.1*** *Possible nitrogen transformations which can occur during composting.*

### 9.1.4.1 Nitrogen

All the microbial transformations of nitrogen indicated in Figure 9.1.1 occur during composting albeit with different levels of importance. Nitrogen-fixing bacteria have been isolated during the mesophilic phases of composting but their activity is inhibited by the presence of ammonia and by high temperatures (de Bertoldi *et al.*, 1983).

The nitrogen content usually decreases during composting mostly through ammonia volatilization. Denitrification occurs very rarely in some microniches where the oxygen supply is very poor and ammonia already has been oxidized to nitrate. The nitrogen loss during the process however can be evaluated only in terms of absolute value. In terms of dry weight, however, there is an increase due to the mineralization of organic matter and consequent loss of $CO_2$ and $H_2O$, so that a decrease in the C/N ratio can be observed through the process (de Bertoldi *et al.*, 1983). In spite of N loss by ammonia volatilization and denitrification, a partial recovery can take place later, due to the activity of nitrogen-fixing bacteria.

Heterotrophic bacteria, fungi and actinomycetes utilize organic nitrogen present in proteins and other compounds with their aerobic catabolism, producing $CO_2$, $H_2O$, ammonium, hydrogen sulfide and energy. Ammonium can be directly assimilated or oxidized by other microorganisms to nitrite and nitrate. Regardless of the composting system used, autotrophic nitrification appears to be absent in the early phases of the process. This type of nitrification was not detected until the completion of the thermopilic phase. The inhibition of ammonium oxidation is mainly due to high temperatures. Many investigations have confirmed the fact that above 40 °C the rate of nitrification is very low (Alexander, 1977). Excessive amounts of ammonia also inhibit the growth of nitrifiers, especially the *Nitrobacter* (Focht and Chang, 1975). Low oxygen availability may furthermore inhibit nitrification in competition with carbon degradation. Heterotrophic nitrification, operated by other bacteria and by fungi (eumycetes), seems less subjected to conditioning by these factors. In fact, production of nitrate in the early phases of composting seems almost exclusively the work of heterotrophic nitrifiers like *Arthrobacter* and *Aspergillus flavus* (Alexander, 1977). Nitrite and nitrate can during composting be assimilated by microorganisms as source of N or lost by denitrification, when conditions are mainly anaerobic.

### 9.1.4.2 Sulfur

Sulfur is required by microorganisms because of its structural role in the amino acids, cysteine and methionine, and because it is present in a number of vitamins (biotin, thiamine and lipoic acid). In composting, during microbial degradation of these compounds, sulfur undergoes a number of chemical transformations.

During composting, a wide variety of microorganisms can use sulfate as a sulfur source and carry out assimilative sulfate reduction converting the HS⁻ formed to organic sulfur. HS⁻ is continuously reformed from the decomposition of this organic sulfur. Hydrogen sulfide, like ammonia, is mainly produced during the initial phase of composting. Chemical analyses of the gases evolved during composting confirmed that, in a feedback-controlled process, the loss of HS⁻ was limited to few days in the thermophilic phase, and the maximum concentration observed in the gases was 2 ppm (v/v; de Bertoldi *et al.*, 1988a).

**Table 9.1.3** *Odorous compounds from composting, primarily from anaerobic microbial niches in the compost.*

| Group | Compound | Origin |
|---|---|---|
| Fatty acids | Acetic, butyric, propionic | Fermentation |
| Aldeydes, chetons, alcohols, esters | | Fermentation |
| Amines | Methylamine, thilamine, dimethylamine, cadaverine, putrescine | Anaerobic degradation of proteins and aminoacids |
| Aromatics and ring structures | Indole, scatole | Anaerobic degradation of proteins |
| Inorganic sulfide | Hydrogen sulfide | Anaerobic degradation |
| Organic sulfides | Mercaptans, dimethyl, diethyl, methyl and propyl sulfide | Anaerobic degradation of proteins and sulfur compounds |
| Ammonia | | Anaerobic degradation of N-compounds |
| Reduced phosphorus compounds | Phosphines | Anaerobic degradation of P-compounds |

### 9.1.4.3 Odor Generation Under Anaerobic Conditions

In composting materials, rich in nitrogen like animal wastes particularly when the process is not controlled, anaerobic microbial processes can start producing reduced molecules. These molecules arising from fermentation or anaerobic respiration can derive from sulfur, nitrogen and other products (Table 9.1.3). Some of them are volatile and can diffuse in the environment with a resulting odor nuisance. For public acceptability reasons, these highly odorous molecules can not be spread in the environment. Consequently exhausted air from composting plants has to pass through biofilters. The microorganisms inside biofilters are able to oxidize these reduced molecules transforming most of them into liquids without any odor.

## 9.1.5 Energy Release and Temperature Development

Each waste material has a certain potential energy which it will release as heat when fully oxidized. Although the biological oxidation is distinctly different from thermal oxidation (combustion) the total heat released when a material is fully oxidized using either route is similar.

In a real situation most of the wastes will have substantial amounts of water associated with them and consequently a substantial part of the generated heat will be used to evaporate not only the water formed by the degradation process but also a significant fraction of the original moisture in the waste. In some cases the loss of water is so high that water must be added to the compost to ensure enough moisture for the microbial processes.

With respect to the temperature development in a practical composting process, it is not the total amount of energy potentially available which is important, but the rate at which it is released; i.e. the degradation rate of the waste materials. Waste composting systems use a wide range of waste materials as feedstock, each of which has different degradation rates. Table 9.1.4 shows heat output rates determined by different workers for a range of waste materials (Shaw, 1996). The more putrescible the waste, the higher the rate of output and as a result the temperature of the composting mass increases more rapidly. As the heat is being generated due to the biological activity it is also being lost to the surrounding environment as radiation loss and convection loss (air passing over or through the compost). Consequently there are three principal states which can exist in a composting mass:

- Heat generated > heat lost → temperature increases.
- Heat generated = heat lost → temperature static.
- Heat generated < heat lost → temperature decreases.

In general the biological processes which take place during composting are temperature related with the rate increasing with temperature up to the 50–60 °C temperature range after which it falls. Figure 9.1.2 shows a typical rate curve with this increase and eventual fall with the rate dropping to zero in the 70–80 °C range.

**Table 9.1.4**    *Heat output from various materials during composting under a range of experimental conditions.*

| Waste materials | Peak heat output (J/kg(VS)/s) | Temperature (°C) | Reference |
|---|---|---|---|
| Straw | 13.39 | 40 | Carlyle and Norman (1941) |
| | 7.39 | 60 | |
| Refuse | 21.31 | 60 | Wiley (1957) |
| Sewage sludge and woodchips | 12.97–15.0 | 50–60 | Miller (1984) |
| | 6.53 | 72 | |
| | 4.94 | 74 | |
| Oak leaves | 1.61 | 50 | Finstein *et al.* (1986) |
| Oak leaves and nutrients | 3.22 | 50 | Finstein *et al.* (1986) |
| Maple leaves | 3.64 | 50 | Finstein *et al.* (1986) |
| Mushroom compost, straw based | 2.33 | 72 | Miller *et al.* (1989) |
| | 8.72 | 55 | |
| | 10.75 | 45 | |
| Rice hulls and rice flour | 11.61 | 40–55 | Hogan *et al.* (1989) |

In full-scale composting plants there is never one single operating temperature in the mass, since all systems have a temperature gradient across the mass. In reactor-based systems recycling the warm exhaust air, this gradient can be as low as 5 °C and in open windrow systems it can be as high as 50 °C. As a result of these gradients not only do we have a wide range of degradation rates but also a range of microbial communities suited to the local temperature. The rate of composting is very dependent on the temperature and getting the 'best rate' for the process depends very much on what the process is required to achieve. In general terms we can characterize the temperature ranges as follows:

- 25–45 °C: Highest biodiversity.
- 45–55 °C: Highest rate of biodegradation.
- > 55 °C: Highest rate of pathogen inactivation (sanitization).

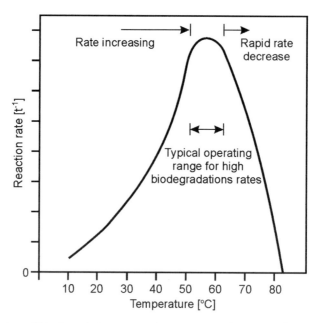

**Figure 9.1.2**    *The variation of biodegradation rate with temperature for a typical material used in composting.*

There are many rate equations which are used to reflect the change with temperature, the most commonly used are of the form shown in Equation (9.1.3) which is based on the Arrhenius equation:

$$k_T = k_{20}(\theta)^{(T-20)} \tag{9.1.3}$$

Where $k_{20}$ and $k_T$ are the rate constants at $20\,^{\circ}\text{C}$ and $T\,^{\circ}\text{C}$ respectively and $\theta$ has a value in the range 1.05–1.1 (1.06 is commonly used). This is in accordance with the rule of thumb that degradation rates approximately double for each $10\,^{\circ}\text{C}$ increase in temperature.

## 9.1.6 Process Factors

In an operational composting system each part of the composting mass is working as a microcomposting unit with each unit being based on an individual particle. The mass transfer rates, both within the particle and at the surface, determine the conditions which exist around the particle. The principal factors involved are: temperature, moisture content, oxygen availability and nutrient availability.

### 9.1.6.1 Temperature

Whilst we have the variation in temperature within the composting mass being dependent on microsite activity, the general variation of temperature with time is represented in Figure 9.1.3. The two temperature/time curves shown in Figure 9.1.3 represent a system with no temperature control (system 1) and one with temperature feedback control (system 2). Despite the apparent difference in the two curves their main features are the same: Increasing temperature, static temperature and decreasing temperature. The difference in the two systems is the middle phase; in the controlled system a relatively constant (generally lower) temperature is maintained for an extended period to maximize the rate of biodegradation (stabilization). In the uncontrolled system there is no attempt to optimize the system and in most cases the high temperatures restrict the rate of degradation, as shown in Figure 9.1.2.

In full-scale systems the designers, manufacturers and operators may have set their systems up to satisfy different local requirements ranging from maximizing the rate of sanitization to minimizing the process time to achieve stability.

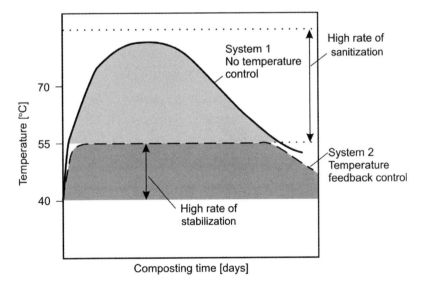

**Figure 9.1.3** *Typical temperature/time profiles for two composting systems. System 1: No temperature control. System 2: With temperature control.*

However, in all cases the same fundamental process of adjusting the heat loss to achieve the required temperature is behind the operation. The heat loss can be controlled by controlling the air flow and the moisture content.

In relation to the heating phase of the process it is worth bearing in mind that if elevated temperatures in excess of 55 °C cannot be reached within one day there is something fundamentally wrong with the system and/or operating protocol.

### 9.1.6.2   Moisture

Water is a vital element in composting as it provides the moist working environment that most of the microorganisms need for moving gases and liquids through their cell membranes. During composting the heat generated evaporates large quantities of water from the composting mass.

In order to prevent the lack of moisture inhibiting composting it is recommended that the moisture content should not fall below 35–40 %. To allow for water loss during the process it is normal to start with values in the initial mixture in the 55–65 % range depending on the materials. By starting at this level then as composting progresses and water is lost, sufficient water still remains, even at the end of the process, to keep it from becoming a limiting factor. Care should be taken during initial mixing to make sure that the value is not too high as water could then fill the interstitial spaces in the structure and cause the system to become anaerobic. In the majority of full-scale systems provision is made by means of the design of the plant and the operating protocols to allow moisture levels to be adjusted during the process.

### 9.1.6.3   Oxygen Availability

Oxygen is required to meet the metabolic needs of the microbial community and it is normally supplied in composting as air. The rate of consumption of oxygen reflects the level of microbial activity in the mass. Figure 9.1.4 shows a typical oxygen demand curve for material being composted. The four zones in Figure 9.1.4 are as follows:

- Zone 1: Consumption rate increasing as the pile temperature increases and the microbial community develops – generally less than 3 days.
- Zone 2: Peak demand – in reactor-based systems with putrescible wastes this period is unlikely to be more than 5–10 days.
- Zone 3: Cooling – the more readily degradable materials have now been used and at the end of this zone the material is approaching stabilization.
- Zone 4: Stabilization is completed and maturation/humification begins.

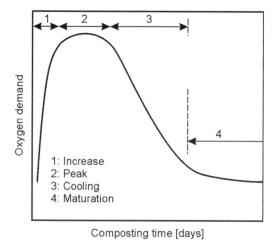

**Figure 9.1.4**  *Typical oxygen consumption curve during composting.*

***Table 9.1.5*** *Range of C/N values for different waste materials.*

| Material | C/N |
|---|---|
| Old, woody, green waste | 100 |
| Conifer mulch | 30–100 |
| Wheat straw | 60–125 |
| Sawdust | 100–500 |
| Bark | 100–130 |
| Prunings | 100–150 |
| Paper/cardboard | 200–500 |
| Fresh, green garden waste | 10–20 |
| Grass clippings | 12 |
| Kitchen waste | 15–23 |
| Fruit residues | 35 |
| Animal manure | 15–25 |
| Mixed municipal solid waste | 20–40 |
| Sewage sludge | 5–15 |

An implicit requirement for achieving the type of oxygen demand profile shown in Figure 9.1.4 is to have the appropriate structure in the mixture which allows the easy passage of air. For example whilst sewage sludge does not compost easily on its own it composts very rapidly when mixed with woodchips. The woodchips not only supply a small amount of carbon but more importantly give the structural 'openness' which facilitates the passage of air into the system and other gases out of it.

### 9.1.6.4  Nutrient Availability

A wide range of nutrients are required to ensure the optimum performance of a particular biological community. In the composting context the issue which raises most concern is the balance of carbon and nitrogen: the C/N ratio. For bacterial cell formation the typical C/N requirement is in the 10–15 range, however, in order to support the energy requirements of the process additional carbon is required. The ideal starting value for the C/N ratio varies depending on the materials used but it is generally agreed to be in the 20–35 range (Haug, 1993). Problems during processing can occur if the initial value of the C/N ratio is outside this range:

- C/N is too high: The rate of degradation is restricted because of the lack of N which is needed by the microorganisms.
- C/N is too low: Excess nitrogen is given off as ammonia which not only represents loss of nutrients but also causes potential odor problems.

Table 9.1.5 shows typical values of carbon and nitrogen for a range of waste materials. In many cases it is necessary to mix materials in order to obtain not only the appropriate starting mixture from a nutrient viewpoint but also to create a suitable structure to allow easy passage of the air. The table shows a range of values for most of the material groups as each of the categories contains within it a whole series of specific materials.

It should be noted that determination of the C/N ratio of a waste requires analytical determination of both C and N, and that the analysis of C determines all organic C, also any recalcitrant organic fraction of the waste. In such cases the optimum C/N ratio for degradation is higher. Where more sophisticated carbon analyzers are not available an approximate value for the C content in waste or sludge can be obtained by dividing the volatile solids content by 1.8.

## 9.1.7  Pathogenic Agents in Composts and Their Control

Waste may contain a huge number of pathogenic agents like bacteria, viruses, fungi and parasites. Table 9.1.6 presents some pertinent pathogens found in organic waste and their related diseases. Waste is not the ideal habitat for many of them, so that their numbers tend to fall steeply with time.

**Table 9.1.6**   *Pathogenic agents likely to be present in compost starting material.*

| Pathogen | Disease |
| --- | --- |
| Viruses | |
| Enterovirus | Gastro-enteritis, heart disease, meningitis |
| Rotavirus | Gastro-entertis |
| Parvovirus | Gastro-enteritis |
| Adenovirus | Respiratory tract infections, conjunctivitis |
| Hepatitis A virus | Viral hepatitis |
| Polio virus | Poliomyelitis |
| Ecovirus | Meningitis |
| Coxsachivirus | Meningitis |
| Bacteria | |
| *Salmonella* (1700 types) | Typhus, salmonellosis |
| *Shigella* | Shigellosis |
| *Mycobacterium tuberculosis* | Tuberculosis |
| *Vibrio cholerae* | Cholera |
| *Escherichia coli* | Gastro-enteritis |
| *Yersinia enterocolica* | Gastro-enteritis |
| *Clostridium perfrigens* | Gangrene |
| *Clostridium botulinum* | Botulism |
| *Listeria monocytogenes* | Meningo-encephalitis |
| Fungi | |
| *Candida albicans* | Systemic and skin mycosis |
| *Trichophyton* sp. | Skin mycosis |
| *Trichosporon cutaneum* | Skin mycosis |
| *Epidermophyton* sp. | Skin mycosis |
| *Microsporum* sp. | Skin mycosis |
| *Aspergilus fumigatus* | Lung mycosis |
| Protozoa | |
| *Entamoeba* | Amebiasis |
| *Giardia lamblia* | Giardiasis |
| *Balantidium* | Balantidiasis |
| *Naegleria fowleri* | Meningo-encephalitis |
| *Acanthameba castellanii* | Meningo-encephalitis |
| Helmints | |
| *Ascaris lumbricoides* | Ascariasis |
| *Ancylostroma* sp. | Ancilostomiosis |
| *Necator americanus* | Necatoriasis |
| *Enterobius vermicularis* | Enterobiasis |
| *Strongyloides stercoralis* | Strongilaidiasis |
| *Toxocara* sp. | Larvae in the viscera |
| *Thrichuris thrichuria* | Thricuriasis |
| *Taenia saginata* | Tapeworm |
| *Hymenolepsis nana* | Tapeworm |
| *Echinococcus granulosus* | Echinococcosi |
| *Echinococcus multilocularis* | Echinococcosi |

Some pathogens, such as parasites and viruses, are incapable of reproducing outside their host; however, they may survive for a long time in waste but without multiplying. Other pathogens, such as bacteria and fungi, may reproduce, feeding on the assimilable organic matter in waste, thereby multiplying in number in time (de Bertoldi *et al.*, 1988b). The latter group of pathogens makes it necessary to provide conditions for compost sanitization, such that these microorganisms not only are reduced in number, but also they are unable to grow and multiply in mature compost.

When pathogens present in compost are spread on land, the homeostatic properties of soil and competition with native microorganisms usually have the effect of reducing their number. The fact remains however, that there is a risk of some pathogens reaching man and, if present in sufficiently high numbers, causing disease or food poisoning. It is therefore extremely important for pathogens to be reduced to a minimum so that the health and hygiene risks are minimized also. Technically speaking this could be done by sterilization, but this would not only be economically impractical, but it would also not eliminate the risk of regrowth with some pathogens (i.e. *Salmonella* spp.), which in absence of microbial competition, could grow rapidly.

It must not be forgotten that the destination of the compost is the soil, which already harbors some pathogens such as *Clostridium tetani* and *Aspergillus fumigatus*, so complete sterilization is unnecessary. It would be superfluous and a waste of money and energy to eliminate pathogens naturally present in soil (WHO, 1981). Composting must therefore guarantee the reduction to below a certain level of all those microorganisms which are foreign to soil and which, above certain concentrations result in the danger of contamination. There is a direct risk of infection when pathogen-contaminated vegetables are consumed by humans or animals without suitable treatment.

All this underlines the importance of reducing the number of pathogens present in waste before their use in agriculture. The two main problems that emerge are first how far to take sanitization and secondly, how to check the degree of sanitization once completed. The former is simply a problem of economics: waste by definition cannot carry heavy overheads for transformation into compost which would make it noncompetitive on the open market. For this reason, pathogen reduction must of necessity be a compromise between the cost of processing and the benefit of a safe sanitized endproduct. With regard to the second point, that is how to check the degree of sanitization, this can be done either during or on completion of the process.

### 9.1.7.1 Sanitization (Temperature – Time)

Sanitization occurs naturally during composting since the composting mass reaches a certain temperature level, which it maintains for some time. Temperature and time span must be selected on the basis of a logarithmic reduction in the main pathogens; what degree of reduction there should be in compost varies from country to country. Some countries have set temperature standards varying from 55 °C to 65 °C, covering a time span of 3 days and 24 h respectively. Before release to market the compost should be tested to check its degree of sanitization.

Indicator microorganisms such as fecal coliforms, fecal streptococci, Enterobacteriaceae, certain viruses and parasite eggs have been used and give reliable results. An indicator microorganism should have the following properties:

- Be present routinely in large quantities in the raw material before composting.
- Must belong to a group with similar reactions to the pathogen with respect to treatment (i.e. thermal changes).
- Must have a higher resistance than the targeted pathogens to allow a safety margin.
- Must be able to be enumerated using simple and cheap tests.
- Must itself not be pathogenic.

One of the most controversial problems under discussion is the quantitative determination of the reduction that indicators should undergo to guarantee hygienicity of end-products. Most research in this field concerns two pathogens: *Salmonella* spp. and infective parasite ova; and two indicators: fecal streptococci and fecal coliforms (or Enterobacteriaceae; see de Bertoldi *et al.*, 1984, 1991). It has been proposed in the EU but not yet accepted that *Salmonella* (preenrichment) should be absent in 100 g sample, fecal coliforms should be $<5 \times 10^2$ colony forming units (cfu)/g in the final product and fecal streptococci $<5 \times 10^3$ cfu/g in the final product.

### 9.1.7.2 Factors Affecting Pathogen Growth in Compost

Some pathogenic bacteria and fungi can multiply in compost but only under particular conditions. Even though viruses and parasites may survive a long time on fresh material, they do not multiply outside the host. There are many factors which influence pathogen growth in compost and some of the key ones are covered in the following paragraphs.

## *Organic Matter*

All pathogens present in composting material are heterotrophic and therefore they need a source of soluble organic matter to metabolize for the growth. Pathogenic bacteria and fungi are generally able to metabolize only readily available organic matter, such as alcohol, organic acids, sugars, etc., but are unable to multiply on complex compounds like cellulose, lignin and humic substances. During composting the easily degradable organic matter is quickly degraded and in the compost the residual organic fraction is composed mainly of stable polymers, often combined together to make up the complex humic fraction. This makes stabilized compost unfavorable for the metabolism and growth of pathogens since they do not have the enzymes to break down these polymers and therefore they cannot feed on them.

## *Moisture*

As with all other biological processes, growth in microorganisms depends on the presence of water. The moisture level in most commonly composted wastes is usually at a level sufficient to support pathogen growth. Even a material with a very high moisture content, such as sewage sludge, does not impair pathogen development, since many (bacteria in particular) are facultative anaerobes, and can therefore grow even in the absence of oxygen. Low moisture levels, below 25 %, slow or halt all forms of microbial growth, pathogens included. Cured and stabilized compost, with a moisture content of 25–30 %, should not therefore create problems concerning the regrowth of pathogens unless local meteorological conditions raise the moisture level.

## *Temperature*

All pathogen microorganisms have a threshold resistance to temperature which varies from one microorganism to another and according to environmental conditions. All sporigenous and nonsporigenous thermophilous bacteria are able to resist temperatures above 100 °C; this means that no composting process is capable of eliminating these microbial forms. Fortunately the most dangerous pathogens for the environment and agriculture in waste and sludge are neither sporigenous nor thermophilic and can therefore be eliminated through conventional heat treatments. The effectiveness of this treatment depends firstly on the temperature reached and the length of time it is maintained. Temperature and time are inversely correlated. Moisture levels affect the impact of the temperature by increasing the thermal conductivity of the mass. Sanitization is the result of high temperatures (65 °C) generated during composting and is more active if the moisture content is high. This lethal effect would be much less marked in dry conditions, for example in a dry endproduct.

The key feature in sanitization by thermal means is that a particular temperature must be reached in all parts of composting mass for a specific time period. This can be very difficult to achieve technically; in some parts of the mass temperatures may rise to above 80 °C while others do not reach a high enough temperature by the end of the process. This gradient of temperature depends on the practical system of composting as was mentioned earlier. It is often forgotten in open systems that if the surface part of the composting material is not insulated it never reaches the hygienization temperature. Turning the composting waste may increase the possibility of having all the material exposed to high temperature for sufficient time.

## *Oxygen Supply*

High temperatures help to produce sanitized mature compost. Temperature elevation in the mass is the direct result of the accumulation of heat which is released during the exothermic reactions in composting. Since these reactions are aerobic and require an adequate oxygen supply, it is easy to see that temperatures will rise only where there is a good supply of oxygen for oxidation. When there are parts of the mass where oxygen is insufficient, anaerobic processes take hold over the aerobic; much lower quantities of energy are liberated, so the temperature in the mass does not rise sufficiently.

## *Microbial Competition*

Microbial competition and antagonism are the second most important factor in pathogen control during composting. An enormous number of saprophytes are involved in composting. This microflora should be considered as native or natural in composting matter. Waste contains a second microbial population, pathogens, which is numerically an insignificant fraction of the total microbial population. When the population density is relative high, antagonism and competition for nutrition sets in, in this heterogeneous microbial community. The native saprophitic population has a considerable advantage over the other population in this competition. Composting matter is not the natural environment for pathogens, so in this ecosystem competition brings about the elimination of the weaker rival.

# References

Adani, F., Confalonieri, R. and Tambone, F. (2004): Dynamic respiration index as a descriptor of the biological stability of organic wastes. *Journal of Environmental Quality*, 33, 1866–1876.

Adani, F., Gigliotti, G., Valentini, F. and Laraia, R. (2003): Respiration index determination: a comparative study of different methods. *Compost Science and Utilization*, 11, 144–151.

Adani, F., Lozzi, P. and Genevini, P.L. (2001): Determination of biological stability by oxygen uptake on municipal solid waste and derived products. *Compost Science and Utilization*, 9, 163–178.

Alexander, M. (1977): *Introduction to soil microbiology*. J. Wiley & Sons, Ltd, New York, USA.

Barberis, R. and Nappi, P. (1996): Evaluation of compost stability. In: De Bertoldi, M., Bert, P. and Tiziano, P. (eds) *The science of composting, part 1*, pp. 175–184. Blackie Academic and Professional, London, UK.

Bardos, R.P. and Lopez-Real, J.M. (1989): The composting process: susceptible feedstocks, temperature, microbiology, sanitisation and decomposition. In: Bidlingmaier, W. and L'Hermite, P. (eds) *Compost processes in waste management*, pp. 179–190. Directorate-General Science, Research and Development, Commission of the European Communities, Brussels, Belgium.

Carlyle, R.E. and Norman, G. (1941): Microbial thermogenesis in the decomposition of plant materials, part II, factors involved. *Journal of Bacteriology*, 41, 699–724.

de Bertoldi, M., Vallini, G. and Pera, A. (1983): The biology of composting: a review. *Waste Management and Research*, 1, 157–176.

de Bertoldi, M., Frassinetti, S., Bianchin, M.L. and Pera, A. (1984): Sludge hygienization with different compost systems. In: Strauch, D., Havelar, D. and L'Hermite, P. (eds) *Inactivation of microorganisms in sewage sludge by stabilization processes*, pp. 64–76. Elsevier Applied Science, London, UK.

de Bertoldi, M., Rutili, A., Citterio, B. and Civilini, M. (1988a): Composting management: a new process control through $O_2$ feedback. *Waste Management and Research*, 6, 239–259.

de Bertoldi, M., Zucconi, F. and Civilini, M. (1988b): Temperature, pathogen control and product quality. *Biocycle*, 28, 56–61.

de Bertoldi, M., Civilini, M. and Manzano, M. (1991): Sewage sludge and agricultural waste hygienization through aerobic stabilization and composting. In: L'Hermite, P. (ed.) *Treatment and use of sewage sludge and liquid agricultural wastes*, pp. 212–226. Elsevier Applied Science, London, UK.

Finstein, M.S. and Morris, M.L. (1975): Microbiology of municipal solid waste composting. *Advances in Applied Microbiology*, 19, 113–151.

Finstein, M.S., Miller, F.C and Strom, P.F. (1986): Waste treatment composting as a controlled system. In: Rehn, H.-J. and Reed, G. (eds) *Biotechnology*, vol 8: Microbial degradations, pp. 363–398. VCH Verlagsgesellschaft mbH, Weinheim, Germany.

Focht, D.D. and Chang, A.C. (1975): Nitrification and denitrification processes related to waste water treatment. *Advances in Applied Microbiology*, 19, 153–186.

Golueke, C.G. (1977): Biological reclamation of solid wastes. Rodale Press, Emmaus, USA.

Haug, R.T. (1993): *The practical handbook of compost engineering*. Lewis Publishers, Boca Raton, USA.

Hogan, J.A., Miller, F.C. and Finstein, M.S. (1989): Physical modeling of the composting ecosystem. *Applied and Environmental Microbiology*, 55, 1082–1092.

Iannotti, D.A., Pang, T., Toth, B.L., Elwell, D.L., Keener, H.M. and Hoitink, H.A.J (1993): A quantitative respirometric method for monitoring compost stability. *Compost Science and Utilization*, 1, 52–65.

Insam, H. and de Bertoldi, M. (2007): Microbiology of the composting process. In: Diez, L.F., de Bertolde, M., Bidolingmaier, W. and Stentiford, E. (eds) *Compost science and technology*, pp. 25–48. Elsevier, Amsterdam, The Netherlands.

Lasaridi, K.E. and Stentiford, E.I. (1996): Respirometric techniques in the context of compost stability assessment: principles and practice. In: de Bertoldi, M., Sequi, P., Lemmes, B. and Papi, T. (eds) *The science of composting, Part 1*, pp. 274–285. Blackie Academic and Professional, Glasgow, UK.

Miller, F.C. (1984): Thermodynamic and matrix water potential analysis in field and laboratory scale composting ecosystems. PhD thesis. M/8424132. Rutgers University, USA.

Miller, F.C., Hogan, J.A. and Macauley, B.J. (1989): Determination of heat evolution and activity in mushroom composting through physical modeling. *International Symposium on Microbial Activity*. 5, Abstract 0-9-7.

Shaw, C.M. (1996): *Computer-based modeling of the composting process*. PhD thesis, The University of Leeds, UK.

Tchobanoglous, G., Theisen, H. and Vigil, S. (1993): *Integrated solid waste management: engineering principles and management issues*. McGraw-Hill, New York, USA.

WHO (1981): *The risk to health of microbes in sewage sludge applied to land*. Euro Reports and Studies 54. World Health Organization – Regional Office for Europe, Copenhagen, Denmark.

Wiley, J.S. (1957): Progress report on high-rate composting studies. In: *Proceedings of the twelfth international waste conference*, pp 596–603. Extension Series No. 94. Engineering Bulletin, Purdue University, vol. XLII, No. 3. Purdue University, Lafayette, USA.

Zucconi, F., Forte, M., Monaco, A. and de Bertoldi, M. (1981a): Biological evaluation of compost maturity. *Biocycle*, 22, 27–29.

Zucconi, F., Pera, A., Forte, M. and de Bertoldi, M. (1981b): Evaluating toxicity of immature compost. *Biocycle*, 22, 54–57.

# 9.2

# Composting: Technology

**Uta Krogmann**

*Rutgers University, New Jersey, USA*

**Ina Körner**

*Hamburg University of Technology, Hamburg, Germany*

**Luis F. Diaz**

*CalRecovery, California, USA*

All composting systems have the common objective to decompose the biodegradable organic fraction of the feedstock under controlled conditions to produce an end product that can be handled, stored, used or disposed without adversely affecting the environment (Golueke, 1977). Composting systems range from backyard composters and simple windrow systems (elongated piles that are regularly mixed) to more complicated enclosed building or in-vessel systems where various parameters such as temperature, moisture and oxygen supply are controlled.

This chapter focuses on large-scale composting facilities (> 2000 t/year). Backyard composting of food and yard waste by residents can be an important part of an integrated waste management system, but is not addressed in this chapter. More information about backyard composting can be found in US EPA (1999) and Kranert (2000). Furthermore, this chapter does not cover the small-scale commercial, institutional and agricultural onsite composting systems (5–100 t/year; e.g., at hospitals, supermarkets, nurseries, and on small-scale farms; German Federal Government, 1998; US EPA, 1999; Krogmann *et al.*, 2006).

Generally, the composting technologies described have been selected on the basis of a variety of key issues such as feedstock, environmental controls, socioeconomic and other nontechnical considerations, and proposed use or disposal of the end product. Aspects regarding mass balances and compost quality are covered in Chapter 9.3.

*Solid Waste Technology & Management*   Edited by Thomas Christensen
© 2011 Blackwell Publishing Ltd

## 9.2.1   Introduction

The key issues in establishing a composting system are: (1) select the waste to be treated (the feedstock), (2) identify the need for environmental controls when composting the selected waste, (3) identify local and regional regulatory and socioeconomic issues relevant for establishing a composting facility, and (4) determine the use or disposal of the product to be produced. If the compost is going to be marketed also information about the size of the market, alternative products and costs may be needed. Having defined the composting system, a range of technologies is available to choose from.

### 9.2.1.1   Feedstock

Generally, all wastes that are biodegradable can be composted. The feedstock may range from very mixed and heterogeneous waste as municipal solid waste (MSW) to separately collected fractions such as yard waste or food waste to specific industrial processing residues. Table 9.2.1 provides a list of potential feedstocks to a composting plant.

Feedstock can be of plant (e.g., vegetable scraps, yard waste) or animal origin (e.g., fish waste, meat waste); feedstock may have already been processed by natural (e.g., animal manure) or by technical processes (e.g., sewage sludge, paper, cardboard, biodegradable plastics). Feedstock is characterized by chemical, physical and biological parameters including moisture content, particle size, organic matter content, nutrient content, heavy metal content, synthetic organics content,

**Table 9.2.1**   *Potential feedstocks for composting.*

| Main waste type | Subfraction |
|---|---|
| Mixed municipal solid waste (MSW) | |
| Residual MSW (fraction of MSW that remains after source separation of food waste, yard waste and dry recyclables) | |
| Yard waste and other green wastes | Grass clippings |
| | Plant residuals |
| | Brush and tree trimmings |
| | Leaves |
| | Cemetery wastes |
| | Christmas trees |
| | Seaweed and other aquatic plants |
| Agricultural wastes | Excess straw |
| | Spoiled hay and silage |
| | Beet leaf residuals |
| | Dead animals (not allowed in some countries) |
| | Solid and liquid manure |
| Biowaste (source separated food and yard waste) | |
| Sewage sludge (biosolids) | |
| Paper products | |
| Market wastes | |
| Processing residuals | Residuals from the pharmaceutical industry |
| | Residuals from the food processing and beverage industry |
| | Residuals from vegetable oil production |
| | Fish processing wastes |
| | Paunch contents from slaughterhouses |
| | Bark |
| | Sawdust and shavings |
| Forestry wastes | Residuals from windbreaks |
| | Logging residuals |

**Table 9.2.2** *Selected characteristics of feedstock for composting.*

| | Moisture (% wet weight) | Volatile solids (% dry weight) | C/N | N (% dry weight) | Structure |
|---|---|---|---|---|---|
| Food waste[a] | 76 | 89 | 22 | 1.9 | Poor |
| Biowaste[b] | | | | | |
| – Downtown | 64–77 | 56–86 | 17–27 | 1.2–2.3 | Poor |
| – Single-family homes | 57–71 | 45–77 | 15–23 | 1.1–2.0 | Good/poor |
| Cardboard[c] | 8 | — | 563 | 0.10 | Medium |
| Newspaper[c] | 3–8 | — | 398–852 | 0.06–0.14 | Medium |
| Paper from residential waste[c] | 18–20 | — | 127–178 | 0.2–0.25 | Medium |
| Residual MSW[d] | — | 40–69 | 23–45 | 0.7–1.8 | Good |
| Municipal yard waste | | | | | |
| – Grass clippings[a] | 70–73 | 60–77 | 13–16 | 2.0–2.8 | Poor |
| – Leaves[a] | 18–49 | 75–92 | 35–56 | 0.9–1.3 | Medium |
| – Grass/leaves[b] | 32–67 | 41–77 | 17–40 | 0.6–1.4 | Medium |
| – Brush[b] | 30–58 | 33–69 | 19–41 | 0.4–1.1 | Good |
| – Mixed[b] | 46–63 | 44–74 | 18–41 | 0.7–1.4 | Good |
| Wood chips[e] | | 65–85 | 400–500 | 0.1–0.4 | Good |
| Straw[c] | 4–27 | — | 48–150 | 0.3–1.1 | Good |
| Horse manure[f] | 35–72 | — | 20–93 | 0.5–1.9 | Good |
| Apple pomace[c] | 88 | — | 48 | 1.1 | Poor |

Ranges are reported if available; otherwise means are reported. The means and ranges should not be considered as true means or ranges, but as representative values.
[a] Krogmann (1994) except description of structure
[b] Woyczechowski *et al.* (1995) except description of structure
[c] Rynk (1992) except description of structure
[d] Kögel-Knabner and Pichler (1999) except description of structure; sampling after mechanical preprocessing
[e] Bidlingmaier (1992)
[f] Krogmann *et al.* (2006) except description of structure

and pathogen content. Examples of feedstock characteristics are presented in Table 9.2.2. More comprehensive data about the characteristics of various feedstocks can be found for example in Bidlingmaier (1992).

Food waste is an example of a feedstock requiring careful selection of the most appropriate composting technology. This feedstock is very moist, easily degradable, and high in proteins; as a result, it is more difficult to handle than other feedstock and requires in many cases more controlled composting systems to avoid odor problems.

### 9.2.1.2 Environmental Controls

During composting, environmental controls may be necessary to appropriately handle liquid emissions (i.e., condensate from active aeration, leachate, run off), air emissions (i.e., odorous emissions, bioaerosols, dust), and noise.

Leachate, and to a lesser extent condensate, contain high concentrations of dissolved organic matter and nutrients which should not be released into surface water and groundwater. These liquid emissions should either be recycled within the facility as a source of moisture or treated prior to discharge.

Odorous emissions released during composting can be a nuisance to neighbors and have resulted in the closure of several composting facilities. Odors at composting facilities are partially controllable. For example, enclosed compost facilities showed a 65–75 % reduction in odor emissions (measured as odor units; see Box 9.2.1 for definition) as compared to odor emissions from windrow composting (Bidlingmaier and Grauenhorst, 1996). Bioaerosols (colloidal particles with bacteria, fungi, actinomycetes or viruses attached) can be a potential health hazard to compost facility workers and neighbors (Millner *et al.*, 1994).

**Box 9.2.1   Biofilters for Treatment of Odorous Air from Composting Facilities. Based on Heining (1998), Schlegelmilch *et al*. (2005), and VDI (2002).**

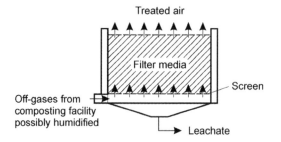

Treatment principle:

- Sorption of odorous compounds on the surface of the filter media.
- Biological oxidation of the odorous compounds.

Biofiltration as air pollution control technology is appropriate under the following conditions:

- Odorous compounds are water soluble and biodegradable.
- Temperature of the gases discharged is between 15 and 45 °C.
- Moist off gases (relative humidity of 95 %, possibly use of humidifier).
- No toxic compounds.

Advantages:

- Relatively inexpensive.
- Relatively low energy consumption.
- Generally no contaminated residues.

Biofilter types:

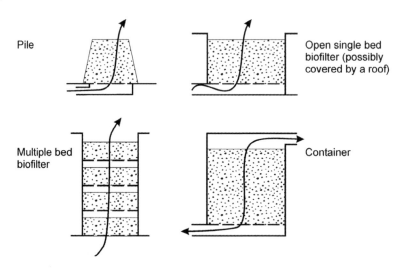

Biofilter media:

- Compost, peat, root wood, wood chips, lava, mixtures of various media.
- Function of media:

- o Suitable environment for microorganisms
- o Sorption sites for odorous compounds
- o Nutrient source
- o Provision of water
- o Buffer and storage of degradation products
- o Porous media ensuring appropriate air flow
- Lifetime of media:
  - o Depending on composition and application, generally 2–5 years, considerably longer if high concentration of inert matter
- Maintenance:
  - o Watering
  - o Removal of plants (or weeds) growing on top of open biofilter
  - o Replenishment and rototilling to prevent cracks and channelling

Sizing (long-term operating experience, but no theoretical model available applicable to composting facilities):

- Pilot test with actual off gas.
- Based on biofilter used for similar application. For example an open single-bed filter (1.5–2.0 m deep) may have a typical gas loading rate of 100–150 m³/m²/h.
- Pilot test with artificial off gas.

Example for sizing based on pilot test:

- Given: Flow rate of off gas ($F$; m³/h), odor concentration ($c_{in}$; OU/m³), target or regulatory odor concentration of treated air ($c_{out}$; OU/m³). Note: An air sample has an odor concentration of 1 odor unit (OU)/m³ if 50 % of an odor panel smells something and 50 % does not. Odor units above 1 indicate the dilutions of the air needed to reach 1 odor unit.
- Conduct biofilter type specific pilot test with a specific filter media and plot the odor degradation rate ($r$; OU/m³/h) over the average logarithmic concentration ($c_M$; OU/m³), see calculation in the figure below).
- Read degradation rate from graph (see figure below).
- Calculate required biofilter volume $V$ (m³, see calculation in figure).

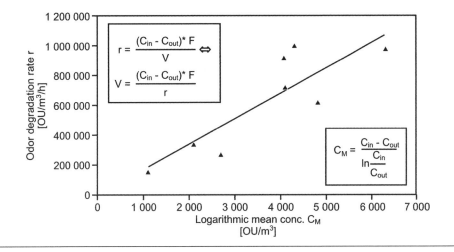

Noise from facilities is inevitable due to trucks delivering the waste, pre- and posttreatment machinery, blowers, turning machines, etc. The significance of the noise is closely related to the siting of the facilities, working hours, types of machinery, and the noise reduction introduced regarding machinery and in terms of noise barriers around the facility.

Environmental controls in composting facilities vary with technology and are key factors in the selection of a composting technology.

### 9.2.1.3   Socioeconomic Considerations

Some municipalities favor smaller decentralized composting facilities over centralized facilities because less transportation is needed and because smaller facilities might be more acceptable to the public. In some municipalities, this advantage might even outweigh higher costs. In urban areas with high property prices and neighbors nearby, generally more controlled systems are installed in comparison to more rural areas. However, it should be noted that also in rural areas, especially adjacent to residential areas, increasing process control is desired.

High population densities in many central European countries have led to the acceptance of higher costs for solid waste management technologies than in other parts of the world. This generally allows the design and construction of more enclosed facilities in this region. Economically developing countries often make use of simpler technologies for economic reasons. As a result, less-developed countries often rely on more labor-intensive technologies than industrialized countries (see Box 9.2.2).

---

**Box 9.2.2    Composting in Developing Countries.**

In developing countries, especially in municipalities located away from the major cities, composting of food and yard waste is a very appealing and cost-effective waste management option. The utility of the compost is particularly high if food and yard wastes are collected separately and not mixed with other wastes.

Municipal solid waste in most developing countries contains a relatively high percentage of putrescible wastes (often up to about 80 % by weight; CalRecovery, 2005). Highly putrescible food wastes particularly in tropical climates cannot be stored for long periods of time without taking special precautions to avoid unpleasant odors and attraction of flies and vermin.

Nuisances during composting from the highly putrescible wastes can be avoided through a combination of appropriate siting of the composting facility (as far away from human receptors as possible), buffer zones, and by processing the putrescible materials immediately upon arrival.

Pre- and postprocessing equipment can be limited to size reduction and screening. Size reduction, a maintenance- and energy-intensive process, is only required if relatively large branches or similar materials are in the waste.

Small throughputs can be managed in windrows turned manually (a person can turn about 1 t/h) or by means of small, generally locally built reactors (CalRecovery, 2005). Relatively successful examples are rotating drums that can be rotated manually if small or by an electrical motor. The laborers, especially when turning the windrows manually, should be properly trained and outfitted with protective clothing. Showers and changing rooms should be provided.

Facilities that receive high throughputs (> 50 t/day) can still rely on windrow composting. However, turning should be conducted by a front-end loader, or aerated static piles should be used. A major disadvantage of windrow composting is the relatively large land area required, which is a problem at sites in or near major cities.

Composting should be conducted on a concrete slab or on compacted clay under a roof (no walls needed) especially in areas with high precipitation (rainfall, snow). Large-scale facilities should incorporate a well thought-out odor management program, which can include buffer zones and coverage of the active windrows with a layer of cured compost or another material as an odor barrier. In the case of aerated static piles, the off gases should be treated in biofilters.

There are many examples where technically complicated compost technologies were implemented in developing countries and failed because replacement parts were not available. Therefore, priority should be on the design of appropriate composting technologies (CalRecovery, 2005).

#### 9.2.1.4   Use and Disposal of the End Products

Most feedstock generates compost that can be used as a soil amendment or as a component of a growing medium. The initial feedstock is the main factor, which determines the main chemical, physical and biological characteristics of the end product including its nutrient and contaminant content. For example, if nutrient-rich compost for use in agriculture is requested, biowaste (source-separated food, yard waste) and manure are more suitable feedstock than most yard wastes. The most efficient way to reduce contaminants (heavy metals, synthetic organics) in compost is the implementation of source separation as opposed to composting mixed MSW. The choice of composting technology also influences the compost's final characteristics. For example, the moisture content can be controlled by aeration, temperature, and pile size; pathogen reduction is affected by temperature; and the salt content in compost is reduced in open windrows due to leaching by rainwater (Krogmann, 1994).

In some European countries residual MSW (the fraction of MSW that remains after source-separation of food, yard waste, dry recyclables) is composted to produce a stabilized and volume reduced material before landfilling to reduce long-term emissions from these wastes in the landfill. In Europe, composting is also employed as a drying process to improve the recovery of recyclables and/or the production of a solid fuel (refuse-derived fuel; RDF) from residual MSW.

The intended final use or disposal of the end products must be clearly defined prior to selection of the technology to be chosen. The economic value of the end product (if any) depends on the intended final use or disposal methods and the geographic location. This may also affect the operational costs of the composting facilities.

### 9.2.2   Technological Process Factors and Their Control

To enhance the composting process and to minimize environmental impacts the microbial decomposition is controlled during the composting process. In comparison to a homogeneous, liquid substrate system where microorganisms are completely dispersed in an aqueous solution containing the soluble substrate such as an activated sludge process, composting is a heterogeneous, solid, and limited moisture substrate system (Haug, 1993). The substrate is in solid form, moisture is limited to that associated with the solids, and the microorganisms are attached to solids and dispersed in the water associated with the solids. Homogeneous systems are normally modeled using Monod kinetics, assuming that the mass transport of the substrate is not limiting substrate use and the kinetics (i.e., reaction rates) are controlled by the concentration of a limiting substrate. In a heterogeneous system like composting, many additional mass transport processes such as transport of both oxygen and soluble substrate to the cell are limiting. Since these processes are very complex, not understood in detail, and not easily modeled, the main influencing factors are typically controlled based on the scientific understanding of the living conditions of the microorganisms in the waste or compost. The main factors are:

- Biodegradability.
- Moisture content.
- Oxygen content, material structure, particle size and aeration.
- Temperature and hygiene.
- Nutrients.
- pH.

#### 9.2.2.1   Biodegradability

The degradation of organic matter during composting, measured as the percentage of the initial mass of volatile solids that is lost, depends on the composition of the feedstock, the efficiency of the technology, and the length of the composting period including curing. In a typical composting facility, food wastes are degraded more than 60 %, biowaste (source separated food and yard waste) about 50 %, and lignocellulytic plant materials about 35–45 % (Krogmann, 1994). A feedstock with a low degradation rate can cause insufficient self-heating especially during cold seasons. For example, Haug and Ellsworth (1991) showed this for a raw sludge (degrading only 48 % of VS within 60 days) amended with pine sawdust (degrading only 11 % of VS within 90 days). To ensure sufficient self-heating, a more biodegradable amendment (e.g., straw) had to replace the pine sawdust.

## 9.2.2.2   Moisture Content

The optimum moisture content is feedstock-specific and varies between 40 and 70 %, with higher optimum moisture contents for feedstocks that are coarser and have higher water holding capacities (Bidlingmaier, 1983). If the moisture content is below 12–25 % (Golueke, 1977; Bidlingmaier, 1983), only very limited biological activity takes place. Moisture contents above optimum reduce the availability of oxygen which favors anaerobic degradation accompanied by the formation of odorous gases. During composting, the maximum tolerable moisture content of coarser feedstocks (e.g., wood and bark 74–90 %) exceeds the tolerable moisture content of less structured feedstocks (e.g., paper 55–65 %, food waste and grass 50–55 %; Bidlingmaier, 1983). At the end of the composting process, the finished compost should not have a moisture content greater than 35–45 % to avoid storage, transport, and handling problems. Compost with a moisture content lower than 35–45 % may increase the release of dust.

For some dry wastes ($< 45$), the addition of water or of a feedstock with a higher moisture content may be necessary. Generally, feedstocks tend to be too wet rather than too dry. The addition of a dry bulking agent like wood chips, shredded bark, sawdust, or recycled compost is a common practice to lower the initial moisture content, which, however, can increase the compost mass significantly (see Box 9.2.3). Biodrying has been suggested to reduce the amount of bulking agent needed (Richard, 1998). The initial composting phase is operated like a sequential batch reactor with sequential addition of the wet feedstock, which is dried by the already composting feedstock.

---

**Box 9.2.3   Bulking Agent Requirements.**

One of the key properties of a bulking agent is its structural strength. A bulking agent with a high structural strength (e.g. wood chips) increases the porosity of the feedstock which, in turn, improves aeration during composting (except for bulking agents with very small particle sizes such as sawdust). Furthermore, the addition of a dry bulking agent like wood chips, shredded bark, sawdust, or recycled compost lowers the moisture content of an incoming feedstock that is too wet, for example sewage sludge, grass clippings, or food waste. In addition to its structural properties, the bulking agent usually is drier, has a higher C/N and a lower biodegradability than the initial feedstock. Furthermore, the optimum moisture content for the feedstock is usually raised through the addition of the bulking agent.

However, the addition of a bulking agent may increase the compost mass significantly. The amount added depends on the moisture content of the feedstock and that of the bulking agent and on the desired moisture content of the mixture. The recycling ratio R is the ratio of the mass of the bulking agent and the initial feedstock to adjust the moisture content of the feedstock to a set point moisture content.

$$R = G_B/G_F$$
$$G_B = (M_F * GF - M_M * G_F)/(M_M - M_B)$$

Where:

$R$   = Recycling ratio
$G_B$  = Wet mass of bulking agent (g)
$G_F$  = Wet mass of initial feedstock (kg)
$M_B$  = Moisture content of bulking agent (%)
$M_F$  = Moisture content of initial feedstock (%)
$M_M$ = Set point moisture content of mixture of bulking agent and initial feedstock (%).

Example:
The initial feedstock mass has a moisture content of 75 % ($M_F$=75 %) and the considered bulking agent a moisture content of 30 % ($M_B$=30 %). For optimum composting the moisture content of the incoming feedstock should be

adjusted to 50 % ($M_M = 50\%$).

$$G_B = (75 * G_F - 50\,G_F)/(50 - 30) = 1.25 * G_F$$
$$R = G_B/G_F = 1.25 * G_F/G_F = 1.25$$

This means that the addition of the bulking agent increases the initial feedstock mass by 125 %. If the set point moisture content is 60%, the initial feedstock mass increases by only 50 %.

The initial moisture content of the material changes during composting. Small amounts of water (e.g., 0.6 g/g $C_6H_{12}O_6$ completely mineralized) are generated as a metabolic end product and organic matter is degraded resulting in an increase in the moisture content. More importantly, the increased temperature during composting reduces the moisture content via evaporation. Evaporation is the major energy release mechanism during composting. How the moisture content changes during composting is especially dependent on the temperature throughout the compost and the aeration rate.

To compensate for the moisture loss during composting, water is added directly – most efficiently during mixing. The added water could be leachate, condensate, rain water, or tap water. Sometimes the aeration air is saturated with water to increase overall moisture levels in the compost. However, the effect of aeration with air saturated with water is limited since the temperature of the aerating air increases as it moves through the compost and the air does not remain water-saturated at elevated temperatures and can still potentially dry the compost. In addition, since evaporation is the major energy release mechanism during composting the aeration air's ability to control the temperature is reduced.

Forced aeration (blowing air into the compost) or vacuum-induced aeration (suction of air through the compost) can generate a vertical moisture gradient in the compost material that increases from the air inlet to the exhaust air outlet because the aeration air is already saturated with water as it moves towards the outlet. This can slow the microbiological degradation near the air inlet if the moisture decreases below optimum. The moisture gradient can be changed by reversing the direction of the active aeration, frequent mixing or recirculation of large amounts of aeration air.

If the water-holding capacity of the compost is exceeded, leachate is released. Leachate should not be released to ground and surface water without treatment (Krogmann and Woyczechowski, 2000).

To monitor the moisture content during composting, the moisture content can be measured in compost samples in the laboratory (US Composting Council, 2004). Experienced operators are able to assess the moisture content based on a visual inspection or a hand squeeze test (i.e., too dry if sample breaks apart, too wet if water is released during squeezing). Using in-vessel systems, the moisture content can be calculated based on the initial moisture content of the feedstock and the humidity and temperature of the inlet air and the exhaust air (off gases).

### 9.2.2.3 Oxygen Content, Material Structure, Particle Size and Aeration

Regardless of the feedstock or the selected technology, a minimum free pore space of 20–30 % is recommended for a sufficient supply of oxygen to the waste (Bidlingmaier, 1983; Haug, 1993). Grinding and shredding reduce the structure and porosity of the feedstock but increase feedstock surfaces, which enhances the microbiological degradation. If the feedstock has a stable structure, the feedstock can be finer without adversely affecting the oxygen supply (woody material < 1 cm, food waste > 2.5–5.0 cm; Golueke, 1977).

The stoichiometric need for oxygen to degrade the organic matter can be calculated (see Table 9.2.3). However the air supply should be much higher in order to ensure good distribution of the air, but also to allow for drying of the compost and for regulating the temperature. A sample calculation (Table 9.2.3) shows that drying can require tenfold more aeration air than would be stoichiometricly necessary for aerobic biological degradation. In most cases, temperature control requires about the same aeration rates as drying (Haug, 1993). The recommended oxygen content in the exhaust gas leaving the compost varies between 5 % (Strom et al., 1980) and 18 % (de Bertoldi et al., 1983).

Table 9.2.3 presents the total air requirement during composting. An average air flow rate (m³ air/m³ compost/h) can be calculated with a known moisture content (% w/w), compost bulk density (kg compost/ m³ compost), and retention time. At the beginning of the composting process the air flow rate has to be higher due to the initial intense microbial degradation and, therefore, the peak air flow rate exceeds the average air flow rate.

***Table 9.2.3***   *Total air requirements based on different objectives of aeration (calculated examples).*

| Objective | Total air demand (l air/g dry feedstock) |
|---|---|
| Supply according to the stoichiometric oxygen demand | 2.54 |
| Supply according to the stoichiometric oxygen demand, 50 % utilization rate | 5.08 |
| Drying (initial moisture content 65 %) | 11.40 |
| Drying (initial moisture content 80 %) | 27.60 |

Assumptions:
Organic matter (initial): 65 %
Degradation rate: 54.3 %
Stoichiometric oxygen demand: 2 g $O_2$/g degraded organic matter

Moisture content (end): 35 %
Air: 1.2 g/l at 25 °C, $10^5$ Pa, 23.4 wt% $O_2$
Air temperature (inlet): 20 °C
Air temperature (outlet): 60 °C

Air is provided by active aeration (forced aeration, vacuum-induced aeration, or a combination of the two), by natural ventilation (diffusion, convection) and to a lesser extent by mixing and turning. Natural ventilation is not as effective as active aeration. Figure 9.2.1 shows an example on how fast $O_2$ concentrations can decrease in a windrow after turning. Vacuum-induced aeration has the advantage that the odorous off gases are already collected and therefore easier to treat. However, the power consumption for forced aeration is lower than for suction (Miller *et al.*, 1982; Haug, 1993).

If active aeration is used, the air flow rate (e.g., controlled by restricting air flow with a baffle or installation of fan with variable rotation frequency), frequency and length of aeration periods, the direction (forced, vacuum-induced), type (fresh, exhaust air), and condition (temperature, humidity) of the aeration air can be varied. The effect of the frequency and length of aeration periods on the oxygen concentration in the compost is shown in Figure 9.2.2.

### 9.2.2.4   Temperature and Hygiene

Most composting studies have concluded that the optimum temperature during the high-rate decomposition period is about 55 °C. At temperatures over 60 °C, the diversity of the microorganisms is greatly reduced. At 70 °C the total biological activity is 10–15 % less than at 60 °C, whereas, at 75–80 °C, no significant biological activity was detected (Strom, 1985). During curing, the optimum temperature is around 40 °C (Jeris and Regan, 1973). For example, the optimum temperature for nitrification which occurs during curing is about 30 °C.

It is not very common, but in some cases spontaneous combustion of yard waste compost piles was reported (Buggeln and Rynk, 2002). Biotic and then abiotic processes (i.e., oxidation of metabolic byproducts, pyrolysis) combined with limited heat transfer increase the temperatures in the composting piles until they ignite at 120–145 °C. Spontaneous combustion can be prevented by limiting the compost pile height and ensuring that the piles are moist.

Elevated temperatures reduce pathogenic organisms in the compost. In the United States, pathogen reduction regulations require temperatures above 55 °C for three days in aerated static piles and in in-vessel facilities and above 55 °C for 15 days and five turnings during this time in windrow facilities. While these requirements are developed for sewage sludge (US EPA, 1993), these pathogen reduction requirements are also applied to other feedstocks (e.g., manure used in composts for organic fruit and vegetable production). In Europe, the pathogen reduction requirements vary from country to country. For example in Germany, pathogen reduction regulations require temperatures above 55 °C for 14 days or above 65 °C in open and above 60 °C in in-vessel facilities for 7 days (German Federal Government, 1998). However, after the foot and mouth disease and bovine spongiforme encephalopathie (BSE) outbreaks, the European Union enacted a directive for animal byproducts (EC, 2002) with various amendments (EC, 2003) that address the handling of animal byproducts. Animal byproducts include biowaste containing waste fractions of animal origin and manure. Certain animal byproducts (e.g., animal parts suspected of carrying infectious agents; internationally traded catering waste from airplanes or cruise ships) are not allowed for composting.

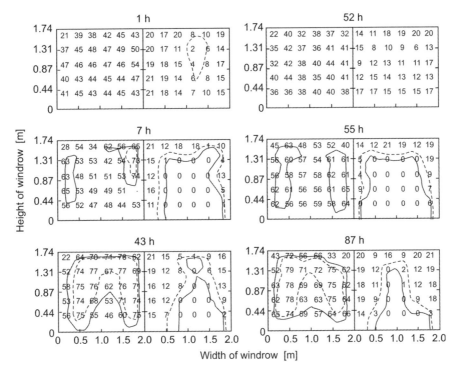

***Figure 9.2.1*** *Oxygen concentrations (right part of windrow) and temperatures (left part of windrow) in a windrow over a period of 87 h (hour 0 = 7 days after beginning of composting, turning at hour 52 (Miller et al., 1989). Reprinted with permission from Australian Journal of Experimental Agriculture (now Animal Production Science), Field examination of temperature and oxygen relationships in mushroom composting stacks – consideration of stack oxygenation based on utilisation and supply by F.C. Miller, E.R. Harper and B.J. Macauley, 29, 5, 741–750 © (1989) CSIRO PUBLISHING.*

In many large-scale composting facilities with active aeration, the temperature is controlled via active aeration during high-rate degradation (Finstein *et al.*, 1986). A temperature sensor in the compost enables feedback control of the blowers. However, with temperature control, measures need to be taken to prevent excessive feedstock drying. The drying effect is more prominent with weekly compared to daily moisture addition (Figure 9.2.3).

### 9.2.2.5  Nutrients

With the exception of nitrogen, biodegradable wastes generally contain enough macronutrients (sulfur, phosphorus, potassium, magnesium, calcium) and micronutrients to sustain the composting process. Very uniform feedstock can create exceptions. For example, MSW with a high paper content was too low in phosphorus (Golueke, 1977; Brown et. al., 1998).

Suitable carbon/nitrogen ratios (C/N) at the beginning of the composting process are between 20 and 30 for most wastes. Of greater importance is the actual availability of the carbon and nitrogen which is often neglected as a factor (Körner and Stegmann, 2002). Carbon in lignin, some aromatics and cellulose bedded in lignin are resistant to degradation. For woody feedstocks which contain a significant portion of lignin, C/N of 35 to 40 is considered optimum (Golueke, 1977). With the exception of keratin (structure protein, for example in hair) and a few similarly resistant components, nitrogen is considered very easily degradable. Wastes with lower or higher C/N can be composted, but too high C/N slow down the microbial degradation and too low C/N result in the release of nitrogen as ammonia. The most important method of controlling the C/N is by varying the composition of the feedstock.

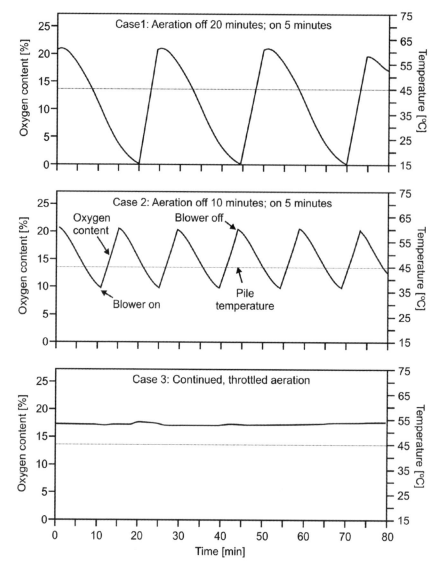

***Figure* 9.2.2**   *Effect of frequency and length of aeration periods on the oxygen concentration in compost (US Composting Council, 1994). Reprinted from Composting facility operating guide © (1994) US Composting Council.*

### 9.2.2.6   pH

For composting, the optimum pH range is between 7 and 8. Low molecular organic acids produced as anaerobic intermediate products in collection containers can reduce the pH in biowaste to 5 (Krogmann, 1994). During composting, the pH increases due to the degradation and volatilization of organic acids and the production of bases like ammonium, pyridine, and pyracine (Finstein and Morris, 1975). Fernandes *et al.* (1988) enhanced the initial degradation of fat-containing feedstocks at a pH less than 6 by the addition of CaO. However, in most cases, the addition of alkaline materials is not considered to be necessary.

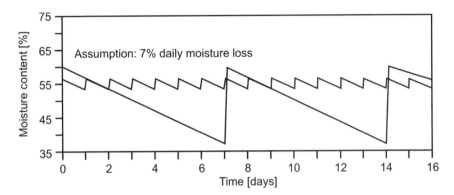

**Figure 9.2.3** *Effect of daily and weekly moisture addition on the moisture content in an actively aerated pile (US Composting Council, 1994). Reprinted from Composting facility operating guide © (1994) US Composting Council.*

### 9.2.2.7   Process Control

In a composting facility, the previously discussed influencing factors are manually or automatically controlled according to predetermined set points which vary with process stage. Generally, the composting process is technically divided into three degradation phases: high-rate degradation, stabilization, curing. High-rate degradation is the thermophilic part of the composting process where the feedstock is degraded to a 'fresh compost'. During the high-rate degradation phase, volume and mass are reduced by degradation of the easily degradable organic matter that is usually responsible for both vector attraction (disease-spreading organisms like flies, rats, etc.) and the most intense odor emissions released from a composting facility. Additionally, pathogens are destroyed due to the thermophilic temperatures. During stabilization the 'fresh compost' is degraded to 'stabilized compost'. As the temperature decreases, the degradation continues and the organic matter is further stabilized. During curing, the compost matures to 'cured compost'. Ambient temperatures are reached and the humification started earlier continues. However, it should be noted that even 'cured compost' is not completely stable; it still has an oxygen demand.

Process control is facility-specific; however, a generic example for composting biowaste in an enclosed building or in-vessel system is presented in Figure 9.2.4. During feedstock preparation the moisture content is adjusted to 45–60 % by the addition of a bulking agent which increases the porosity of the feedstock and the C/N. During composting, especially during high-rate degradation and stabilization moisture, aeration and temperature are controlled. Pathogen reduction takes places over 3 days at the beginning of the high-rate degradation at temperatures over 55 °C or 60 °C at reduced aeration rates complying with United States pathogen reduction standards. To comply with German pathogen reduction standards the temperature would need to be raised and the duration of the pathogen reduction phase extended. This pathogen reduction phase might also be moved to a later time during high-rate degradation or stabilization. For maximum degradation during the remaining high-rate degradation and stabilization, the aeration rate controls the temperatures between 40 and 50 °C. This results in the highest aeration rates during high-rate degradation. If the compost is moved or screened the aeration rate is increased to cool and dry the compost. During curing, no temperature control is required. The pH change is not controlled.

The discussed example presents an ideal process control, which in most cases cannot be accomplished by simpler technologies. However, even if process control is limited a quality end product can be produced, but the retention time might be longer and/or more emissions are released.

## 9.2.3   Composting Systems

Thousands of composting systems have been established around the world. Table 9.2.4 shows estimates of the number of facilities and the type of feedstock they used in the United States, Germany, and the region of Flanders in Belgium. Composting systems differ with respect to the product they produce and the technology they apply.

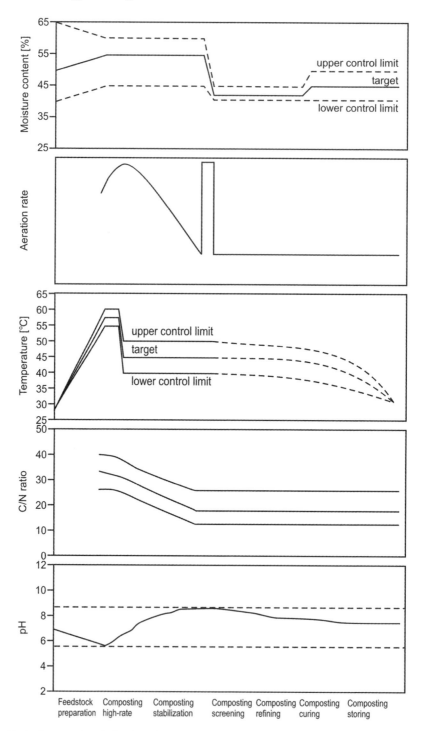

**Figure 9.2.4**   *Typical control or behavior of moisture content, aeration rate, temperature, C/N ratio and pH during composting (modified after US Composting Council, 1994).*

**Table 9.2.4** *Number of composting facilities for different feedstock in selected countries and regions.*

| | USA | Germany | Belgium (Flanders)[g] |
|---|---|---|---|
| Population | 292 000 000 | 82 000 000 | 5 900 000 |
| Land area [km$^2$] | 9 158 960 | 348 950 | 13 510 |
| Yard waste | | | |
| – Number | 3846[a] | | ~30 |
| – throughput (t/year): | | | |
| (a) per facility | — | | 5 000–50 000 |
| (b) total | 8 703 000[i] | | |
| Biowaste (residential) | | | |
| – Number | 5[b] | ~700–900[d] | 9 |
| – Throughput (t/year): | | 1200–87 600 | |
| (a) per facility | 1200–74 000[b] | (mean: ~ 14 000)[e] | |
| (b) total | 76 500[h] | ~ 7 000 000–8 000 000[d] | 20–60 000 |
| Biowaste (commercial, institutional, industrial) | | | A few biowaste composting plants also treat this waste in low quantities (max. 10 %) |
| – Number | 118 | | |
| – Throughput (t/year): | | | |
| (a) per facility | 45–79 500[c] | | |
| (b) total | — | | |
| Mixed MSW | | | |
| – Number | 12[b] | 0 | 0 |
| – Throughput (t/year): | | | |
| (a) per facility | 2 400–74 000[b] | — | — |
| (b) total | 394 000 | — | — |
| Residual MSW | | | |
| – Number | 0 | 23[f] | 0 |
| – Throughput (t/year): | | | |
| (a) per facility | — | 16 000–220 000 | — |
| (b) total | — | 1 803 000 | — |

[a]Goldstein and Madtes (2001)
[b]Goldstein (2003), calculated assuming 270 operating days
[c]Glenn and Goldstein (1999), throughput only for food residuals, facility might have larger throughput
[d]ECN (2004)
[e]Wiemer and Kern (1998)
[f]LAGA (2003) for the year 2005
[g]Devlieger, personal communication
[h]Only four facilities
[i]Only 25 of 50 states reported

Composting biodegradable organic wastes to produce marketable compost includes steps for preprocessing of the initial feedstock and postprocessing of the final compost, as illustrated in Figure 9.2.5. However, if residual MSW is processed with the objective of production a stable product before landfilling or a fuel for thermal processing different pre- and postprocessing steps are needed, as illustrated in Figure 9.2.6. Individual facilities may omit some processing steps or arrange them in a different order.

### 9.2.3.1  Classification of Composting Technologies

The composting technology itself also covers a broad spectrum of approaches, as shown in Table 9.2.5. The main difference is if the high-rate degradation takes place in the open, in an enclosed building or in a reactor. In open technologies, the exhaust gas of the composting process in most cases escapes to the surrounding environment without

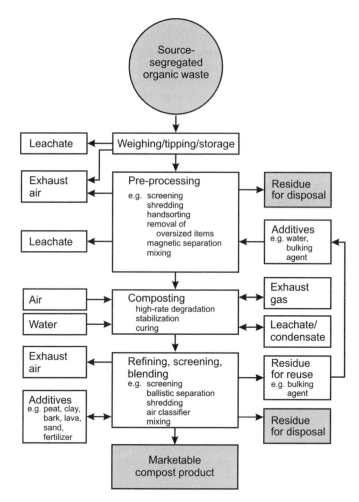

***Figure 9.2.5***   *Pre- and postprocessing to produce a marketable product.*

deodorization. Enclosed technologies and reactor technology enable the treatment of the exhaust gas. In a reactor, the free air space above the compost is minimal which lowers the quantity of the exhaust gas compared to that in an enclosed building system. During the past 15 years, enclosed technologies and reactor technologies have become the choice for municipalities composting biowaste in central Europe with high population densities and residents sensitive to unpleasant odors. Smaller open systems are found in rural areas in Central Europe.

Another distinction between composting technologies is how and in which direction the waste moves through the composting system. In static technologies, the waste is not moved after placement. In dynamic technologies the waste is continuously moving. Dynamic technologies include only a few reactors including towers (vertical flow) and rotating drums (horizontal flow). In agitated technologies, the waste is at rest for most of the time but is moved or turned at certain time intervals for homogenization, fluffing and to a lesser extent for aeration. In some technologies, the compost is agitated in place, in other technologies the agitation is combined with movement of the compost.

Due to higher biological activity and higher odor emissions, high-rate degradation and stabilization require more control compared to curing. Therefore, enclosed technology and reactors are in many cases used only for high-rate degradation and stabilization, while curing takes place in the open.

The composting technologies are presented in detail in Sections 9.3.4–9.3.6.

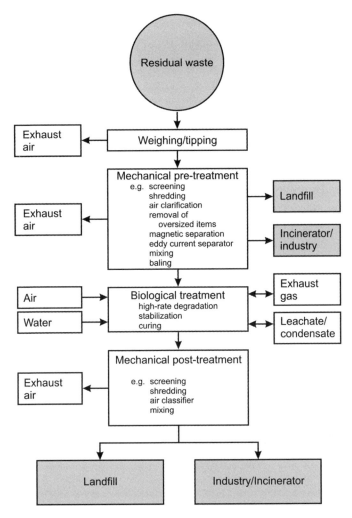

**Figure 9.2.6**  *Pre- and postprocessing of residual MSW to produce stable product before landfilling and fuel for thermal processing.*

**Table 9.2.5**  *Classification of composting technologies.*

|            | Open technology           | Enclosed technology | Reactor technology |
|------------|---------------------------|---------------------|--------------------|
| Static     | Aerated pile              | Aerated pile        | Box reactor        |
|            | Naturally vented pile     | Brikollari          | Container reactor  |
|            |                           |                     | Tunnel reactor     |
| Agitated   | Aerated windrows          | Aerated pile        | Box reactor        |
|            | Naturally vented windrows | Channel             | Container reactor  |
|            |                           |                     | Tunnel reactor     |
|            |                           |                     | Tower reactor      |
| Dynamic    |                           |                     | Rotating drum      |
|            |                           |                     | Tower reactor      |

### 9.2.3.2   Pre- and Postprocessing Technology

Various mechanical treatment technologies are available for pre- and postprocessing. Most of these unit operations are not unique to composting facilities and are used in other recycling or thermal treatment facilities. Mechanical treatment technologies are described in Chapter 7.1.

The mechanical unit operation technologies will be presented in a logical sequence as they appear in a composting facility. In addition, not all unit operations are implemented in each composting facility. Variations based on feedstock will be addressed. In small facilities, mobile equipment might be employed.

*Preprocessing Technology*

As a first step, incoming waste is weighed and unloaded in the tipping area. Generally, small yard waste composting facilities do not have scales and quantities of yard waste are calculated based on volume estimates and estimated bulk densities. The most common tipping area is a flat tipping floor. Another option includes a reception conveyor. Relatively putrescible organic wastes like biowaste need to be processed within a day, while other wastes like tree stumps might be stored for several months before further processing.

In a composting facility that produces a marketable product (Figure 9.2.5), the purpose of the feedstock recovery and preparation step is to remove noncompostable materials to improve the final compost quality (reduction of visible inert material and chemical contaminants) and to prepare the waste to adjust its biological, chemical and physical properties. For example, biowaste contains noncompostable materials that need to be separated (e.g., 0.5–7.0 % determined in a pilot project; Krogmann, 1994), food wastes are commonly too wet and yard wastes are too bulky.

Since MSW contains up to 45 % of noncompostable materials, residual MSW processing facilities need to have more extensive unit separation processing to remove noncompostable fractions, recyclable materials (e.g., metals), and high calorific fraction (e.g., plastics; Figure 9.2.6).

In composting facilities, the mechanical separation units include size fractionation and sorting. For size fractionation, screens separate the waste into large materials (film, plastics, large paper, cardboard, miscellaneous), mid-sized materials (recyclables, most organics, miscellaneous), and fines (organics, metal fragments, miscellaneous). If further preprocessing is required, size fractionation eases the subsequent unit operations. Screens used in composting facilities include vibrating, trommel and disc screens. If the waste is delivered in bags, the trommel screen might be equipped with blades to open the bags. Sorting may involve sorting of plastics and paper by air classification, metals, glass, gravel by ballistic separation, ferrous metal by magnetic separation, nonferrous metals by eddy current separation, recyclables and inerts by hand sorting, and oversized items by front-end loader (Richard, 1992).

The mechanical equipment used to producing sufficient feedstock surface area are shredders and grinders. The preference for certain shredders and grinders varies with country. High-speed hammer mills that have swinging hammers and require the materials passing through a grate and slow-speed shear shredders that consist of counter-rotating knives or hooks are probably the most ubiquitous particle size reducing devices used in composting facilities. Generally, high-speed devices such as hammer mills are very effective, however, they have a high energy consumption, are high in maintenance (e.g., repairing of hammers) and are prone to explosions. In the United States, tub grinders (rotating tub that feed a horizontal hammer mill) are the most common devices in yard waste composting facilities. Slow-speed screw mills (counter-rotating horizontal screws) are popular in biowaste composting facilities in Germany because they separate the different fibers for example in wood, providing surfaces for the microorganisms. The rotating drum, which is discussed as a composting technology, also acts as a particle size reducing device. Dense and abrasive items in the waste such as stones grind the softer materials such as paper.

The last stage of preprocessing is usually the adjustment of moisture and C/N and the addition of a bulking agent. Mixing trommels and pug mills can be used for homogenizing and mixing of two or more feedstocks. In smaller facilities mixing is combined with other unit operations such as size reduction and screening.

To ensure worker health and safety due to the release of odors (i.e., etheric oils and anaerobic metabolic products from waste storage) and bioaerosols, the preprocessing areas in many composting facilities are vented. In many cases, the vented air is used as aeration air for the composting process.

## *Postprocessing Technology*

The selection and order of different unit operations for postprocessing depend on the physical and chemical characteristics of the feedstock, the preprocessing measures and the proposed end use. Controlling the particle size can be accomplished by size fractionation and by shredding. In most facilities that produce a marketable end product (Figure 9.2.5), the compost is screened to remove larger, nondegradable particles and to produce a compost with specified particle sizes according to the proposed use (e.g., growing media 0–10 mm, mulch 10–30 mm). If the oversize materials (materials that are larger than a particular screen opening) in a facility for source-separated materials contain only small amounts of inert materials, the oversize materials can be reused as bulking agent. In addition, other inert material can be removed by ballistic separation or air classification. For example air classification of plastics is more efficient in this process stage, because the compost is drier than the initial feedstock. For some uses (e.g., growing media), the compost is blended with additives (e.g., organic additives like peat or bark and inorganic additives like clay, lava, or sand).

If residual MSW is processed (Figure 9.2.6) to produce a stabilized waste to be disposed in a landfill postprocessing might include additional steps to separate a high-calorific fraction.

## *Odor Controls*

Off gases from composting facilities are always odorous. The measures to reduce odor emissions include short storage periods for a feedstock that is very putrescible, good feedstock preparation (e.g., moisture, structure, C/N) and good control of the composting process. In open windrow systems, excessive moisture in the piles needs to be avoided and turning should only be conducted under appropriate climatic conditions (i.e., wind direction and wind speed). The composting facility should also be located at a certain distance from residential areas and at a location where the main wind direction is away from the residential areas. Best odor control can be achieved when enclosed building and in-vessel composting systems are selected. In this case, the off gases are collected and treated. The most common air pollution control technology in composting facilities is the biofilter. Details are presented in Box 9.2.1. Other air pollution control technologies are less common and are discussed by Haug (1993).

### 9.2.3.3   Design and Space Considerations

A few general guidelines should be followed when planning the layout of a composting facility.

The tipping area, the yard waste drop-off area for residents if applicable, the compost storage/pickup area, the collection area for residuals for disposal and the parking area for visitors should be kept separate (Kranert, 2000). To disentangle the traffic, preferably, the visitor parking lot should be located outside the facility and the yard waste drop-off area for residents adjacent to the scale. The tipping area should be close to the scale, while the preprocessing area and composting area can be further away from the scale.

If an enclosed building facility or in-vessel composting facility is designed, special design measures are required to handle the corrosive, warm and humid air in the enclosure. For example, structural elements are located at the outer parts of the building or certain building materials are treated with special coatings (e.g., epoxy resin).

Space considerations for composting facilities are case-specific. Therefore, space requirements need to be determined for a specific facility and a specific location. Calculations for space requirements for windrow facilities are provided by Diaz *et al.* (2002). General recommendations for biowaste composting facilities were given by Kranert (2000). Space requirements for windrow facilities for biowaste have a range of 1.2–2.5 m$^2$/t/year depending on operation and throughput and for enclosed and in-vessel facilities of 1.0–2.2 m$^2$/t/year for a throughput of 12 000 t/year. Due to reduced traffic areas, enclosed and in-vessel facilities with 50 000 t/year throughput require only 0.4–0.8 m$^2$/t/year. These space requirements can be significantly reduced when curing is not performed.

## 9.2.4   Composting Technologies: Open Technologies

### 9.2.4.1   Open Composting: Windrow Composting

Windrow composting is the oldest and simplest composting technology (Figure 9.2.7). Windrows, which are elongated piles, can be used for the entire composting process or for curing only. Windrow composting requires frequent turning by

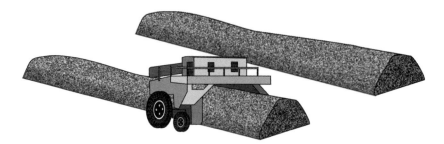

***Figure 9.2.7***  *Windrow with windrow turner. Not to scale.*

specialized equipment. Windrows are naturally ventilated as a result of diffusion and convection. Less often, windrows are aerated by forced or vacuum-induced aeration similar to aerated static piles. The aeration pipes are placed in a bed under the windrow if turning is also performed.

A turning machine is used to increase the porosity, to break up clumps and to homogenize the compost and thereby equalize moisture and temperature gradients in the windrow. The turning machine lifts, turns, reforms and sometimes moistens the windrow. The effect of turning on the oxygen supply of the windrow is not very efficient unless the pile is turned constantly. A simple, however, not very effective turner is the front-end loader, which is used in small-scale facilities. Straddle turners are commonly used. They straddle the windrow and simply turn it over. Some equipment achieves the turning by means of a rotating drum with welded scrolls or teeth; the rotating drum spans over the side frame of the machine at ground level. Others use a wide, back inclined steel plate conveyor that also spans over the frame of the machine. A turning machines drives self-powered over the windrow or is powered by a vehicle that drives next to the windrow.

Only a few facilities operate without turning. Turning frequency decreases from high-rate degradation to curing and during curing turning is often omitted. In most cases, a higher turning frequency leads to a decrease in retention time but also to an increase in operating costs. For example, the retention time to produce a 'stabilized compost' made from leaves, grass clippings and brush was reduced from 4–5 months to 2–3 months when the turning frequency was increased from once per month to seven times per month (Michel *et al.*, 1996).

The turning equipment and the aeration type (natural versus active) determine the windrow dimensions like shape (e.g., triangle or trapezoid), height, and width. For example, the bases of naturally ventilated, triangular windrows of biowaste vary between 3.0 and 4.0 m and the height between 1.0 and 2.5 m (Kern, 1991). The height of naturally ventilated windrows of leaves is recommended to be limited to 1.5 m when not turned and 2.0 m when turned once (Finstein *et al.*, 1986). With active aeration, heights of between 2.5 and 3.0 m are feasible. The length of the windrow depends on the quantities of waste and the available space. For frequently turned, naturally ventilated windrows of biowaste, retention times of 12–20 weeks are reported (Kern, 1991), while for windrows of yard waste 12–72 weeks are found.

### 9.2.4.2  Open Composting: Static Pile Composting

The main difference between the windrow and the static pile technology is that static piles are not agitated or turned. The lack of agitation in static pile composting requires the maintenance of adequate porosity over an extended period of time even more than in windrows. Either the feedstock itself needs to have sufficient porosity (e.g., certain yard waste feedstocks) or a bulking agent needs to be added to feedstocks without any structure (e.g., food waste).

In most cases, the static pile has the shape of a truncated pyramid (Figure 9.2.8). Typical dimensions are between 12 and 15 m at the base with a height of 3 m.

In the United States, the static aerated pile is one of the most common sewage sludge composting technologies. The technology was developed in 1970 in Beltsville, Maryland (USA). Piles are often covered with a layer of 'matured compost' to prevent heat loss from the upper layer and provide a minimum odor treatment (US EPA, 1981). The timer-controlled blowers maintained an oxygen level of 5–15 %. The realization of unfavorable temperatures in the static pile resulted in the development of the Rutgers process, which adopted temperature-controlled blowers (in most cases the

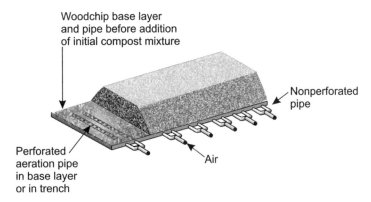

Woodchip base layer
and pipe before addition
of initial compost mixture

Nonperforated
pipe

Perforated
aeration pipe
in base layer
or in trench

Air

**Figure 9.2.8**   *Aerated static pile (WEF, 1995). Not to scale. Reprinted with permission from Operation of Municipal Wastewater Treatment Plants, Manual of Practice FD-9 © (1995) Water Environment Federation, Alexandria, Virginia.*

temperature is below 60 °C in the pile; Finstein *et al.*, 1986). In the initial and the final composting phases, the temperature feedback needs to be overridden to ensure minimum aeration if the temperatures in the pile are below the set point (Lenton and Stentiford, 1990). The aeration of the initial process in Beltsville was vacuum-induced while the Rutgers process used forced aeration. Typical retention times in aerated static piles are 21 days followed by 6–8 weeks of curing in windrows.

In piles without active aeration, a few facilities try to enhance the natural ventilation by placing aeration pipes (loops or open-ended pipes) in the piles (Fernandes and Sartaj, 1997). The pipes enhance the natural ventilation inside the pile.

In another modification, the feedstock can be stacked in open composting cells. To compensate for vertical moisture and temperature gradients in the piles, the compost is moved from one cell to another.

More recently, in some cases aerated or unaerated static piles are covered by fleeces containing air-permeable membranes (Goldstein and Satkofsky, 2001). The fleece overcomes some of the disadvantages of this relatively inexpensive composting method (odors, run-off, leachate). The fleece allows gas exchange but prevents rain from entering. Water vapor condenses on the inside of the fleece.

Another more recent inexpensive modification of the aerated static pile is an aerated pile in a patented plastic bag system which is primarily used in the United States. These plastic bags were originally used as silage bags and are modified for composting application. The company recommends composting in the bags for 8 weeks followed by 1–3 months of curing (Chiumenti *et al.*, 2005).

### 9.2.4.3   Open Composting: Other Approaches

Other less often used, special applications of the previously discussed composting systems include mat composting and vermicomposting.

Mat composting is almost more a storage and preprocessing measure than a composting technology (Kern, 1991). It is a simple process to mix different yard waste feedstocks like grass clippings, branches, leaves, and hedge cuttings which are delivered to the facility over a period of 3–12 months. The first 0.5 m mat layer consists of mostly bulky yard wastes. Additional yard waste layers are spread on this initial layer up to a final height of about 1.5–3.0 m. A front-end loader drives regularly over the previous layer, fluffs and mixes the yard waste with a fork-like device. Due to the lack of effective turning of the entire mat, the bulky particle size of the feedstock and the permanent addition of only a thin layer of fresh yard waste, temperatures in the mat reach only 40–50 °C. To continue stabilization and curing the yard waste, the mat is reformed to windrows.

Vermicomposting is a simple, very unique technology. Biodegradable organic wastes are inoculated with compost worms (i.e., *Eisenia foetida*) which break down and fragment the waste. Low, medium, and high technology vermicomposting systems – all without agitation – are available. Vermicomposting is relatively labor-intensive and, for large-scale production, requires large land areas to cool the waste to avoid temperatures which are detrimental to earthworms.

The temperature suitable for growing compost worms ranges between 22 and 27 °C (Wesemann, 1992). As a result, in traditional open vermicomposting systems, the waste is placed in beds or windrows only up to a height of about 0.5–1.2 m. At elevated temperatures, worms are only found in a fairly localized zone towards the outside of the compost pile. Sometimes vermicomposting is only implemented for curing where temperatures are low. Another reason for combining high-rate degradation (3–15 days) without worms with vermicomposting for stabilization and curing is the need for pathogen reduction. At the end of the process, the worms are separated from the castings by screening. The end product is a fine, peat-like material. Retention times of 6–12 months are reported for techniques relying solely on vermicomposting (Wesemann, 1992).

## 9.2.5   Composting Technologies: Enclosed Technologies

The major difference between open technologies and enclosed technologies is that in the latter case composting takes place in an enclosed building. The main advantage of an enclosed building composting technology is that the off gases of the composting process can be collected and treated thereby reducing odor emissions from the composting facility. However, the warm, humid off gases of the composting process condense on the cooler building roof, walls, and pipes and can lead to corrosion.

### 9.2.5.1   Enclosed Composting: Channel and Cell Composting

In an enclosed building, the feedstock is placed in triangular or trapezoidal windrows, however, in most cases the feedstock is divided by walls which are used as tracks for a turning machine (Figure 9.2.9). In these uncovered channels, the feedstock is stacked up to a height of about 2.0–2.5 m (Haug, 1993). The length (i.e. 50, 65, or 220 m) and the number of channels depend on the capacity of the facility and the proposed retention time. A facility can start with only two channels and additional channels can be added as the capacity of the facility increases. Short channels (about twice the width) are called cells.

Both active aeration (forced or vacuum-induced) and turning are used to control the composting process. Air is supplied from manifolds below each channel and aeration rate and water addition are controlled separately for each channel. In many cases, each channel is divided into several aeration areas each aerated by its own fan.

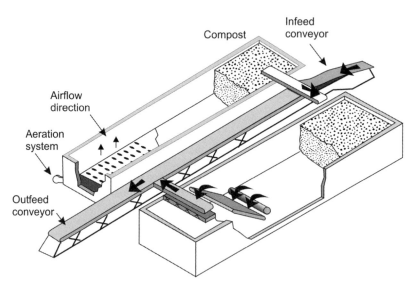

*Figure 9.2.9*   *Channel composting (US EPA, 1989). Reprinted with permission from US EPA (1989): In-vessel composting of municipal wastewater sludge. EPA 625/8-89/016. Center for Environmental Research Information, US EPA, Cincinnati, USA. © (1989) US Government.*

Due to frequent turning, the compost moves from the beginning to the end of the channel. Turning frequencies of every day or every other day are reported. Some technologies compensate for the volume loss during turning to keep the height of the compost constant while other systems compensate for the volume loss when restacking the compost in another channel. Typical retention times in these channels are 6–8 weeks for all composting phases (Schmitz and Meier-Stolle, 1995). However, some facilities compost for 21 days in channels followed by 6 months curing outside in static piles (Haug, 1993).

### 9.2.5.2  Enclosed Composting: Aerated Pile Composting with Automatic Turning Machines

The difference between channel composting and aerated pile composting with automatic turning machines is that the feedstock is not placed in channels which can be individually controlled but in one large pile with heights up to 1.8 and 3.3 m (Kugler *et al.*, 1995). The frame for the turning machine spans over the whole width of the composting hall (e.g., about 35 m in facility shown in Figure 9.2.10).

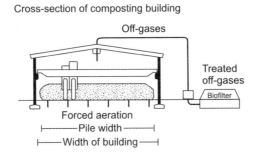

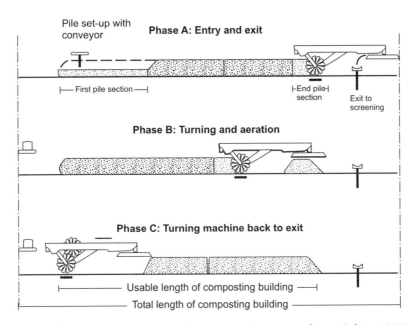

***Figure 9.2.10*** *Aerated pile composting systems with automatic turning machines (Schön, 1992). Reprinted from Schön, M. DasWendelin-Verfahren zur Bioabfallkompostierung (TheWendelin process for biowaste composting; in German). In: Wiemer, K. and Kern, M. (eds) Verfahrenstechnik der Bioabfallkompostierung, pp. 129–144. M.I.C. Baeza-Verlag, Witzenhausen, Germany. © (1992) Baeza-Verlag.*

A forced and/or vacuum-induced aeration system is placed under the compost bed similar to the aerated static pile bed. The composting hall is subdivided into several aeration areas that can be separately aerated and moistened depending on the progress of the composting process. In most cases, the feedstock is conveyed from the preprocessing area to one end of the composting hall, moves through the composting hall via a turning machine and exits the composting hall at the opposite site. The volume loss is compensated for in most systems via the turning machine. The aeration rate can be reduced from $18\,000\,m^3/t$ biowaste for forced aeration to $7000\,m^3/t$ biowaste for a combination of vacuum-induced and forced aeration with a heat exchanger (Kugler *et al.*, 1995).

In some technologies the feedstock is stacked in one area in the composting hall, where it remains for the whole retention time. In these systems, the height of the pile decreases during composting.

In most cases, water is added during turning. Turning machines vary with the manufacturer, but they are typically floor-independent and movable over the entire composting hall (e.g., bucket wheel, diagonally working screws, double spindle agitators, or vertically working screws). The turning frequencies range from once a day to once a week which, of course, affects the moisture gradient in the pile. A few systems do not turn for the first four weeks (Zachäus, 1995). Retention times range between 4 and 12 weeks, followed by windrow composting for the shorter retention times.

### 9.2.5.3   Enclosed Composting: Brikollari Composting

The Brikollari process is a unique type of static composting used in some facilities in Germany (Linder, 1995). Biowaste, well mixed and amended with a bulking agent, is compressed into blocks (30–60 kg) that are stacked crosswise on pallets. Channels pressed into the surface of the blocks provide aeration via natural diffusion. The coarse structure of the blocks ensures aeration even within the block.

The compaction requires more electrical energy than other preprocessing steps which in turn is compensated for by the low space requirements. An automatic transport system moves the stacks with a mass of 1.2–1.4 t each to a multifloor high-rack warehouse for high-rate and stabilization degradation. Dividers separate the warehouse in several areas which can all be ventilated separately depending on the progress of the composting process.

Retention times in the warehouse range between 5 and 6 weeks. At this time the moisture content in the blocks is about 20 %. The blocks can be marketed as 'stabilized compost' after grinding, can be stored before being processed further due to the low moisture content or can be immediately cured after the addition of water. Retention time for curing in windrows is about 8–10 weeks.

## 9.2.6   Composting Technologies: Reactors

In comparison to enclosed technologies, reactor or in-vessel technologies have minimal free air space above the compost and this reduces the volume of exhaust gases that requires treatment. Additionally, the aeration system can be better controlled (i.e., exhaust air recirculation, conditioning of the aeration air).

### 9.2.6.1   Reactor Composting: Tunnel Composting

Tunnel reactors include static and agitated composting technologies with different levels of process control. They have been used for composting of MSW, sewage sludge, and manure for many years. During the past ten years, there has been special interest in tunnel reactors used for composting manure in the mushroom industry because these reactors are better controlled than previous MSW tunnel composting reactors and their design and operation are based on many years of experience. Several manufacturers adapted this concept for composting biowaste. A characteristic of this system is that it maintains a relatively homogeneous temperature and moisture profile over the height of the compost due to large amounts of recycled exhaust gases (Figure 9.2.11). As a result, less turning is needed to homogenize the compost.

The number of tunnels is chosen according to the proposed facility capacity. Typical tunnel reactor lengths are 30–50 m and widths and heights 4–6 m. Each tunnel reactor is separately controlled and, depending on the process control, fresh air, recycled exhaust gases, or a mixture of both are supplied from below the bed. Tunnels are often equipped with nozzles in the ceiling that are used to moisten the compost with process waters (condensate, leachate). The tunnels are provided

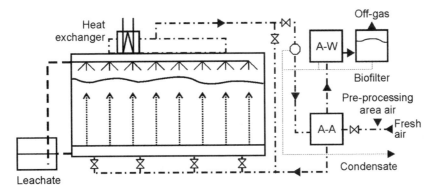

**Figure 9.2.11** *Example of typical process control in tunnel or box systems (A-A: air–air heat exchanger; A-W: air–water heat exchanger).*

with hatches or doors to feed or remove the compost. Adapted from the mushroom industry, special plastic slide nets are used to empty the tunnel with a movable net winder with integrated skimming roller. In most cases, feeding is carried out automatically by band conveyor systems. If the retention times in the tunnel system are short (i.e., 1 week), agitation may not be carried out. One way of turning is to empty the tunnel and to backfill another tunnel which also compensates for the volume and mass loss. Turning can also be accomplished within one tunnel by means of turning devices movable into each tunnel – for example on rails.

The retention time of the compost in the different tunnel systems varies between 1 and 7 weeks (Zachäus, 1995). If composting in the tunnel system is of short duration, additional curing in windrows is required.

### 9.2.6.2   Reactor Composting: Box and Container Composting

In box and container composting, the reactor units are very similar to tunnels, however shorter. Composting boxes with volumes of about 50–60 m³ (7 × 3 × 3 m) up to volumes of 250 m³ are found. The entire front area of a box reactor consists of a door.

In comparison to composting boxes, the smaller composting containers with volumes of about 20–25 m³ (5 × 2 × 2 or 6 × 2 × 2) are movable. Next to the preprocessing area, the containers that can be opened at the top are filled. A truck or crane moves the container back to the composting area where each container is connected to the aeration, the air conditioning and the leachate collection system with quick-couplings.

The boxes and containers are filled by a front-end loader or automatically by specialized conveyor belts. In the composting area, the aeration and air conditioning system are individually controlled for each reactor. Forced aeration, vacuum-induced aeration or a combination of both are possible. A homogeneous moisture and temperature profile is accomplished due to air recirculation and the addition of fresh air. Some reactors have segmented aeration floors which enables control of the different aeration areas individually. Some composting boxes pull air from a box filled with fresh feedstock and force it through a composting box with more mature feedstock.

The retention times in container or box reactors are usually between 7 and 14 days followed by curing for about 12 weeks in windrows, resulting in total retention times of 13–14 weeks (Zachäus, 1995). The retention time is shorter if the compost after the first run is backfilled twice to a box for a second and a third run with a total retention time of 6–7 weeks. In some systems, agitation is carried out between 1 and 4 times a week with retention times of 14–50 days in the box. Curing in windrows may follow. An example of a box composting technology can be found in Section 9.2.7.

### 9.2.6.3   Reactor Composting: Rotating Drum Composting

Composting in a rotating drum is a dynamic process; this technology has been used for composting MSW all over the world. The rotating drum (Figure 9.2.12) acts like a ball mill and the constant movement of the rotating drum ensures

*Figure 9.2.12*   *Rotating drum.*

homogenization and fluffing. As an outcome of the resulting thorough aeration, odor emissions from rotating drums are low. The removal of inert materials is more effective after leaving the rotating drum, because the compost is drier and well homogenized.

Rotating drums are inclined to move the material and volumes vary between 30 and 500 m$^3$ (Haug, 1993; Kasberger, 1995). The reactors are not completely filled to ensure sufficient mixing. Feedstock inlets and compost outlets are located on opposite ends of the rotating drum and the compost tumbles slowly through the reactor. To prevent possible short-circuiting of compost stages resulting in some material undergoing less processing, the rotating drum can be divided into sections.

The rotational speed (i.e., 0.1–1.0 rpm) and whether the rotation is continuous or intermittent (i.e., 45 min in 24 h on the average for biowastes; Kasberger, 1995) can be controlled. Especially for wet feedstocks like biowaste, high rotational speeds can result in compaction. In most cases, aeration is provided by a fan, less often the rotating drum is perforated allowing aeration via natural diffusion. In the last case, the rotating drum is vented to minimize odor emissions.

The retention time in the rotating drum system ranges between 1 and 10 days (Kern, 1991; Kasberger, 1995). At lower retention times only a portion of the high-rate degradation is performed while at higher retention time high-rate degradation and stabilization are conducted in the rotating drum. In any case, additional windrow composting between 2 and 3 months will be required (Haug, 1993; Kasberger, 1995; Chiumenti *et al.*, 2005).

### 9.2.6.4   Reactor Composting: Tower Composting

Vertical flow reactors, mostly cylindrical towers and less often rectangular reactors are agitated or dynamic composting technologies. They are often used to compost sewage sludge and less often MSW and biowaste. Some towers are divided into separate vertical compartments by interior floors. Tower volumes of 400–1800 m$^3$ with total material depths of 6–9 m can be found (Haug, 1993).

Vertical reactors are commonly fed at the top on a continuous or intermittent basis as the feedstock in the tower moves slowly from the top to the bottom during composting. If interior floors are used, the material is transported vertically to the next floor by flaps or movable grates. The floors enhance agitation during the movement down through the reactor. Without the floors, the movement through the reactor is more like a plug flow.

In most cases, the reactors are aerated by forced aeration counter-current to the compost flow. Large compost depths result in a high pressure drop in the compost which affects the efficiency of the aeration system. In addition, due to the lack of effective agitation, the compost can become compacted. This can potentially result in anaerobic niches. In particular, odor problems have led to the closure of a significant number of tower facilities.

After movement through the vertical composting reactor, the material is either cured in open windrows or filled into another composting tower for a second run. Typical retention times in vertical reactors range between 14 and 20 days (Haug, 1993). Stabilization and curing in windrows follow.

## 9.2.7 Examples of Composting Facilities

In the following section, three composting facilities are schematically presented, each processing a different feedstock. The purpose of these illustrative examples is to show how the feedstock affects the selection and operation of a composting facility, to present the process sequence in a composting facility and to summarize the information presented in this chapter. These featured facilities are only examples since there are many other technologies that can be used to compost these feedstocks and there are other ways to operate these facilities. The selection of these three facilities is not an endorsement of these specific technologies over others. It is recommended that detailed information from various vendors be obtained and visiting reference facilities can also be useful when planning a composting facility.

### 9.2.7.1 Example 1: Biowaste Composting Facility in Cloppenburg-Stapelfeld, Germany (Status Summer 2004)

The biowaste composting facility in Cloppenburg-Stapelfeld, Germany, is illustrated in Figure 9.2.13 and has the following specifications:

- Feedstock: Biowaste from residents (BF), yard waste from residential, institutional and commercial sources (Y), potato processing wastes (P).
- Design capacity: 20 000 t/year.
- Current capacity: 17 000 t/year (BF: 13 400 t/year, Y: 3400 t/year, P: 200 t/year).
- Beginning of operation: July 1994.
- Total area of composting facility: ~ 1.25 ha (Note: Offices, maintenance shop, and scale are not included, and the majority of the compost is marketed as stabilized compost).
- Composting system: System Herhof composting boxes (50 m³ each).
- Mobile equipment: Two front-end loaders (70 kW, 2.5 m³ bucket), one Rudnick and Enners MTS 5000 trommel screen (diesel/electric, 31 kW, throughput material dependant, approximately 40 m³/h with 20 mm screen openings), one self-built air classifier, one AZ Jenz 55 hammer mill (200 kW, 110–220 m³/h).
- Stationary equipment: One overhead magnet at exiting conveyor from hammer mill, one hand-sorting room, 12 System Herhof composting boxes, two System Herhof biofilter boxes (50 m³ each), various belt conveyors.
- Power consumption: 450 000 kWh/year.
- Number of employees: 3.5–4.0.
- Process control in composting box: Temperatures in the off gases are lower than the temperatures in the composting boxes and are initially controlled at 45 °C for optimum degradation and then at 60 °C for 3 days for pathogen control. The German regulatory authorities approved this technology with the above described temperature control as a technology being equivalent to a technology complying with the German temperature requirements for hygienization, as described in Section 9.2.2. Before emptying the composting box, the temperatures are lowered to about 30 °C, so that the compost is less odorous when moved out of the boxes. The box is divided into six different aeration segments to enhance the equal distribution of the aeration air. Leachate is recirculated to keep the moisture content of the biowaste at 50 % by weight and to avoid discharging highly contaminated water (COD, $BOD_5$) into the sewer at high costs.

(1) Tipping area for yard waste
(2) Reception area with chain conveyer
(3) Hand sorting cabin
(4) Mobile shredder
(5) Composting box
(6) Biofilters
(7) Curing area
(8) Screen

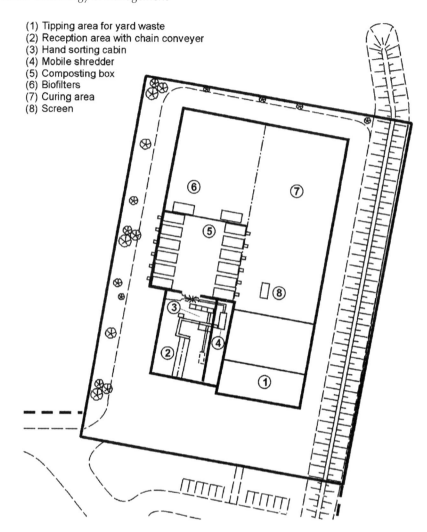

**Figure 9.2.13** *Layout of composting facility in Cloppenburg-Stapelfeld/Germany (courtesy of BioKomp Verwertungsgesellschaft mbH, Cloppenburg, Germany). Reprinted from Biokomp © (2010) BioKomp.*

- Each box has two blowers, blower 1 for the entering fresh air ($1800\,\text{m}^3/\text{h}$ at $1080\,\text{Pa}$, $1.1\,\text{kW}$) and blower 2 for the off gas and air recirculation ($1800\,\text{m}^3/\text{h}$ at $1940\,\text{Pa}$, $2.2\,\text{kW}$). In the off gas mode, blower 1 pushes fresh air in the box and blower 2 draws off gas from the box into the biofilter. The recirculation mode ensures pathogen control requirements ($61\,°\text{C}$ in the off gas). In the recirculation mode, blower 1 is off and blower 2 draws air from the box and pushes it back into the box. If the set point temperature of $61\,°\text{C}$ or a $CO_2$ concentration of $5\,\%$ in the recirculating air is exceeded, the off gas mode is operated for a short period.
- Retention time: BF and P: 6–7 days in composting boxes plus 2 weeks in windrows (end product is fresh compost), Y: 10–12 weeks in windrows (end product is stabilized to cured compost). Note: Retention time longer when compost cannot be marketed immediately, e.g. in winter.
- Environmental controls: Distance to neighboring houses is 500 m. Off gases of the composting boxes are treated in two biofilters. Condensate from the off gases and the biofilter is collected in a cistern ($10\,\text{m}^3$) and is used to adjust

the moisture content of the windrows (Y waste). All curing piles are covered by a roof to reduce run off and to avoid that the piles get too wet. Since 2000 the facility is operated in a mode that results in no excess process water being discharged from the facility.

- Investment costs: ~ € 6 000 000. Operating costs: € 40/t. Tipping fee: € 5–10/m$^3$ for Y waste (funded by residents and businesses via 'solid waste fees' for BF waste).
- End products: Fresh compost from BF waste, 0–30 mm (~ 7200 t/year); stabilized/cured compost made from Y waste, 0–20 mm (~ 1800 t/year); residuals sent to landfill (~ 250–300 t/year).
- End market: Agriculture markets for fresh compost (grain, corn, free transport and spreading); landscapers and the general public (garden centers) for stabilized/cured compost.
- Heavy metal content in ppm (median for 10 samples from 2002, normalized to a compost with 30 % organic matter as required by German compost quality guidelines):– BF and P: Pb 22.8, Cd 0.30, Cr 22.3, Cu 51.0, Ni 6.80, Hg 0.06, Zn 147– Y: Pb 24.4, Cd 0.33, Cr 24.8, Cu 31.8, Ni 6.25, Hg 0.07, Zn 121.
- Potential upgrades: Expansion of curing and storage area.
- Description of processing sequence: Public works and private customers tip the Y waste in a tipping area (1) and the BF and P wastes on a 3-m wide reception bunker with chain conveyor (2) at 10–100 m/h. One staff person inspects the load before the collection vehicle leaves. Loads that are contaminated are rejected. If BF and P wastes are too wet, up to 15–20 % Y waste and oversize materials (> 30 mm) from screening finished compost are added. BF and P wastes are transported by several covered conveyors including a vibrating conveyor that fluffs and distributes the waste over the width of the conveyor to an elevated hand sorting area (3). In the elevated hand sorting area, impurities are removed manually by one or two persons. The rejects are transported via a conveyor to a container that is regularly hauled to the landfill. From the hand sorting room, the BF waste is conveyed to a mobile hammer mill (4) where it is shredded. Then, ferrous metals are separated by an overhead magnet and the feedstock is conveyed to the composting area. There, a front-end loader fills the composting boxes (5) where BF waste is composted for 1 week. The off gases of the composting boxes are treated in two biofilters (6). After 1 week the composting boxes are emptied and the compost is moved to a curing area which is covered by a roof (7). In the curing area, the compost is stabilized in 2-m high piles for about 2 weeks. Then, a front-end loader loads the fresh compost into the trommel screen for screening (8). Plastic contaminants are removed from the oversize material by an air classifier. The oversize materials are returned to the input if they are clean, otherwise they are landfilled. The compost is stored until marketed. The Y waste is not processed in the composting boxes. After grinding in the hammer mill (4), piles (max. 3 m high) of shredded Y wastes are formed in the curing area. The piles are turned every month by a front-end loader and wetted by sprinklers with condensate from the cistern or with rainwater from the roof collected in a second cistern. To ensure pathogen reduction in Y waste piles, temperatures in the pile center need to reach 65 °C for 2 weeks (1 week before and 1 week after turning, measured daily). After 10–12 weeks the compost is screened.

### 9.2.7.2 Example 2: Yard Waste Composting Facility in Dickerson (Status Summer 2004)

The yard waste composting facility in Dickerson (Montgomery County, Maryland, USA) is illustrated in Figure 9.2.14 and has the following specifications:

- Feedstock: Yard waste from residential, commercial and institutional sources (~ 50 vol% leaves, 50 vol% grass clippings, includes small amounts of other yard trimmings; delivered from transfer station where yard waste was preshredded to break up collection bags made of kraft paper).
- Design capacity: 80 000 t/year.
- Current capacity: 77 000 t/year (based on legal agreement with the community).
- Beginning of operation: 1984 (expanded once).
- Total area of composting facility: 46.7 ha, including 22.3 ha asphalt for compost operation.
- Composting system: Windrow composting.
- Mobile equipment: three Scarab 18 windrow turners (5.5-m drum, 340 kW), four farm tractors (41–63 kW), five rubber tire loaders (two Caterpillar 966 with 6.8-m$^3$ Balderson roll out buckets, 185 kW; three Komatsu WA450 with 6.8-m$^3$ Rockland roll out buckets, 185 kW), two Diamond Z tub grinders at transfer station (735 kW, 912 kW),

**Figure 9.2.14**   *Picture of Montgomery County composting facility in Dickerson. Courtesy of Montgomery County, Department of Public Works and Transportation, MD, USA.*

two trommel screens (Royer 616, electric, 4.9-m trommel with magnet; McCloskey Brothers 833, 9.2-m trommel, 56 kW), two irrigation pumps.

- Stationary equipment: two Amadas bagging machines (1.5 kW), two stretch wrapping machines (Wolftec, Lantech).
- Power consumption: 232 000 kWh/year.
- Number of employees: 9.
- Process control: Based on experience, hand squeeze test and manually measured temperatures that decline too early, windrows that are too dry are spread out and irrigated by a sprinkler. Watering is generally needed from May to September. The time when the compost is ready for screening and curing is based on the Solvita compost maturity test and other determinants such as experience, smell and consistency.
- Retention time: 18–24 months. Note: The leaves are mainly stored for the initial 6 months. The compost could be marketed after 12 months. However, the retention time is longer because of the volume of material processed and the time it takes to coordinate activities on such a large scale and because of the preference of soil blending market of a very stable end product. The retention time includes storage of compost when it cannot be marketed immediately (e.g., in winter).
- Environmental controls: Distance to closest neighbor is 400 m. Pieces of plastic from bags that were not appropriately source-separated are collected from perimeter fence. The facility has three sediment ponds; one for each asphalt compost pad. The ponds are permitted through the Maryland Department of Environment (MDE) under the National Pollutant Discharge Elimination System (NPDES). Runoff from each of the three composting pads is separately drained through a 'forebay' sediment basin into a holding pond. The water is stored on the site in the holding pond for future use as irrigation water for the piles. After proper testing and assurance that permit requirements are met, remaining water is sometimes released through a monitored process to tributaries of the Monocacy River. The permit allows a discharge of a $BOD_5$ of 100 mg/l, total suspended solids of 100 mg/l and a pH of 6–9.
- Investment costs: Not available. Operating costs: $ 28.60/t (no debt service, revenues subtracted), Tipping fee: Funded by municipalities via 'solid waste fees'.
- End products: Cured compost ($< 0.95$ cm) in bulk (37 250 m$^3$/year) and bags (12 150 m$^3$/year); 'ground oversize materials' (shredded screen overs) as mid-grade mulch. Residuals (plastic bag pieces collected from fences) are sent to the county transfer station and ultimately to the adjacent MSW incinerator.

- End market: Top soil blenders, landscapers, and nurseries.
- Heavy metal content in ppm (one sample for January 2004, for comparison with example 1, numbers in parentheses are normalized to a compost with 30 % organic matter): Pb 38 (70), Cd 2.0 (3.7), Cr 24 (44), Cu 50 (92), Ni 20 (37), Hg < 1.0 (< 1.84), Mo 2.0 (3.7), Se 4.0 (7.4), Zn 143 (263).
- Description of processing sequence: All preshredded yard waste delivered from the transfer station is weighed on the scale. In the fall, approximately 60 m long windrows of leaves are formed by a rubber tire loader. A Scarab windrow turner mixes the pile once and then larger storage piles are formed (5.5 m high by 9 m wide) until the grass season starts in early April. Beginning of March, smaller windrows are formed. Grass clippings are deposited next to the leaf windrows and mixed with the leaves initially by a rubber tire loader and then by a windrow turner. The first week, the 60 m long and 2.7 m high windrows are turned twice per week and afterwards weekly. The piles are spread and watered when needed. Volume reduction due to occurring degradation is compensated by combining piles when they are smaller. After about three to four months, based on the Solvita compost maturity test and other determinants as specified earlier, the compost is cured for 3 months in 5.5-m high piles. Before marketing and bagging, the compost is screened in a 0.8-ha covered processing building (see compost storage building in Figure 9.2.14). If the piles are too wet (> 30–40 %) they are spread for drying before screening. The oversized fraction from screening is ground into mulch. The bagging takes place in the processing building.

### 9.2.7.3   Example 3: Residual MSW Composting Facility in Bassum, Germany (Status Summer 2004)

The Residual MSW composting facility in Bassum, Germany is illustrated in Figure 9.2.15 and has the following specifications:

- Feedstock: Residential, commercial and institutional MSW 37 000 t/year, dry industrial MSW 29 000 t/year, bulky waste (i.e., old furniture) 7000 t/year, residuals from construction and demolition (C&D) waste processing 2000 t/year.
- Design capacity: 112 000 t/year (preprocessing), 12 000 t/year (anaerobic digestion), 40 000 t/year (composting including 7500 t/year residuals from anaerobic digestion).
- Current capacity: 75 000–80 000 t/year (preprocessing), 9000 t/year (anaerobic digestion), 40 000 t/year (composting).
- Beginning of operation: Fall 1997.
- Total area of facility: $\sim$ 3 ha.
- Composting system: Aerated pile system with Wendelin automatic turning machine.
- Mobile equipment: one Liebherr A 902 grapple excavator (74 kW), two front-end loaders (CAT, 928G, 98 kW; KOMATSU, WA270, 103 kW).
- Stationary equipment (anaerobic digestion not included): two roller shredders (Doppstadt DW 3080 E, 360 kW; Doppstadt DW 2560 E, 280 kW), one Husmann stationary compactor MP 1600 (12 kW), four belt magnets (Andrin, 3–5 kW), two trommel screens (screen 1: Bühler, ZSR-30V, trommel length 12 m, 80-mm openings, 2 × 24 kW; screen 2: Bühler, ZSR-30V, trommel length 12 m, 40-mm and 80-mm openings, 2 × 24 kW), onw biofilter (two closed chambers), one automatic turning machine 'Wendelin', one mixing trommel (will be taken out of commission because mixing of anaerobic digester residuals and fraction < 80 mm not effective), various belt conveyors.
- Power consumption: $\sim$ 30 kWh/t (2 400 000 kWh/year at a capacity of 80 000 t/year).
- Number of employees: 11 (including the anaerobic digester).
- Process control of composting system: The pile in the composting building is divided into eight pile sections and five aeration sections. Each aeration section is controlled individually (widths of all sections: 35 m; lengths: 14.45 m, 12.50 m, 25.20 m, 31.50 m, and 9.45 m, respectively). Both forced aeration and vacuum-induced aeration are available. The blowers are sized as follows: 9000 m³/h at 290 Pa, 9000 m³/h at 290 Pa, 15 800 m³/h at 250 Pa, 9300 m³/h at 190 Pa, 4600 m³/h at 250 Pa. Each blower has a fixed aeration rate and the length of each aeration period is controlled by a timer. During weekly turning the waste is watered if necessary.
- Retention time: 8 weeks in composting system.
- Environmental controls: The building air (up to 96 700 m³/h = $\sim$ 1.5× building volume/h) is treated in a biofilter with two chambers (each chamber: 3.3 × 26.5 × 10 m shredded root wood, gas loading rate: 198 m³/m/h or 60 m³/m³/h). The biofilter is operated from top to bottom. Before entering the biofilter the building air is humidified by two

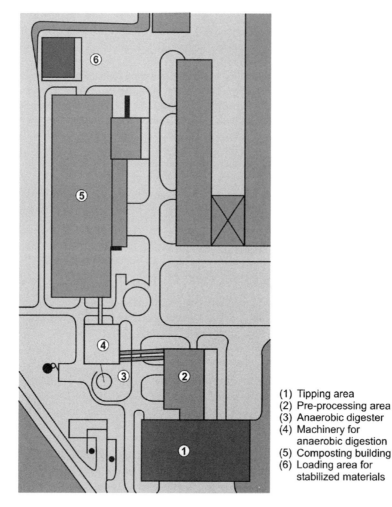

(1) Tipping area
(2) Pre-processing area
(3) Anaerobic digester
(4) Machinery for
    anaerobic digestion
(5) Composting building
(6) Loading area for
    stabilized materials

**Figure 9.2.15**    *Layout of residual MSW treatment facility in Bassum/Germany (courtesy of Abfallwirtschaftsge-sellschaft mbH, Bassum, Germany). Reprinted from http://www.awg-bassum.de/virthos.php?-pg=474 © (2010) AWG-Bassum.*

spray nozzle reactors (humidifiers), each processing 48 350 m³/h building air and recirculating 18 m³/h water. The off gases of the biofilter are discharged through a 18-m high stack for better dispersion. Condensate from walls of the composting building and the vacuum-induced aeration system and leachate are collected and treated in the leachate treatment facility of a nearby landfill.

- Investment costs: ∼ € 25 000 000, Operating costs: € 70/t, Tipping fee: € 60/t.
- End products: Fraction > 80 mm (35 000 t/year, from preprocessing dry industrial and C&D waste, sent to a MSW incinerator or as solid recovered fuel to power plants and cement industry), biogas (1200 t/year), biologically stabilized waste (25 000 t/year, sent to landfill), ferrous metals (1000 t/year, sent to steel industry).
- Regulatory requirements for end products: Solid recovered fuel (e.g., requirements concerning heavy metals; for details see RAL, 2001); biologically stabilized waste [among other criteria compliance with criteria assessing the stability such as TOC (total organic carbon) < 18 % total solids, lower heating value < 6 MJ/kg, oxygen consumption rate < 5 mg/kg total solids, TOC (extract) < 250 mg/l (German Federal Government, 2001)].

- Description of processing sequence: Bulky waste and bulky fractions of dry industrial wastes are delivered to flat tipping area I (1) where these wastes, if necessary, are preshredded by one of the roller shredders. The preshredded waste is screened in a Bühler ZSR-30V screen (80-mm openings). The fraction > 80 mm is compacted, possibly after additional shredding, to be sent to an MSW incinerator or to industry as solid recovered fuel. The other feedstocks are delivered to flat bunker area II (1) and then conveyed to the other BÜHLER ZSR-30V in the preprocessing area (2). The fraction < 40 mm is conveyed to the anaerobic digester (3, 4), the fraction 40–80 mm together with the fraction < 80 mm from tipping area I to the composting area (5) and the fraction > 80 mm to the stationary compactor. During preprocessing, ferrous metals are separated by several overhead magnets. The fraction sent to the composting area is mixed with the residuals from the anaerobic digester, currently in a mixing trommel but in the future on a conveyor belt. The mixed wastes are set up as a 2.7-m high pile (designed for a maximum of 3.3 m) in the 40-m by 132-m large composting building. The pile (35 × 93 m) is turned weekly by the automatic turning machine and watered during turning when needed. Condensate and leachate from the composting building are stored in a tank and either used for watering the waste or discharged. Additional water for watering comes from a rainwater retention basin that is also used for fire emergencies or two tanks (600 m$^3$) storing surface water. After 8 weeks the stabilized waste is sent to the landfill (6).
- Upgrades in the near future: Some modification of the screening process is planned to improve the quality of the solid recovered fuel. The fraction < 40 mm from residential waste will be treated in the anaerobic digestion unit, the fraction < 60 mm will be processed in the composting unit, the fraction 60–300 mm will be used as solid recovered fuel and the fraction > 300 mm will be sent to a MSW incinerator. Further curing of the biologically stabilized waste is necessary to fulfill the regulatory requirements for the biologically stabilized waste. Curing for 4 weeks in a composting pile under a roof is planned. The biofilter of this facility is not efficient enough (e.g. total organic carbon in the off gases) to comply with regulations, therefore, a regenerative thermal oxidation unit will be installed to treat the off gases of the high-rate degradation of the composting process.

# References

CalRecovery (2005): *Solid waste management*. CalRecovery Inc./UNEP International Environmental Technology Centre, Osaka, Japan.

de Bertoldi, M., Vallini, G., Pera, A. and Zucconi, F. (1983): Comparison of three windrow systems. *BioCycle*, 2 (3/4), 45–50.

Bidlingmaier, W. (1983): Das Wesen der Kompostierung von Siedlungsabfällen (Basics of MSW composting; in German). In: Hösel, G., Bilitewski, B., Schenkel, W. and Schnurer, H. (eds) *Müll-Handbuch*, vol. 4. KZ 5305. Erich Schmidt Verlag, Berlin, Germany.

Bidlingmaier, W. (1992): Charakteristik fester Abfälle im Hinblick auf ihre biologische Zersetzung (Characterization of solid wastes regarding their biodegradability; in German). In: Hösel, G., Bilitewski, B., Schenkel, W. and Schnurer, H. (eds) *Müll-Handbuch*, vol. 5. KZ 5303. Erich Schmidt Verlag, Berlin, Germany.

Bidlingmaier, W. and Grauenhorst, V. (1996): Geruchsemissionen von Kompostierungsanlagen (Odor emissions of composing facilities; in German). In: Stegmann, R. (ed.): *Neue Techniken der Kompostierung*. Economica Verlag, Bonn, Germany.

Brown, K.H., Bouwkamp, J.C. and Gouin, F.R. (1998): Carbon:phosphorus ratio in MSW composting. *Compost Science and Utilization*, 6 (1), 53–58.

Buggeln, R. and Rynk, R. (2002): Self-heating in yard trimmings: conditions leading to spontaneous combustion. *Compost Science and Utilization*, 10 (2), 162–182.

Chiumenti, A, Chiumenti, R., Diaz, L.F., Savage G.M., Eggerth L.L. and N. Goldstein (2005): *Modern composting technologies*. J.G. Press, Emmaus, USA.

Diaz, L.F., Savage, G.M. and Golueke, C.G. (2002): Composting of municipal solid wastes. In: Tchobanoglous, G. and Kreith, F. (eds.): *Handbook of Solid Waste Management*, 2nd edn., pp. 12.1–12.70. McGraw Hill, New York, USA.

EC (2002): Animal by-product regulation No. 1774/2002. *Official Journal of the European Communities*, L 273, 1–95.

EC (2003): Commission regulation (EC) No 808/2003 of 12 May 2003 amending regulation (EC) No 1774/2002 of the European Parliament and of the Council laying down health rules concerning animal by-products not intended for human consumption. *Official Journal of the European Communities*, L 117, 1–9.

ECN (2004): *EU country reports – Germany. European composting network.* Located on the internet: 31 July 2007. http://www.compostnetwork.info/index.php?id=37.

Fernandes, F., Viel, M., Sayag, D. and André, L. (1988): Microbial breakdown of fats through in-vessel co-composting of agricultural and urban wastes. *Biological Wastes*, 26, 33–48.

Fernandes, L. and Sartaj, M. (1997): Comparative static pile composting using natural, forced and passive aeration methods. *Compost Science and Utilization*, 5 (4), 65–77.

Finstein, M.S. and Morris, M.L. (1975): Microbiology of municipal solid waste composting. *Advanced and Applied Microbiology*, 19, 113–151.

Finstein, M.S., Miller, F.C. and Strom, P.F. (1986): Waste treatment composting as a controlled system. In: Rehm, H.-J. and Reed, G. (eds): *Biotechnology*, vol. 8, pp. 363–398. VCH Verlagsgesellschaft, Weinheim, Germany.

German Federal Government (1998): *Verordnung über die Verwertung von Bioabfällen auf landwirtschaftlich, forstwirtschaftlich und gärtnerisch genutzten Böden (Bioabfallverordnung – BioAbfV)* [Ordinance about land application of biowastes on soils in agriculture, forestry and horticulture (Biowaste ordinance – BioAbV); in German]. Das Bundesgesetzblatt Teil I – BGBl, s. I 2955. Bundesministerium der Justiz, Berlin, Germany.

German Federal Government (2001): *Verordnung über die umweltverträgliche Ablagerung von Siedlungsabfällen und über biologische Abfallbehandlungsanlagen* (Ordinance concerning the environmentally sound landfilling of MSW and concerning biological waste treatment facilities; in German). Das Bundesgesetzblatt Teil I - BGBl, s. I 305. Bundesministerium der Justiz, Berlin, Germany.

Glenn, J. and Goldstein, N. (1999): Food residuals composting in the U.S. *BioCycle*, 39 (8), 30–36.

Goldstein, N. (2003): Solid waste composting trends in the United States. *BioCycle*, 43 (1), 38–44.

Goldstein, N. and Madtes, C. (2001): The state of garbage. *BioCycle*, 41 (12), 42–53.

Goldstein, N. and Satkofsky, A. (2001): Enhanced fabrics spur composting innovations. *BioCycle*, 42 (11), 49–52.

Golueke, C.G. (1977): *Biological reclamation of solid waste.* Rodale Press, Emmaus, USA.

Haug, R.T. (1993): *The practical handbook of compost engineering.* CRC Press, Inc., Boca Raton, USA.

Haug, R.T. and Ellsworth, W.F. (1991): Measuring compost substrate degradability. *BioCycle*, 37 (1), 56–62.

Heining, K. (1998): *Biofilter und Biowäscher – Die Geruchselimination bei Kompostwerken* (Biofilter and bioscrubber – odor treatment technology in composting facilities; in German). Hamburger Berichte 14. Economica Verlag, Bonn, Germany.

Jeris, J.S. and Regan, R.W. (1973): Controlling environmental parameters for optimum composting. *Compost Science*, 14 (1/2), 10–15.

Kasberger, P. (1995): Das dynamisch gesteuerte LESCHA-Verfahren (The dynamically controlled LESCHA-process; in German). In: Wiemer, K. and Kern, M. (eds): *Abfallwirtschaft Neues aus Forschung und Praxis – Herstellerforum Bioabfall – Verfahren der Kompostierung und anaeroben Abfallbehandlung im Vergleich*, pp. 120–131. M.I.C. Baeza-Verlag, Witzenhausen, Germany.

Kern, M. (1991): Untersuchungen zur vergleichenden Beurteilung von Kompostierungsverfahren: Technik-Umweltrelevanz-Kosten (Comparative evaluation of composting processes: technology – environmental relevance – costs; in German). In: Wiemer, K. and Kern, M. (eds.): *Abfallwirtschaft: Bioabfallkompostierung – flächendeckende Einführung*, pp. 235-278. M.I.C. Baeza-Verlag, Witzenhausen, Germany.

Kögel-Knabner, I. and Pichler, M. (1999): *Humifizierungsprozesse und Huminstoffhaushalt während der Rotte und Deponierung von Restmüll* (Humification processes and humic substance balance during composting and landfilling of residual MSW; in German). Final research report as part of the BMFT project 'Mechanisch-biologische Behandlung von zu deponierenden Abfällen', TU Munich, Institute of Soil Science, Freising-Weihenstefan, Germany.

Körner, I. and Stegmann, R. (2002): N-Dynamics during composting – Overview and experimental results. In: Insam, H., Riddech, N. and Klammer, S. (eds.), *Microbiology of composting*, pp. 143–154. Springer-Verlag, Berlin, Germany.

Kranert, M. (2000): Aerobe Verfahren (Kompostierung) [Aerobic processes (composting); in German]. In: Bidlingmaier, W. (ed.), *Biologische Abfallverwertung*, pp. 56–119. Eugen Ulmer GmbH and Co., Stuttgart, Germany.

Krogmann, U. (1994): *Kompostierung – Grundlagen zur Einsammlung und Behandlung von Bioabfällen unterschiedlicher Zusammensetzung* (Composting – basics of collection and processing of biowastes of different compositions; in

German). PhD Thesis, Technical University of Hamburg-Harburg. Hamburger Berichte Band 7. Economica Verlag, Bonn, Germany.

Krogmann, U. and Woyczechowski, H. (2000): Selected characteristics of leachate, condensate and runoff released during composting of biogenic waste. *Waste Managment and Research*, 18, 235–248.

Krogmann, U., Westendorf, M.L. and Rogers, B.F. (2006): *Best management practices for horse manure composting on small farms*. Rutgers Cooperative Extension Bulletin E307. Rutgers, the State University of New Jersey, USA. Located on the internet: 13 November 2006. http://www.rcre.rutgers.edu/pubs/publication.asp?pid=E307.

Kugler, R., Hofer, H. and Leisner, R. (1995): Das Wendelin-Tafelmieten-Kompostierungsverfahren (The Wendelin pile composting process; in German). In: Wiemer, K. and Kern, M. (eds) *Abfallwirtschaft Neues aus Forschung und Praxis – Herstellerforum Bioabfall – Verfahren der Kompostierung und anaeroben Abfallbehandlung im Vergleich*, pp. 13–23. M.I.C. Baeza-Verlag, Witzenhausen, Germany.

LAGA (2003): Umsetzung der Abfallablagerungsverordnung (Implementation of the landfill ordinance; in German). In: Hösel, G., Bilitewski, B., Schenkel, W. and Schnurer, H. (eds): *Müll-Handbuch*, vol. 4. KZ 4509. Erich Schmidt Verlag, Berlin, Germany.

Lenton, T.G. and Stentiford, E.I. (1990): Control of aeration in static pile composting. *Waste Management and Research*, 8, 299–306.

Linder, H. (1995): Das Brikollare-Verfahren (The Bricollari process; in German). In: Wiemer, K. and Kern, M. (eds) *Abfallwirtschaft Neues aus Forschung und Praxis – Herstellerforum Bioabfall – Verfahren der Kompostierung und anaeroben Abfallbehandlung im Vergleich*, pp. 170–182. M.I.C. Baeza-Verlag, Witzenhausen, Germany.

Michel, F.C., Forney, L.J., Huang, A.J.-F., Drew, S., Czuprenski, M., Lindeberg, J.D. and Reddy, C.A. (1996): Effects of turning frequency, leaves to grass mix ratio and windrow vs. pile configuration on the composting of yard trimmings. *Compost Science and Utilization*, 4 (1), 26–43.

Miller, F.C., MacGregor, S.T., Psarianos, K.M., Cirello, J. and Finstein, M.S. (1982): Direction of ventilation in composting wastewater sludge. *Journal of the Water Pollution Control Federation*, 57, 111–113.

Miller, F.C., Harper, E.R. and Macauley, B.J. (1989): Field examination of temperature and oxygen relationships in mushroom composting stacks – consideration of stack oxygenation based on utilisation and supply. *Australian Journal of Experimental Agriculture*, 29, 741–750.

Millner, P.D., Olenchock, S.A., Epstein, E., Rylander, R., Haines, J., Walker, J., Ooi, B.L., Horne, E. and Maritato, M. (1994): Bioaerosols associated with composting facilities. *Compost Science and Utililization*, 2 (4), 8–57.

RAL (2001): *Solid recovered fuels – quality assurance RAL-GZ 724* (translated version). RAL Deutsches Institut für Gütesicherung und Kennzeichnung e.V., St. Augustin, Germany.

Richard, T.L. (1992): Municipal solid waste composting: physical and biological processing. *Biomass and Bioenergy*, 3, 163–180.

Richard, T.L. (1998): Composting strategies for high moisture manures. In: *Proceedings of the manure management in harmony with the environment and society conference, Ames, Iowa*, pp. 135–138. The Soil and Water Conservation Society, West North Central Region, Iowa, USA.

Rynk, R. (ed.) (1992): *On-farm composting handbook*. Northeast Regional Agricultural Engineering Service, Ithaca, USA.

Schmitz, T. and Meier-Stolle, G. (1995): Das Biofix- und Kompoflex-Verfahren (The Biofix and Kompoflex processes; in German). In: Wiemer, K. and Kern, M. (eds) *Abfallwirtschaft Neues aus Forschung und Praxis – Herstellerforum Bioabfall – Verfahren der Kompostierung und anaeroben Abfallbehandlung im Vergleich*, pp. 183–192. M.I.C. Baeza-Verlag Witzenhausen, Germany.

Schlegelmilch, M., Kleeberg, K., Stresse, J. and Stegmann, R. (2005): *Geruchsmanagement – Ein Anwenderhandbuch mit beispielen aus der Lebensmittelindustrie* (Odor management – A user handbook with examples from the food industry; in German). Hamburger Berichte 24. Verlag Abfall aktuell, Stuttgart, Germany.

Schön, M. (1992): Das Wendelin-Verfahren zur Bioabfallkompostierung (The Wendelin process for biowaste composting; in German). In: Wiemer, K. and Kern, M. (eds) *Verfahrenstechnik der Bioabfallkompostierung*, pp. 129–144. M.I.C. Baeza-Verlag, Witzenhausen, Germany.

Strom, P. (1985): Effect of temperature on bacterial species diversity in thermophilic solid-waste composting. *Applied and Environmental Microbiology*, 50, 899–905.

Strom, P.F., Morris, M.L., Finstein, M.S. (1980): Leaf composting through appropriate, low-level technology. *Compost Science/Land Utilization*, 21 (11/12), 44–48.

US Composting Council (1994): *Composting facility operating guide*. Composting Council, Alexandria, USA.

US Composting Council (2004): Test methods for the examination of composting and compost. Located on the internet: 13 November 2006. http://tmecc.org.

US EPA (1981): *Composting processes to stabilize and disinfect municipal sewage sludge*. EPA 430/9-81-011. Office of Water Program Operations, US EPA, Washington, D.C., USA.

US EPA (1989): In-vessel composting of municipal wastewater sludge. EPA 625/8-89/016. Center for Environmental Research Information, US EPA, Cincinnati, USA.

US EPA (1993): 40 CFR Part 503 – Standards for the use or disposal of sewage sludge. *Federal Register*, 58, 9387–9401.

US EPA (1999): *Organic materials management strategies*. EPA 530-R-99-016. Office of Solid Waste, Municipal and Industrial Solid waste Division, US EPA, Washington, D.C., USA.

VDI (2002): Biologische Abgasreinigung-Biofilter, VDI 3477 (Biological air pollution control, VDI 3477; in German). In: *VDI/DIN Handbuch, Reinhaltung der Luft, Vol. 6*, V Beuth Verlag, Berlin, Germany.

WEF (1995): *Wastewater residuals stabilization manual*. Manual of practice FD-9. Water Environment Federation, Alexandria, USA.

Wesemann, W. (1992): Wurmkompostierung (Vermicomposting; in German). In: Amlinger, F. (ed.) *Handbuch der Kompostierung – Ein Leitfaden für Praxis, Verwaltung, Forschung*, pp. 106–110. Bundesministerium für Land- und Forstwirtschaft, Bundesministerium für Wissenschaft und Forschung, Vienna, Austria.

Wiemer, K. and Kern, M. (eds) (1998): *Kompostatlas 1998/99 – Referenzhandbuch für Kompostierungsanlagen, Anaerobanlagen, Mechanisch-biologische Restabfallbehandlungsanlagen sowie Aggregate* (Composting handbook 1998/99 – Handbook for composting facilities, anaerobic digestion plants, mechanical-biological residual MSW treatment facilities and equipment; in German). M.I.C. Baeza Verlag, Witzenhausen, Germany.

Woyczechowski, H., Krogmann, U., Arndt, M. and Stegmann, R. (1995): Optimierung des Betriebs von Kompostierungsanlagen (Optimization of the operation of composting facilities; in German). Final research report. Deutsche Bundesstiftung Umwelt, Technical University of Hamburg-Harburg, Department of Waste Management, Hamburg, Germany.

Zachäus, D. (1995): Kompostierung (Composting; in German). In: Thomé-Kozmiensky, K.J. (ed.) *Biologische Abfallbehandlung*, pp. 215–353, EF-Verlag für Energie- und Umwelttechnik GmbH, Berlin, Germany.

# 9.3

# Composting: Mass Balances and Product Quality

**Alessio Boldrin and Thomas H. Christensen**

*Technical University of Denmark, Denmark*

**Ina Körner**

*Hamburg University of Technology, Hamburg, Germany*

**Uta Krogmann**

*Rutgers University, New Jersey, USA*

While the basic processes involved in composting of waste are described in Chapter 9.1 and the main composting technologies are presented in Chapter 9.2, this chapter focuses on mass balances, environmental emissions, unit process inventories and the quality of the compost produced. Understanding these issues and being able to account for them is a prerequisite in compost engineering and for establishing and running a successful composting facility.

Of specific importance is the final use of the compost product. Use in agriculture is described in Chapter 9.10 and the use of compost in soil amendment products are presented in Chapter 9.9.

## 9.3.1 Mass Balances

The overall mass balance for a composting facility geared to produce a compost product for use on land depends on many factors. Most important is the type and composition of waste being composted, because this determines the amount of undesirable items to be removed from the waste and the degradability of the waste. Second most important is the intended quality of the final product: the better the quality and the more stable the compost is, the less the mass that ends up in the final product. The composting technology may also influence the mass balance, but having the feedstock and the

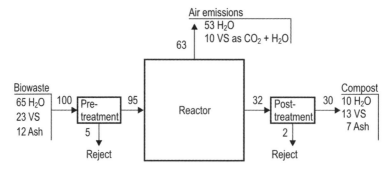

***Figure 9.3.1***    *Mass balance for typical composting plant for mixed biowaste.*

end product defined, the choice of technology (see Chapter 9.2 for a wide variety of technologies) has only a secondary influence on the overall mass balance.

Mass balances can be made with different purposes, approaches and degree of details. The more complex and detailed the mass balance is, the more correct it is, but this may be at the expense of clarity.

### 9.3.1.1    Balances of Volatile Solids, Water and Wet Mass

Figure 9.3.1 shows a mass balance of a composting facility treating biowaste (source-separated organic household waste) focusing on the fate of the waste as characterized by its content of water, volatile solids (VS) and ash (non-volatile solids) (after Bidlingmaier and Müsken, 1993). The biowaste is relatively clean and about 5 % of the weight (foreign items and some organics sticking to the items) is removed in the pre-processing. During the composting nearly two-thirds of all mass is lost. Out of 63 units lost, however, 53 units are water and only 10 units are volatile solids actually being degraded. The postprocessing removes additionally 2 % of the weight, showing that about 30 % of the incoming waste ends up as a compost product. The balance also shows that about 7 % of the waste is removed for other treatment or disposal and about 63 % is removed as air emissions. The final compost contains about 33 % water, 43 % volatile organic matter and 24 % ash. This mass balance clearly shows the fate of waste, but it does not show any air ventilation and pumping of water, or any internal recirculation of materials.

Figure 9.3.2 shows a mass balance for another type of composting facility also composting biowaste. In this example, straw and lime are added to the incoming waste to prevent acidic conditions and a significant fraction of the compost is

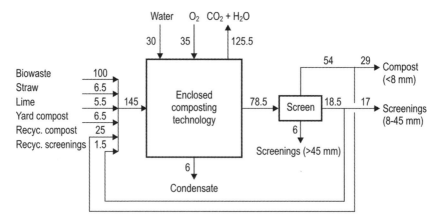

***Figure 9.3.2***    *Mass balance for a typical composting plant for a wet biowaste containing a high percentage of household kitchen waste.*

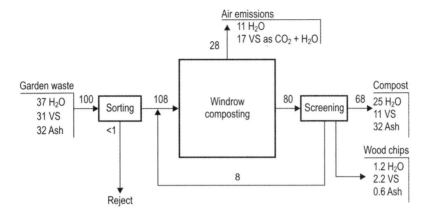

**Figure 9.3.3** *Mass balance for a typical garden waste windrow composting plant.*

recirculated in the enclosed composting facility. As in the previous case, the end product makes up about 30 % of the incoming waste. The reject from the postprocess screening of the compost makes up 17 %, which is further processed by the nearby composting facility for garden waste. Also some garden waste compost is mixed in with biowaste to provide structure.

Figure 9.3.3 shows a mass balance for a garden waste composting facility in terms of average values for a year (Boldrin *et al.*, 2009a). The content of foreign materials in terms of plant pots, tiles, etc. is very low, less than 1 %. The garden waste is shredded and composted in open windrows with occasional turning over a 12-month period; no water is added to the process. A bit less than one-third of the weight is lost during the process: 11 units of water and 17 units of water and degraded volatile solids. The compost is sieved before use and the large fraction is partly recirculated back to the windrows (8 units) and partly sent to thermal treatment (4 units). Since garden waste composting facilities often use open technologies, the moisture content of the compost, which may have been stored over longer periods, may also vary considerably (e.g. due to the addition of rain water) making mass balancing uncertain. In this example, the final compost consists by weight of about 68 % of the initial feedstock and it contains 37 % water, 16 % volatile organic matter and 47 % ash. The latter suggests that the garden waste contained a significant fraction of soil. The degradation of volatile solids was 55 % in this example.

### 9.3.1.2  Balances of Elements

The elemental composition of the waste changes during composting primarily with respect to carbon and nitrogen. Both of these elements are partially converted and lost, primarily with the off gasses. Some may also be lost with the leachate, but most facilities recirculate this liquid and avoid discharging any liquid. This means that also carbon and nitrogen in the leachate are returned to the compost and therefore in practice all other elements entering the composting process are conserved and will appear in the final product, except if removed physically in the postprocessing. The only exception is mercury that would volatilize to a large extent at the elevated composting temperatures. However, the mercury content in most organic waste is already very low and this has no practical implications.

Table 9.3.1 illustrates in a general sense the balancing of N, P and K in biowaste undergoing composting. Elements like N, P and K, which are important nutrients in compost, are usually quantified on the basis of dry matter (VS + ash), so any change in water content has no implication on the measured concentrations. However, the fact that organic matter is being mineralized induces an apparent increase in the concentrations of K and P after composting. If the loss of nitrogen relatively speaking is less than the loss of solids, the concentrations of nitrogen will also increase during the composting. Box 9.3.1 discusses the various losses of nitrogen during composting. It should be noted that the loss of nitrogen during composting is hard to quantify and currently only limited data exist (see below regarding ammonia).

***Table 9.3.1***    *Mass (Total solids, VS, ash) and element (N, P, K) changes during composting on a dry matter basis.*

| | Organic fraction of biowaste | | Loss during composting | | Organic fraction of compost | |
|---|---|---|---|---|---|---|
| | Concentration | Content per kg input | % | Mass | Concentration | Content per kg input |
| Total solids | 100 % | 1000 g | 40 | 400 g | 100 % | 600 g |
| VS | 85 % | 850 g | 47 | 400 g | 75 % | 450 g |
| Ash | 15 % | 150 g | 0 | 0 | 25 % | 150 g |
| Nitrogen, N | 14 g/kg dw | 14 g | 30 | 4.2 g | 16 g/kg dw | 9.8 g |
| Phosphorous, P | 1.2 g/kg dw | 1.2 g | 0 | 0 | 2 g/kg dw | 1.2 g |
| Potassium, K | 5.6 g/kg dw | 5.6 g | 0 | 0 | 9 g/kg dw | 5.6 g |

**Box 9.3.1    Nitrogen Mineralization and Losses During Composting.**

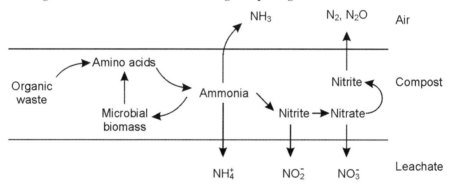

In the biowaste the majority of nitrogen is found as organic nitrogen (measured as Kjeldahl-N). As degradation proceeds proteinous compounds mineralize and nitrogen is released as dissolved organic nitrogen and as ammonium. Some of the nitrogen will be used in the build up of microbial biomass in the system but the excess nitrogen will be mineralized to ammonium. Ammonium ($NH_4^+$) will dissociate according to:

$$NH_4^+ \rightarrow NH_3 + H^+$$

The acid dissociation constant for this reaction is $pK_a = 9.24$ at 25 °C, but 8.04 at 70 °C as easily reached in a composting system. Since pH in the composting system after the initial phase increases and may reach 7.5–8.0, the equilibrium is pushed to the right. The venting of the system (induced or natural) provides a constant removal of degradation gases and hence removal of $NH_3$, which again enhances the dissociation. Ammonium, $NH_4^+$ in solution may bind to particle surfaces in the waste by ion exchange. The actual release of $NH_3$ thus will depend on the amount of nitrogen in the waste, the degradability of the waste, the temperature in compost, the pH of the compost and the amount of air and gas released through the compost.

As the compost stabilizes, oxidation of ammonium slowly starts. In very old and well matured compost, part of the nitrogen eventually ends up as nitrate ($NO_3^{2-}$) in the moisture. However, the compost is ready as a product long before and in finished compost, nitrogen will be available primarily as organic N (in decaying microbial biomass and in humic compounds as most nitrogen-containing compounds in the original biowaste are easily degradable during composting), some ammonium and only limited nitrate.

The microbial oxidation of ammonium to nitrate as well as the microbial processes reducing nitrate are complex in composting since the redox conditions locally may change with the availability of degradable organic matter and moisture content. Intermediates such as nitrite ($NO_2^-$) and nitrous oxide ($N_2O$) appear, the first dissolved in the compost moisture and the second as a gas. The quantities are highly variable with respect to operational conditions and measured values fluctuate significantly. Quantitatively these nitrogen compounds are of minor importance with respect to the nitrogen mass balance, but in particular $N_2O$ is of environmental concern because its global warming potential is about 298 times that of $CO_2$.

If leachate is generated during the composting process, high concentrations of organic nitrogen (Kjeldahl-N minus ammonium) and ammonium (measured as the sum of $NH_4^+$ and $NH_3$ depending on the sample handling) are found. Typical levels: 100–1000 mg Kjeldahl-N/l; 10–500 mg $NH_4^+$-N/l, 1–100 mg $NO_3$-N/l, and <1 mg $NO_2$-N/l.

## 9.3.2 Environmental Emissions and Controls

The environmental emissions are primarily liquid emissions and air emissions.

### 9.3.2.1 Liquid Emission

During composting, depending on the feedstock and the chosen composting technology, leachate (water that percolates through the compost and exits at the bottom) and condensate (water that evaporates from the compost and condensates in condensers or on cooler building surfaces such as roofs and walls) are collected as excess water. Water in the composting process originates from the moisture in the waste itself, reaction water that is generated during the composting process, rain water in open facilities without a roof, and water that is added to adjust the moisture content of the feedstock. An additional part of the water balance is the evaporated water that is released from open windrows or through biofilters directly to the ambient air. In addition to leachate and condensate, the water balance of a composting facility contains run off from traffic areas and windrow surfaces. The run off from traffic areas and windrow surfaces does not pass through the pile and therefore is, in most cases, less contaminated than leachate and condensate (Table 9.3.2). Since condensate is less contaminated than leachate, enclosed and reactor-based facilities recycle leachate, if possible, and only discharge condensate. Some facilities even cover the entire facility with a roof and reuse condensate for watering of windrows and therefore do not discharge any processing liquids.

In most cases, leachate, condensate and run off from composting facilities need to be treated or handled appropriately onsite or offsite before release into surface water or groundwater. The regulatory requirements concerning liquid emissions being discharged into a receiving water body or a sewer vary among countries. Therefore, general guidelines on how to handle the liquid emissions cannot be provided without knowing the local regulatory requirements.

### 9.3.2.2 Air Emissions

Air emissions from composting facilities are primarily related to $CH_4$, $NH_3$, $N_2O$, odours, dust and bioaerosols. Table 9.3.3 shows a typical, but not complete, composition of off gasses from a composting plant prior to any treatment. Controlling air flow and ventilation and collecting and treating all air prior to discharge will significantly reduce the air emissions.

While carbon dioxide emitted from composting is considered neutral in terms of global warming potential (the organic materials constituting the organic waste were produced recently by photosynthesis), emissions of methane ($CH_4$) and nitrous oxide ($N_2O$), generated in less aerobic pockets in the feedstock, do contribute to global warming. The global warming potential on a weight basis of $CH_4$ is 25 times higher and of $N_2O$ 298 times higher than the global warming potential of carbon dioxide (Solomon *et al.*, 2007). This suggests that even minor emissions of these gases should be avoided. Monitoring data on $CH_4$ emissions from actual plants are few. $CH_4$ have been reported at high concentrations in the voids of compost windrows, but how much is emitted without being oxidized is not well documented. Occurrence and magnitude of $CH_4$ emissions depend on the feedstock (C content, degradability, structure) and the technological level of

**Table 9.3.2**   Chemical characteristics of liquid flows from composting facilities. Conductivity in mS/cm, other parameters in mg/l.

| | Mixture | Leachate | | Condensate | Run off | |
| --- | --- | --- | --- | --- | --- | --- |
| | Krogmann and Woyczechowski (2000) | Cole (1994) | Loll (1994) | Loll (1994) | Krogmann and Woyczechowski (2000) | Loll (1994) |
| pH | | 8.35–9.2 | 5.8–8.6 | 8.0–8.6 | 6.62–8.47 | 7.0–8.1 |
| Waste type[a] | B | G | B, G | B, G | B | B, G |
| Conductivity | 9.37–27.94 | 3.89–6.4 | 4.1–14.7 | 1.8–2.5 | — | 1.3–8.8 |
| BOD$_5$ | 8–11 571 | — | 10 000–45 000 | 100–1000 | < 2–513 | 100–1200 |
| COD | 2 434–31 812 | — | 20 000–100 000 | 500–2000 | 56–1768 | 500–2500 |
| TOC | — | — | 5000–18 000 | < 50–500 | — | < 50–500 |
| VFA | 118–9535 | — | — | — | — | — |
| NH$_4^+$-N | 98–558 | 5.1–10.5[b] | 50–800 | < 5–100 | 2.0–46.0 | 15–300 |
| NO$_3^-$-N | — | 3.6–5.8[c] | < 5–190 | < 1 | < 0.1–96.4 | <5–150 |
| NO$_2^-$-N | — | — | — | — | < 0.1–0.80 | — |
| TKN | 250–1602 | — | — | — | — | — |
| Total N | — | — | — | — | — | — |
| Total P | — | — | 50–150 | < 1 | — | < 1–50 |
| K$^+$ | — | 1629–2323 | 1075–7280 | — | — | — |
| Ca$^{2+}$ | — | 386–462 | — | — | — | — |
| Mg$^{2+}$ | — | 213–290 | — | — | — | — |
| Na$^+$ | — | — | — | — | — | — |
| Mn | — | 1.5–3.5 | — | — | — | — |
| Fe | — | 16–52 | — | — | — | — |
| S | — | — | — | — | — | — |
| Cl$^-$ | 1 514–5 254 | — | 2000–10 000 | — | 106–445 | 30–500 |
| As | — | — | — | — | 0.001–0.044 | — |
| Cr | — | 0.02–0.06 | — | — | — | — |
| Ni | — | — | 0.07–2.6 | < 0.04 | — | < 0.05–1.0 |
| Cu | — | 0.25–0.38 | — | — | — | — |
| Co | — | — | 0.01–0.2 | < 0.05 | — | < 0.05–0.2 |
| Zn | — | — | 1–8 | 0.2-1.6 | 0.011–2.4 | < 1–2 |
| Cd | — | < 0.1 | 0.01–0.2 | < 0.02 | < 0.001–0.172 | < 0.05–0.2 |
| Hg | — | — | — | <0.0005 | — | — |
| Pb | — | < 0.25 | 0.01–0.2 | < 0.1 | < 0.001–0.500 | < 0.1–0.2 |

[a]B = Biowaste; G = Garden waste
[b]NH$_4$
[c]NO$_3$

the facility (turning frequencies, ventilation, size and design of the piles, compaction), as described in detail by Boldrin *et al.* (2009b). Literature data reported in Table 9.3.4 suggest that 0.77–3.0 % of the carbon degraded during the composting process is emitted as CH$_4$ or about 0.5–10 kg CH$_4$/t dry biowaste.

Data on N$_2$O emission are very rare; the emission of N$_2$O may be in the order of 0.13–1.8 % of the initial N content or about 0.15–1.0 kg/t dry biowaste (Table 9.3.4). N$_2$O formation occurs in conditions with low oxygen concentrations. Therefore, emissions depend on N content, aeration, temperature and to some extent age of the material.

Ammonia is not a greenhouse gas but a significant nutrient and an odorous compound. As discussed in Box 9.3.1 a significant fraction of nitrogen is lost during composting and this primarily as NH$_3$. Data from full scale plants are limited; potentially 10–35 % of all nitrogen in the biowaste could be emitted as NH$_3$, corresponding to 1–5 kg/t biowaste. Table 9.3.4 suggests ammonia volatilization in the range of 0.6–11.0 % of the initial nitrogen content of the feedstock depending on its availability, temperature, pH and aeration.

**Table 9.3.3** *Chemical characteristics of air flows from composting facilities (adapted from Ødegård et al., 2005). The unit is ppm if no other information is given.*

|  | Organic waste | Garden waste |
| --- | --- | --- |
| $CO_2$ | 0–50 % | 0–50 % |
| $CH_4$ | 2–47 % | 0–900 |
| $N_2O$ |  | 1–12 |
| $NH_3$ | 300 | 10–100 |
| NO | 10–26 |  |
| CO |  | 120 |
| $H_2S$ | 0–500 |  |
| Methanol | 5–25 |  |
| Ethanol | 0.002 |  |
| Acetone | 3.3 |  |
| Toluene | 0–0.009 | 0.0005–0.0008 |
| Benzene | 0–0.028 | 0.0004 |
| Ethylbenzene | 0.2 |  |
| $CCl_4$ |  | 0.0001 |
| $CH_2Cl_2$ |  | 0.0003–0.0005 |
| $CH_3Cl$ |  | 0.001 |

Odorous compounds during composting include low molecular alcohols, aldehydes and ketones, low molecular carbonic acids, nitrogen-containing compounds, sulfur containing compounds and etheric oils. Odorous emissions in a composting facility can be separated into three types: waste specific, biogenic and abiogenic odours. For example, a waste specific odour is limonene released by orange and lemon peels. The incoming waste itself might also release biogenic odours especially at low waste collection frequencies during the summer. However, biogenic odours are mainly released during the thermophilic phases of composting. Biogenic odorous emissions include anaerobic processes odour emissions, process-specific odour emissions (e.g., ammonia) and odour emissions from aerobic–anaerobic transition processes (Schildknecht *et al.*, 1979). The latter are released because 100 % aerobic composting is unlikely. In a composting pile that is not actively aerated, there are always some niches where aerobic and anaerobic processes coexist. Many of the odorous emissions, which are anaerobic metabolic products, are degraded within the pile, unless the pile is turned or moved releasing the volatile compounds. The anaerobic process odours are generated because of insufficient aeration or a high moisture content of the feedstock.

Dust can be expected when dry materials are processed at the following processing locations: shredding of yard waste, moving and turning of dry compost, screening during posttreatment, ballistic separation, loading of compost and traffic areas. Enclosures can also reduce the dust emissions. Fences and dense vegetation around the facility are helpful.

Generally, in presence of dust also bioaerosols (biological particulate matter) are expected to be found. In particular, the highest bioaerosol concentrations are found in areas where the highest dust concentrations are measured. Therefore, dust generation needs to be minimized. The highest concentrations of bioaerosols are measured during windrow turning. In composting facilities, appropriate measures need to be taken to reduce the exposure of personnel to bioaerosols.

When composting is performed in enclosed buildings, a common and not expensive treatment is the filtration of air in biofilters. The efficiencies of biofilters depend on air flow, load, residence time, materials and design (Chung, 2007; Chung *et al.*, 2007; Powelson *et al.*, 2006). In some cases biofilters represent a source of some of the gaseous compounds. It is the case, for example, of $N_2O$: ammonia volatilized from compost is absorbed on the biofilter and part of it is transformed into $N_2O$ (26 % according to Clemens and Cuhls, 2003).

### 9.3.3 Unit Process Inventories

While mass balances are useful for understanding the performance of the composting process and the features of specific facilities, unit process inventories are useful for establishing green accounts, bench marking and as a basis for performing a

**Table 9.3.4**   Gas emissions from organic waste composting (total N is total amount of N in the feedstock).

| Waste type[a] | Methane | Nitrous oxide | Ammonia | Data | Source |
|---|---|---|---|---|---|
| H, G mix | 2.1–3.6 % of C emissions | 0.95–1.25 % of total N; 14–15 % of emitted N | 6.04–6.97 % of total N; 85–86 % of emitted N | Home-composting | Amlinger and Peyr, 2002 |
| H, G mix | 0.77–2.5 % of C emissions | 0.13–0.64 % of total N; 7–17 % of emitted N | 0.63–8.15 % of total N; 83–93 % of emitted N | Windrow composting | Amlinger and Peyr, 2003 |
| G | 1.5–2.0 % of total C | 0.5 % of total N | 1.2 % of total N | Batch and field experiments | Hellebrand, 1998 |
| H | 3 % of C emissions | 1.8 % of total N | 11 % of total N | Closed system | Gronauer et al., 1997 |
| H | 2.4 % of C emissions | | 1.6 % of total N | Closed reactors | Marb et al., 1997 |
| H | | 0.4–0.6 % of Total N | 1–4 % of Total N | Laboratory reactors | Beck-Friis et al., 2000a |
| G | 0.7–1.3 % of total C; 2.24 % of C emissions | 1.2 % of total N; 50.4 % of N-loss | | Windrow composting | Andersen et al., 2010 |
| H | 195 g/t waste | 101 g/t waste | 27 g/t waste | Closed system | Saft and Van Ewijk, 2004 |
| H | 6–12 kg $CH_4$/t waste | 1.44–378 g $N_2O$/ t waste | 18–1150 g $NH_3$/t waste | Measurement at biofilter | Clemens and Cuhls, 2003 |
| H, G mix | 1.9 kg $CH_4$/t TS | 0.35 kg $N_2O$/t TS | 1.3 kg $NH_3$/t TS | Literature value | Pitschke et al., 2004 |
| H | 0–119 g $CH_4$/m$^2$·dag | 1–1464 mg $N_2O$/m$^2$·dag | | Flux chambers technique | Beck-Friis et al., 2000b |
| G | | | 4.5 g $NH_3$-N/kg TS | Laboratory-scale reactors | Komilis and Ham, 2006 |
| K | | | 41 g $NH_3$-N/kg TS | Laboratory-scale reactors | Komilis and Ham., 2006 |
| G | | 1.18 % of total N | | Full-scale plant | Ballestero and Douglas, 1996 |
| O | 4 kg $CH_4$/t waste; 10 kg $CH_4$/t TS | 0.3 g $N_2O$/kg waste; 0.6 g $N_2O$/kg TS | | | IPCC, 2006 |

[a]H = Household waste; K = Kitchen waste; G = Garden waste; O = organic waste

broader environmental assessment. The inventory lists, in principal, all inputs and outputs of a facility and by normalizing with the amount of waste treated by the facility the unit process inventory is prepared. In particular the use and eventual production of energy by the facility – electricity, fuel, heat, etc. – is important information in the inventory. Also the use of chemicals may in some cases appear significant.

Unit process inventories for three composting facilities are presented in Table 9.3.5. The inventories suggest that composting of waste requires a significant input of energy, but hardly any chemicals and alike are used.

## 9.3.4   Quality of Compost

The composition of the compost reflects the composition of the waste, the composting technology and the final use of the compost. Table 9.3.6 provides examples of the composition of compost from various sources. In addition to the chemical composition of the compost, also quality in terms of compost maturity and compost standards are important.

***Table 9.3.5***  *Unit process inventory for composting of 1 t of waste (wet weight).*

| Per tonne of waste composted | Unit | Biowaste composting in static boxes[a] | MSW composting in open windrows[b] | Garden waste composting in open windrows[c] |
|---|---|---|---|---|
| Inputs | | | | |
| Kitchen waste | kg | 790 | 610 (15 % food, 65 % paper, 20 % other organic) | 0 |
| Garden waste | kg | 200 | 185 | 1000 |
| Other | kg | 10 | 205 (glass, tins, aluminium, other metals) | 0 |
| Diesel | l | ? | 1.13 | 30.4 |
| Electricity | kWh | 26 | 149 | 0.2 |
| Water | l | ? | Up to 50 % of waste | — |
| Outputs | | | | |
| Compost | kg | 530 | 654 | 680 |
| Screening reject | kg | 0 (recycled) | 141 | 40 |
| Metals | kg | <10 | — | 0 |
| Reject[d] | kg | 15 | 141 | <10 |
| Wastewater | m³ | Recirculated | ~ 0 | Recirculated |
| Off gasses | m³ | Not measured | Not measured | Not measured |

[a]Cloppenburg-Stapelfeld, Germany
[b]Komilis and Ham (1999)
[c]After Boldrin *et al.* (2009a)
[d]Reject for incineration or landfilling

### 9.3.4.1  Compost Maturity

Compost maturity expresses the stability of the compost and is a key parameter with respect to both the management of the composting facility (Is the compost ready?) and the markets for the final product (The compost is of good quality every time). The main problems encountered if the compost is not mature are:

- Odour problems when spreading.
- Odour problems during storage.
- Negative effects on plant growth when compost is mixed with soil.

Numerous tests have been developed over the years. The main principles for the tests, which all are surrogate tests, are to:

- Measure low microbial activity, which signals that degradation now is slow and that the compost is mature.
- Measure something that does not appear until the compost is well degraded and mature.

Common approaches are:

- C/N ratio: The C/N ratio has traditionally been used to determine the 'compostability' of the waste and a significant decrease in the ratio indicates that degradation has proceeded. However, it is not possible in general to set absolute values for the C/N ratio to monitor maturity (Garcia *et al.*, 1992). The problems are related to the facts that the chemical analysis of C also measure C that is not degradable and that N also is lost during the composting process. In some cases the C/N ratio in a water extract (*L/S* 20 l/kg for 2 h) is used as an indicator of maturity.

**Table 9.3.6**   *Chemical composition of different types of compost across Europe.*

| Parameter | Unit | Sweden Eklind et al. (1997) | Germany BGKeV (2006) | Switzerland Schleiss (2007) | Holland Schokker (2007) | Denmark Carlsbæk (2007) | Denmark Christensen (1998) | Germany BayLFU (2003) | Denmark Boldrin et al. (2009a) |
|---|---|---|---|---|---|---|---|---|---|
| Waste type[a] | | K | B | B | B | K, G mix | K | B | G |
| Moisture | % | | 27.1 | 32.8–67.2 | 30.2 | 40–75 | 64–77 | 64.1 | 29-44 |
| VS | % TS | 56–72 | 38.9 | 24.2–67.2 | 35.1 | 35–75 | 37–50 | | 21-28 |
| Ash | % TS | 28–44 | 61.1 | 32.8–65.8 | | 25–65 | 50–63 | | 72-79 |
| Density | t/m³ | | 0.651 | | | 0.4–0.6 | 0.4–0.5 | 327 | 0.67-0.82 |
| pH | | | 7.06 | 6.9–8.5 | 7.44 | 7.2–8.5 | 6.6–9.0 | | |
| Conductivity | m S/cm | | | 11–65 | 3.88 | 150–400 | 160–200 | | |
| Maximum size | mm | | 5 | | 5 | 2 | | | 4 |
| Total C | % TS | 33–47 | | | | | 19–25 | 37.9 | 10-19 |
| TKN | % TS | | | | | | | | |
| Total N | % TS | 2.3–2.8 | 1.39 | 0.9–2.6 | 1.34 | 0.8–1.7 | 1.7–2.1 | 1.6 | 0.7-0.9 |
| Total P | % TS | 0.48–0.7 | 0.29 | 0.18–0.44 | 0.63 | 0.19–0.33 | 0.25–0.38 | 0.06 | 0.15-0.23 |
| Total K | % TS | 1.3–2.3 | 0.96 | 0.53–1.89 | 1.91 | 0.41–0.66 | 0.87–0.99 | 0.72 | 1.52-1.94 |
| C/N | | 13–20 | | 12.2–42.2 | | | 9.0–14.7 | 24 | 11.1-27.1 |
| Ca | % TS | 3.1–4.8 | 3.17 | 3.45–9.56 | | | 2.0–6.7 | 3.42 | 2.17-2.89 |
| Cl⁻ | % TS | | | | 0.31 | | | | 0.001-0.09 |
| Fe | % TS | | | | | | | | 0.98-1.57 |
| Mg | % TS | 0.17–0.46 | 0.43 | 0.42–1.54 | 0.47 | 0.12–0.19 | 0.25–1.0 | 0.60 | 0.29-0.37 |
| Na | % TS | | | | | | | | 0.56-0.79 |
| Mn | % TS | | | | | | | | 0.03-0.05 |
| S | % TS | | | 0.057–0.253 | 0.21 | 0.12–0.15 | 0.18–0.29 | | 0.091-0.132 |
| As | mg/kg TS | | | | 3.80 | | 2.9–5.0 | | 3.32-4.04 |
| Cd | mg/kg TS | 0.4225 | 0.20–0.58 | 0.15–0.53 | 0.41 | 0.36–0.70 | 0.15–0.46 | 0.17 | 0.237-0.528 |
| Co | mg/kg TS | | | | | | — | | 2.81-3.44 |
| Cr | mg/kg TS | 21.1 | | | 20.23 | 9–26 | 7.6–18 | 8.8 | 23.3-37.7 |
| Cu | mg/kg TS | 45.8 | 22–88 | 20–78 | 36.78 | | 39–73 | 32.6 | 28.4-49.6 |
| Hg | mg/kg TS | 0.11 | <0.05–0.10 | 0.09–0.16 | 0.10 | 0.02–0.30 | <0.03–0.22 | 0.04 | 0.055-0.086 |
| Ni | mg/kg TS | 13.1 | 2.6–2.7 | 7.1–40.6 | 9.95 | 6.0–14.0 | 6.9–11.0 | 2.8 | 6.37-8.34 |
| Pb | mg/kg TS | 37 | 2.6–5.1 | 9.8–63.9 | 58.83 | 15–50 | 20–60 | 9.3 | 20.3-54.2 |
| Zn | mg/kg TS | 169 | 81–136 | 32.6–188.3 | 173.41 | 115–290 | 94–160 | 74.7 | 111-206 |
| DEHP[b] | mg/kg TS | | | | | | 15–24 | | |
| LAS[c] | mg/kg TS | | | | | | 23–190 | | |
| NPE[d] | mg/kg TS | | | | | | 1.1–8.9 | | |
| PAH[e] | mg/kg TS | | | | | | 0.52–0.83 | 1.5 | |

[a]H = Household waste; K = Kitchen waste; G = Garden waste; O = organic waste; n.d. = not detected
[b]DEHP: di (2 ethylhexyl) phthalate
[c]LAS: linear alkylbenzene sulfsulfonates
[d]NPE: Nonylphenol
[e]PAH: Polycyclic aromatic hydrocarbon

- Respiration test in various versions measuring the oxygen demand of a compost sample for example during 4 days. The lower the value the more mature the compost. Different methods for respiration tests have been previously presented in Chapter 9.1, Box 9.1.4.
- Self-heating test, where the ability of a small sample to increase its temperature reflects microbial activity in the sample. The less self-heating the more mature the compost.
- Colour development when the compost sample is suspended in a standardized buffer solution with a pH indicator. The more acids present the less mature the compost is and a pH-sensitive stick submerged in the solution will appear yellow/orange. If the compost is mature, less acid will be present and the stick will appear blue and purple (e.g. the Solvita test).

- Water soluble nitrate. Since nitrate will not appear until late in the composting process, a high concentration of nitrate in a water extract indicates a mature compost.
- Growth of specific fungi, e.g. *Chaetomium gracilis*, which requires high redox potentials and cannot compete when easily degradable substrate is present.
- Germination test since an elevated temperature in the compost will limit germination of seeds.

### 9.3.4.2 Compost Standards

Several countries have introduced standards for compost in order to provide product information to the market and to ensure sufficient quality for the various applications from an environmental point of view. While the first issue focuses on physical and growth related parameters (e.g. nutrient content, pH, maturity), the second issue addresses heavy metals and organic pollutants. Examples of quality standards for selected countries are presented in Table 9.3.7 and Table 9.3.8, for heavy metals and organic contaminants, respectively. In some cases, a country can have several quality standards, depending on what is the final utilization of compost. Table 9.3.7 also presents European Union (EU) quality standards, as proposed in the draft biowaste directive. Such proposal is elaborated accounting for long-term accumulation of heavy metals in soil due to repeated application of compost on land, as presented by Amlinger *et al.* (2004).

**Table 9.3.7** *Heavy metals limits for some compost standards (mg/kg dm), from Hogg et al. (2002). Hogg, D., Barth, J., Favoino, E., Centemero, M., Caimi, V., Amlinger, F., Devliegher, W., Brinton, W. and Antler S. (2002): Comparison of compost standards within the EU, North America and Australasia. Waste and Resources Action Programme, WRAP, Banbury, UK.*

| Country | Regulation | Cd | Cr$_{tot}$ | Cr(VI) | Cu | Hg | Ni | Pb | Zn | As |
|---|---|---|---|---|---|---|---|---|---|---|
| Austria | Compost ordinance: Quality Class A+ (organic farming) | 0.7 | 70 | — | 70 | 0.4 | 25 | 45 | 200 | — |
| | Compost ordinance: Quality Class A (agriculture; hobby gardening) | 1 | 70 | — | 150 | 0.7 | 60 | 120 | 500 | — |
| | Compost ordinance: Quality Class B (landscaping; reclaimed land) limit value | 3 | 250 | — | 500 | 3 | 100 | 200 | 1800 | — |
| | Compost ordinance: Quality Class B (landscaping; reclaimed land) guide value | — | — | — | 400 | — | — | — | 1200 | — |
| Germany | Bio waste ordinance (I) | 1 | 70 | — | 70 | 0.7 | 35 | 100 | 300 | — |
| | Bio waste ordinance (II) | 1.5 | 100 | — | 100 | 1 | 50 | 150 | 400 | — |
| Netherlands | Compost | 1 | 50 | — | 60 | 0.3 | 20 | 100 | 200 | 15 |
| | Compost (very clean) | 0.7 | 50 | — | 25 | 0.2 | 10 | 65 | 75 | 5 |
| EC | Draft W.D. biological treatment of biowaste (class 1) | 0.7 | 100 | | 100 | 0.5 | 50 | 100 | 200 | |
| | Draft W.D. biological treatment of biowaste (class 2) | 1.5 | 150 | | 150 | 1 | 75 | 150 | 400 | |
| | 2092/91 EC – 1488/98 EC (organic farming) | 0.7 | 70 | 0 | 70 | 0.4 | 25 | 45 | 200 | — |
| UK | UKROFS 'Composted household waste' | 0.7 | 70 | — | 70 | 0.4 | 25 | 45 | 200 | — |
| | Composting association quality label | 1.5 | 150 | | 200 | 1 | 50 | 150 | 400 | |
| USA | Texas TNRCC Grade 1 compost | 16 | 180 | — | 1020 | 11 | 160 | 300 | 2190 | 10 |
| | Texas TNRCC Grade 2 compost | 39 | 1200 | — | 1500 | 17 | 420 | 300 | 2800 | 41 |

**Table 9.3.8**   *Limit values for organic contaminants in compost, from Hogg et al. (2002). Hogg, D., Barth, J., Favoino, E., Centemero, M., Caimi, V., Amlinger, F., Devliegher, W., Brinton, W. and Antler S. (2002): Comparison of compost standards within the EU, North America and Australasia.Waste and Resources Action Programme, WRAP, Banbury, UK.*

|  | Austria | Denmark | Luxemburg |
|---|---|---|---|
|  | Mixed MSW compost only | Biowaste compost (1 analysis per year) | Guide values for fresh and matured compost |
| PCB | 1 mg/kg dm |  | 0.1 mg/kg dm (4 analyses per year) |
| PCCD/F |  |  | 20 ng/kg dm (4 analyses per year) |
| Dioxins | 50 ng I-TEQ/kg dm |  |  |
| PAH | 6 mg/kg dm | 3 mg/kg dm | 10 mg/kg dm (2 analyses per year) |
| AOX | 500 mg/kg dm |  |  |
| Hydrocarbons | 3000 mg/kg dm |  |  |
| LAS |  | 1300 mg/kg dm |  |
| NPE |  | 30 mg/kg dm |  |
| DEHP |  | 50 mg/kg dm |  |

# References

Amlinger, F. and Peyr, S. (2002): *Umweltrelevanz der Hausgartenkompostierung – Klimarelevante Gasemissionen, flüssige Emissionen, Massenbilanz, Hygienisierungsleistung* (Environmental impacts from home composting – global warming potential, leaching, mass balance, sanitation; in German). Technisches Büro für Landwirtschaft, Perchtoldsdorf, Austria.

Amlinger, F. and Peyr, S. (2003): *Umweltrelevanz der dezentralen Kompostirung* (Environmental impacts of decentralized composting; in German). Technisches Büro für Landwirtschaft, Perchtoldsdorf, Austria.

Amlinger, F., Favoino, E., Pollak, M., Peyr, S., Centemero, M. and Caima, V. (2004): *Heavy metals and organic compounds from wastes used as organic fertilisers.* Study on behalf of the European Commission, Directorate-General Environment. Located on the internet: 19 July 2007. http://ec.europa.eu/environment/waste/compost/pdf/hm_finalreport.pdf.

Andersen, J.K., Boldrin, A., Samuelsson, J., Christensen, T.H. and Scheutz, C. (2010): Quantification of GHG emissions from windrow composting of garden waste. *Journal of Environmental Quality*, 39, 713–724.

Ballestero, T.P. and Douglas, E.M. (1996): Comparison between the nitrogen fluxes from composting farm wastes and composting yard wastes. *Transactions of the ASAE*, 39, 1709–1711.

BayLFU (2003): *Kompostierung von Bioabfällen mit anderen organischen Abfällen* (Composting biowaste with other types of organic waste; in German). Bayerisches landesamt für Umweltschutz, Augsburg, Germany.

Beck-Friis, B., Smårs, S., Jönsson, H. and Kirchmann, H. (2000a): Gaseous emissions of carbon dioxide, ammonia and nitrious oxide from organic household waste in a compost reactor under different temperatur regimes. *Journal of Agricultural Engineering Research*, 78, 423–430.

Beck-Friis, B.G., Pell, M., Sonesson, U., Jönsson, H. and Kirchmann, H. (2000b): Formation and emission of $N_2O$ and $CH_4$ from compost heaps of organic household waste. *Environmental Monitoring and Assessment*, 62, 317–331.

Bidlingmaier, W. and Müsken, J. (1993): Biotechnische Verfahren zur Behandlung von Bioabfall. In: Bidlingmeyer, W. (ed.) BioNet Netzwerk Umweltbiotechnologie (Biological–technical plant for the treatment of municipal biodegradable waste; in German). University of Essen. Located on the internet: 19 July 2007. http://www.bionet.net.

Boldrin, A., Andersen, J.K., and Christensen, T.H. (2009a). LCA report: Environmental assessment of garden waste management in Århus Kommune (Miljøvurdering af haveaffald i Århus Kommune). Department of Environmental Engineering, Technical University of Denmark, Copenhagen, Denmark.

Boldrin, A., Andersen, J.K., Møller, J. and Christensen, T.H. (2009b): Composting and compost utilization: Accounting of greenhouse gases and global warming contributions. *Waste Management and Research*, 27, 800–812.

Carlsbæk, M. (2007): Personal communication. Solum, Sengeløse, Denmark.

Christensen, T.H. (1998): Affaldsteknologi (Waste technology; in Danish). Teknisk Forlag, Copenhagen, Denmark.

Chung, Y.C. (2007): Evaluation of gas removal and bacterial community diversity in a biofilter developed to treat composting exhaust gases. *Journal of Hazardous Materials*, 144, 377–385.

Chung, Y.C., Ho, K.L. and Tseng, C.P. (2007): Two-stage biofilter for effective $NH_3$ removal from waste gases containing high concentrations of $H_2S$. *Journal of the Air and Waste Management Association*, 57, 337–347.

Clemens, J. and Cuhls, C. (2003): Greenhouse gas emissions from mechanical and biological waste treatment of municipal waste. *Environmental Technology*, 24, 745–754.

Cole, M.A. (1994): Assessing the impact of composting yard trimmings. *BioCycle*, 35 (4), 92–96.

Eklind, Y., Beck-Friis, B., Bengtsson, S., Ejlertsson, J., Kirchmann, H., Mathisen, B., Nordkvist, E., Sonesson, U., Svensson, B.H. and Torstensson, L. (1997): Chemical characterization of source-separated organic household waste. *Swedish Journal of Agricultural Research*, 27, 167–178.

Garcia, C., Hernandez, T., Costa, F. and Ayosu, M. (1992): Evaluation of the maturity of municipal waste compost using simple chemical parameters. *Communication in Soil Science and Plant Analysis*, 23, 1501–1512.

Gronauer, A., Claassen, N., Ebertseder, T., Fischer, P., Gutser, R., Helm, M., Popp, L. and Schön, H. (1997): *Bioabfallkompostierung, Verfahren und Verwertung* (Composting of biowaste, procedures and recycling; in German). Schriftenreihe Heft 139. Bayerisches Landesamt für Umwelt-schutz, München, Germany.

Hellebrand, H.J. (1998): Emission of nitrous oxide and other trace gases during composting of grass and green waste. *Journal of Agricultural Engineering Research*, 69, 365–375.

Hogg, D., Barth, J., Favoino, E., Centemero, M., Caimi, V., Amlinger, F., Devliegher, W., Brinton, W. and Antler S. (2002): *Comparison of compost standards within the EU, North America and Australasia*. Waste and Resources Action Programme, WRAP, Banbury, UK.

IPCC (2006): *2006 IPCC Guidelines for national greenhouse gas inventories, volume 5: Waste.* The Intergovernmental Panel on Climate Change, IPCC National Greenhouse Gas Inventories Programme, Hayama, Kanagawa, Japan.

Komilis, D. and Ham, R.K. (1999): *Life cycle inventory and cost model for mixed municipal and yard waste composting.* United States Environmental Protection Agency, Washington, D.C., USA.

Komilis, D.P. and Ham, R.K. (2006): Carbon dioxide and ammonia emissions during composting of mixed paper, yard waste and food waste. *Waste Management*, 26, 62–70.

Krogmann, U. and Woyczechowski, H. (2000): Selected characteristics of leachate, condensate and runoff released during composting of biogenic waste. *Waste Management and Research*, 18, 235–248.

Loll, U. (1994): *Behandlung von Abwässern aus aeroben und anaeroben Verfahren zur biologischen Abfallbehandlung* (Treatment of wastewater from aerobic and anaerobic biological waste treatment processes; in German). In: Wiemer, K. and Kern, M. (eds) *Verwertung Biologischer Abfälle*. M.I.C. Baeza Verlag, Witzenhausen, Germany, pp. 281–307.

Marb, C., Dietrich, G., Köbernik, M. and Neuchl, C. (1997): *Vergleichende Untersuchungen zur Kompostierung von Bioabfällen in i Reaktoren und auf Mieten: Emissionen, Qualität und Schadstoffe* (Comparison of biowaste composting in reactors: Emissions, quality and harmful substances; in German). *Müll und Abfall*, 29, 609–620.

Ødegård, K.E., Berg, B.E. and Bergersen O. (2005): *Emisjoner fra kompostering* (Emissions from composting; in Swedish). RVF Raport 2005-13. Svenska Rennhålningsverkförening, Malmö, Sweden.

Pitschke, T., Roth, U., Hottenroth, S. and Rommel W. (2004): *Ökoeffizienz von öffentlichen Entsorgungsstrukturen* (Eco efficiency of public waste disposal; in German). BIfA-Text Nr. 30. Bayerisches Institut für Angewandte Umweltforschung und Technik GmbH (BIfA), Augsburg, Germany.

Powelson, D.K., Chanton, J., Abichou, T. and Morales, J. (2006): Methane oxidation in water-spreading and compost biofilters. *Waste Management and Research*, 24, 528–536.

Saft, R.J. and Van Ewijk, H.A.L. (2004): *Dutch LCA experiences with biowaste management.* Status Paper. IVAM, Amsterdam, The Netherlands.

Schildknecht, H. and Jager, J. (1979): *Zur chemischen Ökologie der biologischen Abfallbeseitigung* (Chemical ecology of biological waste treatment; in German). Report 10302407. Ministry of Internal Affairs, Bonn, Germany.

Schleiss K. (2007): *Average of data collected in year 2005 in some Swiss composting plants.* Personal communication.

Schokker, E. (2007): *Dutch Waste Management Association (DWMA).* Personal communication.

Siebert, S. (2006): *Data on average composition of compost in Germany in 2006.* Bundesgütegemeinschaft Kompost e.V. (BGKeV). Personal communication.

Solomon, S., Qin, D., Manning, M., Alley, R.B., Berntsen, T., Bindoff, N.L., Chen, Z., Chidthaisong, A., Gregory, J.M., Hegerl, G.C., Heimann, M., Hewitson, B., Hoskins, B.J., Joos, F., Jouzel, J., Kattsov, V., Lohmann, U., Matsuno, T., Molina, M., Nicholls, N., Overpeck, J., Raga, G., Ramaswamy, V., Ren, J., Rusticucci, M., Somerville, R., Stocker, T.F.; Whetton, P., Wood, R.A. and Wratt, D. (2007): Technical summary. In: Solomon, S., Qin, D., Manning, M., Chen, Z., Marquis, M., Averyt, K.B., Tignor, M. and Miller, H.L. (eds) *Climate change 2007: The physical science basis. Contribution of Working Group I to the fourth assessment report of the Intergovernmental Panel on Climate Change.* Cambridge University Press, Cambridge, UK.

# 9.4

# Anaerobic Digestion: Process

**Irini Angelidaki**

*Technical University of Denmark, Denmark*

**Damien John Batstone**

*University of Queensland, Australia*

Organic waste may degrade anaerobically in nature as well as in engineered systems. The latter is called anaerobic digestion or biogasification. Anaerobic digestion produces two main outputs: An energy-rich gas called biogas and an effluent. The effluent, which may be a solid as well as liquid with very little dry matter may also be called a digest. The digest should not be termed compost unless it specifically has been composted in an aerated step.

This chapter describes the basic processes of anaerobic digestion. Chapter 9.5 describes the anaerobic treatment technologies, and Chapter 9.6 addresses the mass balances and environmental aspects of anaerobic digestion.

## 9.4.1 Definition of Anaerobic Digestion, Biogas and Digestate

### 9.4.1.1 Anaerobic Digestion

Anaerobic digestion may be defined as a biological conversion process without an external electron acceptor. That is, there is no overall supplied electron acceptor such as oxygen (in aerobic processes such composting oxygen is the overall electron acceptor), or nitrate and sulfate (in anoxic processes such as denitrification and sulfate reduction respectively). This definition determines some of the most obvious technological characteristics of anaerobic digestion:

- Combined endproducts have the same oxidation/reduction state as the waste degraded. The final catabolic endproducts are the carbon dioxide and methane, the most oxidized and least oxidized forms of carbon, respectively. The relative composition of the two gases in the biogas produced in anaerobic digestion is determined by the oxidation state of the waste.

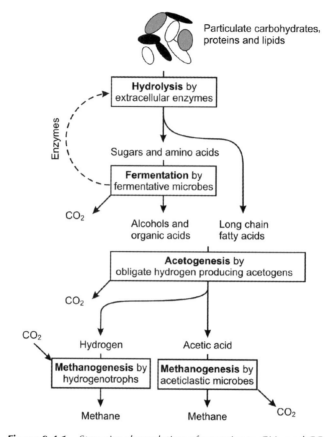

**Figure 9.4.1**   *Stepwise degradation of organics to CH₄ and CO₂.*

- Energy available from anaerobic conversion processes is relatively low compared to aerobic or anoxic reactions. Therefore, biomass produced from anaerobic digestion during anabolism is limited and very little energy is released as heat. Increase in temperature due to anaerobic digestion is not observed in anaerobic digesters; it has only been observed in large well insulated landfills.

Because of the limited energy available and specialized reactions occurring during anaerobic digestion, different reactions within anaerobic digestion are mediated by different groups of microorganisms. This is why anaerobic digestion is often termed a 'structured process'. The different digestion processes are generalized in Figure 9.4.1:

- Hydrolysis.
- Fermentation (acidogenesis).
- Acetogenesis.
- Methanogenesis.

Fermentative organisms convert simple sugars and proteins to organic acids, alcohols, hydrogen, and carbon dioxide, and produce enzymes that hydrolyze complex particulates to dissolved substrates. Acetogenic organisms produce acetate from organic acids, and waste electrons as hydrogen. This hydrogen is converted by hydrogenotrophic methanogens, while the acetic acid is converted to methane by aceticlastic methanogens.

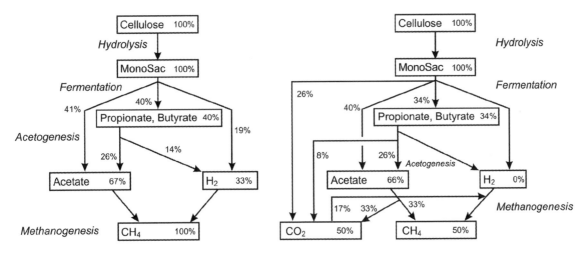

***Figure 9.4.2*** *COD (left) and carbon (right) flow in anaerobic digestion of cellulose. MonoSac = Monosaccharides.*

Within each of these overall steps, there are a number of parallel steps describing conversion of specific substrates. For example both monosaccharides and amino acids are fermented, but different microbes may mediate the process, and different governing factors determine endproducts and rates of fermentation. Additionally, under specific conditions, backward or lateral reactions may occur (e.g., acetate formation from $H_2$ and $CO_2$).

Under special conditions, other functional organisms play a role in anaerobic digestion: At low temperatures, homoacetogenic organisms convert hydrogen and carbon dioxide to acetic acid, and at high temperatures acetate-oxidizing organisms may convert acetate to hydrogen gas and carbon dioxide (Lepisto and Rintala, 1999).

Acetate is the most important final precursor to methane in the anaerobic digestion with 60–70 % of the organic matter passing through acetate and the remaining 30–40 % via hydrogen and carbon dioxide. As an example, the organic matter flow for digestion of cellulose is shown in Figure 9.4.2, with a large proportion of the methane forming directly from acetate.

### 9.4.1.2 Biogas

The main components in the biogas, $CO_2$ and $CH_4$, are determined by the composition of the waste digested. Electron-rich (energy-rich) waste such as fats produce a biogas high in methane, while electron-neutral substrates such as carbohydrates produce biogas with equal amounts of carbon dioxide and methane. $CO_2$ is partially soluble in water and the biogas from anaerobic digestion of waste usually contains 55–65 % $CH_4$ and 35–45 % $CO_2$. Ammonia, $H_2S$ and numerous volatile organic compounds usually constitute only less than 1 % of the biogas.

The energy content in biogas is significant and usually exceeds the amount of energy used technically in running the anaerobic digester. The biogas can be used directly for producing electricity and heat or can be converted to a fuel, making the digester a net energy producer. In addition, the energy originating from recently plant-based materials synthesized by photolysis, can be considered neutral with respect to global warming potentials. However, $CH_4$ escaping the process or the energy utilization unit should be captured, since unconverted $CH_4$ emitted to the atmosphere is a significant green house gas contributing to global warming approximately 25 times more than $CO_2$. Capturing methane is usually done by flaring at the plant and by covering after-storage tanks in order to retrieve residual biogas production. After-storage biogas production has been found to constitute 5–25 % of the total biogas production of the plants (Angelidaki *et al.*, 2005).

### 9.4.1.3   Digestate Residue

The residue from anaerobic digestion can be a solid, if the process is a dry process, or a liquid effluent, if the process is wet. In the latter case the process runs in a mixed suspension of 4–10 % total solids. As the digestion proceeds, the solid content decreases. The composition of the digestate depends strongly on the composition of the waste treated (see Chapter 9.6).

The uses or disposal of the digestate can be many, depending on the overall concept of using anaerobic digestion. Examples are:

- Spread on land as a fertilizer if based on digestion of biowaste.
- Landfilled.
- Incinerated.
- Composted and landfilled.
- Further treated in subsequent step for removing the residual organic matter and nutrient.

## 9.4.2   The Structure of Anaerobic Digestion

The different biochemical processes – hydrolysis, fermentation, acetogenesis, and methanogenesis – active in anaerobic digestion are described in the following sections.

### 9.4.2.1   Hydrolysis

Hydrolysis of particulates to soluble molecules is the first step in anaerobic digestion. Hydrolysis is extracellular, because microbes cannot accept particles. The strict definition of hydrolysis is given in Box 9.4.1. In anaerobic digestion, hydrolysis is used as a lump term for solubilization of solids. There are three main classes of compounds that can be hydrolyzed: Carbohydrates, proteins, and lipids. There are also recalcitrant organic compounds such as lignins, keratin, plastics, waxes and mineral compounds. Inert organic compounds are generally included in mass balances, while mineral compounds often are excluded.

Different groups of fermentative bacteria are capable of excreting the extracellular enzymes that are needed for the hydrolysis of complex polymeric compounds in the waste into oligomers and monomers that can be taken up by the microorganisms. The proteolytic bacteria produce proteases that catalyze the hydrolysis of proteins into amino acids. The cellulytic and xylanolytic bacteria produce cellulases and/or xylanases that degrade cellulose and xylan (both are carbohydrates) to glucose and xylose, respectively. The lipolytic bacteria produce lipases that degrade lipids (fat and oils) to glycerol and long-chain fatty acids.

Carbohydrates are plant material formed from long-chain, branched polysaccharides, and are the main component in food waste, paper, cardboard and wood in municipal solid waste. By definition, they have a generalized empirical

---

**Box 9.4.1   Definition of Hydrolysis.**

The strict definition of hydrolysis is 'a reaction in which a molecule is split, and the hydrogen and hydroxide ions from a water molecule are attached to the separate products' (Brown, 1993). As a biological reaction, hydrolysis is commonly used to describe the lumped reactions in the solubilization of particulate compounds. Most of the individual depolymerization reactions during hydrolysis, such as cleavage of peptide bonds in proteins, ester bonds in lipids, and glycosidic bonds in polysaccharides are true hydrolysis reactions, but others, such as cleavage of disulfide bonds in protein, are not. Additionally, one should distinguish between hydrolysis and pretreatment. Several pretreatment methods (physico-chemical and/or enzymatic/biological) aim at increasing particle surface area or removing lignin to unpack carbohydrates from biofibers and contribute to an increase in the hydrolysis; indirectly by making particulate more accessible for hydrolysis.

Box 9.4.2    **Hydrolysis of Lignocellulose.**

Lignocellulose is a common name for compounds consisting of the three main components of the plants vascular tissue: Cellulose, hemicellulose and lignin. The composition of the lignocellulose differs according to age of the plant and also between crops, softwood and hardwood. Due to the chemical and physical composition of lignocellulose, this complex polymer is very hard to degrade for the microorganisms. The major problem in the degradation is the lignin associated with the cellulose. A physical contact between the cellulases and cellulose has to be established before any degradation can take place. Before the cellulases can function effectively, the cellulose has to swell by uptake of water, but a close association between lignin and the cellulose 'seals off' the cellulose for any water entrance thereby preventing any loosening of the cellulose structure. So lignin acts as a physical barrier making microbial attacks difficult. In fact, lignin is one of the plants defense mechanisms against microbial attack and is practically undegradable at anaerobic conditions. The compound can even be toxic to the microorganisms. Of minor importance for the degradation of lignocellulose are the surface area available for microbial attack and the complexity (crystallinity) of the cellulose.

Since a large fraction of organic waste is composed of lignocellulose, the microbial separation of the cellulose fraction is prerequisite for a satisfactory degradation. The initial hydrolysis proceeds at various speed depending on the nature of the waste and may in some cases be a rate-limiting step for the entire degradation process.

formula $C_n(H_2O)_m$, but are better described as chains of polysaccharides. Carbohydrates are the most common oxidized, generalized particulate substrate in anaerobic digestion, and generally produce a gas with 50% $CH_4$ and 50% $CO_2$. Some simple carbohydrates are easily solubilized (sugars, starch and the like), but those being lignocellulose are hard to hydrolyze. Lignocellulose is a common name for compounds consisting of the three main components of the plants vascular tissue: Cellulose, hemicellulose, and lignin. The composition of the lignocellulose differs according to age of the plant and also between crops, softwood, and hardwood. Due to the chemical and physical composition of lignocellulose, this complex polymer is very hard to hydrolyze (see Box 9.4.2). If cellulose is the main component in the waste, the hydrolysis step is rate-limiting, while the decomposition of acetate to methane is rate-limiting, if the substrate is mainly consisting of easily metabolized material, i.e. dissolved starch.

Proteins and lipids are main components in food waste. Both compounds are relatively more reduced compared to carbohydrates and therefore produce larger amounts of $CH_4$ than $CO_2$. They are also the source for inhibitory compounds; proteins produce ammonia when the amino acids are degraded, and lipids produce long-chain fatty acids. Both ammonia and long-chain fatty acids inhibit aceticlastic methanogenesis. Most proteins are relatively rapidly hydrolyzed. The rate of lipid hydrolysis varies, depending on whether it is a liquid (oil), or a solid (fat). Most animal-based fats are hydrolyzed slowly in comparison to proteins.

The products from hydrolysis are simple sugars, amino acids and long-chain fatty acids. The process is also called acidogenesis, since the main outcome is acids.

### 9.4.2.2    Fermentation

Sugars and amino acids are converted to volatile fatty acids (VFA), alcohols, hydrogen, and $CO_2$ in a process called fermentation. The amino acids release ammonium during fermentation. This process is not obligated to use an external electron acceptor. Electrons from oxidative processes can be directly utilized in coupled reductive processes. This often means that multiple products are produced. Long-chain fatty acids from the hydrolysis of lipids are not converted in fermentation, but instead are oxidized during the following acetogenesis.

Generally, fermentation is a rapid process involving many pathways, intermediates and endproducts. The versatility of the process is illustrated with the fermentation of glucose in Figure 9.4.3. The fermentation of monosaccharides (nominally glucose) has two main branches. One may produce propionate via pyruvate, which overall, is an electron consuming reaction, while the other may produce a variety of products via acetyl-coenzyme A (CoA), including acetate,

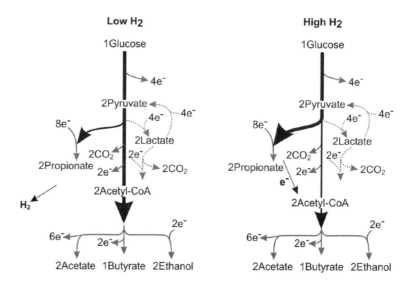

**Figure 9.4.3** *Major degradation pathways for glucose in anaerobic digestion, showing major directions of energy flow under low and high $H_2$ conditions.*

butyrate, and ethanol, which are generally (net) electron producing. If the overall reaction is electron producing, the excess electrons must be wasted to produce hydrogen or formate. Formate and hydrogen are energetically interchangeable as final electron acceptors. Examples of some balanced reactions, with net adenosine triphosphate (ATP) generation, are shown in Table 9.4.1. There are also many alternative products from acetyl-CoA (e.g., see Madigan *et al.*, 2000), but the major products are butyrate and acetate. Ethanol may also be produced under low pH conditions (Ren *et al.*, 1997). Lactate is also an important potential intermediate that may be produced and released with the same overall oxidation state as glucose, but which is very quickly consumed.

Fermentation has under most conditions a considerable energy yield when producing both highly oxidized products such as acetate and more reduced products such as propionate and butyrate. When the reactor is operating under stable conditions, most substrate is converted to acetate and hydrogen directly (see Figure 9.4.3), rather than passing through the reduced products. Production of mainly acetate and hydrogen is an oxidative pathway, as electrons are wasted to hydrogen ions (to form dissolved $H_2$). However, when the reactor is overloaded, either through excessive production of acetate and hydrogen, or pH extremes, fermentation directs to coupled pathways to form larger amounts of less oxidized products such as propionate, butyrate, and ethanol. Proteins may also form higher organic acids such as valerate and aromatic organic acids. These higher organic acids need to be oxidized by organic acid-oxidizing bacteria, which are also subject to hydrogen and pH inhibition, and the reactor can easily move into an organic overload situation.

**Table 9.4.1** *Thermodynamically possible reactions in glucose fermentation.*

| Products | Reaction | ATP/mol glucose | Conditions |
|---|---|---|---|
| Acetate | $C_6H_{12}O_6 + 2H_2O \rightarrow 2CH_3COOH + 2CO_2 + 4H_2$ | 4 | Low $H_2$ |
| Acetate + propionate | $3C_6H_{12}O_6 \rightarrow 4CH_3CH_2COOH + 2CH_3COOH + 2CO_2 + 2H_2O$ | 4/3 | Any $H_2$ |
| Butyrate | $C_6H_{12}O_6 \rightarrow CH_3CH_2CH_2COOH + 2CO_2 + 2H_2$ | 3 | Low $H_2$ |
| Lactate | $C_6H_{12}O_6 \rightarrow 2CH_3CHOHCOOH$ | 2 | Any $H_2$ |
| Ethanol | $C_6H_{12}O_6 \rightarrow 2CH_3CH_2OH + 2CO_2$ | 2 | Low pH |

**Table 9.4.2** *Thermodynamics of reactions for fatty acid oxidation (acetogenesis).*

| Substrate | Reaction | $\Delta G_0$ (kJ/g COD) | $\Delta G'$ (kJ/g COD) |
|---|---|---|---|
| Propionate | $CH_3CH_2COOH + 2H_2O \rightarrow CH_3COOH + 3H_2 + CO_2$ | 0.68 | −0.13 |
| Butyrate | $CH_3CH_2CH_2COOH + 2H_2O \rightarrow 2CH_3COOH + 2H_2$ | 0.30 | −0.16 |
| Palmitate | $CH_3(CH_2)_{14}COOH + 14H_2O \rightarrow 8CH_3COOH + 14H_2$ | 0.55 | −0.16 |

$\Delta G'$ Calculated for $T = 298\,K$, pH 7.0, $p(H_2)$ $1e^{-5}$ bar, $p(CH_4)$ 0.7 bar, 0.1M $HCO_3^-$ and 1 mM organic acid.

### 9.4.2.3 Acetogenesis

The VFA (such as propionate and butyrate) and alcohol (such as ethanol) produced during the fermentation step are oxidized to acetate by obligate hydrogen producing acetogens (OHPA; see Table 9.4.2). Electrons produced from this oxidation reaction are wasted to hydrogen ions to produce $H_2$. By obligate, it is meant that there is no coupled reaction or external electron acceptor available, as wasting electrons by producing hydrogen is generally energetically unfavorable.

The free energy of reaction for fatty acid oxidation is positive at standard conditions and therefore the reaction needs very low hydrogen concentrations to achieve a negative free energy (and thereby yield energy for anabolism). Hydrogen-consuming methanogens, which convert $H_2$ and $CO_2$ to $CH_4$, help in a balanced biomass to keep $H_2$ concentrations low. However, it is only in a very narrow band of $H_2$ concentrations that it is favorable for both $H_2$ producers (acetogens) and $H_2$ consumers (methanogens). $H_2$ is the most important endproduct with respect to determining the free energy of the reaction, since more $H_2$ is produced (stoichiometrically) than other products. Acetate is also important, as it is also an endproduct, but acetate is stoichiometrically produced at lower levels. This is demonstrated in Figure 9.4.4, which indicates the 'operating regions' for different acetogenic reactions, the major aceticlastic reaction to methane, and the methanogenic, $H_2$-consuming reaction. As demonstrated, the range in which the reaction is possible (shaded for propionate) is far smaller for $H_2$ than acetate.

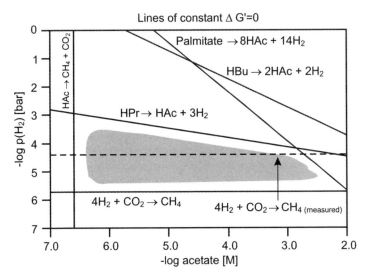

**Figure 9.4.4** *Lines of zero $\Delta G'$ (298 K) for five reactions with the same assumptions (1 mM VFA concentrations, 100 mM bicarbonate, 0.5 bar methane and $CO_2$ in the gas phase). Shaded portion shows regions where all reactions are possible. The measured hydrogen-utilizing threshold and corresponding line (–) are based on Cord-Ruwisch et al. (1988); figure from Batstone et al. (2002). HBu = Butyrate, HAc = Acetic acid, HPr = Propionate. Reprinted with permission from Anaerobic digestion model No. 1 (ADM1) 9781900222785 by D.J. Batstone, J. Keller, I. Angelidaki et al., © (2002) IWA Publishing.*

---

**Box 9.4.3    Ecology of Interspecies Hydrogen Transfer.**

Because of the critical levels of $H_2$ concentration, $H_2$ transfer between the $H_2$ producers and $H_2$ consumers is very important. $H_2$ is a small molecule, with high diffusivity and low solubility. This means that the optimal distance for $H_2$ transfer is quite low (10 µm; Boone *et al.*, 1989). Formate ($HCOO^-$) is thermodynamically and stoichiometrically equivalent to $H_2$, when it is considered that the $CO_2$ concentration and anaerobic digesters are relatively stable. There are three main differences between $H_2$ and formate as electron carriers:

- Formate is more soluble.
- $H_2$ has a higher diffusivity.
- Formic acid is a stronger acid (than $CO_2$).

These differences led Boone *et al.* (1989) to conclude that, when the microbes are separated by more than 10 µm, formate is used for electron transfer (a), and if they are separated by less than 10 µm, $H_2$ is used for electron transfer (b). In practical terms, if $H_2$ is used for electron transfer, there is very little transfer between the microcolony and bulk liquid, and bulk liquid changes in $H_2$ concentration may have a low or delayed influence on $H_2$ producing reactions within the microcolony. From studies in a variety of environments, including anaerobic granules, municipal solid waste digesters, soil and anaerobic aquatic systems, it seems that close co-location of $H_2$ producers and consumers is the preferred orientation, and therefore $H_2$ is used most commonly as electron carrier, although $H_2$ concentrations in the liquid and in the gas may not reflect this directly.

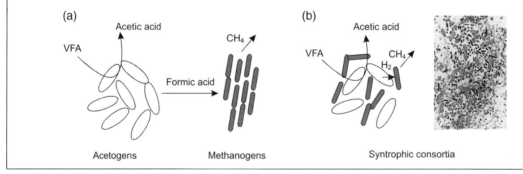

---

Because of the critical levels of $H_2$ concentration, $H_2$ transfer between the $H_2$ producers and $H_2$ consumers is very important and most likely they live in synthrophic consortia (see Box 9.4.3). Because they have no alternative pathways apart from direct oxidation of the organic acids, fatty acid-oxidizing organisms may be referred to as having a linear metabolism.

## 9.4.2.4    Methanogenesis

Methane is generated primarily by two pathways: Hydrogenotrophic methanogenesis, which converts $H_2$ and $CO_2$ into $CH_4$, and the aceticlastic methanogenesis, which converts acetate into $CH_4$ and $CO_2$.

Hydrogenotrophic methanogensis, as previously discussed, takes place in close coexistence with acetogenesis and typically accounts for 30–40 % of the $CH_4$ according to this overall reaction:

$$4H_2 + CO_2 \rightarrow CH_4 + 2H_2O \quad \Delta G_0 = -130\,\text{kJ/mol } CH_4 \qquad (9.4.1)$$

Aceticlastic methanogenesis typically generates 60–70 % of the $CH_4$, and is one of the most sensitive processes in anaerobic digestion to a number of factors. It is also a highly specialized reaction and is mediated by two groups of

Archaea, specifically the Methanosarcinaceae and Methanosaetaceae within a single family Methanosarcinales. The overall reaction is as follows:

$$CH_3COOH \rightarrow CH_4 + CO_2 \quad \Delta G_0 = -31\, kJ/mole\, CH_4 \tag{9.4.2}$$

There are several important differences between the two groups, mainly related to the transport and reaction pathways. Methanosarcinaceae convert 1 mol ATP/mol acetate activated, while Methanosaetaceae convert 2 mol ATP/mol acetate. Consequently, Methanosaetaceae achieve an overall yield of approximately 0.25 mol ATP/mol acetate converted, while Methanosarcinaceae achieves an overall yield of between 1.0 and 1.25 mol ATP/mol acetate converted. As a negative, higher acetate concentration levels are required for the active transport mechanism used by Methanosarcinaceae (Zinder, 1993). In practical terms, this means that Methanosarcinaceae can outcompete Methanosaetaceae at high acetate concentrations, while Methanosaetaceae dominates at low acetate concentrations. It also means that Methanosarcinaceae has a higher maximum growth rate and also a lower minimum acetate level. Additionally, Methanosaetaceae is more susceptible to load shocks and inhibition (as there is less energy available to maintain cellular stasis). Therefore, while Methanosarcinaceae can achieve higher conversion rates, Methanosaetaceae can achieve lower effluent acetate concentrations.

Acetate oxidation is also an alternative route for converting acetate. In this process, acetate is oxidized directly to $CO_2$ and $H_2$, with the same thermodynamic considerations as for acetogenesis. This pathway mainly occurs in thermophilic (high temperature) digesters, where the acetate ion oxidation reaction becomes more thermodynamically favorable (Petersen and Ahring, 1991). It has recently been found that in the absence of Methanosaeta this reaction is the dominant reaction for acetate conversion (Karakashev *et al.*, 2006).

Finally it should be mentioned that acetoclastic methanogenesis may be suffering by competition for acetate if sulfate enters the digester. Oxidation of acetate by sulfate reduction is thermodynamically a more favorable process leading to the generation of $CO_2$ and $HS^-$. The latter may precipitate as sulfides, for example with iron, but may also contribute to odor formation.

## 9.4.3 Estimation of Methane Potential and Gas Composition

Prediction of the amount of biogas produced is a key issue in anaerobic digestion since the gas is the value carrier of energy and the more gas produced the more degraded the waste and the better the economy of the process. In fact, the prime interest is the methane, because methane is the energy rich component of the gas.

The amount of biogas produced and the content of methane in the gas depends on the waste being degraded, both its degradability and its oxidation state: The better degradable and the lower the oxidation state, the more the amount of methane produced.

### 9.4.3.1 Theoretical Methane Potential

If the composition of the feed stock is known and all the organic material is converted to biogas, the theoretical methane potential can be calculated from the following equation:

$$C_nH_aO_b + \left(n - \frac{a}{4} - \frac{b}{2}\right)H_2O \rightarrow \left(\frac{n}{2} + \frac{a}{8} - \frac{b}{4}\right)CH_4 + \left(\frac{n}{2} - \frac{a}{8} + \frac{b}{4}\right)CO_2 \tag{9.4.3}$$

This formula is derived by balancing the total conversion of the organic material to $CH_4$ and $CO_2$ with $H_2O$ as the only external source, i.e. under strictly anaerobic conditions.

The specific methane yield, B usually expressed as (standard temperature and pressure, STP), $l\, CH_4/g$ VS can then be calculated as (STP):

$$B = \frac{\left(\frac{n}{2} + \frac{a}{8} - \frac{b}{4}\right) \cdot 22.4}{12n + a + 16b} \tag{9.4.4}$$

**Table 9.4.3**   Theoretical methane yield for a number of organic compounds.

| Substrate | Composition | CH$_4$ yield (STP l/g VS) | CH$_4$ yield (STP l/g COD) | CH$_4$ (%) |
|---|---|---|---|---|
| Carbohydrate | $(C_6H_{10}O_5)_n$ | 0.415 | 0.35 | 50 |
| Protein[a] | $C_5H_7NO_2$ | 0.496 | 0.35 | 50 |
| Lipids | $C_{57}H_{104}O_6$ | 1.014 | 0.35 | 70 |
| Ethanol | $C_2H_6O$ | 0.730 | 0.35 | 75 |
| Acetate | $C_2H_4O_2$ | 0.373 | 0.35 | 50 |
| Propionate | $C_3H_6O_2$ | 0.530 | 0.35 | 58 |

[a]Nitrogen is converted to $NH_3$

where 22.4 is the volume in liters of one mole of gas at standard conditions. Table 9.4.3 shows the theoretical methane yield for different organic compounds.

### 9.4.3.2   Measured Methane Potentials

Since wastes often are very heterogeneous and contain many different organic components, it is often difficult to estimate the methane potential by theoretical calculations and therefore it may be practical to measure the methane potential in the laboratory. Methods for measuring biogas production are utilizing either measurement of the volume of the produced gas under constant pressure (volumetric), measurement of pressure increase in constant volume (manometric), or measurement of methane formation by gas chromatography (Angelidaki and Sanders, 2004). These methods are briefly described in Box 9.4.4.

The laboratory methane potential measurement should be performed under optimal conditions and for a long time, i.e. until methane formation ceases.

### 9.4.3.3   Practical Methane Yields

Although the theoretical methane potential as well as the measured methane potential gives a rough idea of the amount of biogas production to expect from the waste composition, the practical yield obtained in a biogas reactor will always be lower due to a number of factors:

- A fraction of the substrate is utilized to synthesize bacterial mass, typically 5–10 % of the organic material degraded.
- At a finite retention time a fraction of the organic material will be lost in the effluent, typically 5–10 %.
- Lignin is not degraded anaerobically.
- Often a part of the organic material is inaccessible due to binding in particles or structural organic matter.
- Limitation of other nutrient factors.

Under favorable conditions with mainly water soluble matter, degrees of conversion up to 85–95 % can be achieved. If the organic matter is highly particulate or structural 30–60 % conversion is more normal. Davidson *et al.* (2007) found 75–80 % conversion of VS for biowaste at thermophilic digestion for 15 days in pilot-scale reactors.

### 9.4.3.4   Gas Composition

A prediction of the composition of the gas produced is more complex and depends first of all on the amount of CH$_4$ and CO$_2$ produced, but also on the pH of the reactor content. The CH$_4$ produced is mainly released to the gas phase, but CO$_2$ is partly dissolved in the liquid phase of the reactor or is converted to bicarbonate as a function of the pH. Consequently, the % CH$_4$ in the biogas produced will generally be higher than predicted by the stoichiometric ratio. Methane content in the gas phase is typically 55–65 %. Methane content in the biogas is somehow reflecting the pH in the reactor.

**Box 9.4.4   Experimental Determination of Methane Potentials. Reprinted with permission from Waste Management, Method of determination of methane potentials of solid organic waste by T. L. Hansen, J. E. Schmidt, I. Angelidaki et al., 24, 4, 393–400 © (2004) Elsevier Ltd.**

Basically, three different approaches are used to measure methane potentials of waste in the laboratory (Angelidaki and Sanders, 2004). References cited in this box are found in Angelidaki and Sanders (2004).

## Volumetric Methods

Volumetric methods are based on measurement of gas volume produced during anaerobic degradation of a compound. Either the total biogas volume is measured or only the volume of methane produced is measured, after carbon dioxide is removed through a carbon dioxide trap (strong alkali solution). The volume is either measured by displacement of the piston of a syringe attached to the rubber seal of the closed vials where anaerobic degradation is taking place (Owen *et al.*, 1979), as a part of the reactor (Cohen, 1992) or by water displacement systems (Field and Sierra, 1989; Angelidaki *et al.*, 1992).

Simple tests for measuring the anaerobic biodegradability and biogas potential were first described by Owen *et al.* (1979). They measured biodegradability of the test material in closed bottles, where the test material was mixed with inoculum from a laboratory scale reactor, diluted 20 % v/v with synthetic medium containing important salts. Finally, the bottles were flushed with inert gas ($N_2$:$CO_2$ = 70:30) in order to displace oxygen. The volume of gas produced was measured by displacement of the piston of a glass syringe. The methane content of the produced gas was also determined. The biochemical methane potential (BMP), as they called it, was calculated by subtracting methane production in control vials (vials without the test material), from that in vials with the test material. A slight modification of Owen's test led to a proposed standard for anaerobic biodegradability presented by the American Society of Testing Materials (ASTM).

Volumetric methods are often combined with gas chromatographic (GC) methods, where the gas composition of the head space gas is measure at the end of the experiment.

## Manometric Methods

The manometeric methods originate from Warburg's respirometer, in which the gas produced in constant volume is measured by pressure increase by a differential manometer. The measured gas is discharged, to avoid increase of pressure outside the optimal measuring range of the pressure transducer used (Miller and Wolin, 1974). Many researchers have applied manometric respirometer methods for anaerobic degradation tests (Campos and Chernicharo, 1991; Battersby and Wilson, 1988; Birch *et al.*, 1989; Angelidaki *et al.*, 1998).

The manometric-based biodegradability assays were used by Shelton and Tiedje (1984) who developed the first protocol for the determination of anaerobic biodegradability assays. This work, in 1995, lead to ISO 11 734. With this ISO test, the pressure increase from vials with sample is subtracted from the pressure increase from control vials only containing inoculum. In order to account for solubilized carbon dioxide in liquid phase, dissolved inorganic carbon in the liquid phase is measured at the end of the test. By this, the total biogas produced during the anaerobic degradation process can be accurately determined. In order to keep addition of dissolved inorganic carbon by the inoculum (digested sludge) low, inoculum is washed prior to inoculation. The ISO 11 734 standard suggests a substrate concentration of 100 mg/l and a total solids (TS) concentration of 1–3 g/l.

## Gas Chromatographic Methods

Gas chromatography is used to measure the content of methane and carbon dioxide of the biogas that ends up in headspace of closed vials (Dolfing and Bloemen, 1985). The system may be either kept at constant volume or the produced gas may be released. Soto *et al.* (1993) compared liquid displacement systems to gas chromatography methods and concluded that the latter are more accurate for low methane productions.

Gas chromatographic methods can use either a GC with thermal conductivity detection (TCD) where both methane and carbon dioxide are measured or a GC with flame ionization detection (FID), where only methane is measured. The latter method is very fast and one measurement takes less than 1 min.

In methane potential tests samples are inoculated in closed vials and incubated at constant temperature. Controls with only inoculum added are included in order to account for the biogas produced from organic matter contained in the inoculum. Digested sludge is often the used inoculum. However, in some cases, microorganisms adapted to specific conditions such as high ammonia concentrations are needed.

Hansen *et al.* (2004) describes a method for determination of the methane potential of solid wastes. The method uses 2-l glass bottles supplied with solid waste (up to 10 g on a dry basis) and inoculated with digested manure from a reactor operating at thermophilic temperature (55 °C). Large inoculation volumes are ensuring high microbial activity, low risk for overloading and low risk of inhibition. Thermophilic conditions ensure fast reaction rates. Furthermore, digested manure has been shown to be superior to digested sludge as it is rich to many nutrients, and has a high buffer capacity. Inoculum is degassed before use for a few days to ensure that methane production from inoculum is as low as possible. Sample is diluted with water at different dilutions to ensure that possible toxicity of the sample is not overseen. The test is run in triplicates. Methane production in the headspace of the vials is followed by sampling gas from headspace of the vial with a pressure-tight syringe, and subsequent analysis of methane content by GC (FID detection). Each sampling point takes less than a minute to analyze. The batch set-up used in the above investigation as well as a typical methane formation curve (biowaste and cellulose as a control) are shown below.

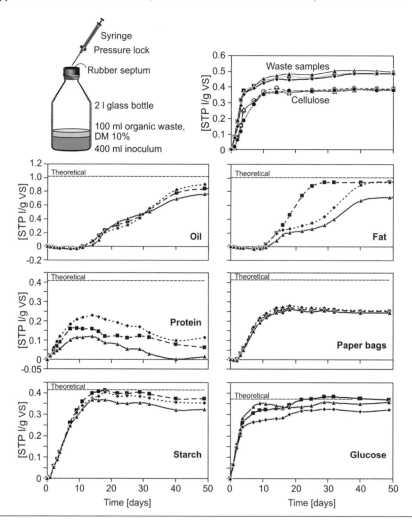

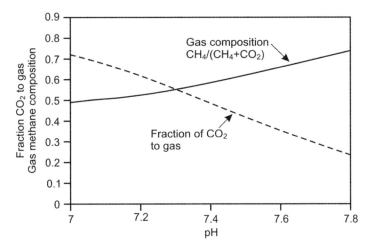

***Figure 9.4.5*** *Approximate gas composition of $CO_2$ to gas and gas composition [% methane/(methane+$CO_2$) at different pH values with mesophilic conditions in a typical digester fed with 30 % inerts, 40 % carbohydrates, 25 % proteins, and 5 % lipids, and with 0.15 M bicarbonate alkalinity in the feed (Batstone et al., 2002)]. This figure is estimated with 5 % solids in the feed, and in solid phase digesters, the impact of pH decreases, and the slope of the gas composition line therefore decreases.*

Figure 9.4.5 shows predictions of the methane content of the gas as a function of pH in the reactor as well as the fraction of the generated $CO_2$ that is found in the gas. The remaining part of $CO_2$ is dissolved in the water.

## 9.4.4   Process Factors

The overall performance of the anaerobic digestion process depends on several process factors of which the most important are nutrients, temperature and inhibitory factors.

### 9.4.4.1   Nutrients

The chemical composition of the cellular material highly reflects which nutrients a microorganism needs. The nutritional requirements in anaerobic digestions is fairly well known from many studies on a variety of substrates. In general, in the solid waste area nutritional limitations for anaerobic digestion are rarely seen.

Since very little biomass is synthesized during anaerobic digestion the requirement for nitrogen is much lower than in composting. Consequently, the C/N ratio of the feedstock is rarely a limiting factor with mixed solid waste.

Calcium, magnesium, sodium, and potassium can in high concentrations be inhibitory, but can result in a stimulation of the fermentation if they are added in low concentrations. Improved digestion by addition of the above-mentioned monovalent and divalent cations in concentrations between 0.01 M and 0.005 M has been shown. An important factor is that there is a synergistic and antagonistic connection between the cations.

Many micronutrients are needed for optimum process rates, especially Ni and Co are important for growth of anaerobic organisms. Nickel is necessary for activating factor F430 (a cofactor involved in the methanogenesis), but in high concentrations nickel can be inhibitory for fermentative as well as methanogenic bacteria.

The addition of iron can stimulate the process by increasing the precipitation of sulfides and phosphates, which could otherwise have precipitated important trace metals.

### 9.4.4.2 Temperature

Temperature has a strong influence on both physicochemical parameters and microbiological processes.

#### *Physicochemical Parameters*

Physicochemical parameters are primarily related to phase distribution, mass transfer rates, and solubility. They include:

- Viscosity decreases with increasing temperature. Increased temperature makes pumping and mixing easier and thus improves on mass transfer as well as on gas–liquid transfer.
- Diffusivity increases with increasing temperature. This slightly decreases the effective threshold for the microbes and improves the gas–liquid transfer rate.
- Thermodynamic influence on equilibrium coefficients. This influences both the acid-base dissociation constants and gas–liquid equilibrium. Generally gases are less soluble at higher temperature, and therefore, more gas is transferred to the gas phase. This can actually be a disadvantage, as more $CO_2$ than $CH_4$ is transferred (the solubility of methane is relatively low), and the gas contains more water vapor. Acid–base dissociation constants generally decrease with temperature (mainly for inorganic acids), so that there is generally more of the base form of the acid–base pair. This is generally of disadvantage to the process, as the base form (e.g., ammonia, sulfides) is the inhibitory form, and inhibits more at higher temperatures.
- Availability of solids is generally increased. This is partly due to increased solubility of solids, but also because particulates such as fats may be melted (and thus emulsified).

#### *Microbiological Processes*

There are three generalized microbial operating regimes, defined by the organisms with optimal operating points in those ranges: Thermophilic (40–60 °C), mesophilic (25–40 °C) and psychrophilic (0–25 °C). The growth rates of methanogens in the three regimes are shown in Figure 9.4.6. For each group of microbes, the growth rate increases with temperature exponentially to the optimum. Beyond the optimum, there is a rapid decrease in growth rate, mainly caused by denaturizing of enzymes, and disruption of cellular stasis.

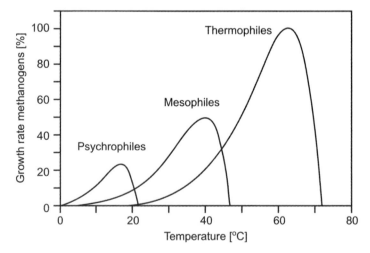

**Figure 9.4.6** *Relative growth rate of psychrophilic, mesophilic and thermophilic methanogens (from van Lier, 1997). Reproduced with permission from Water Science & Technology © (1997) IWA Publishing, London.*

In practice, psychrophilic digestion is not applied by choice, but is normally forced by climate and available energy (for heating). Usually, low technological anaerobic digestion operates at psychrophilic temperatures due to a lack of heating of the digesters. The design decision is normally between mesophilic and thermophilic digestion.

The thermophilic digestion provides a number of advantages compared to mesophilic digestion:

- Reduction of the residence time in the plant.
- Good destruction of pathogenic organisms.
- Improved possibility for separation of solid matter from the liquid phase.
- Better degradation of long-chain fatty acids.
- Less biomass formation compared to the product formation.
- Improved solubility and availability of substrates.

Essential disadvantages of thermophilic digestion compared to mesophilic digestion are:

- Demand of larger amount of process energy, or/and investment costs for heat exchanging and insulation.
- Larger risk of ammonia inhibition.

Destruction of pathogens is a main advantage of thermophilic digestion compared to mesophilic digestion. Process optimization has allowed development of more continuous feeding regimes, and with thermophilic conditions, significant sanitation of the biomass occurs. However, control must always be maintained on the period of time between additions of undigested material and removals from the reactors to insure adequate sanitation time.

Even though the investigations with thermophilic pure cultures and cocultures have shown an optimum temperature near 55 °C, it can be wise especially when the ammonia load is high to keep a lower temperature (52–54 °C) in thermophilic biogas plants. This insures a safety margin compared with the negative effects a temperature increase above 60 °C can give. Thermophilic anaerobic digestion is now becoming more widespread. In Denmark, almost two-third of the anaerobic digesters are operating at thermophilic temperatures.

### 9.4.4.3   Inhibitory Factors

Inhibition is a generalized term that means restriction of biological processes. Speece (1996) divided the two inhibition processes into: (1) Toxicity, which is an adverse effect (not necessarily lethal) on microbial metabolism, and (2) inhibition, which is an impairment of bacterial function. Batstone *et al.* (2002) further clarified these terms as biocidal inhibition, biostatic inhibition and endproduct inhibition.

#### *Biocidal Inhibition*

Reactive toxicity, normally irreversible. By reactive it is meant that the toxic compound reacts with a functional component of the microbial cell, rendering it nonfunctional. This type of inhibition matches Speece's definition of toxicity. Compounds which are generally biocidal to some or all anaerobes include detergents, cyanide and antibiotics. The main other potential toxic inhibitors found in solid waste are xenobiotics such as polyaromatic aromatic hydrocarbons. These also have an adsorption mechanism, but the technical difficulties involved with removal of the xenobiotics generally outweigh those caused by their toxicity to general anaerobic digestion.

#### *Biostatic Inhibition*

Nonreactive toxicity, normally reversible. Compounds that cause biostatic inhibition do not disable functional components, but rather disrupt cellular stasis, or change energy production. Microbes require intracellular conditions with redox potential, pH, and total salts within a small margin. Biostatic inhibiting compounds disrupt these conditions, and the microbe is required to spend energy on maintenance of stasis, rather than anabolism. Free acid and bases (e.g., VFAs, $H_2S$, $NH_3$), salts, and pH changes all cause biostatic inhibition. Biostatic inhibition is also caused by drops in thermodynamic yield caused by accumulation of product (e.g., hydrogen inhibition). Organisms with particularly marginal yields (e.g., acetogenic, hydrogen utilizing, and especially aceticlastic microbes) are particularly susceptible to biostatic inhibition.

pH inhibition is the most important form of biostatic inhibition, and is partly due to inhibition by free acids and bases at low and high pH, respectively, but mainly because a pH well outside the range for energy-limited microbes such as acetogens and methanogens decreases the energy for anabolism and directs it towards maintenance. One of the most common overloads in anaerobic digesters termed an 'acid overload', is pH-related and follows the following self-reinforcing pattern:

- Biostatic inhibitor, overload or other stress causes accumulation of acetate.
- Accumulation of acetate causes drop in pH (normally below 7).
- Drop in pH inhibits aceticlasts, causing further accumulation of acetate.
- Further drop in pH, as well as accumulated acetate inhibits hydrogenotrophs and acetogens (pH drops below 6).
- System is fully inhibited and only acidogenesis can occur. Therefore all the COD is converted to acids instead of methane (pH normally below 5).

It is however important to mention that in digestions (such as digestions of animal slurries or household wastes) with high ammonia load, the pH can almost not decrease below 7, due to the high buffering capacity of ammonia. In these cases the system is not fully breaking down but balances in an 'inhibited steady state' condition, characterized by high VFA concentrations and low methane production yields (Angelidaki *et al.*, 2006). It is very important to detect an inhibition condition at an early stage, when inhibition can be corrected by decrease in loading or removal of the inhibitor causing the inhibition of the process.

Only systems that are fed mainly readily degradable organics are susceptible to overload. Systems fed with a significant fraction of the influent as protein are buffered against pH decreases by ammonia, produced from the proteins. These systems are susceptible to high pH inhibition, which is largely caused by ammonia inhibition.

Free acid and bases, in particular VFAs, ammonia and hydrogen sulfide may at high concentrations, which again depend on pH of the system, also disrupt cellular stasis and hence reduce the performance of the biomass (see Box 9.4.5).

---

**Box 9.4.5   Inhibition by Free Acids and Bases (VFAs, H$_2$S, NH$_3$).**

The interior of a microbe is not in equilibrium with the bulk liquid. Concentrations of ionic compounds (including nutrients) and pH are different from bulk liquid. This non-equilibrium condition is maintained by a semipermeable membrane; the cell wall, with a hydrophobic centre. Charged molecules (including hydrogen ions) cannot pass this membrane readily, but must be actively transported (usually by embedded enzymes, or ion-selective sites). Small, uncharged molecules, however, can pass passively through the cell membrane, and disrupt homeostasis; commonly pH. Because it is the free acid or base that is the inhibitor, the free acid or base inhibition is highly pH-dependent.

The concentration of a free acid or base can be calculated from the pH, the acid–base constant and the total concentration of combined acid and base (normally the measured variable):

$$\text{For acids} : [C_{\text{acid,free}}] = \frac{[H^+][C_{\text{total}}]}{K_a + [H^+]}$$

$$\text{For bases} : [C_{\text{base,free}}] = \frac{K_a[C_{\text{total}}]}{K_a + [H^+]}$$

where $K$ is the acid–base constant ($pK = -\log_{10} K$), $[H^+]$ is the hydrogen ion concentration ($pH = -\log_{10} [H^+]$) and $[C_{\text{total}}]$ is the total concentration of acid + base. For most measurements, including gas chromatography analysis for organic acids, distillation or high performance liquid chromatography (HPLC) measurement of ammonia/ammonium and titration measurement of sulfides, the concentration of combined acid and base is measured.

The start concentrations for inhibition and acid–base coefficients (at 25 °C) for common inhibitory weak acids and bases are shown in the table below. Note that the dissociation constants may be heavily influenced by temperature, but so can the inhibitory concentration.

| Acid/Base | Type | Ionic partner | p*K* (at 25 °C) | KI[a] (mM) |
|---|---|---|---|---|
| VFA[b] | Acid | VFA⁻ | 4.8 | 0.2 |
| NH₃ | Base | NH₄⁺ | 9.25 | 1 |
| H₂S | Acid | HS⁻ | 7.05 | 3–4 |

[a]Concentration of free acid or base at which inhibition generally begins in mesophilic systems. Systems can normally operate at levels at least 2× these levels, with elevated effluent organic acids.
[b]Volatile fatty acids; similar p$K_a$ values.

As can be seen in the table, a system with sulfides is particularly susceptible to weak acid inhibition, because the p$K_a$ value is close to 7 (and therefore 50 % of the total sulfides are as $H_2S$ at neutral pH). Sulfide is also a nuisance because it causes occupational health and safety problems, and may cause corrosion in the gas. Sulfides are normally produced by reduction of sulfate within the digester, and are an issue to be addressed whenever there is sulfate or sulfides in the influent. Methods are precipitation with cationic metals (such as $Fe^{2+}$, $Fe^{3+}$, $Ca^{2+}$), gas stripping, and biological treatment.

Free acid inhibition (and calculated inhibition values) is also considerably complicated by direct saline inhibition by the saline form of the free acid or base. While salts are much less diffusive across the cell membrane, at high concentrations, the driving force allows substantial diffusion, and disruption of cellular stasis occurs.

## *Product Inhibition*

Product inhibition is caused by a drop in free energy available from catabolism, caused by an increase in product concentration.

The most common product inhibition is hydrogen inhibition of acetogens, though acetate can also inhibit the same organisms at high concentrations. This has been discussed earlier in the chapter.

# References

Angelidaki, I,. and Sanders, W. (2004): Assessment of the anaerobic biodegradability of macropollutants. *Reviews in Environmental Science and Biotechnology*, 3 (2), 117–129.

Angelidaki, I., Boe, K. and Ellegaard, L. (2005): Effect of operating conditions and reactor configuration on efficiency of full-scale biogas plants. *Water Science and Technology*, 52 (1/2), 189–194.

Angelidaki, I., Cui, J., Chen, X. and Kaparaju, P. (2006): Operational strategies for thermophilic anaerobic digestion of organic fraction of municipal solid waste in continuously stirred tank reactors. *Environmental Technology*, 27, 855–861.

Batstone, D.J., Keller, J., Angelidaki, I., Kalyuzhny, S.V., Pavlostathis, S.G., Rozzi, A., Sanders, W.T.M., Siegrist, H. and Vavilin, V.A. (2002): *Anaerobic digestion model No. 1 (ADM1)*. Scientific and Technical Report 13. IWA Publishing, London, UK.

Boone, D.R., Johnson, R.L. and Liu, Y. (1989): Diffusion of interspecies electron carriers H2 and formate in methanogenic ecosystems and its implications in the measurement of Km for H2 or formate uptake. *Applied and Environmental Microbiology*, 55, 1735–1741.

Brown, L. (1993): *The New Shorter Oxford English Dictionary*. Clarendon Press, Oxford, UK.

Cord-Ruwisch, R., Seitz, H.J. and Conrad, R. (1988): The capacity of hydrogenotrophic anaerobic bacteria to compete for traces of hydrogen depends on the redox potential of the terminal electron acceptor. *Archives of Microbiology*, 1988, 350–357.

Davidsson, Å., Gruvberger, C., Christensen, T.H., Hansen, T.L. and Jansen, J.I.C (2007): Methane yield in source-sorted organic fraction of minicipal solid waste. *Waste Management*, 27, 406–414.

Karakashev, D., Batstone, D.J., Trably, E. and Angelidaki, I. (2006): Acetate oxidation is the dominant methanogenic pathway from acetate in the absence of Methanosaeta. *Applied and Environmental Microbiology*, 72, 5138–5141.

Lepisto, R. and Rintala, J. (1999): Kinetics and characteristics of 70 °C, VFA-grown, UASB granular sludge. *Applied Microbiology and Biotechnology*, 52, 730–736.

Madigan, M.T. Martinko, M.J. and Parker, J. (2000): *Brock biology of microorganisms*, 9th edn. Prentice Hall, Upper Saddle River, USA.

Petersen, S. and Ahring, B. (1991): Acetate oxidation in thermophilic anaerobic sewage sludge digester: the importance of non-aceticlastic methanogenesis of acetate. *FEMS Microbiology Ecology*, 86, 149–158.

Ren, N.Q, Wan, B.Z. and Huang, J.C. (1997): Ethanol-type fermentation from carbohydrate in high rate acidogenic reactor. *Biotechnology and Bioengineering*, 54, 428–433.

Speece, R.E. (1996): *Anaerobic biotechnology*. Archae Press, Nashville, USA.

van Lier, J. B., Rebac, S. and Lettinga G. (1997): High-rate anaerobic wastewater treatment under psychrophilic and thermophilic conditions. *Water Science and Technology*, 35 (10), 199–206.

Zinder, S.H. (1993): Physiological ecology of methanogens. Chapter 3. In: Ferry, J.G. (ed.) *Methanogenesis. Ecology, physiology, biochemistry and genetics*. Chapman and Hall, New York, USA.

# 9.5

# Anaerobic Digestion: Technology

### Jes la Cour Jansen

*Lund University, Sweden*

Anaerobic digestion is a process used in many different technical applications for treatment of organic solid waste for gas production and waste stabilization. Technologies have been developed for treatment of many different types of organic matter such as wastewater sludge, municipal organic solid waste, high strength industrial wastewater, manure, and mixtures of different types of organic waste. Today the technology is successfully used within all these areas. The technology range from sophisticated highly engineered large capacity plants to simple single-family plants operated by the owner.

This chapter presents digestion technologies for municipal organic solid waste digestion as it is developed in Western Europe. In this context municipal organic solid waste includes household waste together with garden waste, organic waste from restaurants, green markets, food stores and similar types of waste from urban areas. Digestion of other types of organic material, co-digestion with municipal solid waste and low-tech solutions are only briefly presented in Box 9.5.1.

## 9.5.1 Introduction to Planning and Design Factors

A number of different factors influence the technology for anaerobic digestion of solid waste. The technology is young and immature, as the first full-scale commercial plant came into operation in 1986. Development and innovation have taken place in different countries and companies where local political priorities, economical conditions and traditions together with technical factors related to the actual waste have significantly influenced the technological solutions.

*Solid Waste Technology & Management*   Edited by Thomas Christensen
© 2011 Blackwell Publishing Ltd

**Box 9.5.1    Examples of Technologies for Digestion of Different Types of Organic Matter.**

Digestion of solid waste is more widespread than as described in this chapter focussing on municipal organic waste. More widespread use of anaerobic digestion is:

- Digestion of sludge from wastewater treatment plants.
- Small, low-tech solutions in India and China.
- Digestion of manure.
- Digestion of industrial wastewater with high content of organic matter.
- Co-digestion of waste from different sources.

## Digestion of Sludge from Wastewater Treatment Plants

Anaerobic digestion of wastewater sludge is by far the area where the greatest amount of waste is anaerobically digested. Almost all large wastewater treatment plants use the method for reduction of the sludge volume and stabilization of the sludge in order to be able to handle it without odor problems (e.g. as fertilizer or for soil improvement). Finally the methane gas is used for heat or energy production. Typically digestion is performed at mesophilic temperature and in a one-stage process. Thermophilic digestion has got more attention during the recent years as the increased temperature enable better hygienization and a higher degree of degradation of the organic matter. A typical digester with two reactors for sludge treatment is shown below.

## Small, Low-Tech Solutions in China

Especially in China small anaerobic treatment plants serving individual houses or small communities at the countryside are popular. According to IEA (2005) about eight million plants are in operation. The plants are loaded with manure and the organic waste from households. They are normally operated at ambient temperature with a minimum of technology involved. The anaerobic digestion means that the houses are served with cheap energy (methane gas) and at the same time the quality of the manure and household waste for fertilization is improved. Finally the digesters act as storage of fertilizer. The figure below shows the most common type of digester (after Nazir, 1991).

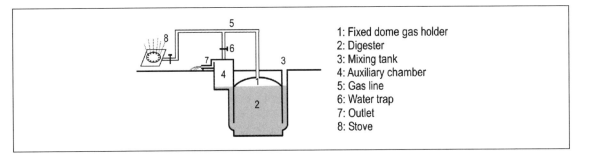

1: Fixed dome gas holder
2: Digester
3: Mixing tank
4: Auxiliary chamber
5: Gas line
6: Water trap
7: Outlet
8: Stove

### 9.5.1.1 Composition and Structure of the Waste

First of all, technologies for the treatment of organic solid waste have to take into account the composition and structure of the waste to be treated. If MSW has to be treated more complicated solutions are needed than if well sorted organic waste is available. Kitchen waste has a composition and structure that differs from the composition and structure of garden waste. The sorting instruction provided to the citizens and the system offered for collection of the waste to a high degree define what is collected and set the requirements for the pretreatment of the waste prior to digestion. If plastic, such as nappies, is accepted in the organic fraction or as bags used for sorting in the households other pretreatment systems are needed than in systems with collection in bins without linings. If garden waste is accepted care has to be taken to remove stone, gravel, and sand in order to protect the installations. If waste from restaurants or fluid industrial waste is accepted the whole reception of the waste have to include facilities for the solid waste as well as fluid waste, as pretreatment of mixtures of such wastes with household waste is difficult. It is important also to consider the typical misplaced items present in the waste as many plants has experienced the greatest problems with items not intended to be part of the organic waste.

Some countries have a long tradition for composting of garden waste and consequently this type of waste is excluded from digestion. In other countries, garden waste that has very low water content compared to kitchen waste and makes up more than 50 % of the waste for digestion. Such significant differences in waste composition and water content call for significant technological differences.

### 9.5.1.2 Degradability of the Waste

The degradability of the waste has significant impact on the technical solutions needed for a proper treatment. Fast and extensive digestion takes place if the waste is homogenized, size-reduced, and consists of easy degradable organic matter. However, municipal solid waste is composed of inorganic material and with respect to anaerobic degradation of easily degradable, moderately degradable, and nondegradable organic matter. The technology has to match the actual waste and a balance between pretreatment (separation of impurities, size reduction) and biological degradation must be sought.

Most anaerobic digesters operate on a mix of organic wastes and information about a likely mix is important when addressing degradability, process stability, and ammonia inhibition.

### 9.5.1.3 Use and Disposal of the End Products

The main output from digestion in terms of value is usually the gas produced and how the utilization is established is a crucial factor for the environmental profile of the plant. Efficient use of the energy produced, e.g. electricity, gas for vehicle fuel, and heat, is as important as the energy that it substitutes.

The digested material also is an important output; in particular from the wet processes where the digested material may contain only 3–5 % solids. Three different routes for its management are available: direct use on land as a fertilizer (e.g. following the routines used for manure application to land), dewatering and composting with the aim of producing a product that can be used as any other compost product (on land, in soil manufacturing, for landscaping, in horticulture), or for landfilling (e.g. as part of a mechanical biological pretreatment prior to landfilling). Finally in some cases very dilute digest is sent to a waste water treatment plant for removal of remaining organic matter and ammonium.

While the use of the gas may define what utilization technology should be available at the plant, the routing of the digest is to a large degree controlling the set-up of the pretreatment, the biological reactor and the post-treatment. Direct use on agricultural land demands the most pure digest while digest going to landfill primarily is characterized by its residual methane potential.

If the digest is going to be used as compost or directly on land, the technology must be able to meet the hygienization requirements. These differ from country to country. Within the European Union (EU) the ambition is to have common rules for organic waste and sludge handling and a proposal has been in the pipeline for many years. The EU draft proposal allows the use of the digest without restrictions from a hygienic point of view only if the waste or digest: (1) has been heated to 70 °C for 30 min + digested at least for 12 days at the mesophilic stage (35 °C), or (2) digested at the thermophilic stage (53 °C) and kept at this temperature as a batch for at least 20 h without admixture or withdrawal of material within the 20 h.

### 9.5.1.4   Economy and Subsidies

Digestion of municipal solid waste has only seldom been established on normal commercial conditions in Western Europe. Development, innovation, and implementation are heavily subsidized in most countries.

The extent and character of the subsidies have been different leading to focus on different aspects of the treatment. This in turns leads to very different technical solutions for collection, treatment, and final utilization of the products/residues from the process.

In many countries the focus has been on diverting organic waste from landfills and subsidies for construction is used very widespread in order to support installation of new capacity for treatment. In some countries electricity produced from the gas is subsidized. In Germany, plants are guaranteed a price (2005) between 0.1133 and 0.0877 €/kWh (depending of plant size) whereas no subsidies in Sweden are available for electricity production based on biomass. However, tax relief is provided for upgrading of the gas to car fuel in Sweden but not in Germany. Such differences in subsidies have a significant impact on the development of technology.

## 9.5.2   Technological Process Factors and Their Control

The influencing factors described above together with technical insight and knowledge have given rise to the development of many different types of full-scale plants for anaerobic digestion of presorted municipal solid waste. Many companies have worked in the field but no single company or technology has got a big market share. From the start in the mid1980s in Western Europe it took more than 10 years to reach 70 plants. The plants ranged from a few hundred tonnes/year up to more than 100 000 t/year (Nyns, 1999; Jansen and Svärd, 2002) The International Energy Agency reports (IEA, 2005) that more than 100 plants are in commercial operation and half of all plants are in Germany. Also Switzerland, Sweden, and Austria have a significant number of plants.

Due to the immature technology and the many different technologies and companies involved in design and construction of a relatively small number of plants, no general accepted design rules for the different parts of anaerobic digestion plants are available. Each company involved seems to use own design criteria and rules. Only for the digestion process itself there seems to be some kind of agreement on a proper organic loading rate and a minimum solid retention time depending on the selected process. Box 9.5.2 states the typical design values.

The systems for anaerobic digestion can be divided technologically according to four characteristics of the digestion process:

- Dry/wet digestion.
- Thermophilic/mesophilic digestion.
- One-stage/two-stage digestion.
- One-phase/two-phase digestion.

---

**Box 9.5.2   Typical Design Values for Retention Time and Organic Loading Rate for Digestion of Municipal Organic Solid Waste.**

Design considerations:

| Temperature | Organic loading (kg VS/m$^3$*day) | Minimum solid retention time (days) |
| --- | --- | --- |
| Mesophilic process | 4–5 | 20 |
| Thermophilic process | 8–10 | 10 |

Calculation of needed volume of an anaerobic reactor for treatment of 10 000 t/year household sorted organic waste: Dry process.
   The following information is provided:

- Dry matter (DM) content of household sorted organic waste: 35 %.
- VS content: 80 %.
- Loss of organic waste in pretreatment: 10 %.
- Daily load: 10 000*1000*0.35*0.8*0.9/365 = 6900 kg/day VS.
- Selected design value of organic loading: 10 kg VS/m$^3$*day.

Needed volume of reactor: 6900/10 ∼ 700 m$^3$
   If no dilution water is used the daily load is 10 000*0.9/365 ∼ 25 t waste ∼ 25 m$^3$ waste.
   Retention time: 700/25 = 28 days

---

## 9.5.2.1   Dry/Wet Digestion

The division into dry or wet processes is simply a question of the moisture content in the biological reactor. A dry process has moisture content of less than 75 % and the biomass looks like a thick slurry. The wet process has moisture content above 90 % and the biomass looks like a liquid.

   The choice of moisture content in the process takes its starting point from the moisture content in the waste. A high proportion of garden waste favors dry processes, whereas liquid industrial wastes call for a wet process. However, also the management of the digest and issues related to ammonia inhibition may play a role. If a wet residue can be accepted as a fertilizer to be used on agricultural land, it favors a wet process. However, if a wet digest is believed to be wastewater, which has to be treated in a conventional wastewater treatment plant, the dry process is favored, as no or almost no wet digest is produced in the dry process. A high nitrogen content in the waste can lead to high and inhibiting ammonia concentrations in the reactor and as water is used for dilution in the wet process such waste is best handled in a wet process.

   Wet and dry processes are equally abundant in Europe. It seems that wet processes are selected in countries with a long tradition for composting of garden waste and where wet digest is accepted as a fertilizer to be used in agriculture.

## 9.5.2.2   Thermophilic/Mesophilic Digestion

The digestion temperature is typically around 35 °C for mesophilic digestion or around 53–55 °C for thermophilic digestion. Selection of process temperature is a balance between several factors. A higher temperature gives faster biological degradation and some kind of material is more easily treated at high temperature. However, the thermophilic process is more difficult to operate and the need for heating and insulation add an extra cost to the treatment. In some countries the requirement for hygienization favors thermophilic processes, as the digestion can be part of the hygienization process whereas other countries requires separate hygienization where the benefit of the high temperature is less. Mesophilic digestion is the most common.

### 9.5.2.3   One-Stage/Two-Stage Digestion

Anaerobic digestion is a complicated and staged process where different groups of bacteria cooperate in order to degrade organic material and produce methane (see Chapter 9.4). The optimal operation condition for the different stages differs and therefore a staged process technology may be favorable. Normally no more than two stages are used, even though more stages could be theoretically interesting. In the first stage the acidification is performed and in the second stage methanogenesis is performed, converting the acids into methane. The prize for the more optimal condition for the different processes is a more complicated construction and normally also a more difficult process to operate. Both one- and two-stage processes are implemented, although one-stage processes seem most common. Two-stage processes are still selected for special applications.

### 9.5.2.4   One-Phase/Two-Phase Digestion

Phased processes are used in combination with staged processes where the biomass after acidification is separated into a solid phase further treated in the acidification stage and a liquid phase, rich in acids, which is routed to the methanogenic stage. Doing so enables a much higher methanogenic conversion rate, typically accomplished in a fixed film system. Phased processes are favored as part of a high rate staged process. Only few plants are operated with phase separation due to problems with separation of the phases and stability in the process.

## 9.5.3   Anaerobic Digestion Systems

Many aspects have to be considered when the technological choices are made for a plant for digestion of organic solid waste. Proper selection of technology in order to have a smooth and safe operation is needed. Economical considerations are essential as the economy in anaerobic digestion in general is poor and legal requirement with respect to handling of the digest, smell and noise from the plants together with requirements to working environment need extra attention. Several full-scale installations have been closed down due to serious mistakes within one or more of these areas.

The seven major technological elements of an anaerobic digestion plant for organic solid waste are presented below and the technologies available presented:

- Reception of waste: Weighbridge, tipping place, bunker.
- Pretreatment: Grinders/shredders, separators, hygienization and treatment of residues from pretreatment.
- Digestion: Feeding, mixing.
- Gas handling: Collection, treatment, storage and utilization.
- Management of digest from digestion.
- Odor control.

### 9.5.3.1   Reception of Waste: Weighbridge, Tipping Place, Bunker

Function: Reception, weighing for payment of the waste received and mass estimate of waste to be handled, storage of waste in bunkers and/or silos to level out variations in delivery, and visual control of waste quality.

Reception of waste is a very simple but also very important part of the whole system. Great variations in waste delivery have to be properly organized in order to be able to receive the waste and to feed the treatment system rather constant as required by the biological processes. Further, the possibility to identify unwanted waste types is important. Many plants have had great troubles with poor design of this part of the system because too little attention is put on handling of all the different kinds of waste that may arrive at the plant. Figure 9.5.1 shows a typical layout of the reception area.

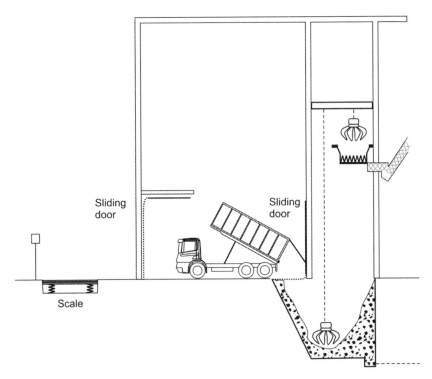

***Figure 9.5.1*** *Typical waste reception area at digestion plant. Reprinted with permission from Hans William Rasmussen © (2009) Hans William Rasmussen.*

### 9.5.3.2   Pretreatment: Grinders/Shredders, Separators, Hygienization, and Treatment of Residues from Pretreatment

Function: Removal of unwanted items, removal of plastic bags used in collection or other items that can disturb the process or the quality of the residues, size reduction of the biomass prior to anaerobic treatment, and hygienization of the biomass, if needed.

The need for pretreatment of organic waste depends on the collected waste and the need set by the other process steps. The pretreatment must be able to remove all kind of misplaced items in the arriving waste. It is not enough to select the pretreatment based on only an evaluation of the sorting criteria and collection system alone.

Further, economical considerations have a significant impact as any separation leads to the creation of a reject stream that can be very costly to eliminate and which contains organic material together with the material that has been separated. Different separation systems have very different separation efficiencies and the need and price for separating missortings have to be balanced with the loss of organic material. Typically, several steps are necessary in order to get sufficient reduction of missortings.

Normally, a grinder or shredder is used as the first step in the pretreatment process in order to open the collection bags and break down the big items. Next, pretreatment systems typically are focused on three different types of foreign items. First, any magnetic material is removed in a magnetic separator. Then any heavy, hard material that can wear down the machinery in the plants is separated out by gravity. Further, plastic or other lightweight material, not accepted as biomass or part of the residue, is sorted out by gravity or by some kind of sieving. Finally, hygienization often is introduced as part of the pretreatment, where the biomass is heated to destroy any pathogens present in the waste prior to the anaerobic digestion. The various technologies are briefly described below:

### *Bag Openers/Size Reduction*

Grinders or bag openers are used to open the bags and to reduce the waste to size fractions that can be handled in the following treatment steps (see Chapter 7.1).

### *Magnetic Separation*

Grinders and shredders are typically followed by magnetic separation where a magnet is placed above a transport conveyer for the size-reduced material. The magnet diverts any magnetic material from the conveyer. Such material makes up a very minor fraction of the waste, but magnetic iron is difficult to sort out and the residue fraction usually contains a lot of organic material.

### *Separation by Gravity in a Pulper*

In wet processes, where the waste has to be diluted with water or recycled wet digest, dilution, separation, and further diminishing typically takes place in a pulper like the one shown in Figure 9.5.2. The waste is diluted in the tank under intense mixing. Heavy materials such as stone and metallic components fall to the bottom and are removed through a sluice system. Light material such as plastic is skimmed off the top. The remaining organic materials are diminished by the strong hydraulic shear caused by the mixing. When the separation is finished and the material has been sufficiently diminished it can be pumped directly to anaerobic treatment or pumped to storage for later treatment.

### *Separation by Size Using Sieves*

In dry processes, where no or only minor dilution water is used, sieves are more common as separation technology, although sieves can be used for wet processes as well. Different sieving principles are used (see Chapter 7.1). Drum sieves are most common. Size of sieve openings determines the quality of the biomass and the amount of waste taken out as reject. The finer the openings the better the quality of the biomass but also the more reject is realized. Sieves with different opening can be used to sort out different reject fraction. Sieves based on the same principles may be used for separation of the final digest after anaerobic treatment.

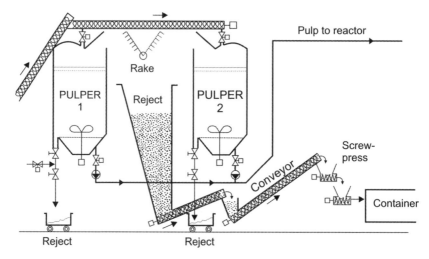

*Figure 9.5.2*   *Pulper used for pretreatment of organic solid waste. Heavy materials and plastic are sorted out by gravity and the organic material is diminished and diluted. Reprinted with permission from Hans William Rasmussen © (2009) Hans William Rasmussen.*

**Figure 9.5.3**  *Presses used for pretreatment of municipal organic solid waste. The green equipment to the right is at a pilot scale (photo: Jes la Cour Jansen).*

### Separation by Size and Gravity on a Disk Screen

The principle of the disk screen is to separate big light particles (typically plastic and paper) from smaller heavy particles. The disc screen is described in Chapter 7.1. Disk screens can be used as a pretreatment before dry as well as wet processes. After bag opening and grinding, the waste enters the screen at one end and big light particles are lifted up by the rotating disks and moved over the edge of the screen and collected as reject. Heavy and small items pass the screen between the disks and are collected for the anaerobic treatment.

### Separation in Presses

The principle for separation in presses is that a great part of the organic material can be separated as a liquid or slurry when the waste is put under pressure. The remaining and dryer material can be taken out as reject. Presses are used for wet processes only. Two different systems are shown in Figure 9.5.3. The one to the left is a continuous operated screw press where the pressure is build up as the waste is screwed through a conical housing and the organic material is pressed out during narrow slits in the pressure house seen on the top of the press. The reject comes out of the end as clearly shown in the figure. The pilot-scale piston press to the right is based on batch operation where one load of waste is pressed at a time in the pressure chamber and the organic material leave the chamber through narrow slits or small holes. Reject is taken out at the end of the pressure cycle.

### Separation by Hand

In a few plans foreign items in the waste are sorted by hand. The diminished material is transported on a conveyer and the foreign items removed by hand.

### Hygienization

Hygienization of the organic waste is needed in order to get a hygienic, safe digest for agricultural use. The requirements to hygienization prior to land application of digest vary among countries. Hygienization can be performed before, during, or after the anaerobic digestion. Typically, hygienization is made before the anaerobic treatment and is normally based on heating a batch of pretreated waste to about 70 °C. The temperature is then kept for 30–60 min, dependent on the local requirement.

### *Management of Residues from Pretreatment*

Three residues normally have to be taken care of from pretreatment of organic waste. The material from magnetic separation normally makes up a very small amount of magnetic material mixed with organic material sticking to the magnetic items. This fraction is normally landfilled. Heavy materials such as stone, gravel, sand and, nonmetallic material from sluices or from the bottom of storage tanks normally also make up a minor part of the residues from the pretreatment. This material is also landfilled. The major part of the residuals is made up of plastic and other missortings together with organic matter sorted out together with the missortings. This material has to be incinerated at least within the EU since the organic content is too high to allow landfilling according to EU regulations.

### 9.5.3.3 Digestion: Feeding, Mixing

Function: The anaerobic digestion shall convert the waste into biogas and digest.
    The digestion system is comprised by the following functions:

- Storage and feeding system.
- Preheating of the biomass.
- Mixing of new biomass and active microorganisms.
- Gas collection system.
- Separation of solid and liquid digest.

The technologies chosen to meet the functions and requirements set by the process layout has a very significant impact on the complicity of the system comprising the reactor and its associated technologies. Figure 9.5.4 shows a reactor for anaerobic digestion. The various technologies are briefly described below.

### *Storage and Feeding System*

In all cases the technology involves storage and feeding systems in order to facilitate controlled feeding of the system in spite of the variation in waste delivery. The feeding system has to secure a proper feeding according to the selected

***Figure 9.5.4***   *Reactor for anaerobic digestion.*

loading and feeding strategy (batch, continuous or semicontinuous operation). In some cases the storage system is used for mixing of the incoming waste with the active biomass.

### Preheating of the Biomass

The waste always needs preheating to obtain the selected process temperature. Normally, a heat exchanger will suffice transferring heat from the outgoing digest to the incoming waste. In case of thermophilic digestion more heat may be needed and excess heat from gas motors or from combustion of gas is used for the purpose.

### Mixing of New Biomass and Active Microorganisms

The digestion takes place in one or more reactors depending on the selected process scheme. In wet processes the reactions takes place in a totally mixed reactor, due to the fluid character of the biomass. New waste is introduced into the reactor and easily mixed with the active biomass by gentle mixing either mechanically or by recirculation of gas inside the reactor. In dry processes, where the biomass is a thick slurry, the flow in the reactor is more or less plug flow and mixing of waste and active biomass takes place prior to introduction in the reactor. In phased processes, where the methanogenic biomass is placed as a biofilm on carrier material, the contact between biomass and the organic material is normally ensured by the hydraulic turbulence in the system without separate need for mixing.

### Gas Collection System

Biogas produced in the digestion process has to be collected. In wet systems the gas is simply released at the surface of the reactor content and can be withdrawn. The mixing in the reactor is sufficient to get a good separation of gas from the waste. In dry systems gas can be entrapped in the biomass and a very gentle mixing is needed in order to get the gas released to the top of the reactor.

### Separation of Solid and Liquid Digestate

After digestion separation of solid and liquid digestate are normally needed. In wet systems part of the wet digest is recirculated for dilution of the incoming waste and part is withdrawn as liquid digestate. In dry systems part of the residue is recirculated as inoculum for new waste, whereas the rest proceeds to the posttreatment. Normally a minor fraction of wet digestate is separated out in order to have a proper product for the following posttreatment of the solid digestate.

## 9.5.3.4 Gas Handling: Collection, Treatment, Storage, and Utilization

Function: The gas storage and treatment system collects and prepares the produced biogas for the final utilization of the gas.

The need for gas treatment depends to a great extent on the final utilization of the gas. The gas coming from the digestion process normally is saturated with water and has a methane content of about 64 %. Further, it contains carbon dioxide and has a minor content of hydrogen sulfide and ammonium. The various technologies are briefly described below.

### Gas Storage

The need for gas storage highly depends on the expected use of the gas. In some cases no storage is needed because the gas is delivered to a power plant or directly incinerated for heat and power production. In other cases a storage is needed in order to get a smooth operation of the gas handling facilities and finally, payment of electricity produced from biogas may vary during the day, meaning that electricity is economically advantageous to produce only during short periods of the day. Storage for about one day is needed.

## Gas Treatment and Utilization

Because of the very different priorities and economic subsidies in European countries, both the utilization of the gas and also the technologies for treatment of the gas are very different. Four main applications with increasing demand for gas treatment are common: In all cases a flare is used as backup if the present gas utilization is out of order:

- Heat production: The simplest use of biogas is to incinerate the gas directly in a boiler for internal use of heat at the plant or in a local district heating network. Such utilization is possible at small plants where the cost for further treatment is too high to be attractive under the actual economic conditions.
- Power and heat production: Power and heat production is the most common application for biogas utilization. The requirement to pretreatment is moderate and simply consists of removal of water and hydrogen sulfide. Then the gas can be utilized in a standard gas engine.
- Vehicle fuel production: Upgrading of gas to vehicle fuel requires higher methane content than obtained in the digestion process and besides water and hydrogen sulfide most of the carbon dioxide must be removed to reach a methane content above 95 %. Different commercial processes exist for such upgrading.
- Upgrading of biogas to natural gas quality: Upgrading of biogas to natural gas quality requires even higher methane content than upgrading to vehicle fuel. Different technologies are under development, but they are expensive and only used in few cases. The upgrading is attractive in areas where a natural gas network exists, as it is quite simple to have a smooth and safe delivery of the gas. However at the moment the process has to be highly subsidized.

### 9.5.3.5   Management of Digestate from Digestion

Function: Brings the digestate from the digestion process into proper conditions for the final utilization or disposal.

The management of the digestate may (depending on the final utilization or disposal of the digestate) require posttreatment in terms of dewatering, wastewater treatment, composting, and storage facilities for digestate. Four main routes of final disposal seem to be preferred:

- Wastewater treatment of liquid residues: Wet residues from digestion have a high content of ammonium and/or soluble organic matter. In cases where use of the digestate on land is not possible, the digestate is subject to wastewater treatment for removal of organic matter and nitrogen. Some plants have internal treatment plants for that purpose and utilize the treated water as dilution water, whereas other plants simply discharge the wet residue to the local wastewater treatment plan.
- Incineration or landfilling without further treatment of dry digestate either as the prime disposal route or because of inadequate quality for producing compost.
- Composting and use of compost on land and in horticulture: The most common method for handling of digestate from anaerobic digestion is composting. In some cases the product can fulfill the requirements for agricultural utilizations and thus the compost is used for soil amendment. In other cases the compost has lower quality but can be used for growth medium in parks, golf courses and similar purposes. In case of wet processes the digestate is normally mixed with other materials in order to promote composting whereas digestate from dry processes often can be composted directly after dewatering.
- Use of digestate in agriculture: In some countries the wet digestate is accepted for use in agriculture without further treatment. Since agriculture can only receive digestate when it fits in to the production schemes, storage tanks may be needed. The digestion process may continue in such storage tanks resulting in release of methane. However, Hansen *et al.* (2006) showed that at least under temperate climatic conditions this was a minor problem.

### 9.5.3.6   Odor Control

Function: Controls and minimizes odor from the raw waste and from the anaerobic digestion and other biological processes at the plant.

Organic solid waste itself can give rise to serious smell problems as the waste sometimes can be up to two weeks old when it arrives at the biogas plant. Especially during warm periods heavy smell problems can be experienced. Further the anaerobic process itself and especially production of hydrogen sulfide can cause problems. Malfunction of other processes such as the composting of digest may also cause problems. Almost all existing plants experience smell problems now and then and some have been closed down due to unsolved smell problems.

Solving odor problems from biogas plants involves a lot of air that needs to be cleaned; the concentration of problematic substances is low but many different smelly substances are present. Further the concentration levels that lead to odor problems are low for many substances. Consequently the solution of odor problems starts with reducing the risk of odor production by keeping a good and smooth handling of the waste. Raw waste shall not stay untreated for long periods and good control of the processes shall be kept. Areas with a risk of odor problems shall be separated from other areas in order to keep the amount of smelly ventilation air down. The next step is to confine any smell in order to protect the surroundings. In many cases, waste is delivered in a closed building and the processes are to the greatest possible extent performed in closed tanks and completely treatment of ventilation air from buildings or tanks is needed. Five main systems for odor treatment are in use:

- Dilution of the ventilation air.
- Biological filters.
- Chemical scrubbers.
- Incineration.
- Catalytic oxidation/regenerative thermal oxidation.

Details about the different techniques are given in Box 9.5.3.

---

**Box 9.5.3  Methods for Odor Control.**

There are typically five methods available for odor control:

The simplest way to reduce odor problems is to *dilute the ventilation air*. It is most easy done by discharging air through a chimney. Only minor odor problems can be solved in this way but a chimney is often used as the last step in combination with some of the other methods mentioned below.

The large quantity of air and the many substances present in low concentrations call for methods that can handle this great variety of substances. *Incineration* is very effective but expensive as a lot of air needs to be treated and the content of substances is low. Incineration are seldom used but can be used for treatment of ventilation air from smaller, very smelling areas.

*Catalytic oxidation* is a new technology for odor control. The addition of a catalyst accelerates the oxidation and enables a lower operating temperature (250–500 °C) than incineration. The metal catalyst normally used is expensive but the operating costs are low do to the low combustion temperature.

*Biological filters* utilize that microorganisms can degrade a great variety of organic substances and convert inorganic substances such as hydrogen sulfide and ammonia to less problematic substances. The basic idea is to create a large surface for growth of microorganisms. If the ventilation air is brought in contact with the microorganisms significant degradation and conversion takes place. The water content plays a significant role, as the microorganisms need water, in order not to dry out. Further the smelling substances need to be solubilized in water before the microorganisms can take them up. Too much water on the other hand can lead to poor oxygen transport to the biological degradation and clogging of the filters. In practice many different types of biological filters are used. Most common are compost or bark filters (or mixture of the two materials). Such filters are inexpensive and provide a good support material for a great variety of microorganisms. The area requirement is significant and the filters need to be operated carefully in order not to get too dry or too wet. Trickling filters of bioscrubbers are more expensive but require less area.

*Chemical scrubbers*, where the air is washed first acidic, then under alkaline conditions, and finally with addition of oxidizing materials (hypochlorite), has shown to be an effective and robust solution but the construction and operational costs are high.

## 9.5.4   Examples of Anaerobic Digestion Facilities

Two actual plants are presented below in order to show how the different technological elements are used in a context.

### 9.5.4.1   Digestion Plant 1: Wet, One Stage, One Phase, Thermophilic Process

The plant is located in Germany and treats 10 000–15 000 t/year of organic waste. Sorted organic household waste is collected in bins without linings and the waste is collected in containers also without lining. Garden waste is accepted as part of the organic waste and make up a significant part. Figure 9.5.5 shows the layout of the plant.

### *Waste Reception*

After weighing, waste is delivered in a sluice system with two parallel lines for waste reception. The waste is tipped in a bunker where a screw mixer diminishes the waste and moves the waste to a conveyer belt for transportation to the pulper/mixing tank.

### *Pretreatment*

No bag opening is needed as no bags are used in the collection system. On the conveyor belt the waste passes a separator for magnetic material. From the two reception lines the waste goes to one of the two mixing tanks ($2 \times 20\,\mathrm{m}^3$) operated in parallel, where the waste is diluted with a mixture of treated wastewater and recirculated process water. Treated wastewater is used in order to reduce the ammonium content in the process. When the tank contains 9–10 % dry matter, the feeding of waste is stopped and the homogenization starts with a high speed stirring (200 rpm) that disintegrates the biomass. A light fraction of e.g. plastic and paper is withdrawn at the top of the tank with a fork system that rakes the material from the upper layer of the mixed biomass. In the bottom of the tank a sluice system collects the heavy fraction of e.g. sand, glass, and metal. Air from the tank is led to a compost filter for odor treatment before it is led out through a high chimney. The building is ventilated and ventilation air is led to the compost filter as well.

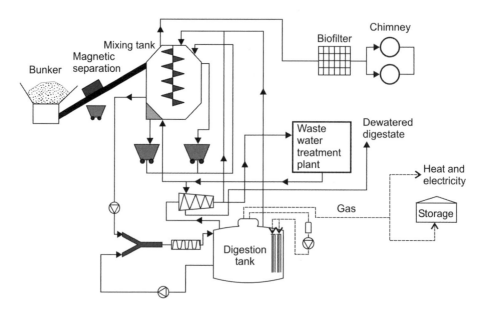

***Figure 9.5.5***   *Plant with wet, one-stage, one-phase thermophilic process (Germany).*

## Hygienization

The plant does not include a separate hygienization step. The thermophilic digestion with a temperature of 60 °C and a retention time of 14–20 days is not sufficient in Germany today as hygienization method for using the digest in agriculture. With the extension of the plant (with two digesters, $2 \times 1000\,m^3$) the retention time of the thermophilic digestion will increase and the hygienization effect will be sufficient for agricultural use of the digest.

## Digestion

The digestion is carried out in one digester ($1500\,m^3$) but the capacity is being increased with two digesters each of $1000\,m^3$ that will be operated in parallel with the first one. The process is therefore wet, one-step, one-phase, and thermophilic with a retention time of more than 20 days after the extension of the plant. The feeding to the digestion tank is semicontinuous as the two mixers are operated in batch mode but with a relatively short cycle time. Gas is recirculated in the tank for mixing.

## Gas Collection Treatment and Utilization

The plant produces approximately $5000\,m^3$/day of biogas collected from the top of the digester. The gas production is estimated to be around $260\,m^3$ biogas/t dry matter fed to the plant. The produced biogas is delivered untreated to the landfill next to the plant where it is mixed with landfill gas and gas from the waste water treatment plant. Six gas engines produce electricity and heat from the gas.

## Treatment of Digestate

The digestate is dewatered in a centrifuge. The wet fraction is partly recirculated to the mixers and partly sent to the municipal wastewater treatment plant. The solid fraction (45 % dry matter) is disposed off at the landfill next to the plant due to insufficient hygienization. After extension of the digestion capacity the hygienization will be sufficient and the municipality will take care of the digestate and is expected to use it after composting as soil amendment in parks and other public green areas.

## Odor Control

Air from the mixing tank and the pretreatment building passes a biofilter (compost filter) with outlet through a high chimney.

### 9.5.4.2   Digestion Plant 2: Dry, One Stage, One Phase, Thermophilic Process

The plant is located in Switzerland and treats about 13 000 t/year of organic waste. It has been in operation since 1996. Household-sorted organic waste is collected in bins without linings and the waste is collected in containers without linings. Garden waste makes up the dominant part of the waste depending on season; even during winter garden waste constitutes about 50 %. Figure 9.5.6 shows the layout of the plant.

## Waste Reception and Pretreatment

When the waste arrives at the plant it is weighed and registered and thereafter tipped into a bunker via an opening in the building. All pretreatment activities are carried out indoors in a ventilated building. No bag opening is needed as no bags are used in the collection system. A crane moves the waste to the primary disintegration step, which consists of low-speed screws. The shredded material is then brought to a belt for manual sorting and inspection and for automatic separation of magnetic material. Only 2 % of the waste is sorted out in the sorting section, but the cost for the manual sorting corresponds to 40 % of all operational costs at the plant. The waste is further disintegrated in a shredder with

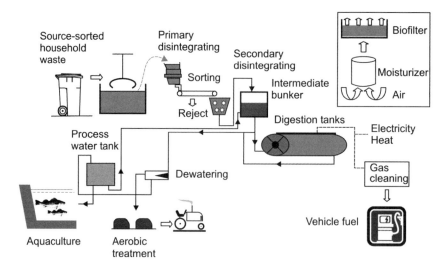

**Figure 9.5.6**    *Plant with dry, one-stage, one-phase thermophilic process (Switzerland).*

rotating cutting disk to a size of 10–20 mm. In the intermediate bunker water is added. Process water is normally used but fresh water has to be added some times if the ammonium content is too high.

### Hygienization

The thermophilic digestion itself is used for hygienization. Continuous plug flow in the reactor guarantees that all biomass stays in the digester for 20 days with an incoming temperature of 55 °C. This type of hygienization has been tested out and is sufficient to allow agricultural use of the digest in Switzerland.

### Digestion

The digestion section consists of one large horizontal concrete digestion tank. The tank is constructed for 10 000 t/year of waste. The biomass is heated to 55 °C and fed to the tank with a plug flow that gives a retention time of about 20 days. The digested biomass is withdrawn at the end of the tank. Part of the digested material is recycled to the front of the tank for inoculation of incoming waste. Centrally assembled sticks that rotate slowly are stirring the biomass in order to secure release of the gas. The stirrers are not moving the biomass forward as the transportation is caused by new incoming waste that pushes old material forward.

### Gas Collection Treatment and Utilization

Gas is collected from the top of the reactor. Gas production is 80 Nm$^3$ biogas/t incoming waste. The methane content is normally 65 %. Heat and power is produced from the gas. Heat is used internal while electricity is sold. Part of the biogas is upgraded to methane content of 96–98 % and is used as fuel in vehicles. Several different companies are interested in buying biogas to their vehicles. Biogas used as vehicle fuel is considered to be the best environmental and economical utilization of biogas today in Switzerland. All kinds of use of biogas are tax-free.

### Treatment of Digestate

The digestate (with a dry matter content of ∼25 %) is dewatered in screw presses, which separate some liquid to get a dry digestate with a dry matter content of ∼45 %. The separated liquid is stored in a process water tank to be used either for diluting the incoming waste (intermediate tank) or at the 'aquaculture'. The aquaculture is a project for treatment of the wet residue in cooperation between a university and the plant. The Swiss government has subsidized half of the investment cost. The aquaculture has water plants resistant to high ammonia concentration. The water plants is harvested and used for feeding chickens and rabbits (2 t/month). The treated liquid is used to irrigate cultivation of e.g. salad and watermelon. The surplus of wet residue is being sent to a wastewater treatment plant.

The dry residue is composted in an aeration bunker with forced aeration and agitation. After 5–10 days, the product, called 'Frischkompost' is stored outdoors under a roof until use. About 2500–3000 t compost (with 45–55 % dry matter) is produced each year. Compost can be collected for free by farmers and the public. There is no market for compost in the area.

### *Odor Control*

Ventilation air from the building is saturated with water and treated in a biofilter before discharged through an outlet just above the roof.

# References

IEA (2005): *Biogas production and utilization*. IEA Bioenergy T37:2005.01. IEA Bioenergy, Task 37, International Energy Agency, New York, USA.

Hansen, T.L., Sommer, S.G., Gabriel, S. and Christensen, T.H. (2006): Methane production during storage of anaerobically digested municipal organic waste. *Journal of Environmental Quality*, 35, 830–836.

Jansen, J. la Cour and Svärd, Å. (2002): *Kortlægning af driftserfaringer fra europæiske biogasanlæg til behandling af husholdningsaffald* (Survey of operational experiences from European biogas plants for treatment of household waste; in Danish). Department of Water and Environmental Engineering, Lund University, Sweden.

Nazir, M. (1991): Biogas construction technology for rural areas. *Bioresource Technology*, 35, 283–289.

Nyns, E-J. and Thomas, S. (1999): *Biogas from waste and wastewater treatment, version 2*. Lior CD-ROM collections on renewable energies series. LOIR International SA, Hoeilaart, Belgium.

# 9.6

# Anaerobic Digestion:
# Mass Balances and Products

**Jacob Møller and Thomas H. Christensen**

*Technical University of Denmark, Denmark*

**Jes la Cour Jansen**

*Lund University, Sweden*

While the basic processes involved in anaerobic digestion of waste are described in Chapter 9.4 and the main digestion technologies are presented in Chapter 9.5, this chapter focuses on mass balances, gas production and energy aspects, environmental emissions and unit process inventories. Understanding these issues and being able to account for them is a prerequisite in digestion engineering and for establishing and running a successful anaerobic digestion facility.

Of specific importance is the final use of the digestate. Use in agriculture as a fertilizer is described in Chapter 9.10 and use after composting of the digestate as a soil amendment product is analogous to issues presented in Chapter 9.9 for compost.

## 9.6.1   Mass Balances

The overall mass balance for a digestion facility geared to produce biogas depends on many factors. Most important is the type and composition of the waste being digested, because this determines the amount of impurities to be removed from the waste and the degradability of the waste. Second most important is the intended use of the digestate and the associated quality requirements. The higher the quality requirements, the less mass will end up in the final product. The digestion technology may also influence the mass balance, but with the feedstock and the end product defined, the choice of technology – Chapter 9.5 presents a wide variety of technologies – has only secondary influence on the overall mass balance.

*Solid Waste Technology & Management*   Edited by Thomas Christensen
© 2011 Blackwell Publishing Ltd

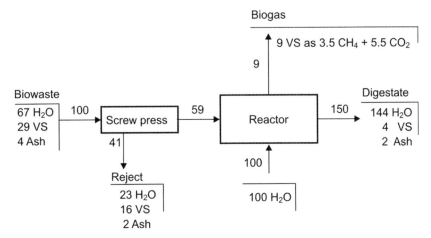

**Figure 9.6.1** *Mass balance for a hypothetical 'wet' anaerobic digestion plant treating (100 mass units) source-separated organic household waste (constructed from Hansen* et al.*, 2006a, b; Davidsson* et al.*, 2007). The biowaste was pretreated by a screw-press that leaves a relative dry reject with high concentration of slowly degradable volatile solids.*

Mass balances can be made for different purposes, using different approaches and degree of details. The more complex and detailed the more correct a mass balance is, but this may be at the expense of clarity.

### 9.6.1.1 Balances of Volatile Solids, Water and Wet Mass

Figure 9.6.1 shows a mass balance of an anaerobic digestion facility treating biowaste (source-separated organic household waste) focusing on the fate of the waste as characterized by its content of water, volatile solids (VS) and ash (nonvolatile solids), using a case constructed from data published by Hansen *et al.* (2006a, b) and Davidsson *et al.* (2007). The biowaste is collected at the source in plastic bags and subsequently undergoes extensive cleaning in a screw press (see Hansen *et al.*, 2006b) where all foreign items and the plastic bags are removed. The residue from the anaerobic treatment, i.e. the digestate, is spread directly on agricultural land without any posttreatment. This requires that the biowaste is very clean. As a consequence a major part of the biowaste is lost together with the reject (53 % of the VS). A significant amount of water is added to bring the waste in suspension corresponding to a solid content of about 9 %. The digestion is thermophilic with a retention time of 15 days. About 350 m³ of $CH_4$ is produced per tonne of VS coming into the reactor, corresponding to a production of about 75 m³ of biogas with a methane content of 63 %/t waste entering the plant. The gas produced corresponds to about 70 % of the VS actually fed into the digester. The digestate is dilute (about 4 % solids), but still contains a potential for further production of biogas if kept anaerobic.

Figure 9.6.2 shows a mass balance for a full-scale thermofilic 'dry' digestion facility in Braunschweig, Germany, treating 20 000 t/year of source-separated biowaste (Kranert and Hillebrect, 2000). The biowaste is pretreated by mechanically shredding and fed into the horizontal digester along with presswater from a dewatering unit. A biologically active 'inoculum' from older digested material is also added. The dry matter content in the reactor is > 20 % and because the reactor is horizontal, new and old material in the reactor is not mixed thereby preventing hydraulic short-circuiting. The retention time of the organic material in the reactor is about 22 days. After dewatering and adjusting the water content at about 65 %, the digestate is further treated under aerobic condition for 7–10 days before it is used in agriculture. The biogas is utilized in a nearby power station that produces electricity to the grid, but does not recover the produced heat. During a monitoring period of one month the plant produced on average 94 Nm³ biogas/t biowaste – with a methane content of 54 %. The net electricity production based on the electricity produced at the power station and subtracted the electricity used at the plant for heating the reactor and operating the pumps etc. amounted to 100 KWh/t waste. This corresponds to 20 % of the total energy of the produced biogas.

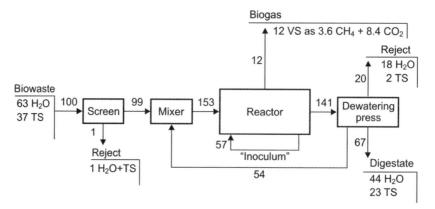

**Figure 9.6.2**   *Mass balance for the 'dry' anaerobic digestion plant 'Braunschweig-WatenBüttel' treating (100 mass units) source-separated organic wastes. Based on Kranert and Hillebrecht (2000).*

### 9.6.1.2   Balances of Elements

The elemental composition of the waste changes during digestion primarily with respect to carbon and nitrogen. Both of these elements are converted, but very little is lost since biogas as well as digestate are collected in confined technical systems. This also suggests that in a practical context all other elements entering the digestion process are conserved and will appear in the final product, except if removed physically in the pre- or postprocessing. The only main losses of C and potentially also N could happen if the digestate, after removal from the reactor, is not fully controlled. If the temperature of the digestate remains high the microbial processes will continue and if the gas is not collected this may yield emissions of $CH_4$, $CO_2$ and potentially also $NH_3$. Davidson *et al.* (2007) showed that the digestate after being removed from the tank still had the potential to produce an additional 35–55 $m^3$ $CH_4$/t VS. Hansen *et al.* (2006c) showed that, in a temperate climate such as the Danish and with spreading of the digestates on land in April, the actual additional production from the storage tanks probably was only of the order of 2 $m^3$ $CH_4$/t VS relative to the waste fed to the digester. The low winter temperature was the main controlling factor.

The C in the reject is primarily organic C adhering to the foreign items removed during pre- and posttreatment. C is also found in any plastic items removed with the reject. Of the C entering the digester, about 45 % leaves as $CH_4$, 30 % as $CO_2$ and 25 % as suspended and dissolved C in the digestate. The N in the incoming waste is primarily organic N as proteins. Only a minor part of N is lost with the gas and in the digestate about half is ammonium and half is organic N. This means that a large part of N in the digestate is easily available for plants, but could also evaporate as ammonia if not treated correctly.

### 9.6.2   Gas

The gas produced in the digester contains in addition to $CH_4$ and $CO_2$ also small amounts of other gases, e.g. $N_2$, $H_2S$, $NH_3$ and various organic compounds produced by the anaerobic degradation or compounds volatilized from the original waste. Water vapor will also be present and raw biogas directly from the reactor can be considered saturated by water vapor. The water content is therefore temperature dependent and can be estimated from psychrometric tables. Table 9.6.1 shows a typical main component composition of biogas from an anaerobic digester. $N_2$ is present in low concentration probably of atmospheric origin trapped in air pockets in the feedstock or from denitrification of small amount of nitrate initially present in the waste. The concentration of ammonia in the biogas inside the reactor is controlled by the pH- and temperature-dependent $NH_4^+/NH_3$ equilibrium in the liquid phase and the subsequent $NH_{3liquid}/NH_{3gas}$ equilibrium, as described by Henry's law. When the gas is transported from the reactor to e.g. storage tanks or gas engine, the gas temperature drops and water vapor condenses, thereby trapping water-soluble compounds such as ammonia. As a

**Table 9.6.1** *Typical composition of biogas from anaerobic digestion plants (adapted from Persson* et al., *2006). Reprinted with permission from Biogas upgrading to Vehicle Fuel Standards and Grid Injecti9on Task 37 – Energy from Biogas and Landfill Gas by M. Persson, O. Jonsson and A. Wellinger © (2006) IEA Bioenergy.*

| Biogas composition | Unit | Value |
|---|---|---|
| Energy content | MJ/Nm$^3$ of methane | 37 |
| Density | kg/Nm$^3$ | 1.2 |
| Methane | vol% | 63 (range 53–70) |
| Carbon dioxide | vol% | 37 (range 30–47) |
| Nitrogen as $N_2$ | vol% | 0.2 |
| Hydrogen sulfide | ppm | < 1000 (from 0 to 10 000) |
| Ammonia | ppm | < 100 |
| Hydrogen | vol% | ~ 0 |
| Carbon monooxide | vol% | ~ 0 |
| Oxygen | vol% | ~ 0 |
| Water[a] | | — |

[a]Water vapor is present in raw biogas at saturation concentration, depending on gas temperature.

consequence, the handling of the condensate from gas cooling can be of some environmental concern. Hydrogen, carbon monoxide and oxygen are, if detectable at all, only present in very small concentrations. It should be stressed that the gas composition in Table 9.6.1 represents average values that could be different as a result of other feedstock composition and the specific technology at the treatment plant.

The concentration of trace compounds in the biogas from two full-scale plants is presented in Table 9.6.2. The volatile organic carbon (VOC) was measured in the gas that had just left the reactor tank. The very high concentration of terpenes was reflected in the waste composition that had a high fraction of fruits, but interesting enough the terpenes found in the biogas were different from the original compounds in the waste (Smet *et al.*, 1999). From an environmental point of view the composition of the raw biogas is only of interest regarding fugitive losses of gas or condensate. Moreover, any drying or cleaning of the biogas to improve its quality for combustion results in changes of the composition of the gas and most of the organic compounds disappear during the combustion process.

**Table 9.6.2** *Examples of trace compounds in raw biogas (unit is mg/m$^3$).*

| Groups | Compounds | Value |
|---|---|---|
| Inorganic compounds | $H_2S$[a] | 170 |
| | $NH_3$[a] | 18 |
| | $NO_x$[b] | 29 |
| | $SO_2$[b] | 44 |
| VOC[a] | Terpenes | 2060 |
| | Alcohols | 44 |
| | Carbonyl compounds | 28 |
| | Organic sulfur compounds | 17 |
| | Esters | 3.1 |
| | Ethers | 3.0 |
| | Other volatile organic compounds | 12 |
| Heavy metals[b] | Cd | < 0.0018 |
| | Ni | < 0.0039 |
| | As | < 0.0079 |
| | Hg | < 0.0083 |

[a]Data from a thermophilic combined anaerobic/aerobic composting plant in Brecht, Belgium, treating source-separated vegetable, fruit, garden and paper waste (Smet *et al.*, 1999).
[b]Data from a one-step mesophilic wet fermentation digester near Münster, Germany treating source-separated MSW (NSCA, 2002).

***Table 9.6.3***   *The significance of units in assessing gas production from biowaste: Volume of methane and biogas produced by the same batch of biowaste, based on theoretical calculations, experimental determinations and calculated back to the amount of biowaste collected (based on Davidsson et al., 2007; Hansen et al. 2006b).*

| Gas production by anaerobic digestion | Unit | Methane $(CH_4; Nm^3)$ | Biomass $(CH_4 + CO_2; Nm^3)^a$ |
|---|---|---|---|
| Theoretical, chemical composition-based (C, H, N, O) | Tonne $VS_{con}{}^b$ | 634 | 1006 |
| Theoretical, component-based (fat, protein, carbohydrates) | Tonne $VS_{con}$ | 521 | 827 |
| Measured potential in batch experiment | Tonne $VS_{in}{}^b$ | 428 | 679 |
| Measured yield in pilot-scale experiment | Tonne $VS_{in}$ | 342 | 542 |
| Measured yield in pilot-scale experiment (dry matter) | Tonne $TS_{in}$ | 291 | 462 |
| Measured yield in pilot-scale experiment(wet waste) | Tonne $ww_{in}$ | 90 | 143 |
| Measured yield in pilot-scale experiment (collected waste) | Tonne ww | $60^c$ | 95 |

[a] Assumed 63 % $CH_4$ and 37 % $CO_2$ in the biogas.
[b] The index 'con' means converted or fully degraded and 'in' means feed to the reactor.
[c] The waste was pretreated using a disc screen that removed approx. 34 % of the waste mass.

The amount of gas produced is a key factor at any anaerobic digestion plant, both environmentally and economically. In most cases a full-scale digestion plant works on many different waste types and the overall specific gas production data ($m^3$/t waste received) from such facilities may not represent the amount of gas that can be produced from biowaste. Furthermore it is extremely important to be precise about the definition of gas and of the tonne that it is related to. Table 9.6.3 shows the volume ($m^3$) of $CH_4$ and biogas ($CH_4 + CO_2$) per tonne of waste, representing different ways of estimating the amount of gas and using different tonnes of waste. The difference for the same waste of $CH_4$ ($m^3$; and also for biogas) produced per tonne is about a factor of 10 depending on approach: The highest amount is calculated theoretically on the basis of the elemental composition (C, H, N, O) and full conversion of VS ($634\,m^3$ $CH_4$/t $VS_{con}$), while the lowest amount is deduced from measurements done in long-term pilot-scale experiments with 15 days retention time and related to the amount of wet waste collected at the source ($60\,m^3$ $CH_4$/t wet waste collected). A significant fraction (one-third) of the collected waste is here assumed lost as a reject during pretreatment, which is a major factor for the amount of gas produced. Davidson *et al.* (2007) reported on 15 long-term pilot-scale experiments with thermophilic digestion of source-separated biowaste from different sources in Denmark and Sweden and subjected to different types of pretreatment and found that the $CH_4$ yield per tonne of VS fed to the reactor in all cases were in the range 300–400 $m^3$ $CH_4$/t $VS_{in}$ (see Figure 9.6.3).

Table 9.6.4 shows gas production at European biogas plants treating organic household waste containing kitchen waste and garden waste in various proportions.

## 9.6.3   Energy

The amount of gas produced represents a significant source of energy. Typically about 70 % of the energy content of the waste fed to the digester is converted to biogas. The gas can be utilized in many ways, as described in Chapter 9.5. However, the digestion plant also uses energy in terms of electricity (pumps, ventilators, mixers, etc.), fuels (diesel for trucks and machines) and heat (eventually based on fuel or electricity). The major consumption of energy is to heat the reactors, which are mesophilic (35 °C) or thermophilic (53–55 °C).

For anaerobic digestion the overall change of enthalpy, $\Delta H$, i.e. the heat released or absorbed by the reaction, is strongly dependent on waste composition because the main organic components carbohydrates, fats and proteins, have very different $\Delta H$ (Lindorfer *et al.*, 2006). Based on this reference it can be calculated that the complete anaerobic degradation of source-separated organic household waste with typical chemical composition is endergonic, i.e. has $\Delta H > 0$ and the reaction will require additional energy to proceed to the finish. At a 'wet' thermophilic digestion plant the energy needed to heat the reactor can be substantial, especially during the winter period.

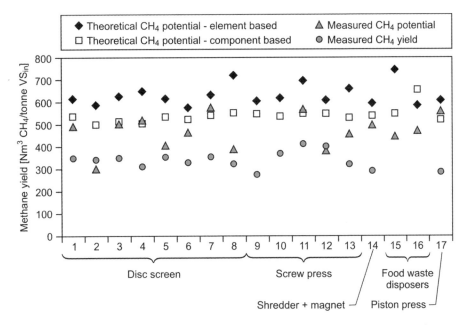

**Figure 9.6.3**   *Methane production from biowaste depending on approach for estimation and measurement for 15 different biowaste samples subject to different types of pretreatment. Note that the measured methane yield always is in the range 300–400 m³ CH₄/t VS$_{in}$ (from Davidsson et al., 2007). Reprinted with permission from Waste Management, Methane yield in source-sorted organic fraction of municipal solid waste by A. Davidsson, C. Gruvberger, T.H. Christensen et al., 27, 3, 406–414 © (2007) Elsevier Ltd.*

**Table 9.6.4**   *Gas production at European biogas plants treating organic household waste (data from Jansen and Svärd, 2002).*

| Plant | Type | Waste type | Capacity (t/year) | Gas production (Nm³/t wet waste collected) | Methane content in biogas (vol%) |
|---|---|---|---|---|---|
| Brecht, Belgium | One-step, dry, thermophilic | 15 % kitchen waste, 75 % garden waste and 10 % paper | 25 000 | 90–120 | 50–60 |
| Münster, Westphalia, Germany | One-step, wet, thermophilic | 100 % source-separated household waste | 10–15 000 | 90 | 45–65 |
| Kirchstockach, Brunnthal, Germany | Two-step, wet, mesophilic | 100 % source-separated household waste | 20–25 000 | 60–100 | 60–65 |
| Tilburg, The Netherlands | One-step, dry, mesophilic | 40 % source-separated household waste, 60 % garden waste | 52 000 | 80 | 55 |
| Bachenbülach, Switzerland | One-step, dry, thermophilic | 50 % source-separated household waste, 50 % garden waste | 10 000 | 100–120 | 65 |
| Otelfingen, Switzerland | One-step, dry, thermophilic | 50 % source-separated household waste, 50 % garden waste | 13 000 | 100–120 | 65 |

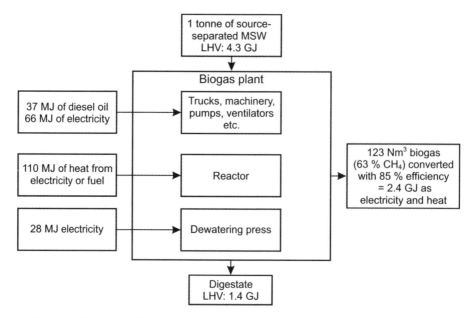

**Figure 9.6.4** *Energy balance for a hypothetical mesophilic anaerobic digestion plant treating source-separated MSW. 'LHV': Lower heating value. Data on electricity use partly from Berglund and Börjesson (2006).*

Figure 9.6.4 shows a simple energy balance for a digestion plant including typical energy uses for pretreatment as well as posttreatment. It is assumed that the waste treated at the plant is source-separated household waste of high quality with a water content of 70 % including food waste, used kitchen tissue and flower cuttings. The gas yield is 70 % of the theoretical methane potential, which is an average value according to Davidsson *et al.* (2007). The biogas is converted to electricity delivered to the grid and to heat for district heating. With an efficient utilization of the energy content of the gas as this, the digestion plant is a net producer of energy. The overall efficiency based on the lower heating value of the biowaste collected and including energy used on the plant is in this example approx. 52 %. In many biogas plants with their own biogas engine it is not possible to connect to a district heating grid and even if the plant uses the heat from combustion of gas to heat the reactor, the surplus heat lost will be substantial. Assuming the heat production to be 46 %, the efficiency of the plant in Figure 9.6.4 would drop to 21 % if the heat could not be recovered.

## 9.6.4 Environmental Emissions and Controls

Since the anaerobic digestion plant does not work with open reactors, the emissions to air and water are few or only appear as indirect emissions. The main emissions may be:

- Air emissions from conversion of biogas.
- Diffuse air emissions from leaks and solid handling.
- Emissions to air, water or soil from digestate (if not considered a product) or digester liquid after dewatering.

### 9.6.4.1 Air Emissions from Gas Combustion

The air emissions from biogas conversion depends on the quality of the raw biogas, any pretreatment (removal of water, hydrogen sulfide, etc.) and the type and quality of the combustion process itself. In addition any gas cleaning of the off

**Table 9.6.5** *Emissions from combustion of biogas in stationary engines < 25 MW (data from Nielsen and Illerup, 2006).*

| Compounds | Emission factor (g/GJ of energy produced) |
|---|---|
| $CH_4$ | 323 |
| $N_2O$ | 0.5 |
| CO | 273 |
| $NO_x$ | 540 |
| $SO_2$ | 19.2 |
| TSP (total suspended particulate matter) | 2.63 |
| PM10 (Particulate matter <10 μm) | 0.451 |
| PM2.5 (Particulate matter <5 μm) | 0.206 |
| Benzo(a)pyrene | $1 \cdot 10^{-6}$ |
| Benzo(b)fluoranthene | $1 \cdot 10^{-6}$ |
| Benzo(k)fluoranthene | $0.4 \cdot 10^{-6}$ |
| Indeno(1,2,3-c,d)pyrene | $1.1 \cdot 10^{-6}$ |

gas may reduce the actual emissions. Table 9.6.5 shows reported off gas composition related to biogas from anaerobic digestion plants. The data covers only emissions from combustion of gas in stationary gas engines typically present at the plants; combustion in large centralized power plants is not represented by the emission factors in Table 9.6.5. It should also be noted that the emission factors are average values from a series of measurements on the off gas from different biogas engines and that the emissions vary substantially between engines from different manufactures.

**Table 9.6.6** *Chemical composition of dewatered digestate and press water from anaerobic digestion of biowaste.*

| | One-stage, thermophilic, 'dry' digestion plant in Kaiserslautern, Germany[a] | | One-stage, thermophilic, 'wet' digestion plant in Elsinore, Denmark[b] |
|---|---|---|---|
| | Dewatered digestate | Press water | Liquid digestate |
| TS | 44 % | | 2.7 % |
| VS | 38 % of TS | | |
| COD | | 45 g/l | |
| pH | 8.3 | | |
| NH4-N | 2.43 g/kg w/w | 3.2 g/l | |
| Total P | | | 215 mg/l |
| Pb | 54 mg/kg TS | 2.61 mg/l | 17 mg/kg TS |
| Cd | 0.76 mg/kg TS | 0.030 mg/l | 0.95 mg/kg TS |
| Ni | 16 mg/kg TS | 1.4 mg/l | 10.7 mg/kg TS |
| Hg | 0.42 mg/kg TS | 0.026 mg/l | 0.24 mg/kg TS |
| Zn | 205 mg/kg TS | 10.8 mg/l | 339 mg/kg TS |
| Cr | | | 9.9 mg/kg TS |
| Cu | | | 76 mg/kg TS |
| AOX (adsorbable organic halogen compounds) | 107 mg/kg TS | | |
| LAS | | | 100 mg/kg TS |
| PAH | | | 0.47 mg/kg TS |
| NPE | | | 5.3 mg/kg TS |
| DEHP | | | 61 mg/kg TS |

[a]From Schmidt *et al.* (2001). Of 1 t biowaste 330 kg was removed during pretreatment; 386 kg and 186 kg ended up as dewatered digestate and press water, respectively.
[b]From Jansen (1996). Approx. 930 l liquid digestate/t biowaste was produced. The plant is no longer operating.

***Table 9.6.7*** *Unit process inventory for thermophilic anaerobic digestion of 1 t waste (ww). The plant is a hypothetical 'wet' plant treating source-separated organic household waste (primarily constructed from data in tables in this chapter).*

| Per tonne of waste anaerobically digested | | Comment |
|---|---|---|
| **Inputs** | | |
| Organic waste | 1000 kg | Source-separated organic household waste including vegetables, meat, fish, garden flowers and kitchen tissue |
| Diesel | 0.9 l | Trucks and machinery |
| Electricity | 18.3 kWh | Pumps, ventilators etc. |
| Heat | 30.6 kWh | To heat the reactor tank to mesophilic conditions (electricity or fuel-based) |
| Water | 2 $m^3$ | |
| **Outputs** | | |
| Gas[a] | 123 $Nm^3$ | 63 % $CH_4$ + 37 % $CO_2$ |
| Electricity[b] | 311 kWh | 39 % engine efficiency |
| Heat[b] | 366 kWh | 46 % engine efficiency |
| Metals | < 1 kg | |
| Digestate | 3000 kg | |
| Fugitive gas loss[c] | 1–3 $Nm^3$ | From pipes and fittings etc. |
| Off gasses | | From combustion in biogas engine |
| $CH_4$ | 124 g | |
| $N_2O$ | 0.19 g | |
| CO | 105 g | |
| $NO_x$ | 208 g | |
| $SO_2$ | 7.4 g | |

[a] The outputs are either gas or energy as listed here. The energy out-put is obtained by combustion of the gas.
[b] Typical values for small biogas engines as presented by manufacturers.
[c] Estimated.

### 9.6.4.2   Diffuse Air Emissions

Diffuse air emission from digestion plants may be very smelly, unless all ventilation air from areas where solids are handled are collected and emitted through a biofilter or tall stack.

### 9.6.4.3   Digestate

Data on digestate composition is scarce and the composition depend on the posttreatment processes, i.e., if the digestate is dewatered before use or the digestate is used in diluted form in agriculture. These differences are demonstrated by Table 9.6.6 that provides data on the composition of dewatered digestate and the water from the dewatering process (press water) as well as untreated liquid digestate.

## 9.6.5   Unit Process Inventories

While mass balances are useful for understanding the performance of the anaerobic degradation process and the features of specific facilities, unit process inventories are useful for establishing green accounts, bench marking and as a basis for performing a broader environmental assessment. The inventory lists, in principal, all inputs and all outputs to a facility and by normalizing with the amount of waste treated by the facility the unit process inventory is prepared. In particular the use and eventual production of energy by the facility – electricity, fuel, heat etc. – is important information in the inventory. Also the use of chemicals may in some cases appear significant.

A unit process inventory for a hypothetical anaerobic digestion plant treating source-separated organic biowaste with energy recover from the produced gas at the plant is presented in Table 9.6.7. The inventory shows that anaerobic digestion in this case is a net energy producer with a very low use of chemicals and alike. The energy for running the plant is in the order of 10 % of the energy recovered from the biogas if the heat from the combustion of gas can be utilized along with the electricity produced. If the digestion process was thermophilic instead of mesophilic, as in this example, a substantial larger amount of energy would have to be spent on heating the reactor vessel. Even though the conversion of biogas to energy is greenhouse gas neutral some greenhouse gases in the form of $N_2O$ and CO are emitted as a result of the combustion process itself. A more significant emission is resulting from unburned $CH_4$ in the biogas engine.

# References

Berglund, M. and Börjesson, P. (2006): Assessment of energy performance in the life-cycle of biogas production. *Biomass and Bioenergy*, 30, 254–266.

Davidsson, Å., Gruvberger, C., Christensen, T.H., Hansen, T.L. and Jansen, J.l.C. (2007): Methane yield in source-sorted organic fraction of municipal solid waste. *Waste Management*, 27, 406–414.

Hansen, T.L., Spliid, H., Jansen, J.l.C., Davidsson, Å. and Christensen, T.H. (2006a): Composition of source-sorted municipal organic waste collected in Danish cities. *Waste Management*, 27, 510–518.

Hansen, T.L., Jansen, J.l.C., Davidsson, Å. and Christensen, T.H. (2006b): Effects of pre-treatment technologies on quantity and quality of source-sorted municipal organic waste for biogas recovery. *Waste Management*, 27, 398–405.

Hansen, T.L., Sommer, S.G., Gabriel, S. and Christensen, T.H. (2006c): Methane production during storage of anaerobically digested municipal organic waste. *Journal of Environmental Quality*, 35, 830–836.

Hansen, T.L., Bhander, G.S. and Christensen, T.H. (2006d): Life cycle modeling of environmental impacts of application of processed organic municipal solid waste on agricultural land (EASEWASTE). *Waste Management and Research*, 24, 153–166.

Jansen, J. la C. (1996): *Nordsjællands Biogasanlæg*. Internal report. Nordsjællands Biogasanlæg, Helsingør, Denmark.

Jansen, J. la C. and Svärd, Å. (2002): *Kortlægning af driftserfaringer fra europæiske biogasanlæg til behandling af husholdningsaffald* (Survey of operational experiences from European biogas plants for treatment of household waste; in Danish). Lunds Tekniska Högskola, Lunds University, Lund, Sweden.

Kranert, M. and Hillebrecht, K. (2000): Anaerobic digestion of organic wastes – process parameters and balances in practice. *Proceedings of Internet Conference on Material Flow Analysis of Integrated Bio-Systems*. CD-ROM. Institute of Advanced Studies, United Nations University, Yokohama, Japan.

Lindorfer, H., Braun, R. and Kirchmayr, R. (2006): Self-heating of anaerobic digesters using energy crops. *Water Science and Technology*, 53, 159–166.

NSCA (2002): *Comparison of emissions from waste management options*. National Society for Clean Air and Environmental Protection, Brighton, UK.

Nielsen, M. and Illerup, J.B. (2006): *Danish emission inventories for stationary combustion plants. Inventories until year 2003*. Research Notes NERI No. 229. National Environmental Research Institute, Denmark.

Persson, M., Jönsson, O. and Wellinger, A. (2006): *Biogas upgrading to vehicle fuel standards and grid injection. Energy from Biogas and Landfill Gas*. IEA Bioenergy, Task 37, International Energy Agency, New York, USA.

Schmidt, S., Welker, A. and Schmidt, T.G. (2001): Vergleichende Untersuchung der Stoffströme bei der Vergärung von Bio- und Restabfall. (Comparative assessment of mass flows related to fermentation of organic waste, in German). *Müll und Abfall*, 33, 456–460.

Smet, E., Van Langenhove, H. and De Bo, I. (1999): The emission of volatile compounds during the aerobic and the combined anaerobic/aerobic composting of biowaste. *Atmospheric Environment*, 33, 1295–1303.

# 9.7

# Mechanical Biological Treatment

**Bernd Bilitewski**

*Technical University, Dresden, Germany*

**Christiane Oros and Thomas H. Christensen**

*Technical University of Denmark, Denmark*

The basic processes and technologies of composting and anaerobic digestion, as described in the previous chapters, are usually used for specific or source-separated organic waste flows. However, in the 1990s mechanical biological waste treatment technologies (MBT) were developed for unsorted or residual waste (after some recyclables removed at the source). The concept was originally to reduce the amount of waste going to landfill, but MBT technologies are today also seen as plants recovering fuel as well as material fractions. As the name suggests the technology combines mechanical treatment technologies (screens, sieves, magnets, etc.) with biological technologies (composting, anaerobic digestion). Two main technologies are available: Mechanical biological pretreatment (MBP), which first removes an RDF fraction and then biologically treats the remaining waste before most of it is landfilled, and mechanical biological stabilization (MBS), which first composts the waste for drying prior to extraction of a large RDF fraction. Only a small fraction is landfilled. The latter technology is also referred to as biodrying. Within each of the two main technologies, a range of variations is available depending on waste received and routing of the RDF fraction.

This chapter offers an introduction to the two technologies. Box 9.7.1 shows the types of MBT plants found in Germany.

## 9.7.1   Mechanical Biological Pretreatment

### 9.7.1.1   MBP: Technology

MBP plants recover a fuel fraction (RDF/SRF) as well as metals and treat the remaining waste prior to landfilling. The first step is mechanical treatment; the second step is biological treatment.

**Box 9.7.1    MBT Plants in Germany Around 2006 (After Kuehle-Weidemeier *et al.*, 2007).**

Germany was one of the first countries to introduce MBT technology as a response to the EU regulation restricting the amount of organic waste going to landfill. By 2006 about 43 plants were in operation and 75 % were using aerobic biological processes, while the remaining ones were using process combinations also involving an anaerobic biological process step. In addition, three plants used a nonbiological drying process. The distribution and capacities of the plants are shown below.

| MBT technology | Number | Treated waste (t/year) |
| --- | --- | --- |
| MBP: Aerobic | 18 | 2 000 000 |
| MBP: Anaerobic/aerobic | 10 | 935 000 |
| MBP: Percolation/anaerobic/aerobic | 3 | 160 000 |
| MBS (aerobic) | 12 | 1 350 000 |

Mechanical processing takes place before biodegradation in order to separate RDF/SRF and recyclables from the remaining waste, which afterwards is biological treated. The basic mechanical operations for material flow separation are shredding, sieving and magnetic separation. After size reduction, the waste stream is usually divided by use of drum screens into oversize and undersize flows. Both flows undergo magnetic separation for ferrous metals recovery. Typical screen sizes are 80 mm, but also smaller screen sizes like 40 mm are used for separation of the fraction for disposal (Thiel 2007). Further, also more sophisticated classification and sorting may be used. The oversize flow may be separated into a light and heavy fraction by means of air classifier or ballistic separators. Nonferrous metals can be segregated by eddy current separators. Near infrared (NIR) is either used for positive sorting of plastics from the waste flow or as negative sorting of PVC for qualitative improvement of the RDF/SRF fraction.

Biological processing can be aerobic treatment (composting) or anaerobic treatment (digestion): the latter may include a percolation step and in most cases the solids are afterwards composted in order to reduce odours and increase stability. In MBP plants using composting for biological degradation, the process often takes place in two steps: intensive composting in contained composting windrows, containers or boxes for 4–5 weeks, and postprocessing in roofed windrows for 9–10 weeks. However, the duration of the biological treatment varies from plant to plant and depends on process intensity and controls (Thiel 2007). Anaerobic digestions for biological treatment are wet or dry fermentation processes with retention times of typically 3–4 weeks. The remaining digestate, where required, is dewatered and afterwards composted (for 4–6 weeks). The anaerobic process may involve percolation washing out soluble components from the organic fraction of the solid waste. The percolate is then used for biogas production in a digestion reactor, while the solids are composted. The biological treatment has to proceed to a level where the stabilized material meets the criteria for landfilling in an MBT landfill (this type of landfill is described in Chapter 10.7).

Figure 9.7.1 presents the technological configuration of a German MBP plant with aerobic treatment. Another example of the technological configuration for mechanical processing in a MBP is given in Chapter 8.7, Figure 8.7.2.

### 9.7.1.2   MBP: Mass Balance

The mass balance of the MBP plant depends on the waste received and the actual configuration of the plant (sieve sizes, composting process, etc.). A typical mass balance is shown in Figure 9.7.2. Five outputs are typical:

- RDF/SRF (30–46 %): This is relatively large pieces of paper, cardboard, wood and plastic that constitutes a relatively dry fraction to be used as a fuel in a power plant or industrial kiln or to be burned in an incinerator.
- Losses (20–24 %) by degradation of the organics and water evaporated during the biological treatment step. Compared to MBS the water loss in the MBP process is less, due to a certain water content, which must be maintained to sustain continued biological degradation. The off gases must be treated prior to discharge (see later).

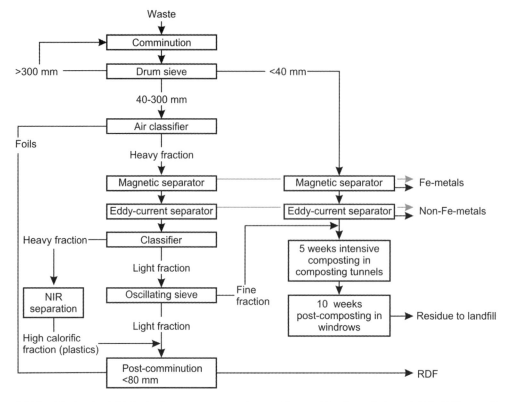

**Figure 9.7.1** *Mechanical biological pretreatment plant with aerobic degradation, MBP Cröbern, Germany (Rößger et al., 2007). Reprinted with permission from Mechanical-biological pre-treatment plant with aerobic degradation, MBP Cröbern, Germany, Rößger et al., © (2007) Erich Schmidt Verlag Germany.*

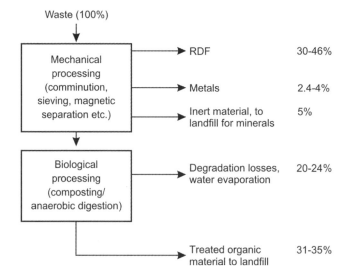

**Figure 9.7.2** *Mass balance of MBP plant (EC 2006; Doedens et al., 2007).*

- Stabilized materials for landfilling (31–35 %) which is like a coarse compost intended for landfilling; the water content of the material is approximately 35 %.
- Inerts for a mineral landfill (5 %) consisting of heavy items like stones, debris and eventually PVC.
- Metals (2–4 %) in terms of magnetic iron scrap and eventually aluminum recovered for recycling.

## 9.7.2 Mechanical Biological Stabilization (Biodrying)

### 9.7.2.1 MBS: Technology

MBS technology aims at conservation of the biodegradable carbon and routing it to the RDF fraction. All the waste is subject to a fast biological process primarily acting as a drying process, thereafter metals are recovered and an inert fraction for landfilling is removed. The main part of the waste ends up in the RDF/SRF fraction. The first step is biological treatment; the second step is mechanical treatment.

Biological processing generates sufficient heat (by degrading a part of the easily degradable organic matter) to evaporate a significant fraction of the water in the waste received. This increases the heating value. Additionally, the waste is stabilized regarding further degradation, as microbial degradation is inhibited by water contents lower than 25 % and ceases at values below 10 % (Mähl, 2005). The target water content for the waste after stabilization is typically below 15 % (Thiel, 2007). The drying also enables a more efficient sorting of metals and minerals. The biological process step typically involves composting for 5–7 days in container composting systems (Thiel, 2007).

Mechanical processing may involve sieves, magnets and eddy current separators. Magnetic iron scrap and eventually aluminum are recovered for recycling. An inert fraction of heavy materials goes to a mineral waste landfill. The RDF/SRF flow may be pelletized prior to energy utilization.

Figure 9.7.3 presents the technological configuration of a German MBS plant (Dresden, Germany; see Figure 9.7.4).

### 9.7.2.2 MBS: Mass Balance

The mass balance of the MBS plant depends on the waste received and the actual configuration of the plant (sieve sizes, composting process, etc.). A mass balance is shown in Figure 9.7.5 for the MBS plant in Dresden, Germany. Four outputs are typical:

- Losses (30–33 %) by degradation of the organics and water evaporated during composting. These off gases must be treated prior to discharge (see later).
- RDF/SRF (approximately 55 %): This is a mix of all waste fractions that constitutes a relatively dry fraction to be used as a fuel in a power plant or industrial kiln or to be burned in an incinerator.
- Inerts for a mineral landfill (9 %) consisting of heavy items like stones, debris and eventually PVC.
- Metals (2–4 %) in terms of magnetic iron scrap and eventually aluminum recovered for recycling.

## 9.7.3 MBT: RDF and Residue for Landfilling

The MBP and the MBS technology both produce a RDF/SRF fraction as well as fractions for landfilling.

The RDF/SRF fraction of the MBS technology is about twice as large as the one from MBP. It is likely that the RDF/SRF fraction from MBP is cleaner than the one from MBS, but little data is available to document this. However, RDF/SRF fractions are in many countries regulated with respect to calorific value, water content, chlorine-content etc. and categorized into various qualities which determine their use. These issues are described in Chapter 8.7.

The inert landfill fractions from both technologies are usually landfilled in a landfill for mineral waste and must in Europe meet the criteria for landfilling of mineral waste (Chapter 10.5). The stabilized landfill fraction from MBP is characterized by its organic content and low biological activity. The criteria valid in some European countries are described in Chapter 10.7.

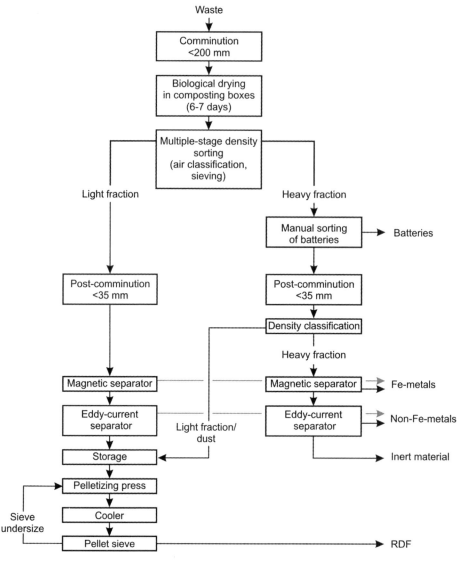

**Figure 9.7.3**  *Mechanical biological stabilization plant, MBS Dresden, Germany (Thiel 2007). Reprinted with permission from Ersatzbrennstoffe in Kohlekraftwerken – Mitverbrennung von Erstazbrennstoffen aus der mechanisch-biologischen Abfallbehandlung in Kohlekraftwerken by S. Thiel © (2007) TK Verlag Karl Thome-Kozmiensky.*

By undergoing mechanical and biological treatment the material for disposal is expected to be have a smaller particle size, and therefore a higher bulk density (up to $1.6 \, kg/m^3$) than untreated MSW (Kühle-Weidemeier *et al.*, 2006). The gas formation potential is reduced by up to 90 % (Stegmann, 2005), resulting in a shortened landfill aftercare period.

## 9.7.4   Environmental Emissions and Controls

MBT plants have two major emissions to control: the off gasses from the treatment process and the leachate and condensate from the biological treatment, respectively.

**Figure 9.7.4**   *The MBS plant in Dresden, Germany. Reprinted with permission from Ersatzbrennstoffe in Kohlekraftwerken – Mitverbrennung von Erstazbrennstoffen aus der mechanisch-biologischen Abfallbehandlung in Kohlekraftwerken by S. Thiel © (2007) TK Verlag Karl Thome-Kozmiensky.*

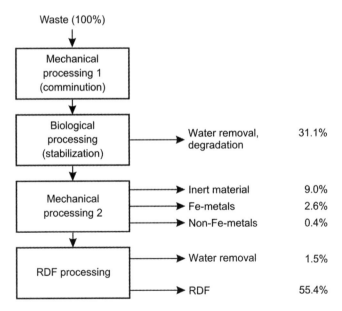

**Figure 9.7.5**   *Mass balance of MBS plant (Dresden, Germany; Thiel, 2007). Reprinted with permission from Mechanical-biological pre-treatment plant with aerobic degradation, MBP Cröbern, Germany, Rößger et al., © (2007) Erich Schmidt Verlag Germany.*

**Table 9.7.1**   *MBT off gas quantities and qualities before treatment compared with German off gas threshold values (Fricke et al., 2005). Reprinted with permission from Waste Management, Comparison of selected aerobic and anaerobic procedures for MSW treatment by K. Fricke, H. Santen and R. Wallmann, 25, 8, 799–810 © (2005) Elsevier Ltd.*

| | Treated off gas | Raw off gas | |
| --- | --- | --- | --- |
| | German guideline[a] | Anaerobic | Anaerobic–aerobic |
| Gas amount | | 5000–9000 | 2000–6000 |
| TOC (mg/Nm$^3$) | 20/40 | 50–200 | 50–200 |
| TOC (g/t) | 55 | 400–800 | 200–600 |
| N$_2$O (g/t) | 100 | <100 | <100 |
| Odour (OU/m$^3$) | 500 | 10 000–30 000 | 10 000–30 000 |
| PCDDF (ng TE/Nm$^3$) | 0.1 | <0.1 | <0.1 |
| Dust (mg/Nm$^3$) | 10/30 | <10 | <10 |
| NH$_3$ (mg/Nm$^3$) | — | 30–100 | 100–300 |

[a]Federal Ministry for the Environment (2001): first value is daily mean, second value is half-hour mean. Reproduced with permission from *Waste Management.* © (2005) Elsevier Ltd.

### 9.7.4.1   Emission to Air and Flue Gas Treatment

The mechanical treatment and in particular the biological treatment generate a significant amount of off gasses; in the order of 2000–10 000 m$^3$/t waste treated. The odorous off gas contains dust, CH$_4$, NH$_3$ and some N$_2$O and volatile organic carbons (VOCs). The latter includes ethanol, BTEX (benzene, toluene, ethylbenzene, xylenes), chlorinated ethenes, etc. (Fricke *et al.*, 2005). In many countries, e.g. Germany, off gas emissions are regulated and all MBT plants are equipped with gas treatment facilities. Table 9.7.1 presents the German threshold values for gas emissions as well as typical ranges of raw gas composition from MBT plants with aerobic biological processes as well as from combined anaerobic–aerobic processes.

In MBT plants off gas treatment involves regenerative thermal oxidation (RTO), acid scrubbers, biofilters and dust filters. Figure 9.7.6 shows the components and flows in a MBT off gas system. Off gas air can be treated and reused depending on the quality of the gas. Less polluted off gas, as it occurs in the waste receiving area or the mechanical processing hall, can be reused for the aeration of the intensive biological process. In contrast, highly polluted off gas from the intensive processing is cooled down and led through a condensate separator and acid scrubber before it is cleaned in the RTO. Off gas from the postprocessing is led as well through a condensate separator, but then reused in the intensive composting process (Warnstedt *et al.*, 2007). Biofilter are used only for the treatment of off gas flows with a low load of

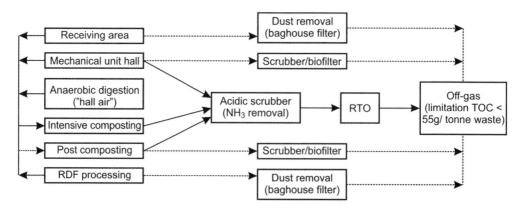

**Figure 9.7.6**   *Typical flue gas treatment system of MBT plants (Wallmann et al., 2006).*

organic compounds, as from the receiving area and mechanical processing (Bisdorf *et al.*, 2007; Doedens *et al.*, 2003), as $CH_4$ is only a little oxidized and $NH_3$ in the off gas may cause generation of $N_2O$.

The main off gas treatment units are:

- **Regenerative thermal oxidation** (RTO). RTO is based on the oxidation of carbon compounds at reaction temperatures of 800–950 °C. The RTO consists of two or three chambers with heat exchangers. The gas to be treated enters the first chamber with a temperature of 30–60 °C, where it is preheated in the heat exchanger and oxidized by the use of auxiliary fuel. The cleaned air passes another heat exchanger, where it cools down to 100 °C (Neese *et al.*, 2007). The off gas flow is routed alternatively through the two heat exchangers, so that the energy can be preserved in the unit. RTO plants are usually designed for exhaust gas preheating efficiencies of 95 %. In the case that the exhaust gas does not possess sufficient heating value, the remaining 5 % have to be added by auxiliary fuels (Dach *et al.*, 2007). For this purpose, mainly natural gas is used, but also the use of landfill gas is possible. The demand of auxiliary fuels can be reduced through aeration management. The higher the carbon load and therefore the heating value of the gas to be treated by the RTO, the less auxiliary fuel is needed. Increasing carbon loads can be achieved by gas/air cycling in the biological process, which also decreases the air volumes to the RTO (Doedens *et al.*, 2003). Another reason for increased fuel consumption is irregular flow through the heat transfer unit. Irregular flows occur when stains on the honeycombs of the heat transfer mass are built up. This is due to organic silicon compounds (silanes, siloxanes), which are released from cosmetic products, packaging residues or other products of the daily life during biological treatment and oxidized to amorphous silica inside the RTO (Dach *et al.*, 2007; Mattersteig *et al.*, 2009). RTO is the only technology that can meet the strict German off gas limits.
- **Acid scrubber**. As RTO cannot achieve any cleaning of dust and ammonia components, an upstream acid scrubber is needed; especially due to the fact that ammonia may generate $N_2O$ or $NO_x$ in the RTO (Dach *et al.*, 2007). By the use of sulfuric acid, nitrous oxide compounds are adsorbed to the scrubber liquor. An ammonium sulfate solution is produced, which can further on be used as fertilizer on agricultural land (Warnstedt *et al.*, 2007).

### 9.7.4.2 Wastewater and Wastewater Treatment

As irrigation is applied during biological processing and waste contains a certain amount of water, wastewater is generated from the composting units, but also as condensate water and acid scrubber water from the off gasses. This water can be cleaned and reused. In some cases, leachate from nearby landfills is also used for irrigation purposes in MBP plants. In case that leachate from MBP cannot be reused in the composting processes and needs to be discharged, usually treatment facilities of nearby landfills are used for this purpose (Kuehle-Weidemeier *et al.*, 2007).

The various types of waste waters that may appear at a MBP plant are presented in terms of their main characteristic in Table 9.7.2. Treatment of leachate is described in detail in Chapter 10.11.

**Table 9.7.2** *Waste waters (raw) from MBT plants (composed after Weichgrebe et al., 2007; Fricke et al., 2005).*

|  | Leachate from intensive composting | Leachate from post composting | Process water from dry digestion | Press water from digestion residue | Scrubber water from off gas | Condensate from off gas |
|---|---|---|---|---|---|---|
| SS (mg/l) | 2000–10000 | 150–500 | 2000–10 000 | 80–2000 | 33–700 | — |
| COD (g/l) | 20–70 | 7–17.2 | 3.5–33.5 | 1.1–7.5 | 0.2–1.2 | 0.4–2.5 |
| BOD (g/l) | 10–12 | 2.5–6.2 | 0.8–2.0 | 0.4–1.6 | — | 0.1–1.0 |
| $N_{tot}$ (g/l) | 2–12 | 1.2–3.2 | 1.4–2.5 | 1.2–2.9 | 40–160 | 0.17 |
| $P_{tot}$ (mg/l) | 13–32 | 3.4–8.5 | 17–85 | 20–56 | 0.1–2.2 | — |
| AOX (mg/l) | 0.4–1.0 | 0.4–0.5 | 0.1–2.1 | 0.5–2.0 | — | <0.1–2.0 |
| Cu (mg/l) | 0.2–3.0 | 0.2–1.2 | 0.2–4.8 | 0.06–0.22 | — | — |
| Zn (mg/l) | 4.1–13.0 | 0.9–4.6 | 0.3–10.0 | 0.2–1.0 | — | 0.2–1.6 |

## 9.7.5   Unit Process Inventories

While mass balances are useful for understanding the performance of the MBT technology, unit process inventories are useful for establishing green accounts, bench marking and as a basis for performing a broader environmental assessment. The inventory list contains, in principal, all inputs and all outputs to a facility normalized with the amount of waste treated by the facility. In particular the use and eventual production of fuels is important information in the inventory (Table 9.7.3).

Regarding the electricity consumption MBS plants show the highest value, due to the operation at higher aeration rates and more sophisticated separation concepts, especially if further RDF processing (size reduction, pelletizing) takes place in the plant. The higher electricity consumptions of MBP with anaerobic treatment are basically depending on the treatment of the digestate (dewatering; Wallmann *et al.*, 2008). Diesel consumptions depend on the transport activity in the plant, especially the feeding systems for mechanical processing and filling of the composting units (automatically or by wheel loaders). The off gas volumes and thereby resulting gas consumption by the RTO, depend on process specific parameters like the aeration rates for aerobic treatment. In case of MBS, aeration is not only required to guarantee a sufficient oxygen supply for microbial degradation, but also to enable 'biodrying' by removing evaporated water via the off gas flow. MBP with anaerobic degradation shows the lowest off gas volumes. The characterization of off gases regarding $N_2O$ and further compounds reflects the emissions from specific plants, but although the emissions vary, the requirements regarding pollution control are fulfilled.

***Table 9.7.3***   *Unit process inventory for two MBP plants and one MBS plant for 1 t (w/w) of waste (EC, 2006; Doedens et al., 2007; Rößger et al., 2007; Thiel, 2007; Wallmann et al., 2008).*

| Per 1 t of MBT-treated waste | Unit | MBP | | MBS |
|---|---|---|---|---|
| | | Aerobic | Anaerobic–aerobic | |
| Inputs | | | | |
| Waste | kg | 1000 | 1000 | 1000 |
| Diesel | l | 1.1[a] | 1.1[a] | 0.4[a] |
| Electricity | kWh | 37[a] | 45[a] | 81[a] |
| Gas | kWh | 53[a] | 50[a] | 59[a] |
| Water | m$^3$ | n.s. | n.s. | 0.1[d] |
| Outputs | | | | |
| RDF/SRF | kg | 370[b] | 370[b] | 555[c] |
| RDF/SRF: LHV | MJ/ tonne RDF | 13.3[a] | 13.2[a] | 14.0[a] |
| Metals | kg | 30[b] | 30[b] | 30[c] |
| Inerts to mineral landfill/ incineration | kg | 50[b] | 50[b] | 90[c] |
| Waste to MBT landfill | kg | 330[b] | 330[b] | 0 |
| Water | l | 140 (recycled)[i] | 170[i] | 1.3[d] |
| Biogas (60% CH$_4$) | Nm$^3$ | — | 9–26[f]/60–65[g,h] | — |
| Off gasses | Nm$^3$ | 5400[a] | 4300[a] | 8700[a] |
| CH$_4$ | g | | | |
| N$_2$O | g | 19.9[d] | n.s. | 1.0[d] |
| TC | g | 29.3[d] | n.s. | 55[d] |

[a]Wallmann *et al.* (2008)
[b]EC (2006), Doedens et al. (2007)
[c]Thiel (2007),
[d]MBS anonymous, Germany,
[e]Rößger *et al.* (2007),
[f]Part flow of biodegradable fraction to digestion, Wallmann et al. (2008),
[g]Full flow of biodegradable fraction to digestion, Wallmann et al. (2008),
[h]85–195 Nm$^3$/t input to digestion reactor, Wallmann et al. (2008),
[i]Fricke et al. (2005).

According to Fricke (2005), about 140 l of waste water is generated per tonne of MSW. But as this water is reused in the composting process, this type of MBP is considered to be wastewater-free. In MBP with anaerobic treatment wastewater generation of 100–170 l/t MSW occurs during the dewatering process of the digestate (Fricke 2005). The water generation for MBS plants does not refer to average numbers, but reflects the situation in one specific plant in Germany.

# References

Bisdorf, R. and Pfliegensdörfer, T. (2007): Customized exhaust gas treatment for MBT. In: Kuehle-Weidemeier, M. (ed.) *Proceedings International Symposium MBT 2007 – Mechanical biological treatment and automatic sorting of municipal solid waste*, pp. 524–537. Cuvillier Verlag, Göttingen, Germany.

Dach, J., Warnstedt, A., Siemion, J. and Müller, G. (2007): Ecobalance of RTO regarding the avoidance of GHG emissions. In: Kuehle-Weidemeier, M. (ed.) *Proceedings International Symposium MBT 2007 – Mechanical biological treatment and automatic sorting of municipal solid waste*, pp. 552–564. Cuvillier Verlag, Göttingen, Germany.

Doedens, H., Kahn, R., Stockinger, J. and Glüsing, J. (2003): Erprobung einer nicht-katalytischen thermischen Oxidation zur Behandlung von Abluft aus der mechanisch-biologischen Abfallbehandlung. BMBF-Verbundvorhaben, Federal Ministry of Education and Research, Germany.

EC (2006): Reference document on the best available techniques (BREF) for waste treatment industries.European Commission, Brussels, Belgium.

Federal Ministry for the Environment (2001): *30th German ordinance (30. Bundes-Immissionsschutzverordnung; BImSchV) on the implementation of the Federal emission control act (Bundes-Immissionsschutzgesetz).* Federal Ministry for the Environment, Nature Conservation and Nuclear Safety, Berlin/Bonn, Germany.

Fricke, K., Santen, H. and Wallmann, R. (2005): Comparison of selected aerobic and anaerobic procedures for MSW treatment. *Waste Management*, 25, 799–810.

Kühle-Weidemeier, M. and Langer, U. (2006): Landfills for waste from mechanical biological treatment (MBT/ MBP) (in German). *Müll and Abfall*, 04/2006, 202–209.

Kühle-Weidemeier, M., Langer, U., Hohmann F. and Butz, W. (2007): Anlagen zur mechanisch-biologischen Restabfallbehandlung (Plants for mechanical–biological waste treatment; in German). Founding Number (UFOPLAN) 206 33 301. *Environmental research program of the Federal Ministry of Environment, Environmental Protection and Nuclear Safety, Germany.*

Mähl, B (2005): Aeration as an instrument for process optimisation of intensive processing in MBT plants. In: Kuehle-Weidemeier, M. (ed.) *Proceedings International Symposium MBT 2005 – Mechanical biological treatment and automatic sorting of municipal solid waste, November 2005*, pp. 281–283. Cuvillier Verlag, Göttingen, Germany.

Mattersteig, S., Brunn, L., Friese, M. and Bilitewski, B. (2009): Organosilicon compounds in the exhaust gas of biological–mechanical treatment plants. In: CISA (ed.) *Proceedings of the Twelfth International Waste Management and Landfill Symposium*. CISA – Environmental Sanitary Engineering Centre, Cagliari, Italy.

Neese, O., Carlowitz, O. and Reindorf, T. (2007): Operational experiences and potential of regenerative thermal oxidation plants in the field of MBT. In: Kuehle-Weidemeier, M. (ed.) *Proceedings International Symposium MBT 2007 – Mechanical biological treatment and automatic sorting of municipal solid waste, May 2007*, pp. 538–551. Cuvillier Verlag, Göttingen, Germany.

Rößger, U., Greif, R., Winden, K.-H., Altepost, G., Kanning, K. and Ketelsen, K. (2007): Construction and startup of MBT Cröbern, experiences from the plant-operator's viewpoint (in German). *Müll and Abfall*, 12/2007, 588–597.

Stegmann, R. (2005): Mechanical biological pretreatment of municipal solid waste. *Proceedings Sardinia 2005, Tenth International Waste Management and Landfill Symposium*. CISA, Environmental Sanitary Engineering Centre, Cagliari, Italy.

Thiel, S. (2007): Ersatzbrennstoffe in Kohlekraftwerken-Mitverbrennung von Ersatzbrennstoffen aus der mechanisch-biologischen Abfallbehandlung in Kohlekraftwerken (RDF in coal-fired power plants; in German). TK Verlag, Nietwerder, Germany.

Wallmann, R., Dorstewitz, H., Hake, J., Santen, H. and Fricke, K. (2006): Operational experiences with the treatment of exhaust air according to the federal emission control regulation (30. BImSchV). *Müll and Abfall*, 06/2006, 304–309.

Wallmann, R., Fricke, K. and Hake, J., (2008): Energy efficiency of mechanical treatment of residual waste (in German). *Müll and Abfall*, 07/2008, 332–339.

Warnstedt, A. Dach, J. and Müller, G. (2007): Two years of experience with new German regulations for MBT plants: View of an MBT operator. In: Kuehle-Weidemeier, M. (ed.) *Proceedings International Symposium MBT 2007, Mechanical biological treatment and automatic sorting of municipal solid waste, May 2007*, pp. 204–219. Cuvillier Verlag, Göttingen, Germany.

Weichgrebe, D., Maerker, S., Böning, T. and Stegmann H. (2007): Process water management for mechanical biological waste treatment plants. *Proceedings Sardinia 2007, Eleventh International Waste Management and Landfill Symposium*. CISA, Environmental Sanitary Engineering Centre, Cagliari, Italy.

# 9.8

# Emerging Biological Technologies: Biofuels and Biochemicals

**Dimitar Karakashev and Irini Angelidaki**

*Technical University of Denmark, Denmark*

Composting and anaerobic digestion are well established technologies, although the latter is a relative recent technology regarding solid waste and full scale plants still are relatively few. However, alternative technologies based on biotechnology are emerging. These technologies are focused around ethanol production, hydrogen production and production of organic chemicals. It is too early to say if these emerging technologies will play a real role in solid waste management in the future, but they may have a potential for treatment of some organic waste streams.

This chapter provides a brief introduction to these emerging biological technologies.

## 9.8.1   Biofuels: Bioethanol

### 9.8.1.1   State of the Technology

Ethanol derived from biomass has the potential to be a sustainable transportation fuel as well as fuel oxygenate. Blending oxygenate such as ethanol is well recognized for causing reduced carbon monoxide levels by improving overall combustion (oxidation) of the fuel. In many countries, the use of bioethanol as an alternative fuel or a gasoline supplement at amounts up to 15 % is highly recommended or even required as an environmentally attractive fuel oxygenate (Mojovic *et al.*, 2006). This liquid energy carrier is produced biologically from two types of biomass – starch/sucrose containing energy crops such as sugarcane and corn (first generation bioethanol technology) and lignocellolosic wastes, which include materials such as agricultural residues (crop straws), herbaceous crops (alfalfa, switchgrass), short rotation woody crops, forestry residues and wastepaper (second generation bioethanol technology). A simple comparison between first and second generation bioethanol technologies is presented in Table 9.8.1. Theoretically, bioethanol production yields 0.51 g ethanol/g glucose, which translates into an energy recovery of approximately 90 %. With current technologies 90–95 % of the theoretical yield is achieved, equivalent to 0.45–0.48 g ethanol/g glucose in the raw material (Hagerdal *et al.*, 2007).

First generation bioethanol technology is conventional and well established and most ethanol worldwide is made by this technology (Lens *et al.*, 2005). Nearly all fuel ethanol is produced by fermentation of corn glucose in the USA

**Table 9.8.1**   Comparison of first and second generation bioethanol technologies.

|  | First generation bioethanol | Second generation bioethanol |
|---|---|---|
| Substrate | Sugar (sucrose) from sugarcane and starch from corn or wheat | Lignocellulosic materials (straw, corn stover, wood, waste) |
| Pretreatment | No chemical/physical pretreatment of biomass before enzymatic hydrolysis | Chemical/physical pretreatment necessary to facilitate enzymatic hydrolysis |
| Enzymes | Not used or commercial enzymes available | Expensive, noncommercial enzymes |
| Status | Commercial scale | Pilot scale |

(five billion liters of ethanol annually) or sucrose from sugar cane in Brazil (12.5 billion liters of ethanol annually), but any country with a significant agronomy-based economy can use current technology for fuel ethanol production. Second generation bioethanol production technology is still under development. Many proposals are mooted to generate ethanol from lignocellulosic biomass, but they are laboratory- or pilot-scale (Table 9.8.2) and not yet full-scale applications. Some of the pilot scale initiatives in Spain, the United States and Denmark are developing into commercial scale in these years (Möller, 2006).

### 9.8.1.2   First Generation Bioethanol Technology: Process Concepts

The basic process in first generation bioethanol technology is hydrolysis and fermentation generally carried out by the yeast *Saccharomyces cerevisiae*. This microorganism has an enzymatic set mediating starch hydrolysis and the subsequent fermentation of resulting sugars to ethanol. In some cases to facilitate the bioconversion process, commercial starch hydrolytic enzymes also can be added. However, the main concern of using primary crops is the high bioethanol production costs. It has been estimated that up to 50 % of the costs in first generation technology is due to feedstock costs (Willke and Vorlop, 2004). One way to overcome this problem and reduce the costs is to use the whole plant crop.

### 9.8.1.3   Second Generation Bioethanol Technology: Process Concepts

The main conceptual steps involved in the second generation biothanol production process are pretreatment, hydrolysis, fermentation and distillation (Figure 9.8.1). Some of the steps can be combined in one depending on the process conditions, substrate characteristics and microorganisms used. A comparison between the first and second generation bioethanol technologies is presented in Table 9.8.1.

**Table 9.8.2**   Pilot scale plants for second generation bioethanol production in the European Union, North America and Japan (adapted from Möller, 2006). Reproduced with permission from CPL Press. © (2006) CPL Press, UK.

| Name | Location | Feedstock | Capacity: Feeding rate (t/day) | Capacity: Ethanol production rate (Mio. l/year) |
|---|---|---|---|---|
| Etek plant | Sweden | Spruce saw dust | 2 | 0.2 |
| Iogen plant | Canada | Straw | 40 | 3 |
| Celunol | USA | Bagasse, energy cane, short rotation wood end poplar, cotton wood, municipal waste | Not mentioned | 6.4 |
| Izumi | Japan | Wood chips | Not mentioned | 0.1 |
| NREL | USA | Various | Not mentioned | Not mentioned |
| Inbicon | Denmark | Straw | 1 | Not mentioned |

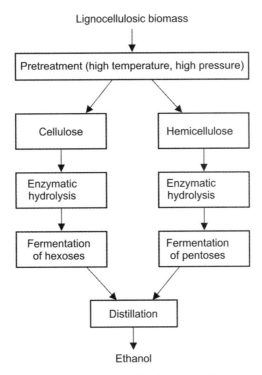

**Figure 9.8.1**   *Schematic diagram of the main steps involved in second generation bioethanol production from lignocellulosic biomass.*

### Pretreatment

Effective pretreatment methods, by which the polysaccharides (cellulose and hemicellulose) are released from the lignocellulosic structure and become available for further conversion (hydrolysis and fermentation), provide the key challenge for lignocellulose-based bioethanol production. The pretreatment is most often made at high temperature and pressure. Different lignocellulose pretreatment methods (steam pretreatment, wet oxidation, thermal hydrolysis) have been investigated. A good pretreatment process is one with a high yield of carbohydrates combined with a low production of fermentation inhibitors.

### Hydrolysis and Fermentation

Pretreatment is followed by an enzymatic hydrolytic step for conversion of cellulose and hemicellulose polysaccharides to fermentable monosaccharides (hexoses, pentoses). There are a few options (Figure 9.8.2) when conducting the hydrolysis and fermentation steps: (1) separate hydrolysis and fermentation (SHF) where hydrolytic enzymes are added before fermentation, (2) simultaneous saccharification and fermentation (SSF) where enzymes are added together with the microorganism, or (3) simultaneous saccharification and fermentation where the microorganisms produce cellylolitic and sugar-fermenting enzymes, a process known as direct microbial conversion (DMC) of lignocellulose. Each of those options has advantages and disadvantages regarding operational conditions, end products inhibition of the hydrolytic process, ethanol yields and ethanol tolerance, and byproduct formation.

In fermentation aspects, the desired traits in a microorganism for commercial ethanol production are broad substrate utilization (ability to use both hexoses and pentoses for fermentation), high ethanol yields and productivity, tolerance to inhibitors presented in the hydrolysates and high ethanol tolerance, cellulolytic activity and ability for sugar fermentation

**Separate hydrolysis and fermentation (SHF)**

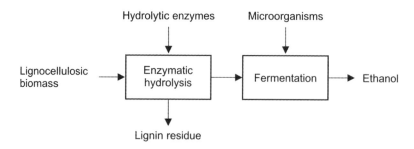

**Simultaneous saccharification and fermentation (SSF)**

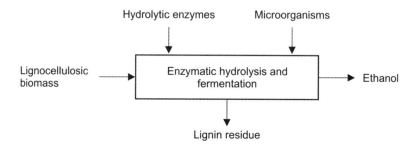

**Direct microbial conversion (DMC)**

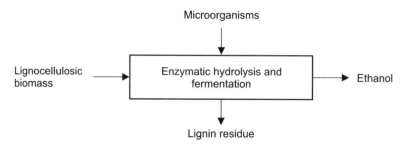

*Figure 9.8.2* *Options for hydrolysis and fermentation steps in second generation bioethanol technology. Reprinted with permission from Cell wall saccharification. Epobio. Realising the Economic Potential of Sustainable Resources – Bioproducts from Non-Food Crops by R.Möller © (2006) CPL Press.*

preferably at high temperatures (for less contamination and decreased cost for ethanol recovery through distillation). The most commonly used microorganism for full-scale ethanol production from sugar cane and corn starch hydrolysates is the yeast *Saccharomyces cerevisiae*. It has two major advantages over other yeasts: high ethanol tolerance and the ability to grow under strict anaerobic conditions and high substrate concentrations. However native *S. cerevisiae* strains have no cellulolytic activity and have limited substrate utilization spectrum including only hexoses and disaccharides (Claasen *et al.*, 1999). Another promising ethanol producer is the bacterium *Zymomonas mobilis* that reaches ethanol yields close to the stoichiometrical value of 0.51 g ethanol/g glucose. The main disadvantage of the native strains of *S. cerevisiae* and *Z. mobilis* is their inability to utilize pentoses (xylose, arabinose). Other microorganisms such as enteric bacteria and the yeasts *Pichia stipitis*, *Candida shehatae* and *Pachysolen tannophilus* are able to ferment xylose to ethanol. However, the ethanol production rates of these microorganisms with glucose as substrate are much lower than that observed with *S. cerevisiae*.

### Distillation

Produced ethanol is separated from the fermentation broth by a distillation or molecular sieve techniques. This product is ready to be used as fuel, either pure or blended with petrol.

### Challenges and Possibilities

Production of ethanol from biomass has a great potential for reducing $CO_2$ emissions and thereby preventing greenhouse effect (global warming). However, the competitiveness of this biologically produced energy carrier, compared to the fossil fuels is low mainly due to the present relatively low price of oil. New technologies are needed for making low price biomass such as lignocellulosic material available for fermentation. New efforts in the area of microbial selection are needed in order to find or create microorganisms with high potential for ethanol production from a wide spectrum of sugars. At the same time improvement of existing strains by genetic manipulations can create a basis for attractive industrial biotechnological applications where strict safety regulations for human health and environment protection should be considered.

## 9.8.2   Biofuels: Biohydrogen

### 9.8.2.1   State of the Technology

In recent years, hydrogen gas is attracting widespread attention as a clean and environmentally friendly fuel that produces water when combusted.

Hydrogen can be produced by both biological and nonbiological methods. The main industrial process to produce $H_2$ involves steam reforming from natural gas and petroleum, a process which depends on fossil fuels and thus is not $CO_2$ neutral. Another source is electrolysis of sea water, which could be sustainable if electricity is generated from renewable resources, but at present there is no significant surplus of sustainable electricity capacity. One of the methods to circumvent the dependence of $H_2$ production from fossil fuels is to utilize the potential of $H_2$-producing microorganisms to derive hydrogen from widely available biomass as renewable energy source. Currently, biologically produced hydrogen is applicable for combustion in internal combustion motors.

Biohydrogen production can be realized by microorganisms using carbohydrate-rich and nontoxic raw materials (Kapdan and Kargi, 2006). Among various processes comprising the biohydrogen production, direct and indirect bio-photolysis, hemoheterotrophic (dark) fermentation, photoheterotrophic (light-driven) fermentation and *in vitro* enzymatic conversion of biomass are important. Currently, the dark fermentation is the most feasible process for biohydrogen production from renewable biomass due to its higher rate of hydrogen evolution in absence of any light sources as well as the versatility of the substrates used.

Although a lot of effort has been made to improve biohydrogen reactor performance, full-scale biohydrogen production has not been developed yet. However, pilot-scale applications have emerged in recent years (Ren *et al.*, 2006).

An anticipated disadvantage of large-scale hydrogen production that needs to be addressed during scale-up is the escape of hydrogen through large plastic enclosures and thin metal sheets that might occur due the high diffusivity of hydrogen. However, as hydrogen becomes more an important fuel in general, full-scale applications will emerge in the future.

### 9.8.2.2   Process Concept

A great number of wastewaters and solid wastes have been studied for dark fermentative hydrogen production at laboratory-scale: Carbohydrate-rich food processing industrial wastewaters, organic fraction of municipal solid waste, sludge from wastewater treatment plants, keratin-containing biowaste, swine manure, cow slurry (Li and Fang, 2007; Zhu *et al.*, 2007; Yokoyama *et al.*, 2007).

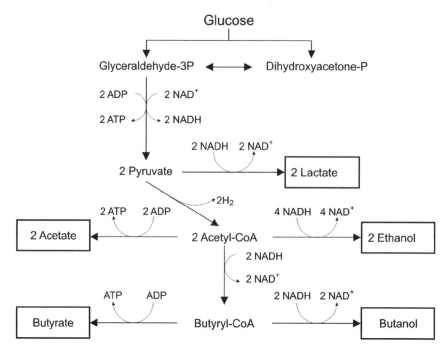

**Figure 9.8.3**   *Schematic pathways for glucose conversion to hydrogen via dark fermentation.*

Fermentative hydrogen production is usually carried out under mesophilic or thermophilic conditions by a wide variety of microorganisms such as strict anaerobes (Clostridia, Ruminococci, Archaea), facultative anaerobes (*Escherichia coli*, *Enterobacter aerogenes*) and aerobes (*Alcaligenes eutrophus*, *Bacillus licheniformis*) when kept under anoxic conditions. Recent attempts for improving biohydrogen production have been made, using hyperthermophilic (70 °C) bacteria and archae (Kanai *et al.*, 2005; Kotsopoulos *et al.*, 2006).

The main biochemical pathways involved in the dark fermentative conversion of the simplest sugar (glucose) to hydrogen and different end products are presented in Figure 9.8.3.

The total hydrogen yields from the different anaerobic glucose degradation pathways are shown in Table 9.8.3. The biochemistry of the anaerobic glucose catabolism shows that the production of propionate, ethanol and lactate is associated

**Table 9.8.3**   *Fermentation end products and hydrogen yields from the main anaerobic glucose degradation pathways.*

| Liquid fermentation end product(s) | Equation for anaerobic glucose degradation pathway | Theoretical hydrogen yield (mol $H_2$/mol glucose) |
|---|---|---|
| Acetic acid $CH_3COOH$ | $C_6H_{12}O_6 + 2H_2O \rightarrow 2CH_3COOH + 4H_2 + 2CO_2$ | 4 |
| Butyric acid $CH_3CH_2CH_2COOH$ | $C_6H_{12}O_6 \rightarrow CH_3CH_2CH_2COOH + 2H_2 + 2CO_2$ | 2 |
| Butyric acid + acetic acid $CH_3CH_2CH_2COOH + CH_3COOH$ | $4C_6H_{12}O_6\ 2H_2O \rightarrow 3CH_3CH_2CH_2COOH + 2CH_3COOH + 10H_2 + 8CO_2$ | 2.5 |
| Ethanol + acetic acid $CH_3CH_2OH + CH_3COOH$ | $C_6H_{12}O_6 + 2H_2O \rightarrow CH_3CH_2OH + CH_3COOH + 2H_2 + 2CO_2$ | 2 |
| Propionic acid $CH_3CH_2COOH$ | $C_6H_{12}O_6 + H_2 \rightarrow 2CH_3CH_2COOH + 2H_2O$ | 0 |
| Ethanol $CH_3CH_2OH$ | $C_6H_{12}O_6 \rightarrow 2CH_3CH_2OH + 2CO_2$ | 0 |
| Lactic acid $C_3H_3O_3$ | $C_6H_{12}O_6 \rightarrow C_3H_3O_3$ | 0 |

with absence of hydrogen generation. In practice, the highest hydrogen yields are associated with the production of acetate, or a mixture of acetate and butyrate, and low hydrogen yields – with the production of butyrate or a mixture of acetate and ethanol as the main liquid fermentation end products. From a thermodynamic perspective, the most favorable products from the breakdown of 1 mol glucose are 2 mol acetates and 4 mol hydrogen. But this stoichiometric theoretical yield is only attainable under near equilibrium conditions, which implies very slow rates and/or at very low partial pressure of hydrogen (Hallenbeck and Benemann, 2002). The practical yield of biologically produced hydrogen is under optimized conditions, approximately 20 % of the electron equivalents in a high-carbohydrate wastewater. The rest of electron equivalents must be utilized in a subsequent step for full recovery of the energy in biomass.

### 9.8.2.3 Challenges and Possibilities

The ultimate goal, and challenge, for fermentative hydrogen research and development are essentially on attaining higher yields of hydrogen. Significant improvement can be expected through rapid gas removal and separation, optimized bioreactor design and genetic modifications in the microorganisms. Nonetheless, it is rather difficult to predict which of the various approaches will ultimately succeed in a substantial enhancement of hydrogen yields so that the dark fermentation process of hydrogen generation becomes commercially competitive.

In order to increase the economical feasibility of the process, recent attempts concentrated on further biological processing (second stage) of the organic acid-rich effluents from dark fermentation hydrogen production. Two possible second stages (Figure 9.8.4) were investigated: photoheterotrophic hydrogen production and methanogenesis (Hawkes *et al.*, 2007; Liu *et al.*, 2006). If technologically and cost-effective photobioreactors were available, the two-stage process combining dark and light-driven hydrogen fermentation would be a very promising method as it has a theoretical maximal molar yield of 12 mol $H_2$/mol hexose converted in the two-stage process (Hawkes *et al.*, 2007). However, some studies indicate that photofermentation is a very inefficient and expensive process with respect to high energy demands for light sources and a requirement for elaborate photobioreactors covering large areas (Hallenbeck and Benemann, 2002). There are good indications that a two-stage process with an acidifying hydrogen-producing first stage and a methanogenic second stage gives rise to more efficient waste treatment and energy recovery than a single-stage methanogenic process (Liu *et al.*, 2006). This process could easily be implemented in existing and new biogas plants, where an additional reactor could be added before the traditional biogas reactor for the production of biohydrogen.

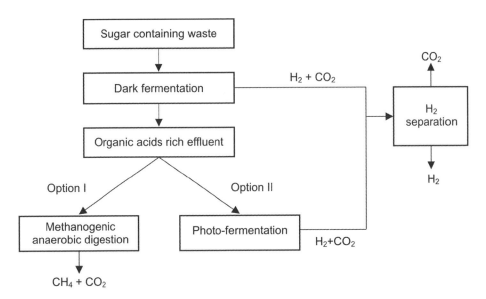

***Figure 9.8.4*** *Possible treatments of the effluent from dark fermentative biohydrogen production.*

It is likely that the food industry and kitchen wastes will prove initially most attractive as substrates. Reactor capital and operating costs are likely to be similar to those already well known for anaerobic digestion. The next major challenge is to determine whether the economics and reliability of dark fermentative hydrogen production are sufficiently attractive for commercial application to be installed.

### 9.8.3   Biofuels: Biodiesel

#### 9.8.3.1   State of the Technology

Biodiesel is defined as the monoalkyl esters of long-chain fatty acids derived from vegetable oils or animal fats, for use in compression/ignition (diesel) engines. Commercial biodiesel production started in 1991 with plants constructed to handle up to 100 000 t/year (Li *et al.*, 2007). The process technology is well understood and established although there are some variants in the technologies used (Marchetti *et al.*, 2007).

Biodiesel has received considerable attention in recent years as a renewable, biodegradable, and nontoxic fuel. Biodiesel production in the European Union (EU) has increased rapidly in recent years, reaching an energy content of 120 PJ in 2005 (EEA, 2007).

Considerable research has been done on producing biodiesel from vegetable oils: Palm oil, soybean oil, sunflower oil, coconut oil and rapeseed oil. Animal fats, although mentioned frequently, have not been studied to the same extent as vegetable oils. Some methods applicable to vegetable oils are not applicable to animal fats because of natural property differences. Oil from bacteria, fungi and algae also has been investigated (Li *et al.*, 2007).

#### 9.8.3.2   Process Concept

Normally, triglycerides present in extracted crude oil are converted into esters through *trans*-esterification (Figure 9.8.5) usually in the presence of a catalyst. Almost all biodiesel is produced using the base-catalyzed technique as it is the most economical process, requiring only low temperatures and pressures and producing over 98 % conversion yield (biofuel production data). Toxicity and/or loss of catalysts and the relatively large amount of water necessary to clean both glycerol and biodiesel product, which represents a polluting burden to the environment, are considered as the main factors responsible for increasing the biodiesel production costs.

#### 9.8.3.3   Challenges and Possibilities

Development of new, clean and effective *trans*-esterification process is a key issue for economically feasible biodiesel production. The main challenge is with respect to the limited availability of fat and oil resources and the relatively high vegetable oil prices. From this point of view, biodiesel can be used most effectively as a supplement to other energy forms, not as the primary source. Biodiesel is particularly useful in mining and marine situations where lower pollution levels are important.

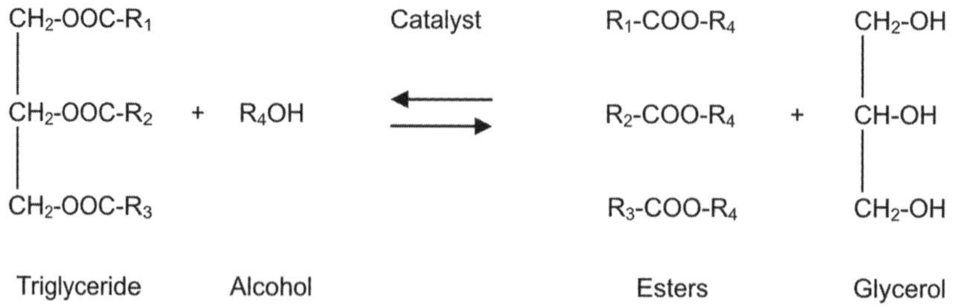

| | | | Catalyst | $R_1$-COO-$R_4$ | | $CH_2$-OH |
|---|---|---|---|---|---|---|
| $CH_2$-OOC-$R_1$ | | | | | | |
| $CH_2$-OOC-$R_2$ | + | $R_4$OH | ⇌ | $R_2$-COO-$R_4$ | + | CH-OH |
| $CH_2$-OOC-$R_3$ | | | | $R_3$-COO-$R_4$ | | $CH_2$-OH |
| Triglyceride | Alcohol | | | Esters | | Glycerol |

***Figure 9.8.5***   *Biodiesel production by trans-esterification of triglycerides with alcohol. Catalysts used in trans-esterification are classified as acid, base or enzyme.*

## 9.8.4   Biochemicals: Butanol, Longer Chain Alcohols and Ketones

### 9.8.4.1   State of the Technology

Acetone butanol ethanol (ABE) fermentation is a process in which the carbohydrate substrate is converted to a mixture of solvents: Acetone, butanol and ethanol, in the approximate ratio 3:6:1, at a total solvent concentration of around 20 g/l (Ezeji *et al.*, 2004). The end fermentation products, acetone, butanol and ethanol, are major basic commodity chemicals consumed in bulk in a variety of ways, e.g. as fuels, fuel additives and solvents. Butanol could be particularly useful; it has a higher octane rating than gasoline, making it a valuable fuel for any internal combustion engine made for burning gasoline. Because it is less hygroscopic, butanol can be shipped through the existing common-carrier pipelines, unlike ethanol.

Before the 1950s, ABE fermentation ranked second to ethanol in its importance and scale of production; but it declined due to increasing substrate costs and the availability of the much cheaper, petrochemically derived butanol. Currently, the ABE fermentation process is operated commercially at full-scale only in China (Karakashev *et al.*, 2007).

Although sugar conversion to solvents was reported to be over 95 % (Ezeji *et al*, 2004) there are three major factors that hamper the economic viability of ABE fermentation: (1) the high cost of the substrates used (e.g. molasses), (2) the low product concentration (due to solvent toxicity) and (3) the high product recovery costs (distillation has been used). Interest in this process was revived after the oil crisis in the 1970s because of the potential utilization of renewable resources (cane molasses, corn, wood hydrolysate, etc). Byproducts from the dairy industry (i.e. whey) and agricultural byproducts seem to be the most promising substrates. Many studies reported that acid and/or enzymatic hydrolysates of cellulosic materials from a variety of biomass sources are potential feedstock for ABE fermentation (Willke and Vorlop, 2004).

### 9.8.4.2   Process Concept

ABE fermentation is mediated by anaerobic saccharolytic microorganism known as solventogenic clostridia. The best known groups are the mesophiles *Clostridium acetobutylicum* and *C. beijerinckii*. During the exponential growth of *Clostridium acetobutylicum* at pH values greater than about 5.6, the major fermentation products from glucose are acetate, butyrate, hydrogen and carbon dioxide (Claasen *et al.*, 1999). The accumulation of organic acids during batch fermentation is responsible for gradual growth inhibition. Solventogenesis requires the induction of a new set of enzymes catalyzing the formation of ABE from glucose and reassimilated organic acids (Figure 9.8.6). The shift to solventogenesis is induced by high intracellular concentrations of acids, low pH, the presence of growth-limiting factors (such as phosphate or sulfate depletion) and high concentrations of glucose and nitrogen compounds.

### 9.8.4.3   Challenges and Possibilities

A promising approach to improve ABE fermentation is the development of genetically modified clostridial strains with increased solvent production due to reutilization of carboxylic acids accumulated during the acidogenic phase of carbohydrate degradation (Qureshi and Blaschek, 2001). Modification in solvent production in genetically manipulated strains of *Clostridium acetobutylicum* ATCC 824 due to induced depression of the solventogenic genes has also been reported (Nair *et al.*, 1999). In contrast to the over-expression of the solventogenic genes, the earlier induction of those genes (deregulated solvent synthesis) resulted in highest solvent production and butanol tolerance reported till now. This strategy appears to be the most promising biotechnological approach for strain improvement in future commercial applications of ABE fermentation.

## 9.8.5   Biochemicals: Biodegradable Plastics

### 9.8.5.1   State of the Technology

Biodegradable plastics are plastics that decompose in the natural environment. Biodegradation of those plastics can be achieved by enabling microorganisms in the environment to metabolize the molecular structure of plastic films to produce an inert humus-like material that is less harmful to the environment. Biodegradable plastics have an extensive range of

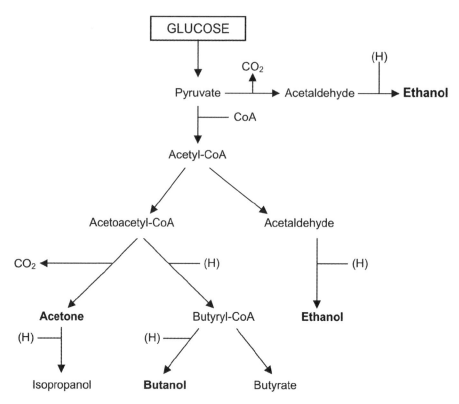

***Figure 9.8.6*** *Overall biochemical scheme of anaerobic breakdown of glucose – metabolic pathways to ethanol, acetone and butanol are shown (modified from Lens et al., 2005). Liquid organic chemicals are presented in bold.*

applications, such as shopping bags, silage wrap, flushable sanitary products, sheet and non woven packaging, bottles, planter boxes and fishing nets, food service cups, cutlery, trays, loose fill foam, etc.

Over the past few years there has been a significant expansion in the range of biodegradable plastics which are commercially available or are under commercial development. The range of biodegradable plastics available include: starch based products, naturally produced polyesters including poly(3-hydroxybutyric acid) (PHB), synthetic aliphatic polyesters, water-soluble polymers such as polyvinyl alcohol and ethylene vinyl alcohol, photo-biodegradable plastics and aliphatic–aromatic copolyesters (Kamm *et al.*, 2006). The most widely studied completely biodegradable plastic is PHB. PHB is a biocompatible, thermoplastic, hydrophobic and stereospecific material belonging to the polyhydroxyalkanoates (PHA) family. This is a material with a unique characteristic among thermoplastic because it presents a complete cycle starting from a waste material and bacterial fermentative synthesis (Kamm *et al.*, 2006). Organic waste materials tested as feed stocks for PHB production include styrene, a toxic byproduct of the polystyrene industry, palm oil mill effluent and wastewater sludge.

### 9.8.5.2   Process Concept

Currently PHB is produced in an aerobic fermentation process in which the sugar carbon source is converted into biopolymer by the microorganism *Ralstonia eutropha* (Kamm *et al.*, 2006). This microorganism accumulates PHB more than other microorganisms studied, such as *Protomonas extraquens, Candida utilis, Azetobacter vinelandii, Synechoccus* sp. and *Rhizobium* sp. (Mercan, 2002; Miyake *et al.*, 2000; Shahhosseini, 2004). The main substrate, sugar, is metabolized to PHB, which accumulates in microbial cells. The growth phase is followed by the polymer biosynthesis stage. At the end of fermentation, the microorganisms are inactivated to prevent polymer loss, because starved microorganisms use PHB as a carbon source in the absence of reducing sugars. The final culture is then ready for the polymer recovery stage including an extraction and purification process.

### 9.8.5.3   Challenges and Possibilities

Currently the main problem which limits the widespread use of PHB is its relatively high cost compared with polypropylene (Shahhosseini, 2004). In the fermentation process, substrates and product recovery are the major costs. However the expectations of the development of industrial PHB production are high. This technology will undergo significant improvements when novel microorganisms are obtained by selection or by genetic engineering.

# References

Claasen, P.A.M., van Lier, J.B., Lopez Contreras, A.M., van Niel, E.W.J., Sijtsma, L., Stams, A.J.M., de Vries, S.S. and Weusthuis, R.A. (1999): Utilisation of biomass for the supply of energy carriers. *Applied Microbiology and Biotechnology*, 52, 741–755.

EEA (2007): Transport and environment: on the way to a new common transport policy. TERM 2006: indicators tracking transport and environment in the European Union. EEA Report No. 1/2007. ISSN 1725-9177. European Environment Agency, Copenhagen, Denmark.

Ezeji, T.C., Qureshi, N. and Blaschek, H.P. (2004): Acetone butanol ethanol (ABE) production from concentrated substrate: reduction in substrate inhibition by fed-batch technique and product inhibition by gas stripping. *Applied Microbiology and Biotechnology*, 63, 653–658.

Hagerdal, H.H., Karhumaa, K., Fonseca, C., Spencer-Martins, I. and Gorwa-Grauslund, M.F. (2007): Towards industrial pentose-fermenting yeast strains. *Applied Microbiology and Biotechnology*, 74, 937–953.

Hallenbeck, P.C. and Benemann, J.R. (2002): Biological hydrogen production; fundamentals and limiting processes. *International Journal of Hydrogen Energy*, 27, 1185–1193.

Hawkes, F.R., Hussy, I., Kyazze, G., Dinsdale, R. and Hawkes, D.L. (2007): Continuous dark fermentative hydrogen production by mesophilic microflora: Principles and progress. *International Journal of Hydrogen Energy*, 32, 172–184.

Kamm, B., Gruber, P.R. and Kamm, M. (eds) (2006): *Biorefineries – Industrial Processes and Products. Volume 1. Status Quo and Future Directions*. Wiley-VCH, Weinheim, Germany.

Kanai, T., Imanaka, H., Nakajima, A., Uwamori, K., Omori, Y., Fukui, T., Atomi, H. and Imanaka, T. (2005): Continuous hydrogen production by the hyperthermophilic archaeon, *Thermococcus kodakaraensis* KOD1. *Journal of Biotechnology*, 116, 271–282.

Kapdan, I.K. and Kargi, F. (2006): Bio-hydrogen production from waste materials. *Enzyme and Microbial Technology*, 38, 569–582.

Karakashev, D., Thomsen, A.B. and Angelidaki, I. (2007): Anaerobic biotechnological approaches for production of liquid energy carriers from biomass. *Biotechnology Letters*, 29, 1005–1012.

Kotsopoulos, T.A., Raymond, J.Z. and Angelidaki, I. (2006): Biohydrogen production in granular up-flow anaerobic sludge blanket (UASB) reactors with mixed cultures under hyper-thermophilic temperature (70 °C). *Biotechnology and Bioengineering*, 94, 296–302.

Lens, P., Westermann, P., Haberbauer, M. and Moreno, A. (eds) (2005): *Biofuels for Fuel Cells. Renewable energy from biomass fermentation*. IWA Publishing, London, UK.

Li, C. and Fang, H.H.P. (2007): Fermentative hydrogen production from wastewater and solid wastes by mixed cultures. Critical Reviews in *Environmental Science and Technology*, 37, 1–39.

Li, X., Xu, H. and Wu, Q. (2007): Large-scale biodiesel production from microalga *Chlorella protothecoides* through heterotrophic cultivation in bioreactors. *Biotechnology and Bioengineering*, 98, 764–771.

Liu, D., Liu, D., Zheng, R. and Angelidaki, I. (2006): Hydrogen and methane production from household solid waste in the two-stage fermentation process. *Water Research*, 40, 2230–2236.

Marchetti, J.M., Miguel, V.U. and Errazu, A.F. (2007): Possible methods for biodiesel production. *Renewable and Sustainable Energy Reviews*, 11, 1300–1311.

Mercan, N. (2002): Production of poly-$\beta$-hydroxybutyrate (PHB) by some *Rhizobium* Bacteria. *Turkish Journal of Biology*, 26, 215–219.

Miyake, M., Miyamoto, C., Schnackenberg, J., Kurane, R. and Asada, Y. (2000): Phosphotransacetylase as a key factor in biological production of polyhydroxybutyrate. *Applied Biochemestry and Biotechnology*, 84/86, 1039–1044.

Mojovic, L., Nikolic, S., Rakin, M. and Vukasinovic, M. (2006): Production of bioethanol from corn meal hydrolyzates. *Fuel*, 85, 1750–1755.

Möller, R. (2006): *Cell wall saccharification. Epobio. Realising the Economic Potential of Sustainable resources – Bioproducts from Non-Food Crops*. CPL Press, Newbury, UK.

Nair, R.V., Green, E.M., Watson, D.E., Bennet, G.N. and Papoutsakis, E.T. (1999): Regulation of the sol locus genes for butanol and acetone formation in Clostridium acetobutylicum ATCC 824 by a putative transcriptional repressor. *Journal of Bacteriology*, 181, 319–330.

Qureshi, N. and Blaschek, H.P. (2001): Recent advances in ABE fermentation: hyper-butanol producing Clostridium beijerinckii BA101. *Journal of Industrial Microbiology and Biotechnology*, 27, 287–291.

Ren, N., Li, J., Li, B., Wang, Y. and Liu, S. (2006): Biohydrogen production from molasses by anaerobic fermentation with a pilot-scale bioreactor system. *International Journal of Hydrogen Energy*, 31, 2147–2157.

Shahhosseini, S. (2004): Simulation and optimization of PHB production in fed-batch culture of Ralstonia eutropha. *Process Biochemistry*, 39, 963–969.

Willke, T. and Vorlop, K-D. (2004): Industrial bioconversion of renewable resources as an alternative to conventional chemistry. *Applied Microbiology Biotechnology*, 66, 131–142.

Yokoyama, H., Waki, M., Moriya, N., Yasuda, T., Tanaka, Y. and Naga, K. (2007): Effect of fermentation temperature on hydrogen production from cow waste slurry by using anaerobic microflora within the slurry. *Applied Microbiology and Biotechnology*, 74, 474–483.

Zhu, J., Wu, X., Miller, C., Yu, F., Chen, P. and Ruan, R. (2007): Biohydrogen production through fermentation using liquid swine manure as substrate. *Journal of Environmental Science and Health, Part B*, 42, 393–401.

# 9.9

# Use of Compost in Horticulture and Landscaping

**Morten Carlsbæk**

*Solum Gruppen, Denmark*

The agricultural sector is potentially, in terms of volume, the largest user of biologically treated waste. There are, however, many other uses for compost besides those in the agricultural sector and these other markets are often prepared to pay a higher price per cubic meter of compost (see Table 9.9.1). Because these sectors require more from compost than merely a high level of plant nutrients, the compost aimed at these markets must be more refined and have higher levels of product documentation. Non-agricultural applications for compost include soil improving and mulching in professional landscaping, the private garden sector and specialized horticultural sectors such as fruit and vine growing, as well as a component in manufactured topsoils for landscaping or in growing media for greenhouses and nurseries.

In the European Union (EU), about 40 % of the compost produced is used in agriculture while 60 % is used for a variety of other purposes (Table 9.9.2).

## 9.9.1 Why Use Compost Products in the Nonagricultural Sector?

There is a large potential for the use of compost in landscaping and private gardening because by applying compost, a number of problem areas common to these sectors can be addressed. Landscape gardeners often have to deal with poor soil that does not promote plant growth. Years without plant cover can lead to very low levels of humus in the topsoil on the site, and unintended mixing of soil with sand from construction activities may be an additional cause. Also, the level of microbiological activity in the topsoil may be less than optimal, for example, caused by long-term storage of the topsoil in large heaps prior to use or by pesticide residues accumulated from many years of applying all-purpose herbicides to the soil. Compost addresses these issues as a soil amendment.

Within urban areas, the demand for topsoils and mulches (dressings) is substantial, particularly for use in infrastructure construction projects, such as the landscaping of newly vegetated areas and road verges. Landscaped areas are often very highly valued areas. They are often high-profile sites occupying important locations and are in many cases subjected to intensive usage and exposed to heavy wear. Initial costs for plant procurement are high. Often, a wide variety of plant species is required and these plants can be difficult to grow under the local climate conditions without optimal soil

*Solid Waste Technology & Management*   Edited by Thomas Christensen
© 2011 Blackwell Publishing Ltd

**Table 9.9.1**   *Volume and value markets for composts and products with compost (modified after Tyler, 1996; Alexander, 2003). European price levels (Carlsbæk and Brøgger, 2002; the price levels still apply in year 2009).*

| ← Increasing market volume.............Increasing product value → | | | | |
|---|---|---|---|---|
| Agriculture – conventional | Soil reclamation Vine and fruit mulch Organic farming Sod farms | Private gardening Erosion control Roadside projects Re-sellers | Landscaping – soil improvers Manufactured topsoils for general landscaping | Sport turf growing media and topdressing Nurseries Growing media for pots and containers |
| € 0–3/m³ | € 0–6/m³ | € 5–20/m³ | € 10–20/m³ | € 10–40/m³ |

conditions. The value of the site makes it economically feasible (and necessary) to optimize plant-growing conditions in order to meet the site's design and utility requirements. Compost low in nutrients, such as yard compost, is very well suited for use in the landscaping sector as mulch or as a component in manufactured soils.

A final benefit of utilizing biologically treated waste in landscaping is the reduction in the labor intensiveness that is a common feature of landscaping. Compost is known for improving the workability of loam and clay soils making easier the preparation of a sowing bed as well as planting in the topsoil. While normal topsoil contains a large number of seeds, it is possible to produce weedfree topsoils by using compost and other constituents. Another benefit is the fact that manufactured topsoils containing a high percentage of sand can be handled easily in rainy periods (which are good times for planting), whereas the common loamy topsoils are very difficult to handle when wet. All of this adds up to possible savings in labor during construction and reduced levels of maintenance afterwards – both of which are important factors to consider in such projects. A summary of these benefits can be seen in Table 9.9.3.

## 9.9.2   Curing, Refinement and Storage of Compost for the Market

When biologically treated waste is to be used in nonagricultural sectors, there is an increased demand for comprehensive product refinement and extended customer service. Landscape gardeners require a wide variety of different products for soil improvement, topsoil replacement and mulching (such as under newly planted shrubbery). In general, only purpose-specific products that have a uniform and consistent high quality from delivery to delivery can penetrate these nonagricultural markets. Curing, refinement and storage of compost products require plenty of space at the facilities producing compost for landscaping and horticulture.

### 9.9.2.1   Curing

Most nonagricultural sectors require stable (mature) products, which can be used in higher quantities and be worked into the ground without causing anaerobic conditions or hampering root development. The ongoing degradation of the compost in the soil must provide a supply of ammonium or better nitrate to the plant roots (nitrate is often regarded as a sign of compost stability). If uncured compost is used, the continued degradation of the compost may even use nitrogen from the soil and thereby reduce the amount of nitrogen originally available in the soil for the plants. A well cured product is mandatory when compost is used in the nonagricultural sector in order to provide good growing conditions for the

**Table 9.9.2**   *European compost market shares by outlet by 2003 (% of total yearly marketed amount). Modified after Hogg et al. (2002) and Barth (2004).*

|  | EU average (approx.) | Individual EU countries |
|---|---|---|
| Agriculture and field horticulture | 40 | 10–70 |
| Landscaping, reclamation, manufactured top soils | 30 | 20–60 |
| Residential/private gardens | 20 | 10–50 |
| Others (greenhouses, nurseries, landfill cover, etc.) | 10 | 5–20 |

***Table 9.9.3*** *Landscaping conditions and benefits from compost utilization.*

|  |  | Site conditions | Compost benefits[a] |
|---|---|---|---|
| Soil | | Low content of humus<br>Low microbial activity<br>Pesticide residues | Supplies organic matter, beneficial microorganisms and nutrients, and has a liming effect. This leads to:<br>• Increased moisture infiltration and permeability of heavy soils – improved drainage and reducing erosion and runoff.<br>• Improved moisture holding capacity of light soils – reduced water loss and nutrient leaching.<br>Binds or degrades specific pollutants. |
| Labor | | Topsoil hauling and leveling<br>Planting<br>Weeding, cutting/pruning, mowing | Manufactured topsoils easier and quicker to handle – improved workability.<br>Manufactured topsoils free of weeds and with slow-release fertilizer – reduced maintenance. |
| Plant and Design | | High-value sites (amenity)<br>Heavy wear/use<br>Expensive plants<br>Foreign climate plants | Improved plant establishment and possible lower need for replanting, because:<br>• Improved soil structure creates a better plant root environment.<br>• The improved soil microbial activity leads to a better plant nutrient uptake and suppresses certain soil-borne diseases. |

[a]In parts modified after Alexander (2003).

plants and in order to avoid any odor problems during its use. Fresh compost from kitchen waste (including structure material) is cured for 2–10 months before being used in general landscaping or as a component in manufactured soil soils or growing media, while fresh compost from sewage sludge (including structure material) is cured for 1–6 months. Cured compost from pure mixed yard waste is produced by windrow composting in 3–12 months often without any clear distinction between the active composting phase and curing. The actual curing time depends on waste mix composition and applied composting technology as well as the foreseen area of application within the landscape and horticultural sectors. The stability (biodegradability) of the compost product should be declared – even though there may be some disagreement as to which analytical methods are best suited for measuring stability. It is often recommended to use two methods when measuring the stability.

The solid digestate from anaerobic digestion has low stability in an aerobic environment and may potentially cause odor problems as well as oxygen depletion in the root zone. The digestate may also have a high ammonia-N content, which may scorch young leaves or roots. Because of these factors, anaerobic digestate is less suited for use in nonagricultural market unless it has been composted after digestion (Carlsbæk and Brøgger, 2002; Fischer and Schmitz, 2004; BGK, 2005).

## 9.9.2.2 Refinement and Mixing

The optimization of internal procedures for product refinement is crucial to keep production costs down. Commonly used screen hole sizes are 10–25 mm for screening of the cured compost and 3–10 mm for blended products. A screen with larger holes (up to 75 mm) may be used for producing mulch or for final mixing of growth media that has peat as its main component.

Blending manufactured topsoils is a two-step process that begins with premixing. The mix is layered and piled up by emptying alternate loads of the constituents (adhering to prescribed ratios). The layered pile is then roughly mixed with the front loader, e.g. by scooping two smaller piles into one larger pile. This roughly mixed pile is screened in a trommel screen. It is completely mixed once the fines slip down the cone-formed heap underneath the conveyer belt. The rotational speed of the trommel and the speed of feeding the premixed material with the front loader must correspond. This avoids the risk of having an empty screen, which would lead to increased levels of coarse material in the final blend, and also prevents the discharge of too much fines into the oversized material which would occur if the trommel became too full. Pictures of the blending process when producing manufactured topsoils are shown in Figure 9.9.1.

**Figure 9.9.1**    *Manufacturing of topsoils for general landscaping.*

When compost, which has already been coarsely screened, is one of the components in a blend, the oversized material from this production may be used in coarser products. For some manufactured topsoils and compost types, the initial screening of the cured compost can coincide with the actual mixing of the topsoil by the trommel screen. In these cases, oversized material may be returned to the incoming waste and used as structural material during the composting process.

Manufactured topsoils are often produced from sand and compost only. But different kinds of sands and composts and different mixing ratios will result in topsoils with very different characteristics regarding water holding capacity, air content, percolation rate (hydraulic conductivity), pH, salinity (electrical conductivity) and nutrient content. Clay or loamy subsoils, peat moss or sedge peat may be added to increase the water-holding capacity of the blend. Softwood/coniferous bark (< 10 % cellulose), perlite, expanded clay or something similar can be added to achieve higher air content in the blend. Hardwood bark or sawdust, shaving and woodchips in general cannot be used without prior composting due to the content of cellulose, an easily degradable compound (US Composting Council, 1996, 2003). Natural topsoil hauled to the composting site from a construction site can be the main constituent of some blends when a weedfree blend is not crucial. The workability of the hauled natural topsoils can be improved by adding compost to the trommel screen when the topsoil is screened to remove stones and other coarse fractions. In addition, the compost supplies nutrients and microbial activity, and as a result, increases the value of the screened topsoil.

Some mixing components (such as certain sands and peat) are expensive and a paved area with retaining walls will facilitate mixing, reduce spillage and minimize possible contamination with stones, etc. from the ground.

### 9.9.2.3 Storage

It is only possible to offer the required variety of individual products to landscapers and gardeners by producing compost in large batches. Stockpiles of all products should be maintained at levels capable of meeting most orders at any time. This places considerable requirements upon having sufficient storage capacity.

Product storage under a roof and with a windscreen may be important for some of the more expensive refined products.

## 9.9.3 Declaration, Guidelines and Marketing

In the nonagricultural sector, a variety of compost products is used in a number of different ways. Declaration of each compost product as well as good guidance regarding its use is mandatory for a successful application as well as for marketing of the product as part of a successful business.

### 9.9.3.1 Declaration

The legal requirements concerning compost quality, declaration and use differ substantially from country to country, depending on the individual country's environmental and agricultural focus. An extensive comparison of compost standards is presented in Hogg *et al.* (2002). In the EU, common European standards are being developed for characterizing treated biowaste and greenwaste (such as compost), sludge and soil. The landscape and horticulture sectors in Europe are asking for common standards – particularly with regards to the parameters impurities, weeds, stability (biodegradability) and phytotoxicity (growth tests) – because these parameters are part of the basis of the current EU eco-label for soil improvers and growing media, which can be awarded to waste-derived products (EC, 2001). Today, different national analysis methods or different methods developed by the industry itself are used to document observance of the criteria in the European eco-label (e.g. BGK 1994). European standards for soil improvers and growing media have been issued for many of the traditionally required parameters like soluble nutrients, electrical conductivity, etc. (CEN, 1999). In the USA the methods in US Composting Council (2002) are commonly accepted.

Traditionally, the different user groups have different requirements and needs regarding compost quality and declaration. Private garden owners require easy to understand user guidelines, but are most likely unable to understand analytical data provided as part of the declaration. However, analytical data are exactly what professional growers usually require. They also request specific and detailed guidelines on the use of the product. Table 9.9.4 shows the frequently declared parameters for compost and products containing compost for use in landscaping, gardening and horticulture. Table 9.9.5

***Table 9.9.4*** *Declared parameters for compost and products with compost used in the landscaping, gardening and horticultural markets.*

| Key:<br>XXX: normally requested<br>YY: often requested<br>I: occasionally requested<br>—: rarely requested | Compost, general use | Mulch compost | Manufactured topsoil, landscaping | Sports turf products: top dressing, growing media | Pots and containers: growing media |
|---|---|---|---|---|---|
| **Input material, user guidelines, etc.** | | | | | |
| Type of facility (windrow/in vessel) and treatment (temperature, turnings) | YY | I | I | I | I |
| Feedstock materials and additives, components in blend | XXX | XXX | XXX | XXX | XXX |
| User guidelines | XXX | YY | YY | XXX | YY |
| Certifications/awards | YY | I | — | I | I |
| **Nutrients, etc.** | | | | | |
| Macro-nutrients (NPK) | XXX | YY | XXX | XXX | XXX |
| Secondary nutrients (Mg, S, Ca) | YY | I | YY | XXX | XXX |
| Trace nutrients | I | — | I | YY | XXX |
| Liming effect ($CaCO_3$) | YY | — | — | YY | — |
| Electrical conductivity (salinity) | XXX | I | YY | XXX | XXX |
| pH | XXX | I | XXX | XXX | XXX |
| Organic matter | XXX | I | YY | XXX | YY |
| **Main physical parameters** | | | | | |
| Screen (mesh hole size) | XXX | XXX | YY | XXX | YY |
| Dry matter | XXX | — | — | I | YY |
| Bulk density | XXX | YY | XXX | XXX | YY |
| Fine particles (< 5 mm or < 10 mm) | I | YY | — | — | I |
| Stones > 5 mm | YY | I | — | XXX | YY |
| **Specific physical analyses** | | | | | |
| Soil texture classification | — | — | XXX | XXX | — |
| Particle size analysis, grading curve | — | — | — | XXX | — |
| Water retention and porosity | — | — | — | YY | YY |
| Percolation rate | — | — | — | YY | — |
| **Guarantee parameters** | | | | | |
| Stability/biodegradability | YY | — | — | — | I |
| Impurities > 2 mm (plastic, glass, metal) | XXX | I | — | YY | XXX |
| Weeds (viable seeds) | YY | — | I | I | XXX |
| Growth tests, references | I | — | YY | YY | XXX |
| Na, Cl | I | — | I | I | XXX |
| Heavy metals | XXX | YY | I | — | — |
| *E. coli, Salmonella* | YY | — | — | — | YY |

shows the characteristic data provided in a declaration of compost for general landscaping. In case the compost is needed for a specific use, it can be necessary to extend the list of declaration parameters.

### 9.9.3.2 Guidelines for Compost Use

In the nonagricultural sector, compost is normally used in large amounts per $m^2$ over a short period of time. Specific usage guidelines are important to ensure that a particular compost is used correctly. All too frequently, good compost is used incorrectly resulting in unsatisfactory plant growth (Carlsbæk and Reeh, 1997).

**Table 9.9.5**  *Characteristic data provided in a declaration of yard waste compost or kitchen waste compost when used for general landscaping.*[a]

| Parameter | Unit | Pure mixed yard waste compost | | Kitchen waste compost, including structure material | |
|---|---|---|---|---|---|
| | | Typical value | Common range | Typical value | Common range |
| Nutrients etc. | | | | | |
| N-total | kg/t | 4.3 | 3.0–6.0 | 11 | 8.0–17.0 |
| NH$_4$-N | kg/t | 0.05 | 0.0–0.4 | 1.2 | 0.1–2.2 |
| NO$_3$-N | kg/t | 0.08 | 0.0–0.1 | 0.1 | 0–0.3 |
| P-total | kg/t | 1.0 | 0.7–1.4 | 2.1 | 1.9–3.3 |
| P-soluble (neutral ammonium citrate) | kg/t | 0.7 | 0.5–1.1 | 1.4 | 1.1–1.9 |
| K-total | kg/t | 2.5 | 1.0–6.0 | 5.3 | 4.1–6.6 |
| Mg-total | kg/t | 0.9 | 0.4–1.7 | 1.5 | 1.2–1.9 |
| S-total | kg/t | 0.6 | 0.4–1.1 | 1.4 | 1.2–1.5 |
| Liming effect (Scheibler) | CaCO$_3$/t | 10 | 5–25 | 20 | 15–25 |
| EC (1 substrate:5 water, v/v)[b] | mS/m | 110 | 50–200 | 260 | 150–400 |
| pH | — | 8.0 | 7.5–8.4 | 8.0 | 7.2–8.5 |
| Organic matter | % of dry matter | 25 | 20–35 | 50 | 35–75 |
| Physical parameters | | | | | |
| Screen (mesh hole size) | mm | 15 | 8–30 | 10 | 8–12 |
| Dry matter | % | 60 | 50–75 | 60 | 40–75 |
| Bulk density | t/m$^3$ | 0.7 | 0.6–0.8 | 0.5 | 0.4–0.6 |
| Fine particles < 5 mm | % of dry matter | 91 | 87–94 | 90 | 83–98 |
| Stones > 5 mm | % of dry matter | 6 | 2–11 | 3 | 1–6 |
| Stability tests | | | | | |
| Oxygen demand in 4 days (Sapromat or OxyTop) | mg O$_2$/g dry organic matter | 12 | 3–17 | 23 | 18–27 |
| Self heating (in a Dewar vessel) | Max. temp. (°C) | 24 | 21–27 | 37 | 30–44 |
| Solvita compost kit | Color no. on scale | 5 | 4–6 | 3 | 1–4 |
| Resulting degree of stability | — | Stable | Fresh – very stable | Fresh | Not ready – stable |
| Guarantee parameters | | | | | |
| Impurities > 2 mm | % of dry matter | 0.1 | 0.0–0.5 | 0.4 | 0.2–1.0 |
| Weeds | Number/liter | 0 | 0–5 | 0 | 0–1 |
| Cd | mg/kg dry matter | 0.36 | 0.22–0.79 | 0.46 | 0.36–0.70 |
| Pb | mg/kg dry matter | 30 | 13–70 | 30 | 15–50 |
| Hg | mg/kg dry matter | 0.12 | 0.01–0.45 | 0.07 | 0.02–0.30 |
| Ni | mg/kg dry matter | 7.0 | 3.0–14 | 12 | 6.0–14.0 |
| Cr | mg/kg dry matter | 9.0 | 5.6–20.0 | 16 | 9–26 |
| Zn | mg/kg dry matter | 140 | 85–210 | 170 | 115–290 |
| Cu | mg/kg dry matter | 50 | 21–85 | 80 | 55–150 |

[a] Modified after background data from the Danish standardized product sheet for compost. The product sheet without data, including calculation of degree of stability, is presented in *Nation-specific supplement 3: Denmark* (Hogg *et al.*, 2002).
[b] Data modified after Devliegher (2005).

Compost is added in large amounts in order to achieve rapid soil improvement. Thanks to the compost's high content of organic material, its varied microbial life and its wealth of different plant nutrients, soil amended with compost quickly becomes better suited to plant growth. Once compost has been mixed into the soil, the resulting content of dissolved plant nutrients and other salts (measured as electrical conductivity EC; see Box 9.9.1) must not be too excessive as this can damage plant roots by scorching. Because of this, 'nutrient-poor' compost types with low levels of soluble nutrients, such

**Box 9.9.1   Electrical Conductivity (Salinity) in Compost Products.**

Electrical conductivity (EC) is a key parameter in characterizing the quality of compost and products with compost. Choice of analytical method for EC determination and the expression of the results should be according to the specific costumers' tradition as well as using more common (international) standards.

The SI unit for EC is S/m (S for Siemen), replacing the old unit mmho, both units having the same value. Commonly used prefixes and their relationships are:

$$1 \text{ mmho/cm} = 10^{-1} \text{ S/m} = 1 \text{ mS/cm} = 1 \text{ dS/m} = 10^2 \text{ mS/m} = 10^3 \text{ } \mu\text{S/cm}.$$

EC values are only fully understandable and possible comparable to others when accompanied by information about unit, extraction (solution, ratio and volume or weight) and a method reference. CEN (1999) and many other EC methods use volume based mixing ratio with dried ($< 75\,^{\circ}$C) crushed substrate $< 2$ mm (stones sorted out) and water.

*Typical EC values in composts (mS/m, substrate:water as 1:5, v/v; CEN, 1999). Modified after Devliegher (2005).*

|  | Typical value | Common range |
| --- | --- | --- |
| Pure mixed yard waste (grass, twigs etc.) compost | 110 | 50–200 |
| Kitchen waste compost, including structure material | 260 | 150–400 |

*General recommendations for EC in growing media and topsoils (mS/m, substrate:water as 1:5, v/v). Modified after Bunt (1988). Specific recommendations must consider soil type, climate conditions, plant species etc.*

|  | Optimum values | Limit values |
| --- | --- | --- |
| Suitable for seeds, young seedlings, salt-sensitive species | 10–20 | 40 |
| Satisfactory for most plants | 30–60 | 80 |
| Suitable for vigorous plants, salt-tolerant species | 50–80 | 120 |
| Reduced growth for most plants | — | > 120 (severe injury common > 160) |

*Cross-reference factors for different EC methods using different extraction ratios – substrate:water, v/v (McLachlan et al., 2004).*

| x = actual used method | y = unknown | | | | |
| --- | --- | --- | --- | --- | --- |
|  | $EC_e$[a] | 2:1 | 1:1 | 1:2 | 1:5[b] |
| $EC_e$[a] | — | 0.8 | 0.3 | 0.2 | 0.1 |
| 2:1 | 1.3 | — | 0.5 | 0.3 | 0.1 |
| 1:1 | 2.7 | 2.1 | — | 0.6 | 0.3 |
| 1:2 | 4.6 | 3.5 | 1.7 | — | 0.5 (0.4)[c] |
| 1:5[b] | 9.2 | 7.1 | 3.5 | 2.0 (2.5)[c] | — |

To cross-reference, multiply method (x) value by the factor in its row to obtain the equivalent result (y) in another method (e.g. if you obtained an EC of 430 mS/m using the 1:2 method and need the equivalent EC value in the ECe method, then 430 × 4.6 (cross-reference factor) = 1978 mS/m in the ECe method).

[a]ECe = saturated paste extract = saturated medium extract = SME. Vacuum extraction of solution.
[b]The methods in CEN (1999), US Composting Council (2002), WRAP (2003), Landscape Institute (2004) for composts, soil improvers, growing media are all based on substrate:water as 1:5 v/v ratio.
[c]Bunt (1988).

as pure yard waste compost (low EC) or sewage sludge compost (low EC but often high content of total phosphorous), are better suited to some landscape applications as opposed to nutrient-rich composts with high levels of soluble nutrients (such as kitchen waste compost with low content of structure material or some types of manure composts). It should be remembered that the aim of growing plants in many areas of this sector is not to produce the maximum possible harvest but rather to produce beautiful, healthy, dense and hardy plants that do not require high maintenance in the form of pruning, cutting back and mowing. Most trees, bushes and perennials are well suited to the slow release of organic nitrogen that stable, mature compost provides.

There are some plant groups, among the many used for landscaping and in horticulture, which have specific needs (BGK, 1992, 2004, 2005; Zentralverband Gartenbau, 2002; WRAP, 2004). The three most noteworthy groups are:

- Acid-loving plants, also named acidophilus or calcifuges: These include among others rhododendrons, azaleas, ericas (such as heathers), and some berries such as blueberry. In general, compost should be used with care when dealing with this group of plants and only in small amounts. This is because the pH of compost is high. Many acid-loving plants furthermore grow best with comparatively low, but constant, levels of nutrients, which also can limit the use of composts.
- Alpine plants, rock garden plants: This varied group of plant species originates from nutrient-poor ecosystems and normally cannot tolerate high levels of nutrients. Apply only small amounts of compost to these plants remembering to take the compost's nutrient content and electrical conductivity (EC) into careful consideration.
- Chloride-sensitive plants, salt-sensitive plants: Several plant species among berries, fruits, bulbs and other ornamental plants are sensitive to chloride ($Cl^-$). Composts produced from materials known to have some chloride content (e.g. kitchen waste or seaweed) should have the content of chloride declared, when marketed to professional growers of these plant groups.

### 9.9.3.3   Marketing and Services

In comparison with, for example, companies marketing fertilizers or peat, the vast majority of composting plants do not actively market their products. Marketing aimed at the landscape, gardening and horticultural sectors will require knowledge about plant growing requirements as well as an understanding of needs of each specific sector. Marketing of compost products requires at least as good declarations and user guidance as those of competing products, but also the potential specific advantages of compost over other products should be pointed out (see Table 9.9.3). Information about the best seasons for compost application with focus on possible terms of delivery is important for production plans at the compost facility and in the marketing of the different products. Landscape gardeners expect that they can be given a price for a large delivery of compost or topsoil either the same day or the day after they make their inquiry. If they do not receive the price information they require, they are unable to fulfill their own obligations – whether submitting tenders or referring directly to a specific construction project that is underway. Landscape gardeners expect to be able to collect small deliveries ($< 100\,m^3$) the same day and often expect to be able to collect large deliveries ($> 1000\,m^3$) within a few weeks of order.

## 9.9.4   Use of Compost Products

Compost and soil mixes containing compost are used for a wide variety of activities within landscaping, gardening and horticulture (US Composting Council, 1996, 2003; Carlsbæk and Reeh, 1997; Zentralverband Gartenbau, 2002; WRAP, 2003; Landscape Institute, 2004).

### 9.9.4.1   Soil Improver

If compost is used as soil improver for upgrading poor soils, it is typical to add a 4-cm layer of nutrient-poor compost or a 2-cm layer of nutrient-rich compost. A 2-cm layer of compost corresponds typically to 140 t/ha or nearly 100 t/ha of dry matter. This is mixed together with the soil down to a level of 10–20 cm using a rotary tiller. Compost normally supplies sufficient phosphorous and potassium for most plants, while vigorous plants (roses, bedding plants, annuals, vegetables, turf-grass, etc.) need additional nitrogen fertilization. Compost and other organic material must not be worked too deeply into the soil as the level of gaseous exchange between soil and air decreases with depth. Even very stable compost should not be mixed to depths over 10–20 cm in heavy soil (soil with a high clay or silt content) as these compact soils have low levels of natural aeration. If the soil is of particular poor quality and lacking in humus, this process is repeated the following year, if possible. If this second application and incorporation is not possible, 2–4 cm of compost may be spread on the ground between plants, keeping it out of direct contact with plant stems and stalks. When carrying out soil improvement of a larger area, such as reclamation of disused quarries, prior contact with environmental authorities should be taken with respect to avoiding groundwater pollution and run off to streams.

### 9.9.4.2   Backfill for Planting

Compost is often used as a backfill mix component when planting shrubberies, bushes and trees. The excavated soil is mixed with compost to the ratio 1 part nutrient-poor compost to 2 or 3 parts soil (or 1 part nutrient-rich compost to 4–6 parts soil). This mixture is filled back into the hole. Unless the topsoil is of poor quality with a very low level of organic matter, a better plant response can be achieved by applying a 2- to 4-cm layer of compost around the plant, rather than mixing compost with the excavated soil (Craul, 1992). Around larger bushes and trees, the layer can be 5–10 cm thick when consisting of nutrient-poor compost. In both cases, it is important to ensure that the compost is kept out of direct contact with trunks, stems and plant stalks.

### 9.9.4.3   Mulching

Laying down a thick layer of (nutrient-poor) compost, bark or wood chips is called mulching. This is done to reduce the occurrence of annual weeds, prevent erosion or to minimize evaporation.

A 5–10 cm layer of mulch prevents the establishment of seedborne weeds but promotes the growth of already established root-propagated weeds (thistles, couch grass, etc.) and these should be dealt with before laying down the mulch. Mulching should not be used on wet soils with poor drainage, as the mulch reduces evaporation (Carlsbæk and Reeh, 1997). Coarse, nutrient-poor compost with low content of fine particles is best suited for mulching especially when the prime objective is to prevent weeds (WRAP 2003; BGK 2005).

Mulching is used on road verges and river embankments as a measure to prevent erosion (see Figure 9.9.2). Mulching used in this way has an equivalent or superior effect to other erosion-prevention measures (Sherman, 2003). If the area is difficult to access with normal machinery, the compost mulch can be spread over the area using specialized blower units. Shredded wood waste may be used directly without prior composting if this is aesthetically acceptable and no grass or ground plant cover is desired for a few years (Lloyd *et al.*, 2002; Demars *et al.*, 2004).

### 9.9.4.4   Manufactured Topsoil in Urban Constructions and Parks

Major construction sites and public works establishing parks, road embankments and other green areas are often short of good topsoils. In such situations, landscape gardeners and engineers often choose to purchase manufactured topsoil (see Figure 9.9.3). It is possible to produce weedfree topsoil by mixing compost with (silty) sand – and clay or peat if necessary. Another possible component is powdered rock/stone, a byproduct of the gravel and quarry industries, which may constitute up to 20 vol% in a manufactured sandy topsoil, if the final blend contains at least 5 vol% organic matter. Common organic matter content of manufactured landscape topsoils are 2–8 % with lowest content in sandy soils for sports fields and heavy, dense soils for general landscaping in need of the highest content. A weedfree topsoil reduces costs for maintenance substantially.

The electrical conductivity (EC) of the final manufactured/improved topsoil must allow plant root growth without scorching (see Box 9.9.1). Beds used for sowing, such as topsoil for grass establishment, should have very low EC and these types of media are often watered prior to sowing. The amount of plant-available nutrients in the compost should be taking into account if commercial fertilizers also are added. Compost normally supplies sufficient phosphorous

***Figure 9.9.2***  *Erosion control using sludge compost along verges and on a ski jump site. Sites will be covered with grass.*

and potassium for most plants, while vigorous plants (roses, bedding plants, annuals, vegetables, turf-grass, etc) need additional nitrogen fertilization.

### 9.9.4.5 Manufactured Topsoil for Sports Facilities

Many golf greens and tees, and sometimes major sports fields (such as football pitches), are established using manufactured topsoils. In these cases, there are highly specific requirements regarding water retention and percolation rate (saturated hydraulic conductivity) in the final topsoil. These requirements, together with specific analytical methods, have been put forward by the United States Golf Association (USGA) to obtain a maximum amount of playing time on the green while maintaining a high-quality grass cover (USGA, 2004). The main component of topsoils for USGA greens is very uniform sand usually mixed with some organic material. Again, mature, uniform, weedfree and fairly nutrient-low compost types are best suited for this application.

The Sports Turf Research Institute (STRI) in Bingley, UK, has set standards for the types of sand (before mixing it with other constituents) used in the construction and maintenance of football pitches and golf greens. In the construction

*Figure 9.9.3*   *Landscaping in urban areas with manufactured soils.*

of football pitches, the median sand particle diameter ($D_{50}$) must be between 0.2 and 0.4 mm, the gradation index ($D_{90/10}$) < 6 and the content of fine particles (< 0.1 mm) must be less than 4 % w/w (other requirements apply; see Baker, 1990).

### 9.9.4.6   Growing Media for Pots and Plant Containers

Compost can be a component in growing media for pot and container plants. Compost can make up to 10–50 % of the total volume depending upon the type of compost and plant species; a common ratio for the hobby market is 33 vol% yard waste compost or 20 vol% kitchen waste compost in a mixture with peat. Specifications for compost used as a component in growing media for pots and containers can be found in Waller (2004) and BGK (2005a, b). American nurseries use a significant amount of composted pine bark in some of their growing media. As a result, they now experience fewer problems with soilborne diseases than previously (Hoitink *et al.*, 1997). Other types of compost have also proved to have disease-suppressive properties for some common pathogens. To maximize the possible disease-suppressive effect that compost can have, the easily degradable cellulose must have been mineralized and the compost must have been recolonized with a wide range of bacteria and fungi once the thermophilic composting stage has been completed. Compost remains disease suppressive for a certain length of time. After this period, lack of microbial available carbon reduces the abundance of the general microflora; the microbial carrying capacity becomes too low and the compost ceases to suppress pathogens (Krause *et al.*, 2001; Boulter *et al.*, 2002).

## References

Alexander, R. (2003): *The practical guide to compost marketing and sales.* R. Alexander Associates, Inc., Apex, USA.

Baker, S.W. (1990): *Sands for sports turf construction and maintenance.* The Sports Turf Research Institute, Bingley, UK.

Barth, J. (2004): Recycling organics in Europe – lessons, success stories and markets. In: European Compost Network (ed.) *Effective compost marketing and compost application in practice.* The fifth international workshop of the European Compost Network. On CD-ROM. European Compost Network, Weimar, Germany. Located on the internet: 19 July 2007. www.compostnetwork.info.

BGK (1992): *Kompost mit Gütezeichen für Ihren Wein- und Obstbau* (Compost with a quality label for your wine and fruit production; in German). Bundesgütegemeinschaft Kompost e.V., Cologne, Germany.

BGK (1994): *Methods book for the analysis of compost*, 3. ed.. Bundesgütegemeinschaft Kompost e.V., Cologne, Germany.

BGK (2004): *Kompost für den Garten- und Landschaftsbau, 2. Auflage.* (Compost for the park and landscape sector; in German). 2.ed. Bundesgütegemeinschaft Kompost e.V., Cologne, Germany.

BGK (2005a): *Qualitätsanforderungen und Güterichtlinien für Komposte: Frischkompost, Fertigkompost, Mulchkompost und Substratkompost.* (Quality criteria and values for composts: fresh compost, stable compost, compost for mulching, compost for growing media; in German). Bundesgütegemeinschaft Kompost e.V.. Cologne, Germany.

BGK (2005b): *Qualitätsanforderungen an flüssige und festige Gärprodukte* (Quality criteria for liquid and solid products from anaerobic digestion; in German). Bundesgütegemeinschaft Kompost e.V.. Cologne, Germany.

Boulter, J.I., Trevors, J.T. and Boland, G.J. (2002): Microbial studies of compost: bacterial identification, and their potential for turfgrass pathogens suppression. *World Journal of Microbiology and Biotechnology*, 18, 661–671.

Bunt, A.C. (1988): *Media and mixes for container-grown plants.* Unwin Hyman, London, UK.

Carlsbæk, M. and Reeh, U. (1997): Application of compost within the park and landscape sector. In: Johansson, C., Kron, E., Svensson, S.-E., Carlsbæk, M. and Reeh, U. (eds.) *Compost quality and potential for use. Literature review and final report.* AFR Report 154, Chapter 1. Swedish Environmental Protection Agency, Stockholm, Sweden.

Carlsbæk, M. and Brøgger, M. (2002): Quality and market issues. Compost and solids from anaerobic digestion. In: Crowe, M., Nolan, K., Collins, C., Carty, G., Donlon, B., Kristoffersen, M., Brøgger, M., Carlsbæk, M., Hummelshøj, R.M., Thomsen, C.D. (eds.) *Biodegradable municipal waste management in Europe. Part 3: Technology and market issues.* EEA Topic Report No. 15/2001, Chapter 2. European Environmental Agency, Copenhagen, Denmark.

CEN (1999): EN 13038:1999 Soil improvers and growing media – Determination of electrical conductivity. CEN/Technical Committee 223, European Committee for Standardization, Brussels, Belgium.

Craul, P.J. (1992): *Urban soil in landscape design*. John Wiley & Sons, Ltd, New York, USA.

Demars, K.R., Long, R.P. and Ives, J.R. (2004): Erosion control using wood waste materials. *Compost Science and Utilization*, 12 (1), 35–47.

Devliegher, W. (2005): Personal communication. VLACO, Belgium.

EC (2001): Commission decision of 28 August 2001 establishing ecological criteria for the award of the Community eco-label to soil improvers and growing media. 2001/688/EC. *Official Journal of the European Communities*, L 242, 17–22.

Fischer, P. and Schmitz, H.-J. (2004): Kompostierte Gärreste als Substratzuschlag (Composted residues from anaerobic digestion as growing medium component; in German). *Deutscher Gartenbau*, 58 (40), 38–40.

Hogg, D., Barth, J., Favoino, E., Centemero, M., Caimi, V., Amlinger, F., Devliegher, W., Brinton, W. and Antler, S. (2002): *Comparison of compost standards within the EU, North America and Australasia*. The Waste and Resources Action Programme (WRAP), Banbury, UK.

Hoitink, H.A.J., Stone, A.G. and Han, D.Y. (1997): Suppression of plant diseases by composts. *HortScience*, 32 (2), 184–187.

Krause, M.S., Madden, L.V. and Hoitink, H.A.J. (2001): Effect of potting mix microbial carrying capacity on biological control of *Rhizoctonia* damping-off of radish and *Rhizoctonia* crown and root rot of *Poinsettia*. *Phytopathology*, 91, 1116–1123.

Landscape Institute (2004): *Compost specifications for the landscape industry*. The Landscape Institute, London, UK.

Lloyd, J.E., Herms, D.A., Stinner, B.R. and Hoitink, H.A.J. (2002): Who gets the nitrogen? Comparing composted yard trimmings and ground wood as mulch. *Biocycle*, 43 (9), 52–55.

McLachlan, K.L., Chong, C., Voroney, R.P., Liu, H.-W. and Holbein, B.E. (2004): Variability of soluble salts using different extraction methods on composts and other substrates. *Compost Science and Unilization*, 12 (2), 180–184.

Sherman, R. (2003): Texas model for erosion control. America's largest compost market. Texas Transportation Department accelerates highway use of compost. *Biocycle*, 44 (7), 24–28.

Tyler, R.W. (1996): *Winning the organics game. The compost marketer's handbook*. ASHS Press, Alexandria, USA.

US Composting Council (1996): *Field guide to compost use*. The Composting Council, Alexandria, Virginia, USA.

US Composting Council (2002): *Test methods for the examination of composting and compost*. CD-ROM. The US Composting Council, Research and Education Foundation, Holbrook, USA.

US Composting Council (2003): *Landscape architect specifications for compost utilization*. The US Composting Council and the Clean Washington Center, Holbrook, USA.

USGA (2004): *USGA recommendations for a method of putting green construction*. United States Golf Association, Far Hills, USA.

Waller, P. (2004): *Guidelines for the specification of composted green materials used as a growing medium component*. The Waste and Resources Action Programme (WRAP), Banbury, UK.

WRAP (2003): *Using compost in landscaping. Compost information package 2*. The Waste and Resources Action Programme (WRAP), Banbury, UK.

WRAP (2004): *Compost use in fruit production. Fact Sheets 08. Using compost in agriculture and field horticulture*. The Waste and Resources Action Programme (WRAP), Banbury, UK.

Zentralverband Gartenbau (2002): *Handbuch. Kompost im Gartenbau* (Handbook. Compost for horticulture, landscaping and gardening; in German). Zentralverband Gartenbau e.V. (ZGV), Bonn, Germany.

# 9.10

# Utilization of Biologically Treated Organic Waste on Land

**Peter E. Holm and Lars Stoumann Jensen**

*University of Copenhagen, Denmark*

**Michael J. McLaughlin**

*CSIRO Land and Water/University of Adelaide, Australia*

Biologically treated organic waste can be recycled back to agriculture and forestry for use on land. The organic waste products considered are processed (composted or anaerobically digested) or unprocessed organic waste from households, gardens, commerce and industry. Agricultural production waste and manure from conventional and organic animal production systems (cattle, pig, poultry slurry) are also used on land, but are not included in this chapter. The use of household wastes in agriculture is not widespread (but is increasing due to urban recycling programs) and most of the regulation in this field arises from the application of wastewater treatment sludges to agricultural land.

Organic wastes can also be beneficially used in urban areas, provided they have been processed in some way, e.g. through composting. Waste composts may be used for soil manufacturing, improvement of degraded urban soils, for soil improvement in parks, gardens and recreational areas, and for revegetation of disturbed soils, e.g. roadways, embankments, etc. These applications are described in Chapter 9.9.

From a historical perspective, land application of waste is as old as agriculture. The Chinese have used human and household waste as a fertilizer for thousands of years. In the beginning of the twentieth century, land application of sludge and waste in Europe and the USA was common. In those times the benefits of the waste were appreciated, and local farmers were glad to use waste or residuals to improve crop yields. However, as soon as concentrated and highly effective sources of plant nutrients became available and affordable in the form of chemical fertilizers, in many situations waste was seen as a disposal problem rather than a resource.

*Solid Waste Technology & Management*   Edited by Thomas Christensen
© 2011 Blackwell Publishing Ltd

## 9.10.1   Introduction

The soil–plant–animal system offers an excellent opportunity for the application and reuse of constituents present in household waste materials, for example in agricultural crop production or in forestry (silviculture). Biologically treated household waste contains organic matter and essential plant nutrients including nitrogen (N), phosphorus (P), potassium (K), sulfur (S), and trace elements. It is therefore potentially useful as a fertilizer and soil conditioner, and may potentially substitute the use of mineral fertilizers to greater or lesser extent. This depends both on the size and type of urban dwelling producing the waste (quantity, quality, concentration of nutrients in the waste) and the type of land use in the area where the waste could be applied (horticultural, arable, grassland, forest). In periurban areas of large cities, with only scattered agricultural or horticultural production, a substantial proportion of regional fertilizer demand could potentially be replaced with nutrients in biologically treated organic waste. However, in areas of intensive agricultural or horticultural production, nutrients in the organic household waste may contribute only a smaller fraction of regional demand, and may furthermore compete with agricultural wastes, such as livestock manures, for available land suitable for land application (see Box 9.10.1 for a more detailed case study).

---

**Box 9.10.1   Case Study Denmark: The Role of Waste as a Potential Nutrient Source for Crops.**

Magid *et al.*, (2006) investigated the potential for use of N, P, and K which can be recovered from urban areas (in sewage and organic household waste) to replace mineral fertilizer currently used in agriculture in the Danish counties. They found that urban organic waste sources could potentially cover the need for 79 % of current agricultural mineral fertilizer N use in the Copenhagen metropolitan area, with approximately 1.5 of Denmark's 5.2 million inhabitants (see figure below). However, in the remaining parts of the country, dominated by intensive agricultural production, organic waste sources would be able to substitute around 10 % of mineral fertilizer N demand.

This was also generally the case for K, but a larger proportion of crop P demand could potentially be covered from organic waste, in the Copenhagen metropolitan area actually as much 170 %, so in excess of local agricultural P demand. On average 22 % of the national P demand could be covered.

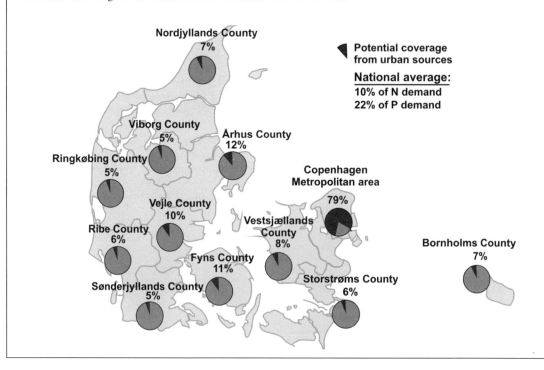

In agricultural production systems, 'closing' the cycling of nutrients is a continuing goal, so that losses of nutrients from the system (to the atmosphere, to water bodies or to sediments) are minimized. Closing the cycle for P is of particular interest, since it is an essential and nonreplaceable plant nutrient, and most of the world's soils have low available P levels, limiting crop production. Fertilization with P is therefore crucial for global food security, with world food demand expected to increase more than 50 % before 2050. However, P is a limited natural resource, with the mineable phosphate-rich rocks used for P fertilizer production projected to be exhausted within the next 60–130 years (Steen, 1998). Therefore, appropriate recycling of P is crucial and will increase in importance in the coming decades.

While recycling of wastes to agricultural systems will assist with closing of the P cycle, use of household wastes may cause environmental problems related to the presence of some constituents in the waste (e.g. household chemicals, pesticides, detergents, etc.) or imbalances between nutrient elements, especially if these cause potential pollution problems and have a negative impact on soil functions. Land application of waste is regulated in most countries to protect human health, the environment and the soil functionality from undesirable chemical constituents such as metals, e.g. cadmium (Cd) and lead (Pb), toxic organic chemicals, e.g. polychlorinated biphenyls (PCBs), eutroficants (e.g. nitrate), and pathogens such as bacteria, e.g. *Escherichia coli* and viruses, e.g. Hepatitis A virus (Epstein, 2003).

Control of use of wastes on land includes regulatory measures concerning quantity (annual and maximum applied quantities) and quality (characteristics, pollutant level) of the waste being applied. Other measures include use of good management practices at land application sites (e.g. season for application, establishment of buffer zones).

Most of the waste products presently applied to agricultural soils are sludges from municipal wastewater treatment plants. In addition, there are other waste products, e.g. sludges from industries and composted organic wastes, which are applied in accordance with individual approvals by environmental authorities. During recent years, a number of initiatives have been taken regarding separation of waste at source. The organic waste products, e.g. food and vegetable waste are composted and reused in gardens, parks or on farmlands. The quality of the compost depends considerably on the efficiency of the separation at the source and on the treatment methods.

Table 9.10.1 presents typical characteristics of waste and manure applied to agricultural land.

**Table 9.10.1**   *Characteristics of anaerobic digestate (household waste), sewage sludge, composts (organic household, green/garden waste) and animal slurries (pig, cattle). Based on numerous sources (e.g. Petersen, 1996; Poulsen, 2009).*

| | Anaerobic digestate (household waste) | Sewage sludge/ biosolids | Compost (organic household waste) | Compost (green/ garden waste) | Pig slurry | Cattle slurry |
|---|---|---|---|---|---|---|
| Total solids, % ww | 1–2 | 2–30 | 40–75 | 50–75 | 2.5–6.5 | 6.5–10.5 |
| Volatile solids, % TS | 65 | 40–80 | 35–70 | 20–40 | 80–90 | 75–85 |
| N-total, % TS | 3.5 | 2–5 | 1.0–2.5 | 0.7–1.6 | 7.5–15.0 | 3.8–6.9 |
| N-NH$_4$, % TS | 1.7 | 0.01 | < 0.2 | — | 5.6–11.3 | 2.3–4.1 |
| N-NO$_3$, % TS | 0.2–1.0 | 0.005 | — | — | 0 | 0 |
| N-organic, % TS | 1.8 | 2–5 | — | — | 1.9–3.8 | 1.5–2.8 |
| P-total, % TS | 0.9 | 0.5–3.0 | 0.2–1.0 | 0.15–0.7 | 1.8–2.8 | 0.9–1.1 |
| K, % TS | 3.7 | 0.1–1.0 | 0.5–2.0 | 0.8–1.9 | 3.8–6.5 | 3.9–6.5 |
| C, % TS | 39 | — | 20–45 | 10–30 | 38 | 36 |
| H, % TS | 5 | — | — | — | nd | nd |
| S, % TS | 0.5 | 0.4–2.0 | 0.06–0.3 | 0.1–0.15 | 1 | 0.6 |
| Cl, % TS | 2.9 | — | 0.3 | 0.001–0.1 | 1.3 | 1.3 |
| C/N-total ratio | 11 | 5–20 | 12–20 | 10–30 | 3–5 | 5–10 |
| C/N-organic ratio | — | nd | — | — | 10–20 | 13–25 |
| Mg, % TS | 0.8–1.1 | 0.1–0.5 | 0.3–1.6 | 0.3–0.4 | 0.8 | 0.6 |
| Ca, % TS | 2–5 | 1–5 | 3–7 | 2.2–3.0 | 2.5 | 1.4 |
| Pb, mg/kg TS | 10–60 | 15–100 | 25–110 | 20–55 | 3.3 | 3.4 |
| Cd, mg/kg TS | 0.3–0.7 | 0.5–5.0 | 0.3–1.0 | 0.2–0.5 | 0.5 | 0.5 |
| Cr, mg/kg TS | — | 10–70 | 15–40 | 15–35 | 10.3 | 2.5 |
| Cu, mg/kg TS | 45–125 | 100–1000 | 28–47 | 28–50 | 325 | 12.5 |
| Ni, mg/kg TS | 8–28 | 7–100 | 8–25 | 6.3–8.3 | 13.8 | 6.5 |
| Hg, mg/kg TS | — | 0.2–3.0 | 0.1–0.3 | 0.05–0.1 | nd | nd |
| Zn, mg/kg TS | 150–300 | 400–1000 | 120–270 | 100–200 | 750 | 150 |

## 9.10.2   Basic Principles of Agricultural and Forestry Land Application

Application of treated organic waste to agricultural and forestry land offers benefits, but may also constitute some problems if not properly constrained. The basic issues are presented below.

### 9.10.2.1   Basic Principles: Benefits

Biologically treated waste contains several plant macronutrients, primarily N and P, some K, and varying amounts of micronutrients, such as boron (B), copper (Cu), iron (Fe), manganese (Mn), molybdenum (Mo), and zinc (Zn). The amounts and ratios of nutrients do not match a well formulated chemical fertilizer, but the soil is usually supplied with supplementary nutrient from other fertilizers to provide the proper amounts of nutrients needed for crop production. Soil incorporation of processed organic solid waste usually results in a positive effect on the growth and yield of a wide variety of crops and may contribute to the restoration of ecological and economic functions of degraded land.

Agricultural uses of municipal solid waste (MSW) have shown promise for a variety of field crops (e.g. maize, sorghum, forage grasses) and vegetable for human consumption (e.g. lettuce, cabbage, beans, potatoes, cucumbers; Shiralipour *et al.*, 1992). The fertilizer value (yield or nutrient uptake per unit of nutrient applied, relative between organic and commercial mineral fertilizer) of MSW and sludges range from the negligible or even negative to more than 50 % of commercial mineral fertilizers, in particular if the residual release of nutrients in subsequent years is included.

Application of biologically treated waste can also improve soil structure. The addition of organic materials like compost to a fine-textured soil can assist the development of structure (aggregation) and can stabilize soil aggregates against dispersion. Improvement of the structure of the soil makes it more friable (lower soil strength, desirable for cultivation and seedbed preparation) and increase the amount of pore space available for plant roots and the entry of water and air into the soil. In coarse-textured sandy soils, organic matter can increase the water holding capacity of the soil and provide exchange and adsorption sites for nutrients. Furthermore, added organic matter may prevent erosion and therefore assist revegetation of sloping land. A major benefit of organic wastes in more arid climates is their use as mulches for water conservation and for modulation of variations in surface soil temperatures. Water requirements and weed infestation of crops can be markedly reduced by mulching with organic wastes (e.g. Preusch and Twokoski, 2003), as can diurnal variations in surface soil temperatures (Dahiya *et al.*, 2001).

As for agricultural land, the advantage of applying organic waste in forestry (silviculture) is the addition of plant macro- and micronutrients. Since forests are often planted on infertile and poorly structured soils (sandy, low pH soils), tree growth is often limited by nutrient deficiencies. Consequently, organic waste can provide a good source of nutrients and organic matter and can be an inexpensive alternative to conventional fertilizers. Because trees are perennials, the scheduling and management of waste applications is not as complex as it may be for annual crops. Application of wastes to forest soils is generally performed either annually or at 2- to 5-year intervals.

Land application of organic wastes may also contribute positively to long-term stabilization of carbon (C) in the soil and hence to decrease atmospheric $CO_2$ levels (Smith *et al.*, 2001). Carbon dioxide from degraded fresh organic matter is considered neutral with respect to the global warming impact, because the plants have recently removed an equal amount of carbon dioxide from the atmosphere during growth. Thus, organic matter from treated waste stored in the soil represents a 'saved' emission of carbon dioxide. This mechanism is referred to as carbon sequestration and carbon is sequestered (protected) in soils as soil organic matter.

Comprehensive model simulations of the soil turnover of applied C in anaerobically treated sewage sludge have shown that 63–86 % is still stabilized in the soil after 10 years, but only 17–37 % after 50 years, and 10–14 % after 100 years (Bruun *et al.*, 2006; see further information in Box 9.10.2). An experimental study of sewage sludge inputs over 30 years (Tian *et al.*, 2009) indicated an average C sequestration rate of approximately 30–35 % of annual applied C, confirming the simulation results. However, C is not sequestered infinitely, and soil C reaches a new equilibrium with decomposition, where no further increase is achieved. However, even with the low net C storage simulated after 100 years (Bruun *et al.*, 2006), this corresponds to an avoided $CO_2$ emission of approximately 180 kg $CO_2$/t dry matter of sewage sludge applied to the soil.

**Box 9.10.2   Environmental Impacts from Soil Application of Processed Organic Waste. Reprinted with permission from Environmental Soil Chemistry by H.C.B. Hansen, Leonardo Da Vinci Programme, Mataj Bel University, Banská Bystrica, Slovakia © (2001) Institut fur Grundvidenskab og Miljø.**

The principles of life cycle assessment (LCA) can be used to quantify and assess the environmental impacts, including any environmental benefits from a substitution of the use of chemical fertilizers. Basically, LCA accounts for all uses of resources and all emissions from the waste application and fertilizer on land, accumulated through the lifetime of the applied waste. Hansen *et al.* (2006) applied such LCA principles for assessing the environmental impacts of application of processed organic waste to agricultural land, including substitution of commercial fertilizer and soil carbon sequestration. Environmental impacts included in the use on land submodel of the EASEWASTE LCA model (Kirkeby *et al.*, 2006) are illustrated in the figure below.

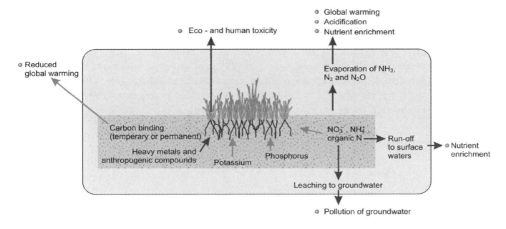

A range of Danish land use scenarios with differing climates, soils, farm types, crop rotations and livestock densities were simulated with a dynamic, mechanistic agroecosystem model (viz *Daisy;* Bruun *et al.*, 2006) in order to determine emission factors, graduated according to local conditions, and these were then applied in the LCA model. The left figure below shows how large an influence e.g. the time frame of the simulated soil C sequestration factor (and hence the amount of waste C stored in soil) has on the global warming potential impact. The right figure illustrates the large influence of soil type and waste decomposition characteristic on particularly the nutrient enrichment potential, leading to eutrophication of the aquatic environment (negative values are savings).

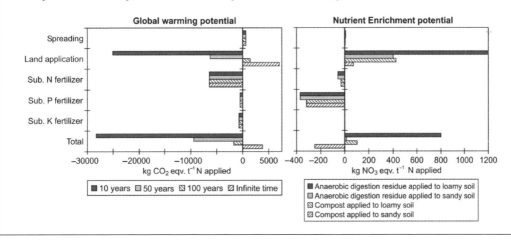

## 9.10.2.2   Basic Principles: Constraints

Although the waste is produced at no cost for the land user, the financial cost of the land application may be a major constraint for reuse. The costs may involve control and regulatory measures and fees, taxes, etc., which may exceed the economic benefits. Also, most organic wastes are bulky, low-grade fertilizers of variable composition and frequently have high water content. Accordingly, they cannot be transported very far before transportation costs exceed the fertilizer value. The fossil fuel-derived $CO_2$ emissions from transport (also termed the carbon footprint) should also be factored into decisions about waste reuse schemes. Many waste-producing facilities lack agricultural and forestry land within economically practical transportation distances.

In addition, application scheduling that is compatible with agricultural and forestry planting, climatic conditions, and harvesting requires careful management. This will often limit application on arable land to certain months, e.g. under temperate, humid climate in autumn or early spring, and may introduce significant costs for storage facilities. Furthermore, some of the nutrients in organic wastes are not readily available for plant uptake, but must be mineralized first.

Organic wastes contain various compounds (metals, organics, see also section on pollutants) that may pose potential environmental and human health risks. In particular, urban wastes may contain relatively high concentrations of potentially toxic metals which may accumulate in surface soils as metals generally bind strongly to soil particles. Potentially toxic metals may extinct or inhibit bacterial species responsible for important soil functions (see Box 9.10.3) and could have significant negative effects on the soil ecosystem to support plant production (e.g. Brandt *et al.*, 2006).

Organic wastes, in particular sewage sludge, may contain a number of organisms pathogenic to humans, e.g. viruses, bacteria and parasites. In addition, organic waste materials may also carry plant pathogenic fungi, bacteria, and weed seeds, which adversely affect crop production. However, various forms of biological treatment of the waste can reduce

---

**Box 9.10.3   Soil Functionality and Land Uses (Compiled by Hansen, 2001).**

Soils are natural entities and represent nonrenewable natural resources at least on a time scale of hundred to thousands of years. About 99 % of the world's food and fiber are produced on soil/land. However, land also has functions or uses other than serving food production.

In general the following functions can be distinguished:

F1: Production of biomass (food, fiber and other nonfood).
F2: Clearing and filtering (mineralization, attenuation of pollutants).
F3: Habitat for plants and animals (maintenance of high biological diversity, gene reserve).
F4: Sites for construction of houses, roads and landfills.
F5: Resource of raw material such as minerals, construction materials (e.g. sand and gravel).
F6: Cultural archive storing remnants of historical activities of man.

The interest in the environmental roles played by soils is not new but has markedly intensified due to the side effects of the technological revolution in the agricultural sector taking place after the Second World War. This revolution introduced the common use of fertilizers and pesticides, and it gave rise to farm units with high animal densities producing large amounts of manures.

There is an increasing awareness of the limits for biomass production set by the other soil functions, in particular F2 and F3. The European community seeks to sustain agricultural production governed by the sustainability criterion. It implies that the quality of the soil resource must not degrade over time and that soil resources shall be managed in a way that minimizes adverse effects in the neighboring aqueous and atmospheric environments.

In addition more interest is now being paid to food quality and health. Therefore the application of pesticides, fertilizers and wastes on land is now under fairly strict control to prevent eutrophication and pollution of the environment and to secure soil functionality and food quality.

these risks significantly (see below), and for some of the pathogens, varying degree of inactivation will also occur shortly after soil application. The risk of transfer of pathogens into the human food chain is greater when the waste is applied to an edible crop, compared to a feed crop used for livestock production, and hence the former is prohibited in many countries.

While reuse of organic wastes in forestry situations eliminates any risks from transfer of contaminants through the human food chain, the risks from nutrient and contaminant runoff and leaching remain. Forest ecosystems are often characterized by relatively low nutrient requirements and have low exports of nutrients (and potential pollutants). Hence, increasing the import of organic matter and nutrients on relatively poor forest soils may increase the leaching of potential pollutants such as metals to groundwater, see e.g. Andersen *et al.* (2002). In many soils, leaching of nutrients, especially nitrate, to groundwater may also limit the application rate.

Many organic wastes contain soluble salts, which can cause problems in their use as fertilizers, particularly in arid regions where soluble salts are already present in irrigation waters (Weggler *et al.* 2004). These can not only cause salt damage to crops through osmotic stress, but may also mobilize contaminants in the soil (e.g. Cd).

Odors and associated nuisances from application of organic waste to land, both real and imagined, create conflict. Often odors are the major obstacle for gaining public and farmer acceptance for waste reuse in agricultural systems, especially those in close proximity to residential areas.

## 9.10.3  Land Use of Biologically Treated Waste

As described above, the use of wastes may cause problems related to their composition because of undesirable chemical and biological constituents. Land application of waste is therefore regulated in most countries to protect human health, the environment and the soil functionality from the possible negative impact of these constituents, refer to Box 9.10.3 (Soil functionality and land uses) for further details. This section focuses on the soil function of biomass production in relation to agriculture and forestry. As stated earlier, the characteristics of a given treated waste product are dependent upon the origin of the waste and the process by which it is produced. A wide range of physical, biological, and chemical characteristics have to be taken into considerations when using waste on land (Table 9.10.2). These characteristics are further described in detail.

***Table 9.10.2***  *Waste parameters used for defining physical, biological and chemical characteristics.*

| Parameter/characteristics | Purpose |
| --- | --- |
| Physical | |
| Color | Aesthetics |
| Odor | Aesthetics, Health, marketability |
| Bulk density | Transportation, handling, application |
| Moisture or dry matter content | Handling |
| Organic matter/volatile solids content | Soil structure and quality |
| Particle size distribution | Handling, aesthetics, soil structure and quality |
| Water holding capacity | Water conservation |
| Biological | |
| Pathogens (bacteria, viruses, prions) | Health, environment, agronomic value |
| Seed germination | Soil quality, crop production |
| Weed seeds | Soil quality, crop production |
| Chemical | |
| Nutrient contents (macro and micro) | Agronomic fertilizer value, soil quality |
| Soluble salts/electric conductivity | Agronomic value, soil quality |
| Heavy metals | Health, environment, agronomic value |
| Organic pollutants | Health, environment, agronomic value |

### 9.10.3.1   Physical Properties

The desired physical characteristics of treated organic waste by farmers are usually dark color, uniform particle size, earthy odor, low moisture content, and high organic matter content.

   The application of biologically treated organic wastes can improve soil structure. The added organic matter enhances both aggregate formation (the binding of fine particles together) and stability (resistance to dispersion). Improvement of the structure of the soil will make it more friable and increase the amount of pore space available for plant roots and the entry of water and air into the soil. In coarse-textured sandy soils, organic matter can increase the water holding capacity of the soil and provide exchange and adsorption sites for nutrients. Furthermore, the organic waste may prevent erosion and therefore assist efforts to revegetate eroded soils.

   In arid regions with limited water supplies, organic wastes used as mulches (surface applications of 5–15 cm depth) are often beneficial in reducing evaporation from the soil surface, thus reducing water requirements and providing a buffer from temperature extremes at the soil surface. This is often beneficial for seed germination and early seedling growth. An additional benefit is that soils in arid regions are often low in organic matter, with low water holding capacities. Addition of organic wastes to these soils is particularly beneficial.

### 9.10.3.2   Biological Properties

Addition of compost or other processed organic waste affects soil biology, fertility and quality (see Box 9.10.4 for further details) in both positive and negative ways. Organic waste application stimulates soil microbial and faunal (especially earthworm) activity, and hence improves soil fertility, as these soil organisms play central role in mineralization and turnover of organically bound nutrients. A recent study (Saison *et al.* 2006) found that compost amendment affected the activity, size and composition of the soil microbial community but the effect was mainly due to the physicochemical properties of the compost matrix and the stimulation of native soil microorganisms by available substrate, rather than to compost-borne microorganisms themselves.

---

**Box 9.10.4   Soil Biology, Fertility, and C and N Cycling. Reprinted with permission from Waste Management and Research, Environmental assessment of solid waste systems and technologies: EASEWASTE by J.T. Kirkeby, H. Birgisdottir, T.H. Christensen et al., 24, 1, 3–15 © (2006) Sage Publications.**

**Soil biota** includes:

– Fauna (animals) in size from earthworms to single-celled protozoan.
– Flora (plants) includes the roots of higher plants as well as microscopic algae.
– Microorganisms including fungi and bacteria which tend to predominate in terms of both numbers, mass and metabolic capacity.

**Soil fertility** refers to the inherent ability of a soil to act as growth medium for plants, including the capacity of the soil to supply nutrients to plants in adequate amounts and suitable proportions. A high level of soil fertility in a soil plant system requires a suitable soil depth for unlimited root growth, suitable mechanical properties, pH, temperature and aeration, high plant available water capacity, and nontoxic concentrations of metals, salts or other contaminants. Finally, fertile soils are able to supply plants with an adequate and a balanced supply of elements throughout the whole growth period. The latter is closely related to the soil biota, as the **biogeochemical cycles of C, N, S and P** are intrinsically linked to biotic activity and associated organic matter turnover.

**Agricultural practices** have different effects on different organisms, and hence on cycling of C, N, S and P, but a few generalizations can be made (Brady and Weil, 2002).

– Monocultures of crops generally reduce the diversity of soil organisms as well as the number of individuals. However, monoculture may increase the population of a few species.

---

- Adding inorganic lime and fertilizers to an infertile soil generally will increase microbial and faunal activity, but largely due to the increase in the plant biomass that is likely to be returned to the soil as roots, root exudates and shoot residues.
- Adding organic fertilizers, such as sludges, biosolids and animal manures, to an infertile soil will however influence microbial and faunal activity directly, as they provide organic components serving as substrate for the microorganisms and fauna.
- Tillage tends to increase the microbial degradation of organic matter in the soil, and the role of bacteria increases at the expense of the fungi and usually decreases overall organism's number as the soil C pool is depleted.

However, organic waste materials may also contain unwanted biota, which include parasites and pathogens of man and animals such as bacteria (e.g. *Salmonella, Campylobacter, Enterococcus*), viruses, and parasites (e.g. *Ascaris*). The bacterium *Escherichia coli* is a normal constituent of the intestinal flora of man and farm animals, but it does not occur in large numbers in uncontaminated soil or water (Stevenson and Cole, 1999), and as such it is commonly used as an easily detectable indicator for fecal pathogen presence. Biological treatment, e.g. biological oxidation (degradation) of organic waste by composting or aeration prior to application to land most often aims to reduce or eliminate these pathogens. Elimination depends on a number of factors, including pH, temperature and retention time of the biological treatment. Figure 9.10.1 illustrates how various combinations of temperature and retention time may be used to safely kill off all relevant pathogens, e.g. 70 °C for < 1 h, 55 °C for > 1 day or 45 °C for > 1 month (Strauch, 1998). The latter temperature–time combinations illustrates that thermophilic digestion in a biogas plant (at 52–55 °C) in many cases may result in hygienization or pasteurization of the waste. Hygienization may also be achieved by increasing the pH to e.g. 12 by liming or other alkaline agents (Carrington, 2001).

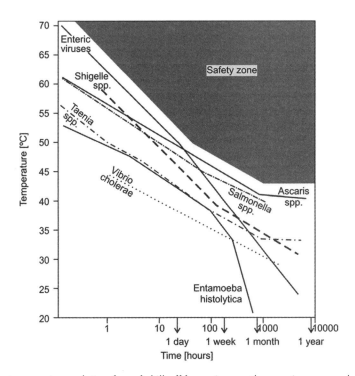

**Figure 9.10.1**   *Time–temperature relation for safe kill-off for various pathogens in sewage sludge (Strauch, 1998). Reprinted from Strauch, D. Pathogenic micro-organisms in sludge. Anaerobic digestion and disinfection methods to make sludge usable as a fertiliser. European Water Management, 1, 12–26.* © *(1998) EWA.*

Once applied to soils, a relatively quick die off of most pathogens occurs, due to competitive advantage of native organisms present in agricultural and forest soils. The survival time for most waste-borne microorganisms following land application is usually very short (hours to days), but a few species, like the persistent *E. coli* O157:H7 has been shown to be able to survive somewhat longer (several months, Avery *et al.*, 2005). Generally, survival depends on a variety of soil and climatic conditions including temperature, moisture content, and pH. Low temperature and high soil moisture result in the longest survival of pathogens.

### 9.10.3.3   Chemical Properties and Fertilizer Value

The agronomic value of applying treated waste is mainly related to its chemical composition, and to its soil physical conditioning value. The three major plant nutrients are nitrogen (N), phosphorus (P), and potassium (K), of which in 2008 approximately 199 million Mg (quantity of $N + P_2O_5 + K_2O$) in inorganic commercial fertilizers are applied throughout the world (FAO, 2008) to ensure sufficient food production and economically profitable crop yields for the farmers. Evaluating the agronomic value of waste on land relies very much on an evaluation of the ability of the waste to supply N, P, and K to crops in terms of commercial fertilizer equivalence.

The nutrient contents of industrial sludges and biosolids from sewage treatment plants vary considerably with ranges (dry matter basis) of 1–200 g/kg for total N, 0.005–70.0 g/kg for $NH_4$-N and 0.002–5.0 g/kg for $NO_3$-N, 1–150 g/kg for total P, and 0.2–25.0 g/kg for K being reported (e.g. Pierzynski and Gehl, 2005). Considering most crop requirement for N, P, and K, biosolids contain a high concentration of P compared to N and K. This is because N, K (and S) to some extent partition to the liquid phase (sewage effluent) during sewage treatment, or are lost from the biosolids during processing (e.g. denitrification for N). Potassium is generally retained relatively strongly (by cation exchange as $K^+$ on soil clays) after application to land, and K is not regarded as a major water pollutant. On the other hand, N and P may be lost from the biosolid or sludge-amended soil by both leaching (N) and surface runoff (N and P). Thus it is critical that loadings of both N and P are chosen with careful consideration of crop demand and potential leaching/runoff losses. In the European Union, the Nitrates Directive imposes a limit for N application in manures and organic waste of 170 kg N/ha/year.

The organic N in sludge is not immediately available for plant uptake, and the mineralization process which converts organic N into plant available ammonium ($NH_4^+$) and subsequently nitrate ($NO_3^-$; see reaction scheme in Box 9.3.1, which is also valid for soil) is slow compared to the timing of plant need for inorganic N for most annual, agricultural crops. This means that only 30–40 % of the N in sewage sludge is mineralized within 3–4 months (Epstein 2003), and hence be available for the crop to which it was applied, whilst the rest of the N is only slowly released during subsequent years. In a humid climate, this may consequently contribute to nitrate leaching to the subsoil and groundwater if released outside the growing season of crops.

The first year fertilizer N equivalent value, i.e. the amount of mineral fertilizer N [in the form of $NH_4NO_3$, $(NH_4)_2SO_4$ or urea $CO(NH_2)_2$] that 100 kg total N in sludge or biosolids may replace, has been found to be around 26–37 % when applied in Spring and only 16–23 % when applied in Autumn, in a number of Danish field experiments (Pedersen, 1999). Similar results in other temperate, humid climates have been found by others (Petersen, 2003; Gilmour *et al.*, 2004). In comparison, somewhat higher first year fertilizer N equivalent values may be achieved for animal manures, ranging from 40 % for solid farmyard manures up to as much as 75 % for liquid slurries, if applied appropriately (minimal N losses and in synchrony with crop demand). This is mainly due to the fact that a higher proportion of manure N is already in the form of ammonium at the time of application.

However, the residual fertilizer N equivalent value (i.e. the extra N availability from mineralization of organic biosolid N in subsequent years) from biosolids has been found to be considerably higher (7–15 % in the second year, 5–7 % in the third year after application) than typically found for animal manures (Pedersen, 1999), again under temperate, humid conditions. The accumulated fertilizer N equivalent value over several years may therefore well reach 45–60 % or even higher as shown by Binder *et al.* (2002), provided losses from leaching prevailing under these climatic conditions can be minimized.

Phosphorus in sludges and biosolids exists in both inorganic and organic forms with the inorganic forms generally predominating (Hedley and McLaughlin, 2005). Organic P must undergo mineralization in the soil before the P is available for plant uptake, similar to organic N, but generally the total P application rates are much higher than crop needs when sludge application is based on the management of N. Compared with commercial mineral phosphate fertilizers, mainly comprised of mono- and diammonium phosphates, biosolid fertilizer P value first of all depends on whether it is derived from a sewage plant with biological treatment only or with chemical biological treatment using coagulants like Fe/Al salts (e.g. Pierzynski and Gehl, 2005). The former typically yields biosolids with a high P availability, corresponding more or less to mineral fertilizer P (McLaughlin and Champion, 1987). Iron and aluminium coagulants, in contrast, yield P precipitates of low solubility, depending on the degree of Fe/Al surplus used in the process. These P salts will only have a low and more long-term fertilizer value.

Sewage sludges are generally considered to be poor sources of K, primarily due to the low concentrations of K in sludges. Despite having low K concentrations, the K in sludges is often a soluble constituent and considered to be fully available for plant uptake, similar to commercial mineral fertilizers such as KCl, $K_2SO_4$, and $KNO_3$.

The application of sludge to land can decrease or increase soil pH, depending on the composition of the sludge. The nitrification of ammonium, hydrolysis of iron (Fe), and aluminum (Al) compounds, and the oxidation of sulfides can all decrease soil pH depending on crop recovery of nitrates and sulfates. The addition of lime in the biological treatment process to stabilize the sludge can increase soil pH, and minerals and organic matter present in the sludge can increase the pH buffering capacity of the soil. Alkalinity may also contribute to lowering sludge P availability.

The concentration of nutrients in compost produced from household, garden and park waste are generally much lower than for sludges and biosolids, in the range 7–28 g/kg for total N, 0.6–7.0 g/kg for total P, and 4–19 g/kg for K, on a dry matter basis. Like for sludges, these composts release the organically bound nutrients (mainly N and S) slowly and thus the main reason for applying such composts may be as a soil conditioner rather than for the supply of nutrients. This applies mainly for N and the literature has documented examples of N fertilizer equivalent values in the order of 10 % or less compared to inorganic fertilizers.

Although sewage sludges, biosolids and compost contain contaminants such as metals, some commercial mineral fertilizers also contain small amounts of contaminants. While N and K fertilizers rarely contain pollutants in any significant amounts, P fertilizers potentially contain contaminants such as Cd, fluorine (F), strontium (Sr), arsenic (As), Pb, and mercury (Hg), depending on the origin of the ingredients, phosphate rock, and sulfuric acid (McLaughlin *et al.*, 1996). For example primary rock phosphate from the Kola peninsula (Russia) has a high Sr content, but it contains less Cd, chromium (Cr), nickel (Ni), and zinc (Zn) than secondary rock phosphates from other locations, e.g. Morocco, the largest phosphate deposits in the world. When assessing ecotoxicological impacts from waste application, the relative substitution of not only plant nutrients, but also possible contaminants as metals, should be included.

## 9.10.4   Potential Pollutants

Organic wastes contain various contaminants (metals, organics) that may be present in concentrations and/or forms which pose potential environmental and human health risks. Chemical residues in solid organic wastes are characterized by their relatively high ability to adsorb to waste particles and by not being fully mineralized during the retention time during processing.

### 9.10.4.1   Heavy Metals and Metalloids

Many metals and metalloids are toxic in the soil environment, even at rather low concentrations ($\mu$mol/kg soil). It is generally accepted that metals having a specific gravity (weight per unit volume) greater than 5 Mg/m$^3$ are termed heavy metals. In soils, these elements include Cd, Co, Cr, Cu, Fe, Hg, Mn, Mo, Ni, Pb, and Zn. Some heavy metals are essential for either plant or animal survival on land (Co, Cr, Cu, Fe, Mn, Mo, Ni, Zn) while others are not essential and are toxic at low concentrations (Cd, Hg, Pb). The toxicity of heavy metals varies from one element to another, and is mainly related to substitution of essential metal cations in enzymes and other vital biomolecules which thereby violates the normal biochemical function.

The heavy metals Cd, Cu, Hg, Ni, Pb, and Zn are of major interest when applying waste to land in the industrialized countries, as these elements are found in most wastes originating from industrialized societies and the exposure to humans in some situations is relatively high. In most countries, limits for these and also the metalloids As, Cr, Mo, and

***Table 9.10.3*** *Guidelines for controlling metal concentrations in soils for reuse of biosolids in USA (EPA 1993) and Europe (CEC 1986), from McLaughlin et al. (2000b).*

| Pollutant | EPA | | | | CEC[a] | | |
| --- | --- | --- | --- | --- | --- | --- | --- |
| | Limit concentration for biosolids[b] (mg/kg) | Cumulative loading limit (kg/ha) | Limit concentration for 'safe biosolids'[c] (mg/kg) | Annual loading limit (kg/ha/year) | Soil MPC[d] (mg/kg) | Biosolid MPC (mg/kg) | Annual loading limit[e] (kg/ha/year) |
| As | 75 | 41 | 41 | 2.0 | | | |
| Cd | 85 | 39 | 39 | 1.9 | 1–3 | 20–40 | 0.15 |
| Cr | 3000 | 3000 | 1200 | 150 | | | |
| Cu | 4300 | 1500 | 1500 | 75 | 50–140 | 1000–1750 | 12 |
| Pb | 840 | 300 | 300 | 15 | 50–300 | 750–1200 | 15 |
| Hg | 57 | 17 | 17 | 0.85 | 1.0–1.5 | 16–25 | 0.1 |
| Mo | 75 | 18 | 18 | 0.90 | | | |
| Ni | 420 | 420 | 420 | 21 | 30–75 | 300–400 | 3 |
| Se | 100 | 100 | 36 | 5.0 | | | |
| Zn | 7500 | 2800 | 2800 | 140 | 150–300 | 2500–4000 | 30 |

[a] Assumes soil pH within range 6–7.
[b] Absolute limit for beneficial use.
[c] Unrestricted use.
[d] MPC: maximum permitted concentration.
[e] Based on 10-year average.

selenium (Se) are set. Table 9.10.3 compares the metal guidelines for controlling metal concentrations in soils for reuse of sludges/biosolids in the USA and Europe.

Solubility and availability/toxicity to organisms of heavy metal cations ($Cd^{2+}$, $Cu^{2+}$, $Hg^{2+}$, $Ni^{2+}$, $Pb^{2+}$, $Zn^{2+}$) decreases as soil pH increases. This is due to the increase in negative charge on variable charge surfaces in soil and the propensity for these metals to precipitate as phosphates, carbonates, and hydroxides as soil pH increases. In contrast, solubility and availability/toxicity to organisms of anionic metals ($CrO_4^{2-}$, $MoO_4^{2-}$) may increase as soil pH increases, due again to increases in surface negative charge on soil particles affecting sorption. Under reducing conditions in soil, many of the metals form insoluble metal sulfides, therefore reducing availability to plants and animals.

Metal retention by soil is often described by a partition coefficient, often termed the distribution coefficient ($K_d$). This is a measure of the amount of metal retained on the soil particles divided by the concentration of metal in soil solution:

$$K_d(l/kg) = \frac{\text{Soil metal (mg/kg)}}{\text{Solution metal (mg/l)}}$$

Thus low $K_d$ values indicate a propensity for leaching and removal from soil (e.g. B, Se), and a high $K_d$ indicates likelihood of high retention time in soil and little transport (e.g. Cr, Pb). Cadmium is most mobile in low pH soil within the range pH 4.5–5.5, whereas in alkaline soils Cd is rather immobile. Based on analysis of 63 Danish soils sampled at three depths, Christensen (1989) found that 72 % of the variation in Cd partitioning ($K_d$) could be explained by variations in pH.

Critical exposure pathways for expression of heavy metal toxicity in soil have been examined as part of regulation governing reuse of sewage sludge on soil (EPA 1993). For some elements, e.g. Cd and Co, food chain transfer is the main risk pathway as these elements are easily accumulated by plants in edible tissues. For other elements, sorption in soil is strong, bioaccumulation by microorganisms low and plant uptake and translocation so low that the dominant risk pathway is (for higher animals and humans) through direct ingestion of soil e.g. Hg and Pb. For the other heavy metals, behavior in soil and bioaccumulation characteristics result in toxicity to plants and microorganisms (phyto- and ecotoxicity) being the dominant risk pathway at high concentrations e.g. Cr, Cu, Mn, Ni, and Zn (McLaughlin, 2002). Thus, prediction of risks from heavy metal pollution of soils through soil testing requires a different emphasis depending on the metal considered (McLaughlin *et al.*, 2000).

Metals which are strongly retained by soil accumulate in the surface soil horizons, and losses by plant uptake/removal and leaching processes, even over decades and centuries, are very small. Holm *et al.* (1998) studied the leaching output from plant pot experiments with 15 contaminated soils which differed in origin, texture, pH (5.1–7.8) and concentrations of Cd (0.2–17.0 mg Cd/kg) and Zn (36–1300 mg Zn/kg). Annual leaching outflows were estimated from soil water concentrations to 0.5–17.0 g Cd/ha/year and 9–3600 g Zn/ha/year per 100 mm of net percolation, corresponding to 0.1 % per year of the total top soil content of Cd and Zn. Assuming a similar removal by plants/crops the average retention time in the top soil for heavy metals like Cd and Zn is several hundred years. For these elements it is critical to ensure that loadings are also small so that build up in the surface soil horizons is avoided. A recent review concluded that the metal content of mechanically segregated MSW compost does not represent a barrier to end use of the product, but that limits on heavy metals in compost may be set to encourage recycling of composted residuals and contaminant reduction measures, which at the same time, also protect the soil and environment from potentially negative impacts caused by long-time accumulation of heavy metals in soil. However, there is no evidence to suggest that risk from metals in MSW composts are lower (per unit metal) than those from metals in biosolids. Most research on risks from metals has been conducted for sewage sludges, and until sound science suggested limits for MSW composts should be otherwise, biosolid-based limits will also be applied to these materials.

### 9.10.4.2 Organic Micropollutants

The potential number of organic pollutants is very high. Wastes applied on land may contain residues from chemicals used in animal production, industries and households. Most of the organic pollutants are non-ionic and nonpolar which means that they show rather low solubilities in water and sorb strongly to organic matter in soils. Unlike the inorganic pollutants these pollutants may be degraded and fully mineralized or they can be incorporated into soil humic substances and loose their toxicity. However, in most cases sorbed organic pollutants are not readily available for microbial degradation. Hence, substances which are less strongly sorbed to the soil solids are more easily degraded provided they are not toxic to the degraders. The less strongly retained substances are also more easily leached to groundwaters. Degradation of organic substances takes place through a number of steps with formation of a suite of metabolites. As most degradation comprises oxidation processes (in aerobic soils) the metabolites are more or less oxidized, hence they are more polar and less strongly bound in the soil than the parent compound. Metabolites may be as toxic as the parent compounds and hence also provide a threat to soil and groundwater quality.

Detergents are used in large amounts in households and industry. The sources of detergents in soils are mainly sewage sludge. Sewage sludges with highest contents of detergents (up to 10 g/kg sludge) come from anaerobic treatment plants as the detergents are rather stable under anaerobic conditions.

Most detergents are rapidly degraded in soils and normally do not represent a threat to soil functions. However, detergents may have some undesired indirect effects in soil, e.g. they can render hydrophobic pesticides, PCBs and PAHs more soluble and thus increase the mobility and bioavailability of these substances.

Drugs and their metabolites are added to soil environments mainly via sludges and animal slurries. Veterinary drugs such as growth promoters contribute the largest amount of drugs added to soils and may cause soil concentrations of > 0.1 mg/kg. Drugs may pass through the animals and storage unchanged or only slightly modified and hence may de disposed to soil in a still-active form. There is much concern about antibiotics in soil due to the risk of inducing resistance to antibiotics among soil bacteria. In general, very few physicochemical properties for drugs have been determined or are available in the literature. Some of these compounds sorb to soils whereas others are mobile.

Polychlorinated biphenyls (PCBs) have been used in electrical equipment due to their insulating and heat conducting properties. Manufacturing stopped in the 1970s. PCBs have also been used as a plasticizer, as lubricants and as fireproofing agents. PCBs in soils mostly originate from sewage sludge. The PCBs are quite resistant to degradation; the more highly chlorinated being the most resistant. In general the compounds are strongly bound in soils.

The information about dioxins in biologically treated organic wastes is very scarce and dioxin originates mainly from chemical processes involving chlorine and from combustion processes.

Polycyclic aromatic hydrocarbons (PAHs) consist of fused benzene rings, and thus only contain carbon and hydrogen. The family of PAHs is very large but the PAHs of environmental concern do not contain more than seven benzene rings. Large amounts of PAHs are formed during combustion processes in power plants, incinerators, open burning, and motor vehicles. Processes producing tars also contain heteroaromatics, i.e. compounds which in addition to C and H also contains N, S, and/or O. Important exposures of PAHs from waste application to land are sewage sludges and composts, but there may be large local variation according to distance to industrial plants, cities, etc.

Many other organics such as pesticides, oils, and chlorinated solvents may end up in the soil environment as a consequence of anthropogenic activities. However, these compounds are not directly related to application of processed organic waste to land and thus not covered in this chapter. One final group to mention is the group of natural toxins. A large number of plants and microorganisms produce metabolites which are toxic. Plants produce such substances in order to compete with other species in the defense against attack by fungi and insects. These toxins and new industrial chemicals, e.g. flame retardants, hormone-like substances, and new persistent pharmaceuticals (including formulated nanoproducts) may also be found in processed organic waste, but the fate of these compounds in the soil environment is not yet known.

## 9.10.5    Quality Requirements and Critical Exposure of Material Utilized

Regulations and practices to control organic waste recycling into agriculture often reflect local political and community opinions on acceptable quality of the material and attitudes to risk. Generally, regulations include limits concerning quantities of waste applied to land and frequency of application (annually and total applied quantities) and quality (chemical, physical, and biological characteristics, ratio of agronomic value to pollutant content). As an example, legislation in Denmark defines the maximum amounts of N and P application per hectare allowed annually to land as sludges and manure. Other measures are voluntary or mandatory pretreatment of waste to improve its quality (e.g. aeration, composting) and the use of good management practices at land application sites (e.g. season for application, establishment of buffer zones).

Often one of the following three basic approaches to setting limits can be distinguished (McGrath *et al.*, 1994):

1.  Comprehensive analyzing the exposure routes of pollutant transfer to selected target organisms and assessing the likely harmful effects that the pollutants may have on the targets.
2.  Setting limits consistent with the lowest observed adverse effect concentrations, which are actual cases of effects due to pollutants.
3.  Attempting (for metals only) to match the metal inputs to soil to the small output or losses of metals due to crop removal, soil erosion, and leaching, the sustainable criterion (see also Box 9.10.3).

These differing approaches have different and sometimes inconsistent numerical limits being set for the same constituents across various jurisdictions.

Table 9.10.4 presents exposure pathways and impacts in relation to environmental pathways of concern in the use of organic waste on land. For agricultural use of wastes, the most important exposure routes are 4–9 in Table 9.10.4, and much research has been conducted on these pathways.

Although standards and legislation for use and application of sewage sludges and other processed organic wastes were created for the safe use of these organic wastes, public concern about the negative effect that the disposal of these solids may have on public health is frequently voiced. However, a number of studies on impacts from land application of waste, in compliance with these standards and regulations, consistently indicate no detrimental effects, rather improvements in soil quality due to the input of organic matter (e.g. Petersen *et al.*, 2003; Gibbs *et al.*, 2006; Smith, 2009). Generally these studies are short term however, and regulations and guidelines must consider time scales of decades or centuries, as has been pointed out by Giller *et al.*, (2009) with respect to long-term sensitivity of soil microorganisms, and that the bioavailability and hence toxicity will be moderated by ageing, leaching and soil type (Smolders *et al.*, 2009).

***Table 9.10.4*** *Exposure routes of concern when using organic waste on land.*

| | Exposure pathway | Description of impact, example |
|---|---|---|
| 1 | Waste – Human | Humans (children) ingesting waste containing pathogens, organic pollutants or metals. |
| 2 | Waste – Soil – Air – Human | Human breathing volatile pollutants from land-applied waste |
| 3 | Waste – Soil – Airborne dust – Human | Tractor operator exposed to dust while ploughing waste-applied land |
| 4 | Waste – Soil – Plant – Human | Humans ingesting crops with increased pollutant content |
| 5 | Waste – Soil – Plant – Animal – Human | Humans eating animals feed with plants containing pollutants. |
| 6 | Waste – Soil – Plant | Yield of crop plants being affected negatively by waste-borne plant pathogens, pests, weed seeds and metal toxicity. |
| 7 | Waste – Soil – Soil organism | Metal or organic pollutant effecting the soil microbe functionality |
| 8 | Waste – Soil – Animal | Toxicity to animals ingesting soil from waste treated land |
| 9 | Waste – Soil – Surface/groundwater – Human | Potable water polluted by surface runoff or leaching from waste-treated land. |

## 9.10.6   Practical Utilization of Organic Waste as a Nutrient Source

Besides the technical, legislative, and environmental considerations of use of treated waste in agriculture, the economics must also be considered. The operation should fully calculate all the operational costs and the information on the economic value of greater crop yields, and/or savings on fertilizer. Some of the most significant economic factors to consider are:

1. The proximity of the agricultural land relative to the waste production site and if dewatering or drying of the waste material is favored.
2. If the waste is not available when needed, can they be stored, are storage facilities available, and who will provide them.
3. The cost and availability of machinery for transporting and applying the waste. Provided waste is properly treated to reduce water content and to attain good physical characteristics (bulk density, aggregation, etc.), no special equipment is needed for application to land and conventional agricultural equipment can be used.
4. Whether the farmer is willing to receive the waste without payment; in some countries/regions alternative treatment cost may enable farmers to demand a 'gate fee' for receiving the waste, as a compensation for any additional trouble with the handling of the waste, even if he saves fertilizer and enhances crop yields.

Whether the material is MSW compost, sewage sludge, or farm manure, several general principles apply to the sustainable sound application of the material to soils. Land application rates of the waste products depend on nutrient and water content of the treated organic waste and the application context (crop type, nutrient demand, etc.). The rate of application is generally governed by the amounts of N or P that the organic material will make available to plants. N usually is the first criterion because it is needed in the largest quantity by most plants, and because excess N often will represent a leaching risk. However, the P/N ratio in most organic wastes is higher than crop requirements. Consequently, if the organic waste material is applied in accordance with crop N requirements, excessive levels of soil P will result. While P is strongly bound in soils, a long-term view is needed and the risk of surface runoff of P is likely. For soils that already have high levels of P, e.g. from inorganic fertilizers, the amount of an organic nutrient source applied may be limited by its P content.

For nutrient-rich waste products like sludge and other biosolids, the application rate will often be limited by regulations on maximum N or P loads (Pierzynski and Gehl, 2005). Typical rates are in the range 1–10 Mg dry matter/ha, corresponding to 3–50 Mg wet weight/ha (Gilmour et. al. 2003). For more nutrient-poor materials like household or green waste compost, application rates may be somewhat higher, e.g. 50–100 Mg wet weight/ha. The nutrient and moisture contents of organic waste materials vary widely among sources, and they also depend on how the material has been stored and treated. Therefore, a representative sample of the material should be analyzed in a laboratory, if this information is not provided by the supplier. Analyses are generally required by laws regulating land application of waste materials.

When land is treated annually with an organic material, the application rate needed will become progressively smaller because, after the first year, the amount of N released from material applied in previous years must be subtracted from the total to be reapplied. This is especially true for composts for which the initial availability of the N is quite low. Instead of making progressively smaller applications, another practical strategy is to use a moderate rate every year, but supplement the N from other sources in the first few years until N release from residual previous applications can supply the entire crop requirement.

In some countries or regions, the controls on agricultural use of wastes have become so strict that options such as incineration or gasification are more viable, if not only for economic reasons, then for security of solutions and independence of farmers willing to receive the waste. As an example, Switzerland in 2003 introduced a ban on land application of sewage sludge, first for fodder crops and vegetables, from 2006 extended to all arable land (Anonymous, 2003), and today nearly all sewage sludge in Switzerland is incinerated.

## 9.10.7   Future Perspectives

It is likely that increasing volumes of waste will be produced by urban centres and by industries. Given the incentives for recycling and full life cycle assessment (LCA) of industrial processes, it is likely that land application and beneficial reuse of wastes will be encouraged. The lack of suitable landfill space in Europe will also provide a strong incentive to develop beneficial reuse options for wastes. However, it will be important that all actual positive and negative impacts are taken in to consideration, e.g. through a LCA. It is important that such an LCA is carried out for the local settings and at the right scale, as the result may be highly influenced by this. As an example, Bruun *et al.* (2006) and Hansen *et al.* (2006) used a land application model dealing with agricultural application of residues from composting or anaerobic digestion of organic MSW and found that the local agricultural conditions (soil type, rainfall, farm type) had a strong influence on the resulting environmental impact and fertilizer N and P substitution (see Box 9.10.2 for further details). Therefore, LCA of alternative options for organic waste disposal, including land application, should always include a careful model analysis of the land use effects, including not only direct emissions, but also secondary effects like changes in crop productivity, soil C sequestration, etc.

Methods to detect contaminants in wastes have improved dramatically, and research into the potential adverse effects of contaminants on humans or the ecosystem is continuously improving the risk assessments relating to land reuse of waste. Much is already known regarding the fate and effects of metals and nutrients. Contaminants are perhaps best controlled at source, where technologies are developed to remove the highest risk contaminants from the waste stream so that land reuse is more feasible. At the same time, waste processing needs to take into account the intended end use of the product, and additional processing (e.g. nutrient addition) may be needed to provide a material that is more suitable and 'fit for purpose'. Nevertheless, the most hazardous wastes will still need to be treated by incineration or land disposal.

An increase in quality (decreases in the concentrations of salt and undesirable elements, e.g. Cd, Hg, Pb) of organic wastes and adequate quality control would make them more competitive with chemical fertilizer and other soil amendments like peat-based growth media and other soil ameliorants. Conservation of the N that usually volatilizes as $NH_3$ during waste treatment would greatly increase the value of many products. Improvements in the physical condition of many waste products for better control and accurate application would also be helpful. Similarly, reduction of odours will assist acceptance by the public and farmers.

Perhaps the greatest gap in our knowledge will be the fate and effects of some of the new industrial chemicals, e.g. flame retardants, hormone-like substances, new persistent pharmaceuticals, personal care products, and nanomaterials. It is important that research understands the fate and effects of these chemicals in the environment prior to any waste containing these contaminants being applied to land.

## References

Andersen, M.K., Refsgaard, A., Raulund-Rasmussen, K., Strobel, B.W. and Hansen, H.C.B. (2002): Content, distribution, and solubility of cadmium in arable and forest soils. *Soil Science Society of America Journal* 66, 1829–1835.

Anonymous (2003): *Ban on the use of sludge as a fertiliser*. Press release of 26 March. Federal Department of Environment, Transport, Energy and Communications, Zurich, Switzerland.

Avery, L.M., Killham, K. and Jones, D.L. (2005): Survival of *E. coli* O157:H7 in organic wastes destined for land application. *Journal of Applied Microbiology* 98, 814–822.

Binder, D.L., Dobermann, A., Sander, D.H. and Cassman, K.G. (2002): Biosolids as nitrogen source for irrigated maize and rainfed sorghum. *Soil Science Society of America Journal* 66, 531–543.

Brandt, K.K., Petersen, A., Holm, P.E., Nybroe, O. (2006): Decreased abundance and diversity of culturable *Pseudomonas* populations with increasing copper exposure in the sugar beet rhizosphere. *FEMS Microbiology Ecology* 56, 281–291.

Bruun, S., Hansen, T.L., Christensen, T.H., Magid, J., Jensen, L.S. (2006): Application of processed organic municipal solid waste on agricultural land – a scenario analysis. *Environmental Modeling and Assessment* 11, 251–265.

Carrington, E.G. (2001): *Evaluation of sludge treatments for pathogen reduction – final report*. Report No. 5026/1. European Commission, Brussels, Belgium.

Christensen, T.H. (1989): Cadmium soil sorption at low concentrations: VIII. Correlation with soil parameters. *Water, Air, and Soil Pollution* 44, 71–82.

Dahiya, R., Malik, R.S. and Jhorar, B.S. (2001): Organic mulch decomposition kinetics in semiarid environment at bare and crop field conditions. *Arid Land Research and Management*, 15, 49–60.

EPA (1993): Part 503 – Standards for the use and disposal of sewage sludge. United States Environment Protection Agency. *Federal Register* 58, 9387–9404.

Epstein, E. (2003): *Land application of sewage sludge and biosolids*. Lewis Publishers, New York, USA.

FAO (2008): *Current world fertilizer trends and outlook to 2012*. Food and Agriculture Organization of the United Nations, Rome, Italy.

Gibbs, P.A., Chambers, B.J., Chaudri, A.M., McGrath, S.P., Carlton-Smith, C.H., Bacon, J.R., Campbell, C.D. and Aitken, M.N. (2006): Initial results from a long-term, multi-site field study of the effects on soil fertility and microbial activity of sludge cakes containing heavy metals. *Soil Use and Management*, 22, 11–21.

Giller, K.E., Witter, E., McGrath, S.P. (2009): Heavy metals and soil microbes. *Soil Biology Biochemistry*, 41, 2031–2037.

Gilmour, J.T., Cogger, C.G., Jacobs, L.W., Evanylo, G.K., and Sullivan, D.M. (2003): Decomposition and plant-available nitrogen in biosolids: Laboratory studies, field studies, and computer simulation. *Journal of Environmental Quality*, 32, 1498–1507.

Hansen, H.C.B. (2001): *Environmental soil chemistry*. Leonardo da Vinci Programme. Mataj Bel University, Banská Bystrica, Slovakia.

Hansen, T.L., Bhander, G.S., Christensen, T.H., Bruun, S. and Jensen, L.S. (2006): Life cycle modeling of environmental impacts of application of processed organic municipal solid waste on agricultural land (EASEWASTE). *Waste Management Research* 24, 153–166.

Hedley, M.J. and McLaughlin, M.J. (2005) *Reactions of phosphorus fertilizers and by-products in soils*. In: 'Phosphorus in Agriculture.' Monograph, American Society of Agronomy/Soil Science Society of America, Madison, USA.

Holm, P.E., Christensen, T.H., Lorenz, S.E., Hamon, R.E., Domingues, H.C., Sequeira, E.M. and McGrath, S.P. (1998): Measured soil solution concentrations of cadmium and zinc in plant pots and estimated leaching outflows from contaminated soils. *Water, Air and Soil Pollution* 102, 105–115.

Kirkeby, J.T., Hansen, T.L., Birgisdottir, H., Bhander, G.S., Hauschild, M. and Christensen, T.H. (2006): Environmental assessment of solid waste systems and technologies: EASEWASTE. *Waste Management and Research*, 24, 3–15.

Magid, J., Eilersen, A.M., Wrisberg, S. and Henze, M. (2006): Possibilities and barriers for recirculation of nutrients and organic matter from urban to rural areas: A technical theoretical framework applied to the medium-sized town Hillerød, Denmark. *Ecological Engineering* 28, 44–54.

McGrath, S.P., Chang, A.C. and Page, A.L. (1994): Land application of sewage sludge: Scientific perspectives of heavy metal loading limits in Europe and the United States. *Environmental Review*, 2, 108–118.

McLaughlin, M.J. (2002): *Heavy metals*. In: Lal, R. (ed.) 'Encyclopedia of Soil Science.' Marcel Dekker Inc., New York, USA.

McLaughlin, M.J. and Champion, L. (1987): Sewage sludge as a phosphorus amendment for sesquioxic soils. *Soil Science*, 143, 113–119.

McLaughlin, M.J., Tiller, K.G., Naidu, R. and Stevens, D.G. (1996): Review: The behaviour and environmental impact of contaminants in fertilizers. *Australian Journal of Soil Research*, 34, 1–54.

McLaughlin, M.J., Zarcinas, B.A., Stevens, D.P. and Cook, N. (2000a): Soil testing for heavy metals. *Communications in Soil Science and Plant Analysis*, 31, 1661–1700.

McLaughlin, M.J., Hamon, R.E., McLaren, R.G, Speir, T.W. and Rogers, S.L. (2000b): Review: A bioavailability-based rationale for controlling metal and metalloid contamination of agricultural land in Australia and New Zealand. *Australian Journal of Soil Research*, 38, 1037–1086.

Pedersen, C.Å. (1999): *Oversigt over Landsforsøgene – Forsøg og undersøgelser i de landøkonomiske foreninger. Landbrugets Rådgivningscenter* (Report on the agricultural field trials conducted by the Danish Agricultural Advisory Service; in Danish). Reports 1993–1999. Located on the internet: 19 July 2007. www.lr.dk/planteavl.

Petersen, J. (1996): *Husdyrgødning og dens anvendelse* (Animal manure and its use; in Danish). SP Rapport no. 11. Statens Planteavlsforsøg, Copenhagen, Denmark.

Petersen, J. (2003): Nitrogen fertilizer replacement value of sewage sludge, composted household waste and farmyard manure. *Journal of Agricultural Sciences*, 140, 169–182.

Petersen, S.O., Henriksen, K., Mortensen, G.K., Krogh, P.H., Brandt, K.K., Sørensen, J., Madsen, T., Petersen, J. and Grøn, C. (2003): Recycling of sewage sludge and household compost to arable land: fate and effects of organic contaminants, and impact on soil fertility. *Soil and Tillage Research*, 72, 139–152.

Pierzynski, G.M. and Gehl, K.A. (2005): Plant nutrient issues for sustainable land application. *Journal of Environmental Quality*, 34, 18–28.

Poulsen, H.D. (2009): *Normtal for husdyrgødning – 2009* (Animal manure norms for nutrient content – 2009; in Danish). Internal report. Faculty of Agricultural Sciences, University of Aarhus, Denmark.

Preusch, P.L. and Tworkoski, T.J. (2003): Nitrogen and phosphorus availability and weed suppression from composted poultry litter applied as mulch in a peach orchard. *Hortscience*, 38, 1108–1111.

Saison, C., Degrange, V., Oliver, R., Millard, P., Commeaux, C., Montagne, D. and Roux, X.L. (2006): Alteration and resilience of the soil microbial community following compost amendment: effects of compost level and compost-borne microbial community. *Environmental Microbiology*, 8, 247–257.

Shiralipour, A., Mcconnell, D. and Smith, W. (1992): Uses and benefits of MSW compost – A review and an assessment. *Biomass and Bioenergy*, 3, 267–279.

Smith, P., Goulding, K.W., Smith, K.A., Powlson, D.S., Smith, J.U., Falloon, P. and Coleman, K. (2001): Enhancing the carbon sink European agricultural soils: including trace gas fluxes in estimates of carbon mitigation potential. *Nutrient Cycling in Agrosystems*, 60, 237–252.

Smith, S.R. (2009): A critical review of the bioavailability and impacts of heavy metals in municipal solid waste compared to sewage sludge. *Environmental International*, 35, 142–156.

Smolders, E., Oorts, K., Van Sprang, P., Schoeters, I., Janssen, C.R., McGrath, S.P. and McLaughlin, M.J. (2009): The toxicity of trace metals in soil as affected by soil type and ageing after contamination: using calibrated bioavailability models to set ecological soil standards. *Environmental Toxicology and Chemistry*, 28, 1633–1642.

Steen I. (1998): Phosphate recovery – Phosphorus availability in the 21st century – Management of a non-renewable resource. *Phosphorus and Potassium*, 217.

Stevenson, F.J. and Cole, M.A. (1999): Cycles of soil. Carbon, nitrogen, phosphorus, sulfur, micronutrients, 2nd edn. John Wiley & Sons, Ltd, New York, USA.

Strauch, D. (1998): Pathogenic micro-organisms in sludge. Anaerobic digestion and disinfection methods to make sludge usable as a fertiliser. *European Water Management*, 1, 12–26.

Tian, G., Granato, T.C., Cox, A.E., Pietz, R.I., Carlson, C. Jr. and Abedin, Z. (2009): Soil carbon sequestration resulting from long-term application of biosolids for land reclamation. *Journal of Environmental Quality*, 38, 61–74.

Weggler, K., McLaughlin, M.J. and Graham, R.D. (2004): Effect of chloride in soil solution on the plant availability of biosolid-borne cadmium. *Journal of Environmental Quality*, 33, 496–504.

# 10
# Landfilling

# 10.1

# Landfilling: Concepts and Challenges

**Thomas H. Christensen**

*Technical University of Denmark, Denmark*

**Heijo Scharff**

*NV Afvalzorg Holding, Assendelft, The Netherlands*

**Ole Hjelmar**

*DHI - Water, Environment & Health, Hørsholm, Denmark*

Landfilling of waste historically has been the main management route for waste, and in many parts of the world it still is. Landfills have developed from open polluting dumps to modern highly engineered facilities with sophisticated control measures and monitoring routines. However, in spite of all new approaches and technological advancement the landfill still is a long lasting accumulation of waste in the environment. Much of current landfill design and technology has been introduced as a reaction to problems encountered at actual landfills. The solution was in many cases sought in isolation of the waste. Although this prevents immediate emission, isolation at the same time is a conservation of potential emission. This potential emission materializes when the isolation fails at some point in time. Therefore it is of importance in the striving for sustainable waste management solutions to understand the concepts, the processes and the long-term aspects of landfilling.

This chapter describes the main conceptual aspects of landfilling. The historical development is presented and key issues of time frames, mass balances and technical approaches are discussed.

The environmental issues of landfilling are described in Chapter 10.2 while specific types of landfilling technology are described in Chapter 10.5 (mineral waste landfill), Chapter 10.6 (reactor landfill) and Chapter 10.7 (pretreated waste landfill).

## 10.1.1   Historical Development of Landfilling

The history of landfilling is the basis for understanding current approaches and technologies as well as the future challenges. The following coarse historical journey through the world of landfills may take its starting point in the 1950s when waste reflected the early industrialization of society and no longer was dominated by organics and ashes. The terms used in the following are descriptive and do not necessarily represent any standardized or legal terms.

- *Open dumps* were often clay and gravel pits or other low-value land filled with what ever waste that might appear, including industrial and hazardous waste: the waste was dumped from the truck where possible. As long as the dumps were small and local waste disposal sites, the impacts and problems were often only local and tolerable, maybe except with respect to smells, rodents and local fires. However, as urbanization developed, the dumps grew bigger and urban areas often moved in close to the disposal sites making nuisances and esthetic issues important.
- The *sanitary landfill* offered a more orderly appearance by limiting access to the site (fences), organizing the disposal activities and often covering of the waste with soil. Where land was plentiful, the sanitary landfill could be trenches dug into the ground. Less contact with the waste, less litter and organized rodent controls were a sanitary improvement. However, issues of gas and leachate were not very much in focus. But large sanitary landfills would after some years start causing damages on vegetation and uncontrollable fires due to gas and pollute nearby streams by seeping leachate. The sanitary landfill site designed with consideration of leachate generation and migration was also called a 'dilute and attenuate' landfill.
- The *controlled landfill* offered controls on leachate and gas by introducing liners, collection systems and treatment facilities for leachate and gas. The protection of groundwater and surface water had high priority in many countries as cases of groundwater pollution around sanitary landfills appeared more frequently and the cost of remediation became high. Much focus was on how to construct liners and leachate drainage systems and how to treat the strongly polluted leachate. The gas issues were more local and controlling the gas led to flaring of the gas and soon also utilization. However, over the years it was experienced that liners may not stay impermeable, leachate collection systems may clog, and leachate from the methanogenic phase is difficult and expensive to treat prior to discharge. It was also learned that gas collection was not complete and landfill covers were needed to ensure efficient gas utilization. Leachate recirculation was in some cases introduced as a treatment option for leachate and an approach to enhance gas generation making collection and utilization more economic. Hazardous waste was not allowed in the controlled landfill, but other restrictions on the waste were few. The controlled landfill and its later developments were also named 'containment landfills'.
- The *dry tomb* was a reaction to all the shortcomings of traditional technology. The concept was to cover the landfill completely in order to prevent water infiltration and leachate generation and to support full collection of landfill gas. However, the dry tomb leads to less degradation of the organic waste and less gas generation, and hence slower stabilization of the organic content because of low moisture content. Leaching is also minimum, thus leading to a conservation of the waste and a need for potentially eternal maintenance of the top cover.
- Various *bioreactor landfill* technologies were suggested and in some cases introduced to enhance degradation and shorten the time for stabilization of the waste: the semiaerobic landfill (primarily in Japan) and the flushing bioreactor (primarily in the UK); the latter based on the understanding that nitrogen potentially is the most persistent pollutant in leachate and therefore removal of nitrogen should be enhanced.
- In recent years quality requirements to the waste being landfilled and mandatory plans for closing of the landfill have been introduced in some countries as a consequence of the fact that potential impacts from landfilled waste may continue longer than most technical measures installed for their control may last. Financial deposits to pay for future remediation are often requested. This new type of landfill has still not gained a common name.

## 10.1.2   Emissions and Time Frames

The two emissions of main environmental concern are gas and leachate. Most engineering of landfills is focused on controlling these emissions to the environment.

### 10.1.2.1 Landfill Gas

Landfill gas ($CH_4$ and $CO_2$ primarily) is generated by the anaerobic degradation of organic waste (Chapter 8.4). Top covers, gas wells, pumps, flares and engines/turbines for utilization are part of the control of the gas (Chapter 10.10). The driving force is to reduce the environmental issues (explosion and fire hazards, greenhouse gas emission, odours, and vegetation damages; Chapter 10.2), but also the utilization of the gas for energy recovery is of importance at least at large landfills.

The amount of gas generated depends on the amount and composition of the organic content of the waste. The rate of gas generation depends on the composition of the organic waste and the biochemical environment in the landfill (plenty of water, no inhibitors present). The amount of gas released is linear with the amount of waste put into the landfill per year and the filling horizon of the landfill. A landfill receiving waste for 25 years may have sections where gas generation rate is on the decline while other sections may hardly have begun generating gas. The gas moves out of the waste by the pressure that it builds. The geometry of the landfill and the compaction of the waste affect how easy it is to collect the gas.

In a modern landfill the majority of the gas is generated within the first 30 years after disposal, but a long tailing of the gas generation is expected. At a certain point in time the flux of gas is so low that methane oxidation in the soil top cover may be able to covert all methane to carbon dioxide and herby remove the main problem associated with gas. If a synthetic liner is used as top cover, this oxidation may be more difficult to obtain. However, this suggests that the gas issue has a limited time period. Based on model predictions and laboratory experiments the gas issue is insignificant 100 years after disposal of the waste. This does not mean that all organic carbon has been removed as gas, but that the flux of methane is balanced by the oxidation in the top cover. In a mixed waste landfill it is likely that half of the biogenic carbon still is in the landfill after 100 years (little from kitchen organics, some from cardboard and paper, significant fractions of wood; Chapter 10.6).

The landfill gas issue is challenging in terms of efficient collection and utilization of the gas and of establishing long-lasting top covers offering methane oxidation. But presuming that the issue does not exceed 100 years, existing well designed and well maintained technology can last to manage the problem. However, the gas issue still suggests that control and maintenance after closing of the landfill is needed.

### 10.1.2.2 Landfill Leachate

Landfill leachate is generated by the infiltrating rain percolating through the waste and leaching of contaminants. Top-covers, bottom liners, drainage systems, pumps and treatment plants are part of the control of the leachate (Chapters 10.3, 10.8, 10.9, 10.11). The driving force is to reduce the risk of groundwater and surface water pollution and to clean collected leachate prior to discharge. Drainage water from the landfill that has not been in contact with waste is not defined as leachate.

The rate of leachate generation depends primarily on the net infiltration, the top cover and the area of the landfill (Chapter 10.3). In addition, the depth of the landfill and the bulk density of the waste determine the amount of leachate per tonne of waste per year. The amount of leachate collected further more depends on the efficiency of the bottom liner and the drainage system. The leachate moves by gravity and in the drainage system also by pumping. For a fixed time frame, a shallow landfill will generate more leachate per year per tonne of waste than a deep landfill. Thus the geometry and the top cover design significantly determine how fast the waste is leached. A shallow landfill with a low quality soil top cover may leach 20–50 times faster than a deep landfill with a good soil cover.

Minerals, salts and hydrolyzed organic matter are dissolved in the leachate. Dissolution of some inorganic substances is limited by their solubility, dissolution of others is limited by their availability. The content of dissolved organic matter depends on the level of organic waste degradation. In the acidic phase, volatile fatty acids are dissolved in the leachate reflected by the high BOD and COD content of the leachate, while in the methanogenic phase the volatile acids have been converted to $CH_4$ and $CO_2$, reducing the organic load of the leachate and increasing its pH.

Manfredi and Christensen (2009) modelled the leaching from a typical mixed waste landfill (20 m deep, 250 mm leachate/year) for a 100-year period and found when compared to the original content of the waste that 60–85 % of N, 30–50 % of Cl and 98.7–99.9 % of heavy metals (Pb, Zn, Cd, Cr, Hg) still were present in the landfill. This suggests that the leaching may continue long after 100 years, so the critical issue is the concentration or flux of the pollutants that continue to leach out. In a real landfill, 100 years may correspond to a liquid:solid (*L/S*) ratio of 1–2 m$^3$/t; and laboratory and lysimeter experiments suggest that significant concentrations continue to appear even after *L/S* = 5, which may be 500 years of leaching in a real landfill. The actual landfill technology (bioreactor, flushing bioreactor, semiaerobic) does not significantly change the problem (Manfredi and Christensen, 2009). It is not likely that bottom liners and drainage systems will fully function that long and, unless we want to reinstall all our landfills with regular intervals of 100–200 years, we need to take advantage of the following opportunities:

- Locate the landfill in an area where the environment by natural attenuation (dilution, degradation, sorption and precipitation) can accept the leachate load when the technical measures stop functioning, e.g. after 100 years.
- Limit the waste accepted at the landfill to types or treated types of waste that show leaching behaviors that will not lead to excessive concentrations for example after 100 years.
- Enhance the stabilization of the waste in the landfill by active methods.

## 10.1.3   Concepts in Landfilling

Accepting that leaching of pollutants from the landfilled waste may last for many centuries and probably longer than the engineered means that we introduce for collecting the leachate, several pragmatic concepts have emerged trying to compensate for our limited ability to perform accurate long-term prediction of leaching and technical barrier performance. The early concepts were frameworks reminding us about the long-term issues and that we, in spite of engineering shortcomings, should do our best, while the later concepts – although still on a simple level – offer some quantitative approaches for addressing the long-term perspectives. The main concepts are briefly described below.

### 10.1.3.1   The Multibarrier Concept

The multibarrier concept, which has been introduced in various forms [Stief (1986) and Ryser (1989) were probably some of the first presentations], is a commonsense concept stating that since our predictions of the long-term aspects of leaching and leachate containment are very uncertain we should in our planning and design of landfills introduce many barriers that each could contribute to the reduction of the problem. Some of the barriers are prophylactic while others are actual technical installations. While all barriers should be considered, not all barriers would be feasible in every case. The barriers that should be considered are:

- Site landfill distant from important groundwater resources and vulnerable surface waters.
- Site landfills in geological stable strata avoiding physical damage to the landfill body and its installations.
- Site landfills in geological strata providing subsurface attenuation capacity (e.g not in fractured rock) in case minor leakage should take place.
- Place the landfills above ground with natural drainage to visible ditches that will reveal inadequate functioning of the leachate collection system.
- Limit the types of waste accepted at the landfill and/or demand pretreatment of the waste prior to landfilling. Enforce the limitations by inspecting the waste entering the landfill.
- Install bottom liners and drainage collection systems that are adequately designed, constructed and maintained.
- Install top covers that are adequately designed, constructed and maintained.
- Operate the landfill so that the stabilization processes can proceed.
- Monitor leachate generation and surrounding groundwater to control that the landfill leachate is collected and does not enter the groundwater.
- Prepare a contingency plan that specifies immediate investigations and possible remediation initiatives if unexpected leachate appearance or behavior is observed.

**Table 10.1.1** *Leachate composition at final storage quality ($C_{FSQ}$) and estimated time it takes to reach these values according to data presented by Belevi and Baccini (1989). Long-term behaviour of municipal solid waste landfills. Waste Management and Research, 7, 43–56.*

|  | $C_{FSQ}$ (mg/l) | Years to reach $C_{FSQ}$ |
|---|---|---|
| $C_{org}$ | 20 | 500–1700 |
| $N (NH_4^+)$ | 5 | 55–80 |
| F | 1 | < 10 |
| P | 0.4 | 100–700 |
| S | 30 | < 10 |
| Cl | 100 | 100–150 |
| Fe | 10 | < 10 |
| Cu | 0.1 | < 10 |
| Zn | 0.6 | < 10 |
| Pb | 0.5 | < 10 |
| Cd | 0.05 | < 10 |

### 10.1.3.2   Final Storage Quality

The concept of 'final storage quality' was first introduced in Switzerland (EKA, 1986). The Swiss waste management guideline contained issues about sustainability before the Brundtland Commission introduced the general concept world-wide in 1987. The Swiss guideline suggested that a landfill reaches final storage quality when it meets the quality of the surrounding geological strata. This is specified in terms of reaching fluxes of elements so low that they will not significantly alter the surrounding environment. Although the concept is general, focus is on leachate. The fluxes were related to the leaching in terms of leaching tests, and originally the limit values obtained in the leaching test were similar to the Swiss discharge limits to surface water. The final storage quality concept thus targets the leachate reaching a quality that will not do unacceptable harm to the environment. The concept tacitly accepts that some kind of dilution will happen after the leachate has exited the landfill.

Belevi and Baccini (1989) estimated based on modelling and leaching test on solid waste samples taken in landfills how long it would take to reach final storage quality. The leachate quality to reach was set by the authors as ten times the Swiss quality standards for running surface water [for dissolved organic carbon (DOC), nitrogen (ammonium) and heavy metals]; but for fluoride, chloride and sulfate the original values of the quality standard were chosen because these anions generally show higher mobility than the other substances. Phosphorous was set as tenfold the upper phosphate level in oligothrophic lakes. The target values and the estimated time to reach these values are shown in Table 10.1.1. It was concluded that, for a typical Swiss landfill, it would take many centuries to reach final storage quality. Dissolved organic carbon (500–1700 years), ammonium (55–80 years), phosphorous (100–700 years) and chloride (100–150 years) were considered the most critical of the investigated substances. They concluded that only treatment of the waste prior to landfilling could make it possible to reach final storage quality within the lifetime of a generation.

### 10.1.3.3   Site-Specific Risk Assessment

Site-specific risk assessment can be used in the evaluation of the environmental aspects of establishing a new landfill or in assessing if an existing landfill has reached a level of stabilization that allows its surrender to nature or – in a legal sense – releases its owner for his obligations to continued aftercare. Regarding existing landfills a risk assessment guideline has by issued in the UK by the Environment Agency (2005).

Site-specific risk assessment is the structured and transparent process of identifying and quantifying risks and assessing the significance of these risks in relation to other risks in a specific area with an existing or planned specified landfill. This implies that emissions of leachate and gas from the specific landfill can be quantified with an acceptable uncertainty for the lifetime of the landfill. For existing landfills actual monitoring data can form part of the data for the assessment.

Risk assessment in general builds on the source–pathway–receptor methodology. This means that if there is no source of pollutants or no pathway along which pollutants can travel or no receptor for the pollutants, then there is no risk. All three elements need to exist for there to be a risk. In the case of landfills, we know that gas and leachate will appear for decades, but their quantification in terms of individual substances is needed to quantify the source term. Considering that gas and leachate contains a large range of substances the level of effort put into assessing each substance must be proportionate to the magnitude and complexity of the risk. The risk assessment may follow the following steps:

***Conceptual model:*** The first step in a site-specific risk assessment often is developing a conceptual model of the proposed landfill and its environmental setting. The preparation of a robust conceptual model is a critical element in successfully evaluating environmental risks since, if the landfill and environmental context of the site are not understood, any subsequent risk assessment will be flawed. The model should (Environment Agency, 2005):
- Describe the design, construction and operation of the landfill.
- Identify possible sources, pathways and receptors.
- Identify the processes that are likely to occur along each source–pathway–receptor linkage.
- Incorporate geological, hydrogeological, topographical, analytical and landfill development information into a single coherent model.

***Risk screening:*** The second step in site-specific risk assessment is risk screening (sometimes referred to as Tier 1), which based on assessing the likelihood and the consequences of effects, prioritizes the risks such that any subsequent, more detailed risk assessment can focus on those risks identified as important. Risk screening involves:
- Identifying possible source-pathway-receptor linkages from the conceptual model. The receptors may be humans, flora, fauna, air, water, land, buildings/structures. Especially airports, hospitals, schools, protected natural habitats, classified surface waters and groundwater reservoirs for drinking water may be sensitive receptors.
- Identifying the most appropriate baseline level for the substances in the environment. The concentrations in the environment resulting from an emission can then be compared against these baseline levels. This allows the significance of emissions to be assessed and a decision to be made on whether the impact of the landfill on air or water quality is acceptable (i.e. does the emission pose a pollution risk).
- An initial assessment of the likelihood and magnitude of any effects that could be associated with each source–pathway–receptor linkage.

***Quantitative risk assessment:*** The third step is to quantify the critical risks in order to provide a basis for deciding that the risk is acceptable, the landfill should not be accepted, or that acceptance is conditioned, for example, in terms of technical improvements in the design or restrictions in the waste accepted. This quantification often involves further data collection and advanced modelling of the source–pathway–receptor linkage. Since the time horizons are long, maybe 100 years, the dynamics of the boundary conditions must also be taking into consideration. Examples are that the general groundwater flow direction may shift due to changes in groundwater abstraction and surface water regulations, and that dwellings with time may be established closer to the landfill as a consequence of urban development.

## 10.1.3.4    Waste Acceptance Criteria for Landfilling in the EU

In the EU waste acceptance criteria have been developed for granular waste to be landfilled in terms of leaching criteria (CEC, 2003). The concept is that the leaching should not lead to an unacceptable increase in concentration of key pollutants in the groundwater downstream the landfill. The methodology used to develop the leaching criteria for acceptance of waste is in essence a general source–pathway–receptor approach and has been described in detail by Hjelmar *et al.* (2001, 2005a).

The procedure consists of a series of consecutive steps. Due to time constraints in the legislation procedure, only the impact on groundwater was considered. The methodology may easily be expanded to include impacts on surface water bodies.

First, a decision must be made concerning the primary target(s) or point(s) of compliance (POC), e.g. the downstream point(s) where the groundwater quality criteria must be fulfilled. For setting of the criteria for mineral waste, the EU considered the groundwater quality 20 m downstream of the landfill.

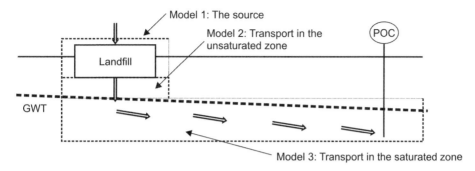

**Figure 10.1.1** *Cross-section showing the principle of three coupled source and transport models used for the forward impact calculation at a landfill scenario used in setting the EU leaching based waste acceptance criteria.*

Based on existing European groundwater/drinking water legislation, primary quality criteria were then set for the peak concentrations in the groundwater of a number of contaminants, and the physical characteristics of the landfill scenario and the environment scenario were selected and described. The environment scenario includes the general net rate of infiltration and a hydrogeological description of the unsaturated and saturated (aquifer) zones upstream, below and downstream of the construction application. The source of the various contaminants was subsequently described in terms of the flux of contaminants as a function of time based on leaching data (which were transformed into an analytical description based on a simplified continuously stirred tank reactor (CSTR) model: $C = C_0 \exp[-(L/S)\kappa]$ and the hydraulic scenario defined. The description of the source term may be improved by using material-specific $\kappa$ values (a kinetic constant describing the rate of decrease of the concentration as a function of $L/S$ for a given material and a given component) over limited $L/S$ ranges. From the above, the amount of contaminant, $E$ (in mg/kg), released over the period of time it takes for $L/S$ to increase from 0 l/kg to the value corresponding to $C$, can be derived ($E = C_0/\kappa \{1 - \exp[-(L/S)\kappa]\}$).

Then the migration of the contaminants from the base of the landfill through the unsaturated zone into the groundwater and through the aquifer to the POC was modelled including only reversible, sorption-based contaminant/subsoil interaction processes and using proven and accomplished flow and transport models (see Hjelmar *et al.*, 2001, 2005a). Selected $K_d$ values were used for each contaminant to calculate and incorporate the retardation factors (assuming linear sorption isotherms).

Using the source terms shown above in conjunction with the migration models, 'forward' calculations were performed to determine so-called 'attenuation factors' (the ratio between the source peak concentration and the peak concentration as modelled at the groundwater POC) for each contaminant. The principle of using the three coupled source and transport models is illustrated in Figure 10.1.1.

The attenuation factors were then used for a 'backward' calculation of the permissible values of the source term corresponding to the selected groundwater quality criteria for each contaminant at the selected POC. In the calculations of the EU waste acceptance criteria the background concentration of the contaminants in the upstream groundwater was not taken into account (this could, however, easily be done).

The resulting source term criteria were then each transformed into limit values for a specific test (a column or batch leaching test) corresponding to a specific L/S value (this allows the use of leaching tests performed at different *L/S* values, each with a specific set of limit values). To find the leaching limit value in terms of released amount (mg/kg) of a given component, the maximum allowable concentration found above was transformed to maximum allowable released amount using the formula for $E$ shown above. The risk-based calculated limit values finally underwent some pragmatic and 'political' adjustments. The final regulatory limit values for leaching, which have been set within the EU for waste to be accepted at landfills for inert waste, landfills for nonhazardous waste receiving stable, nonreactive hazardous waste and landfills for hazardous waste are shown in Chapter 10.5 (CEC, 2003).

It should be noted that the criteria-setting procedure described above involves simplifications and generalisations of complex and diverse physicochemical processes, e.g. neglecting waste–waste interactions. This may be justified by the need to have an operational and relatively simple system, which can be used for the development of general criteria. If needed, it is always possible to apply other or more sophisticated models and to adapt them to other general or site-specific conditions without changing the principle of the calculations. The 'forward' part of the methodology may also be applied for site specific assessments of the potential environmental impacts caused by formation and migration of leachate from a landfill.

## 10.1.4   Challenges in Landfill Technology

The main goal of waste management is to protect human health and the environment. The load to the environment is determined by mass flows. A mass flow is the combination of a concentration ($kg/m^3$) and the volumetric flow of the medium: either gas to the atmosphere or water to soil, groundwater and surface water ($m^3/year$). The volumetric flow can also be related to the surface of the landfill in this case it becomes $m^3/m^2/year$. It can be stated that an acceptable mass flow, that does not compromise human health or the environment, does not depend on the waste but on the environment. In many countries protection measures on landfills are based on the waste categories that the landfill accepts. This may not be the optimal approach to determine how a sustainable landfill is designed; e.g. how should the landfill be capped or should *in situ* stabilization measures be implemented after waste placement.

Locating the landfill in an area where natural attenuation can take over when the technical measures stop functioning, limiting waste acceptance based on leaching behavior, and enhancing the stabilization of the waste in the landfill lead to several challenges:

- We must make engineering predictions of the natural attenuation capacity of a site.
- We must establish what is an acceptable load to the environment.
- Although we treat the waste prior to landfilling it is likely that we need to landfill waste types that in terms of concentrations will show high leaching also after for example $L/S = 2$ ($m^3/t$).
- If leachate concentrations are high we can reduce the flux by limiting the amount of leachate, but at the same time we extend the time period of the problem.

There are various possibilities to determine what emission is acceptable to a surrounding environment. The most detailed and site-oriented approach would entail toxicological knowledge and site-specific soil and groundwater data. Acceptable emission can also be defined in more general terms. In European regulations it is stated that inert waste does not give rise to environmental pollution or harms human health. Consequently no capping and aftercare are required. This implies that the emission associated with the leaching limit values (see above: Waste acceptance criteria for landfilling in the EU) for inert waste is considered acceptable. For landfills for inert waste an infiltration of 300 mm/year was assumed. This in turn implies that, if the infiltration on a landfill by means of a better cover can be sustainably reduced to for instance a third (100 mm/year), the leaching could be three times higher before exceeding the same acceptable emission. Table 10.1.2 shows likely infiltrations through various top covers.

The development of risk-related waste acceptance criteria, e.g. as described above for the EU for granular waste, represents several challenges. One is to ensure that the result of the required leaching test and the associated limit values

**Table 10.1.2**   *Cover constructions and indicative infiltration at 800 mm/year rainfall.*

| Construction | Indicative infiltration (mm/year) |
|---|---|
| 1. Cover soil > 1 m | 300 |
| 2. Optimized cover soil > 1.5 m | 120 (60–150) |
| 3. Capillary barrier | 60 (40–80) |
| 4. Single mineral liner | < 20 |
| 5. Geomembrane + mineral liner (combination liner) | < 5 |
| 6. Geomembrane with leak detection | < 1 |

actually are related to the anticipated risk against which they are meant to provide protection. Another challenge is to reach a reasonable compromise between science and administration: scientists want the truth and administrators want it to be simple. In this case the resulting compromise was reached through modelling and simplification. It is far from perfect and should be further developed as new insights are reached. Another challenge is to develop similar, general waste acceptance criteria for landfilling of monolithic waste, see e.g. Hjelmar *et al.* (2005b).

Many countries require that aftercare of a landfill shall be carried out for at least 30 years. Some countries require that aftercare shall be carried out for as long as may be required by the competent authority, taking into account the time during which the landfill could present a hazard. This is a problem since competent authorities may not have any guidance on how to assess potential hazards or how to make a decision concerning the end of aftercare. Competent authorities tend to avoid risks. In some countries competent authorities can without substantiation increase the period for aftercare. In order to release a landfill from aftercare competent authorities need to be convinced with data on the current condition that the risk of a landfill is minimized. In addition to that they need to be convinced that this situation is stable and not likely to change at any point in the future.

Both for a substantiated decision on the most suitable cap and for release from aftercare it is essential to be able to assess the leaching potential of a waste body several years after placement. A standardized assessment method is currently not available. It is however conceivable that such method can consist of:

- An assessment of the historical and current data on landfill gas and leachate quality and quantity and the hydrology of the waste body.
- Leaching and biodegradation tests on waste samples from cores drilled into the waste body to provide data for modelling of long term emission.
- Statistical analyses (including confidence intervals) and modelling of the long-term emission potential.
- Inclusion of local soil and groundwater conditions in the assessment.

A recent development in modelling (Scharff *et al.*, 2007; van Zomeren *et al.*, 2007) indicates such an approach is feasible. It is a challenge for the landfill community to develop criteria for the end of aftercare and a methodology to assess having reached these criteria.

The end of aftercare, however, is not only related to emission to soil and groundwater. The emission to the atmosphere should also be included. Experience indicates that landfill gas emission reaches a low level after decades, whereas emission into soil and groundwater could last for centuries. It is common that 30–50 years after waste processing the emission of a landfill has reached levels below $10\,m^3\,CH_4$/ha/h. In situations where a more permeable cover is acceptable from a soil and groundwater protection perspective, atmospheric emission can be controlled by means of methane oxidizing covers. For different types of top covers oxidation rates of $10–100\,m^3\,CH_4$/ha/h have been reported (Chapter 10.9). This already has led several regulators to issue guidance that requires no measures or an oxidizing biocover when the emission rate is below $5–10\,m^3\,CH_4$/ha/h (Finnish Environment Institute, 2001; Bour, 2005; Stegmann, 2006). Progress in methane oxidation research could be used to define oxidation rates for specific cover types in specific climatic conditions.

# References

Belevi, H. and Baccini, P. (1989): Long-term behaviour of municipal solid waste landfills. *Waste Management and Research*, 7, 43–56.

Bour, O., Couturier, C., Berger, S. and Riquier L. (2005) *Evaluation des risques liés aux émissions gazeuses des décharges: propositions de seuils de captages*. INERIS-DRC-05-46533/DESP-R01, France.

CEC (2003): Council Decision 2003/33/EC of 19 December 2002 establishing criteria and procedures for the acceptance of waste at landfills pursuant to Article 16 of and Annex II to Directive 199/31/EC. *Official Journal of the European Communities*, L11, 27–49.

Environment Agency (2005) *Guidance on landfill completion and surrender*. Environment Agency, Bristol, UK.

EKA (1986): *Leitbild für die schweizerische Abfallwirtschaft* (Guidelines for the waste management in Switzerland; in German). Schriftenreihe Umweltschutz Nr. 51, BUS, 3003. Eidg. Kommission für Abfallwirtschaft, Bern, Switzerland.

Finnish Environment Institute (2001) *Environment Guide 89: Guide for closing landfills* (in Finnish). ISSN 1238-8602, ISBN 952-11-1021-X and 952-11-1022-8 (PDF). Finnish Environment Institute, Helsinki, Finland.

Hjelmar, O., van der Sloot, H.A., Guyonnet, D., Rietra, R.P.J.J., Brun, A. and Hall, D. (2001): *Development of acceptance criteria for landfilling of waste: An approach based on impact modelling and scenario calculations.* In Christensen, T.H., Cossu, T. and Stegmann, R. (eds): Proceedings of the eighth waste management and landfill symposium, vol. 3, pp. 771–721.

Hjelmar, O., Holm, J., Hansen, J.G. and K. Dahlstrøm (2005a): *Implementation of the EU waste acceptance criteria for landfilling in Denmark.* In: Cossu, R. and Stegmann, R. (eds) Proceedings of the tenth international waste management and landfill symposium, Sardinia. CD-ROM. CISA – Environmental Sanitary Engineering Centre, Cagliari, Italy.

Hjelmar, O., Bendz, D., Wahlström, M., Suér, P., Laine-Ylojoki, J. and Baun, D.L. (2005b): *Criteria for acceptance of monolithic waste at landfills.* In: Cossu, R. and Stegmann, R. (eds) Proceedings of the tenth international waste management and landfill symposium, Sardinia. CD-ROM. CISA – Environmental Sanitary Engineering Centre, Cagliari, Italy.

Manfredi, S. and Christensen, T.H. (2009): Environmental assessment of solid waste landfilling technologies by means of LCA-modeling. *Waste Management*, 29, 32–43.

Ryser, W.H. (1989): Control of reactor landfills by barriers. In: Bacini, P. (ed.) *The landfill. Lecture notes in earth science, No 20.* Springer-Verlag, Berlin, pp. 117–130.

Scharff, H., Jacobs, J., van Zomeren, A.and van der Sloot, H.A. (2007): *Inorganic waste landfill and final storage quality.* In: Cossu, R. and Stegmann, R. (eds) Proceedings of the 11th international waste management and landfill symposium, Sardinia. CD-ROM. CISA – Environmental Sanitary Engineering Centre, Cagliari, Italy.

Stegmann, R., Heyer, K-.U., Hupe, K. and Willand, A. (2006): *Deponienachsorge –Handlungsoptionen, Dauer, Kosten und quantitative Kriterien für die Entlassung aus der Nachsorge.* IFAS, Hamburg, Germany.

Stief, K. (1986): Das Multibarrierenkonzept als Grundlage von Planung, Bau, Betreib und Nachsorge von Deponien (The multi-barrier concept as basis for planning, design, operation and after care at landfills; in German). *Müll und Abfall*, 18, 15–20.

van Zomeren, A., Scharff, H., van der Sloot, H.A. (2009): *Environmental risk assessment for determination of long-term emissions of landfills.* In: Cossu, R. and Stegmann, R. (eds) Proceedings of the 13th international waste management and landfill symposium, Sardinia. CD-ROM. CISA – Environmental Sanitary Engineering Centre, Cagliari, Italy.

# 10.2

# Landfilling: Environmental Issues

**Thomas H. Christensen, Simone Manfredi and Peter Kjeldsen**

*Technical University of Denmark, Denmark*

Waste disposed of in a landfill is by its nature different from the material found in the surroundings of the landfill and thereby the landfill may potentially affect the surrounding environment. This may be in terms of attracting or repelling flora and fauna from the area and through the emission to air, soil and water caused by the processes stabilizing the waste in the landfill. The main factors controlling the actual environmental impacts from the landfilling are: the nature and amount of the waste landfilled, the geological and hydrological setting of the landfill, the landfill technology, the extent and quality of the technical environmental protection measures introduced, the daily operation and the timescale.

This chapter describes the main potential environmental impacts from landfills. The modern landfill is able to avoid most of these impacts. However, in the planning and design of landfills it is important to understand the potential environmental impacts, which must be avoided.

The emissions of landfill gas and leachate causing most of the environmental risks are described in detail in the chapters addressing specific landfill types: Chapter 10.5 (mineral waste landfill), Chapter 10.6 (reactor landfill) and Chapter 10.7 (pretreated waste landfill).

## 10.2.1  Potential Environmental Impact

Figure 10.2.1 shows the potential environmental impacts from waste landfills categorized according to air, soil and water. The individual potential impacts are discussed further in the following sections. Figure 10.2.1 also indicates the distance of influence around the landfill. Most of the potential impacts are fairly local, i.e. within a few kilometers from the landfill. However, two impacts are of a global nature and have no geographical relationship to the actual landfill: Global warming and stratospheric ozone depletion.

Most of our current understanding of the potential environmental impacts from landfills originates from observations during the past 40 years of impacts from old dumps with very little engineering measures introduced to control the impacts. Our understanding of the impacts from landfills thus originates from badly operated landfills that had received an unknown mixture of waste from many different sources. These observations have fostered the introduction of extensive technical measures at new landfills, which have much less impact on the environment than the old dumps had, and, as

*Solid Waste Technology & Management*   Edited by Thomas Christensen
© 2011 Blackwell Publishing Ltd

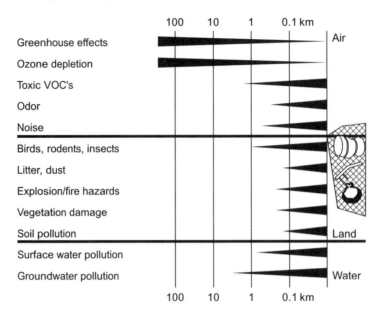

**Figure 10.2.1**   *Potential environmental impacts from landfills with estimated zone of influence of each potential impact.*

a recent thing, quality criteria for waste to be landfilled. Modern landfills should be able to avoid the potential impacts shown in Figure 10.2.1.

### 10.2.1.1   Greenhouse Effects

Landfills containing organic waste produce landfill gas containing $CO_2$ and $CH_4$, which are greenhouse gases. $CO_2$ emitted from a landfill is usually not counted as a greenhouse gas, since $CO_2$ originates from the degradation of recently photosynthesized organic matter. Methane originates from the same organic matter, but has a much higher radiating force than carbon dioxide and therefore counts as a greenhouse gas potentially contributing to global warming.

Methane emitted with landfill gas is oxidized very slowly in air and increased concentrations are observed in the atmosphere. This increases the reflection of the infrared radiation from earth and may lead to increased temperature. The radiation force of $CH_4$ is in a 100-year perspective about 25 times higher than $CO_2$ on a weight basis [the global warming potential (GWP) of $CH_4$ is 25 kg $CO_2$ equivalents/kg $CH_4$]. Box 10.2.1 explains these issues in more detail and provides GWP for several greenhouse gasses. Overall methane from landfills is on the order of 5 % of total anthropogenic greenhouse gas emission (Bogner *et al.*, 2008).

Landfill gas methane is emitted with raw landfill gas through the open landfill face, cracks in top covers and technical vents. However, methane may also be oxidized to carbon dioxide, whereby it looses its potential as greenhouse gas. This may be done technically, e.g. by collection and flaring or utilization in an engine. However, a small percentage of methane is believed even in technical systems to pass through as unburned methane. Methane may also be oxidized biologically either in dedicated filters installed in the top cover to promote methane oxidization or naturally in the top cover soil. Where the flux of methane is limited, e.g. in the late stages of stabilization, and the top soil porous and biologically active, methane oxidation may be complete. These issues are further described in Chapters 10.9 (top covers) and 10.10 (gas production, extraction, utilization).

Biogenic carbon remaining in the landfill due to incomplete degradation constitutes a saving with respect to global warming. This is a direct consequence of assuming that $CO_2$ of biogenic origin is neutral when emitted to the atmosphere (Christensen *et al.*, 2010). Manfredi *et al.* (2010) estimated that, after 100 years, still 30–50 kg $C_{biogenic}$ may be left per tonne of waste landfilled. The GWP of stored biogenic C is 44/12 or 3.67 kg $CO_2$ equivalents/kg C stored, suggesting that bound biogenic carbon may be a significant saving with respect to global warming.

**Box 10.2.1  Global Warming Potentials of Gases. Reproduced with permission from *Environment Centre Northern Territory*, Australia. © (2010) ECNT.**

Many gases present in the atmosphere trap infrared heat which without the presence of the gases would have left the earth's atmosphere. Carbon dioxide has this effect and is the greenhouse gas with the largest manmade emission. However, other gases are also greenhouse gases and their relative strength is expressed in $CO_2$ equivalents. Those gasses relevant within waste management are shown in the table below.

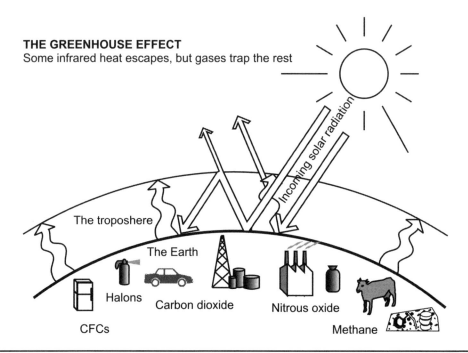

**THE GREENHOUSE EFFECT**
Some infrared heat escapes, but gases trap the rest

| Greenhouse gas | Concentrations in 2005 | Radiative forcing (W/m$^2$) | Lifetime (years) | Global warming potential for given time horizon (kg CO$_2$-eq./kg gas) | | |
|---|---|---|---|---|---|---|
| | | | | 20 years | 100 years | 500 years |
| Carbon dioxide (CO$_2$) | 379 ppmv | 1.66 | — | 1 | 1 | 1 |
| Methane (CH$_4$) | 1774 ppbv | 0.48 | 12 | 72 | 25 | 7.6 |
| Nitrous oxide (N$_2$O) | 319 ppbv | 0.16 | 114 | 289 | 298 | 153 |
| CFC-11 (CCl$_3$F) | 251 pptv | 0.063 | 45 | 6730 | 4750 | 1620 |
| CFC-12 (CCl$_2$F$_2$) | 538 pptv | 0.17 | 100 | 11 000 | 10 900 | 5200 |
| HCFC-22 (CHClF$_2$) | 169 pptv | 0.033 | 12 | 5160 | 1810 | 549 |
| HFC-134a (CH$_2$FCF$_3$) | 35 pptv | 0.0055 | 14 | 3830 | 1430 | 435 |
| HFC-245fa (CHF$_2$CH$_2$CF$_3$) | — | — | 7.6 | 3380 | 1030 | 314 |
| Sulphur hexafluoride (SF$_6$) | 5.6 pptv | 0.0029 | 3200 | 16 300 | 22 800 | 32 600 |

In the future, the importance of greenhouse gases from landfills will change. In the industrialized world, the organic content of waste going to landfill tends to decrease through recycling and pretreatment of the waste, leading to less methane generation in new landfills. However, in economically developing countries, higher waste generation and the abandoning of open dumps for the benefit of modern landfills are expected to lead to an increase in methane emissions from landfills.

### 10.2.1.2   Stratospheric Ozone Depletion

Landfill gas contains chlorinated and fluorinated hydrocarbons (freons). They originate primarily from solvents, spray cans, insulation foams and compressors. Primarily older products contain these compounds. The compounds are very volatile and stable in air. When reaching the stratosphere, chlorine and fluorine atoms are released and forms radicals that cause ozone to break down. The formation of 'ozone holes' allows more UV-B radiation to reach life on earth increasing for example the risk for skin cancer. No account exists of the contribution of landfills to stratospheric ozone depletion.

The chlorinated and fluorinated hydrocarbons are potentially degradable in the strongly anaerobic environments of landfills (e.g. Kromann *et al.*, 1998; Scheutz *et al.*, 2006), but their extreme volatility, when released from the waste components, make them escape rapidly with the gas (e.g. Kjeldsen and Christensen, 2001). However, Kjeldsen and Jensen (2001) have shown (Box 10.2.2) that release of fluorinated hydrocarbons (primarily CFC-11 and CFC-12) from insulation foam from old refrigerators may last decades, suggesting that these compounds will be found in landfill gas many years after they no longer are used in products entering landfills.

The control of emissions of volatile compounds from landfills contributing to stratospheric ozone depletion is by limiting the disposal of waste containing these compounds and by the controlled combustion of the gases from the landfill. Incomplete combustion may generate other unwanted emissions. Scheutz *et al.* (2006) showed that there is a potential for significant reduction of the emission of fluorinated and chlorinated compounds due to natural attenuation processes taking place in the landfilled waste and the top cover. However, no investigation has quantified the importance of natural attenuation processes for a full-scale landfill.

---

**Box 10.2.2    Release of Chlorinated and Fluorinated Compounds from Insulation Foams in Landfills.**

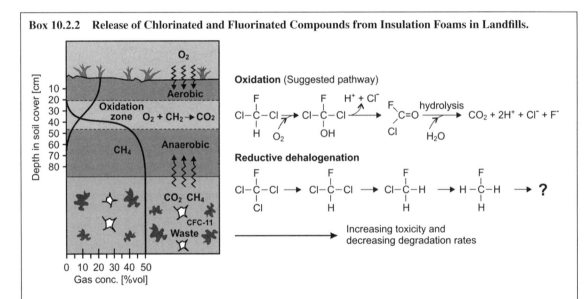

A potential exists for natural attenuation of CFC-11 released from insulation foam in landfills:

- CFC-11 is released from insulation foam waste within the landfill.
- Released CFC-11 is degraded within landfilled waste by reductive dehalogenation which may lead to the formation of more toxic degradation products.
- CFC-11 and degradation products may be transported with generated landfill gas into the soil cover where additional reductive dehalogenation may continue in the lower anaerobic part of soil cover.
- The degradation products are co-degraded along with the methane in the methane-oxidizing zone introduced by oxygen diffusing in from the atmosphere above the soil cover.

### 10.2.1.3 Toxic Gases

The main components of landfill gas (e.g. $CO_2$ and $H_2S$) are potentially toxic to humans, but most concern is generally assumed to be related to organic trace components in the gas. The critical compounds are believed to be vinyl chloride and benzene, due to their carcinogenic effects, but also dioxins and furans, formed during incomplete combustion of landfill gas in flares, may be of concern (Eikman, 1996).

The toxic trace gases are released by routes similar to landfill gas in general. The level of toxic trace gases of course depends on the waste landfilled. Benzene was a common solvent and trace in many types of organic chemical waste, while vinyl chloride is a degradation product from chlorinated hydrocarbons, primarily PCE (tetrachloroethylene) and TCE (trichloroethylene) during anaerobic conditions. Both vinyl chloride and benzene are easily degraded under aerobic conditions, e.g. in the top cover of the landfill. Scheutz and Kjeldsen (2005) showed that a landfill cover in good conditions could degrade as much as $180 \, mg/m^2/day$ of vinyl chloride, and $185 \, mg/m^2/day$ of benzene. The potential environmental impacts of toxic trace gases are believed to be moderate and limited to early stages of landfilling.

### 10.2.1.4 Odor

Landfills may cause odor problems constituting a severe nuisance for people living or working in the vicinity of the landfill. Odor problems often have been reported at landfills (Frechen, 1996). Box 10.2.3 describes different approaches for quantifying odors at a landfill. Many components contribute to the odors emitted by landfills: Compounds in the waste and compounds formed during the degradation of the waste. In Box 10.2.3 common odorous compounds detected at landfills are listed. The most dominant odorous compounds at landfills are probably sulfides, such as hydrogen sulfide, dimethyl sulfide and mercaptans (thiols) (DHHS, 2001). Hydrogen sulfide is emitted at the highest rates and concentrations. Landfilling of construction and demolition (C&D) waste tend to increase the formation of reduced sulfur compounds due to the biological conversion of sulfate from disposed gypsum drywalls (Buske *et al.*, 2005; Kim, 2006). Measurements carried out at two Korean landfills, an active opened in 1999 and a passive closed in 1999, gave information about the concentration of reduced sulfur compounds in landfill gas. Table 10.2.1 presents the concentrations monitored by using gas chromatography.

---

**Box 10.2.3  Odor Measurement at Landfills.**

The measurement of odors from municipal solid waste (MSW) landfills is usually a requirement for compliance monitoring, site expansion and review of operational practices. Monitoring of odor is, however, complicated since odors vary in threshold, intensity and hedonic tone. Misselbrook *et al.* (1993) states that measuring odor intensity alone, is insufficient to assess the human perception of odor.

The measurement can be accomplished in several ways, the three main methods being dynamic dilution olfactometry, electronic methods and chemical analyses.

**Dynamic Dilution Olfactometry**

Dynamic dilution olfactometry tests are commonly performed. However, an objective characterization of odor is difficult to achieve since the perception of odor is inherently subjective. Standard practice is to use a panel of human odor assessors that can determine odor concentration or intensity in terms of odor units. The number of odor units represents the dilution required to reach a detection threshold for a gas. For example, an odor concentration of 1000 odor units implies that the gas must be diluted by a factor of 1000 to reach its detection threshold. The detection threshold is defined as the concentration at which only half of the panel can detect the odor and defines the value of 1 odor unit. The inherent subjectivity of odor panel assessments can lead to large variability in odor detection thresholds (Ruth, 1986), as the threshold for different people can be several orders of magnitude different. The method will also not identify individual odors and will not give any hedonic information. For identification of individual compounds the methods described below may be used.

---

The picture below shows an example of an olfactometric measurement using four persons (Frechen, 2003).

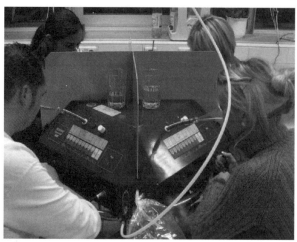

Reprinted with permission from State of the art of odour measurement by F. B. Frechen, Department of Sanitary and Environmental Engineering, University of Kassel, Germany © (2003) F. B. Frechen.

## Electronic Methods

Electronic noses are intelligent instruments that are able to classify and quantify different gasses/odors. This technique uses an array of solid-state sensors combined with neural network software to analyze odor signals (Reinhart, 2007). This makes it possible to detect thousands of chemical species. Current design of the system provides a versatile platform for remote control. Once a connection is established commands can be sent to the instrument (see picture below) in order to execute sampling, getting the current values of sensors or even reprogram the instrument (Perera *et al.*, 2001). The advantage of this method is its ability of detecting a large number of chemicals very rapidly; a drawback is that it does not allow for direct measurement of odor intensity nor odor quality.

Reprinted with permission from IEEE Sensors Journal, A portable electronic nose based on embedded PC technology and GNU/Linux: hardware, software and applications by A. Perera, T. Sundic, A. Pardo et al., 2, 3, 235–246 © (2002) IEEE.

## Chemical Analyses

A third method for odor monitoring is by using gas chromatography. A sample of air is injected into a column which separates the compounds based on their relative vapor pressures and polarities (Brewer and Cadwallader, 2003). The method is very sensitive and allows for classifying and quantifying the different compounds in the gas. A problem associated with this method is that it is cumbersome since air samples must be taken and brought to the laboratory. Additionally the method needs an extraction or preconcentration step in order to separate the compounds of the gas from the bulk of the atmospheric gases and water vapor (Brewer and Cadwallader, 2003). Like the electronic nose the method is not subjective, but it can be used in combination with the dynamic dilution olfactometry tests thereby extending the area of results.

## Different Odor Components

The table below lists common landfill gas components, their smell and odor threshold. The large variability in odor threshold confirms the inherent subjectivity of odor panel assessments.

| Component | Odor description | Odor threshold (ppb) |
|---|---|---|
| 1-Butanethiol | Skunk or garlic | 0.2–1.9[a] |
| Ammonia | Pungent acidic or suffocating odor | 1000–5000[b] |
| Benzene | Paint-thinner-like odor | 840[b] |
| Dichloroethylene | Sweet, ether-like, slightly acrid odor | 85[b] |
| Dichloromethane | Sweet, chloroform-like odor | 205 000–307 000[b] |
| Dimethyl sulfide | Rotten or boiled cabbage | 1[a] |
| Ethanethiol | Rotten egg or leek | 1[a] |
| Ethylbenzene | Aromatic odor like benzene | 90–600[b] |
| Hydrogen sulfide | Strong rotten egg smell | 0.5–1.0[b] |
| Tetrachloroethylene | Sweet, ether-or chloroform-like odor | 50 000[b] |
| Toluene | Aromatic odor like benzene | 10 000–15 000[b] |
| Trichloroethylene | Sweet, chloroform-like odor | 21 400[b] |
| Vinyl chloride | Faintly sweet odor | 10 000–20 000[b] |

[a]Beredskabsstyrelsen (2006)
[b]DHHS (2001)

Ammonia is another odorous landfill gas emitted, but it does not cause as many nuisances as the sulfides, since humans tend to be less sensitive to its odor compared to the odor of sulfides (DHHS, 2001).

The main odor problems at landfills exist during the actual tipping of waste at the landfill and during the acetogenic phase of the degradation of the organic waste (see later). The odor problem is difficult to manage, and some landfills have facilities for dosing aerosols with fragrance at the landfill tip. At the later stages, when the landfilled is covered and the gas control systems installed the odor problems are usually few, maybe primarily related to drops in atmospheric pressure suddenly releasing large quantities of landfill gas. Sironi *et al.* (2005) indicate in their study of seven MSW Italian landfills using dynamic dilution olfactometry that the odor concentration at a restored landfill parcel is less than

**Table 10.2.1**  *Concentrations of odorous compounds measured at an active and a passive landfill, respectively (after Kim, 2006). ppb: Parts per billion. Reprinted with permission from Atmospheric Environment, Emissions of reduced sulfur compounds (RSC) as a landfill gas (LFG): A comparative study of young and old landfill facilities by K.-H. Kim, 40, 34, 6567–6578 © (2006) Elsevier Ltd.*

| Component | Concentration active landfill (ppb) | Concentration passive landfill (ppb) |
|---|---|---|
| Hydrogen sulfide ($H_2S$) | 139 070 | 3.4 |
| Methyl mercaptan ($CH_3SH$) | 577 | 0.1 |
| Dimethyl sulfide ($(CH_3)_2S$) | 395 | 1.2 |
| Carbon disulfide ($CS_2$) | 106 | 0.9 |
| Dimethyl disulfide ($(CH_3)_2S_2$) | 82 | 0.5 |

10 % of the original odor concentration from the freshly tipped waste. The odor concentration of temporarily covered waste is around 20 % of the original concentration.

### 10.2.1.5   Noise

Noise is a common nuisance at landfills in operation. The sources of noise are trucks traveling forth and back, the hydraulic emptying of the trucks, the compactors and earth moving equipment, and in some cases also pumps and blowers, and birds (seagulls) feeding at the tipping front. This suggests that most noise problems are associated with the operational phase of the landfill.

Control of the noise issue may involve technical measures as soil barriers and noise reduction measures on compactors and dozers, as well as proper design of the tipping front, instruction of workers and drivers in reducing noise from engines and reduction in opening hours to avoid operations during the neighbors' leisure time in gardens and surrounding parks.

### 10.2.1.6   Birds, Rodents, Insects, etc.

Various animals are attracted by the landfill for feeding and breeding. In particular, landfills receiving organic waste host several kinds of birds (e.g. seagulls), rodents (e.g. rats) and insects (e.g. flies). Since many of these animals may act as vectors (disease transmitters), their presence may constitute a potential health problem and disinfestations campaigns are usually needed at landfills. Birds, in particular seagulls, are considered a problem, because of the noise they make, their littering around the landfill and their danger to aircrafts at nearby airports. Seagulls suddenly taking of from a landfill may be sucked into the jet engines of an aircraft and may cause accidents. Seagulls are very difficult to avoid at landfills, because the unloading of mixed waste is attractive and within less than a hour they may eat what they need for a whole day. Bird nets, shooting, falconers and alike have all been used in order to reduce the problem of birds at the tipping front of landfills. None of them have yet proven very effective. In siting of landfills, a minimum distance of 10 km to airports is often suggested (US EPA, 2002; US DT, 2006).

### 10.2.1.7   Litter and Dust

Wind-blown litter and dust may be significant nuisances at landfills when spreading into the neighboring areas. Dust may also constitute a contamination issue if originating from ashes or contaminated soil handled on the landfill.

The control of litter and dust involves, in addition to the physical layout of the landfill, a small tipping front, quick covering of landfilled waste, removable catch screens at the tipping front and fences around the landfill (removing the litter regularly so that the fence does not tip in a storm) and water spraying on roads and dry waste. In addition a neat operation will reduce these nuisances.

### 10.2.1.8   Fire and Explosion Hazards

Landfill gas may burn and may explode due to its content of methane. Landfill gas from ash disposal sites may, although in much smaller quantities, produce hydrogen that also may burn and explode. The explosive range for methane is 5–15 % in air at atmospheric pressure and ambient temperature. The explosive range is only slightly affected by the presence of other constituents (Gendebien *et al.*, 1992). If landfill gas is vented directly to the atmosphere, no explosion hazard exists, but methane may sustain fires.

The explosion risk exists where landfill gas migrates through cracks in foundations and pipes into confined space like basements and sewers. Accumulated methane within the explosive range and an energizer (spark from electric appliances, striking of match, etc.) create the explosion. The damage may be significant. Gendebien *et al.* (1992) refers to more than 55 cases in the UK, United States, Canada and Germany reported in the literature involving explosions, fire and human injury. It appears that the main risks exist during extreme metrological conditions: A drop in atmospheric pressure and heavy rain forces large quantities of landfill gas to migrate horizontally out of the landfill and the gas may collect in basements, below floors, and in sewers in the vicinity of the landfill. Box 10.2.4 describes the event at the Skellingsted Landfill in Denmark, where two people were killed in 1991. A similar metrological situation seemed to cause the accident in Loscoe, UK (Williams and Aitjkenhead, 1991). It this case, several houses were located close to the landfill, but the stratified geological conditions forced the gas to collect in a house several hundred meters from the landfill.

**Box 10.2.4   Landfill Gas Explosion at the Skellingsted Landfill, Denmark (After Kjeldsen and Fischer, 1995).**

The Skellingsted Landfill was in operation 1971–1990 receiving a total of 300 000 t of waste, primarily residential waste. The landfill was originally a dump in an old, 10-m gravel pit and had neither a leachate nor a gas collection system. After closing, the landfill was covered with 0.8 m of soil.

On March 21, 1991, a house located 20 m off the landfill boundary exploded and two people were killed. Prior to the explosion it rained significantly and the barometric pressure dropped as low as 995 mbar, as shown on the graph. The high water content of the top cover prevented the landfill gas from migrating out through the top of the landfill and the pressure drop created an increased flow of landfill gas out of the landfill into the surrounding strata comprised of sandy soil. The old house close to the landfill had an old, dry clay layer below the wooden floor, and here the gas collected. Supposedly a match ignited the gas. The two people were killed as the construction collapsed.

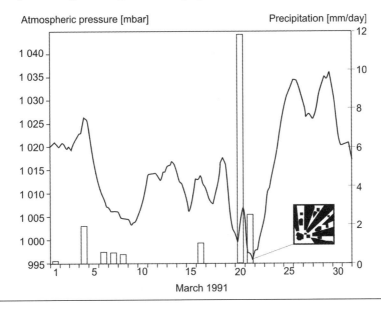

Atmospheric pressure [mbar]     Precipitation [mm/day]

March 1991

The control of gas at the landfill is the main factor in limiting the fire and explosion hazards at landfills. This may involve reduction in the amount of organic waste received, proper collection of landfill gas and liners against migration of gas into the unsaturated zone of the surrounding strata. Kjeldsen (1996) describes how landfill gas migrates. Where houses are at risk, proper drainage around the foundation, forces aeration of confined spaces and gas alarms may be possible approaches to reduce the risk.

### 10.2.1.9   Vegetation Damage

Landfill gas may affect vegetation on the landfill and in the vicinity of the landfill as far as a couple of hundred meters away from the landfill border. Landfill gas may by its flow expel the original soil air, usually containing oxygen, or through oxidation of methane in gas cause removal of oxygen from the root zone. This leads to asphyxia of the roots, eventually killing the plants. In moderate cases landfill gas may lead to dwarf growth or shallow root systems that cannot stand draught or storms, whereby the vegetation also suffers. It is common to see trees at landfills with limited or damaged foliage. Some plant species may also be affected by carbon dioxide and trace gases in the landfill gas (Neumann and Christensen, 1996).

Vegetation damages outside the landfill may be avoided by avoiding landfill gas migration into surrounding strata. Vegetation damages on the landfill also depend on the design of the top cover, as discussed in Chapter 10.9.

**Box 10.2.5   Groundwater Pollution at the Vejen Landfill, Denmark (Based on Kjeldsen, 1993; Lyngkilde and Christensen, 1992a, 1992b; Baun *et al.*, 2003).**

The Vejen Landfill was in operation 1962–1981 and received in total 350 000 t of primarily residential waste. The landfill started out as a dump in a former peat excavation area. The landfill is about 5 m deep and has no collection of leachate and gas. The landfill is covered with 0.5 m of soil of varying quality.

From the central part of the landfill leachate is migrating into the aquifer just below the landfill. The pore flow velocity close to the landfill is about 30–50 m/year but increases to 200 m/year further away as the bottom of the upper aquifer rises downstream of the landfill. The plume was mapped around 10 years after the landfill closed. As illustrated by the figure, the chloride plume stretches about 350 m down-gradient from the landfill, while ammonium and xylene (as an example of an organic compound found in the leachate) have migrated only 100 m and 50 m, respectively. Ammonium is attenuated by ion exchange with calcium on the aquifer material and maybe also by oxidation, while xylene degrades under strongly reducing conditions in the plume.

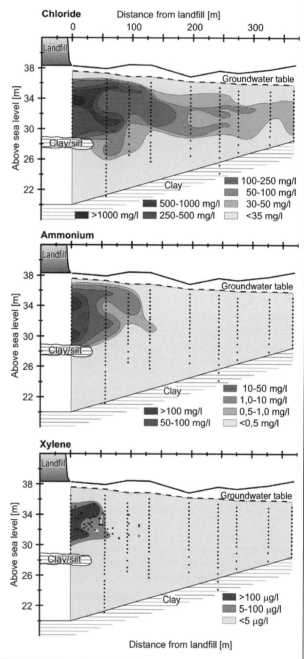

## 10.2.1.10 Soil Contaminations

Contamination of the surrounding land may be caused by spills from waste trucks, dust blowing of the landfill and contaminated soil, ashes and dust carried away by storm water and surface run off. These potential problems can easily be avoided by proper design of the landfill, surface water ditches and a tidy operation.

## 10.2.1.11 Surface Water Pollution

Surface water pollution by leachate from landfills has definitely been observed, but relatively few cases have been reported in the literature. Main effects to be expected are oxygen depletion in part of the surface water body, suffocation of young fish by growth of iron oxides on the gills, changes in bottom fauna and flora, and ammonia toxicity to fish that cannot escape the leachate contamination.

Proper siting and careful control of leachate and surface run off should be able to prevent cases of surface water pollution by leachate.

## 10.2.1.12 Groundwater Pollution

The risk of groundwater pollution is considered to be one of the most important environmental risks at landfills and has led to strict criteria in siting of landfills, and to the installation of liners, leachate collection systems and treatment plants. Christensen *et al.* (1994) reports on the characteristics of leachate pollution plumes described in the open literature: The plumes are relatively short (mostly < 1 km) and are clearly subject to natural attenuation. The reported cases all deal with old landfills with no measures to control leachate and most cases are leachate plumes in sandy aquifers.

The attenuation processes in leachate plumes are reviewed by Christensen *et al.* (2001). At old landfills with organic waste, the leachate plumes develop strongly anaerobic conditions close the landfill. A redox gradient from methanogenic and sulfate-reducing, over iron and manganese reducing conditions to zones with nitrate reduction and aerobic conditions has been observed in well monitored cases. This gradient is apparently beneficial for the degradation of many organic pollutants in the leachate. Chlorinated pollutants may loose the first chlorine during strongly anaerobic conditions and the degradation continues for less chlorinated compounds and daughter products at higher redox conditions. Heavy metals are usually not a problem because of very strong attenuation in the landfill and the plume. Ammonia may appear to be the most critical component in the long run. Box 10.2.5 describes the plume observed at the Vejen Landfill, Denmark. Common ions and nutrients (Cl, Na, N, K, etc.) are found in high concentration in leachate, but also in the general infiltration through the land surface. This is illustrated in Box 10.2.6.

---

**Box 10.2.6    Fluxes of Nutrients and Salts in a Watershed with an Old Dump Contaminating the Groundwater (After Nielsen *et al.*, 1998).**

All land areas contribute to the flux of compounds into the groundwater, partly in terms of compounds applied to the surface by deposition or by human activities, partly from the weathering and conversion processes in the soil and unsaturated zone above the groundwater. The flux of nutrients and salts from different area uses were measured in two small Danish watersheds, both including an old dump. The compound concentrations were measured in the upper 20 cm of the groundwater (where the vertical groundwater component in the center of the area insures that the water concentrations are determined by vertical infiltration) and combined with estimates of the net infiltration to determine the fluxes.

The watershed containing the Vejen landfill is approximately 800 ha. Agriculture is 78 % of the area, forestry 17 %, the landfill 0.9 % and the remaining is roads and civil structures. The landfill contributes (calculation shown below for potassium) to 8 % of the flux of sodium (Na), potassium (K) and magnesium (Mg) and less than 5 % for the other compounds.

$$\text{Flux of potassium}: \left( \frac{300 \cdot 800 \cdot 0.009}{300 \cdot 800 \cdot 0.009 + 40 \cdot 800 \cdot 0.78 + 4 \cdot 800 \cdot 0.17} \right) \cdot 100\% = 7.8\,\%$$

| Flux (kg/ha/year): Vejen | N | K | Mg | Ca | Na | Cl | S |
|---|---|---|---|---|---|---|---|
| Old landfill | 300 | 300 | 150 | 430 | 700 | 1150 | 150 |
| Roads (de-icing) | 0 | 0 | 0 | 0 | 2500 | 4200 | 0 |
| Agriculture (pigs, cattle) | 85 | 40 | 20 | 140 | 65 | 175 | 35 |
| Forestry | 2 | 4 | 7 | 6 | 80 | 130 | 35 |

The watershed containing the Grindsted landfill is approximately 5500 ha. Agriculture is 72%, forestry 18%, the landfill 0.2%, moor 5% and only a little is used for roads and civil structures. The landfill contributes, as shown below, to 3% of the flux of nitrogen (N), but less than 1% of the other compounds.

| Flux (kg/ha/year): Grindsted | N | K | Mg | Ca | Na | Cl | S |
|---|---|---|---|---|---|---|---|
| Old landfill | 480 | 150 | 80 | 700 | 480 | 880 | 50 |
| Roads (de-icing) | 0 | 0 | 0 | 0 | 2230 | 3740 | 0 |
| Agriculture (pigs, plant cultivation) | 45 | 50 | 20 | 160 | 45 | 155 | 35 |
| Forestry | 10 | 13 | 7 | 5 | 65 | 180 | 40 |

The current knowledge about landfill leachate plumes are based on studies of old landfills with mixed waste, that have created plumes in sandy aquifers with long retention times and a favorable redox gradient. This may not necessarily be the case at new landfills with a different waste composition, located in a karst area and run as a modern operation. However, the studies indicate that a minor leaking of leachate may not necessarily be a threat to regional groundwater resources.

Groundwater pollution is controlled through the siting of the landfill, the type of waste received, the quality of the top cover and the efficiency of the bottom liner and leachate collection system established.

# References

Baun, A., Reitzel, L.A., Ledin, A., Christensen, T.H. and Bjerg, P.L. (2003): Natural attenuation of xenobiotic organic compounds in a landfill leachate plume (Vejen, Denmark). *Journal of Contaminant Hydrology*, 65, 269–291.

Beredskabsstyrelsen (2006): *Alfabetisk indeks* (Alphabetic index; in Danish). Located on the internet: 25 July 2007. http://www.kemikalieberedskab.dk/ikidxindl.shtml.

Bogner, J., Pipatti, R., Hashimoto, S., Diaz, C., Mareckova, K., Diaz, L., Kjeldsen, P., Monni, S., Faaij, A., Gao, Q., Zhang, T., Ahmed, M.A., Sutamihardjo, R.T.M. and Gregory, R. (2008): Mitigation of global greenhouse gas emissions from waste: conclusions and strategies from the Intergovernmental Panel on Climate Change (IPCC) Fourth Assessment Report. Working Group III (Mitigation). *Waste Management and Research*, 26, 11–32.

Brewer, M.S. and Cadwallader, K.R. (2003): Overview of odor measurement techniques. In: *Proceedings of the University of Illinois Pork Industry Conference*, pp. 59–74. Illinois Pork Producers Association, Springfield, USA.

Buske D., Lannan M., Riegert M. and Laurila J. (2005): Odor impacts from using C&D as landfill cover. In: *Solid waste and recycling conference with trade show. 2005 Proceedings*. Federation of New York, New York, USA. Located on the internet: 25 July.2007 http://www.nyfederation.org/PDF2005/8Buske.pdf.

Christensen, T.H., Kjeldsen, P., Albrechtsen, H.J., Heron, G., Nielsen, P.H., Bjerg, P.L. and Holm, P.E. (1994): Attenuation of landfill leachate pollutants in aquifers. *Critical Reviews in Environmental Science and Technology*, 24, 119–202.

Christensen, T.H., Kjeldsen, P., Bjerg, P.L., Jensen, D.L., Christensen, J.B., Baun, A., Albrechtsen, H.-J. and Heron, G. (2001): Biogeochemistry of landfill leachate plumes. *Applied Geochemistry*, 16, 659–718.

Christensen, T.H., Gentil, E., Boldrin, A., Larsen, A.W., Weidema, B.P. and Hauschild, M.Z. (2010) C balance, carbon dioxide emissions and global warming potentials in LCA-modeling of waste management systems. *Waste Management and Research*, 27, 707–715.

DHHS (2001): *Landfill gas primer: an overview for environmental health professionals.* Department of Health and Human Services and Agency for Toxic Substances and Disease Registry and Division of Health Assessment and Consultation. Located on the internet: 25 July 2007. http://www.atsdr.cdc.gov/HAC/landfill/html/toc.html.

Eikman, T. (1996): Health aspects of gaseous emissions from landfills. In: Christensen, T.H., Cossu, R. and Stegmann, R. (eds) *Landfilling of Waste: Biogas*, pp. 143–154. E&FN Spon, London, UK.

Frechen, F.B. (1996): Landfill gas odours. In: Christensen, T.H., Cossu, R. and Stegmann, R. (eds) *Landfilling of Waste: Biogas*, pp. 199–213. E&FN Spon, London, UK.

Frechen, F.B. (2003): *State of the art of odour measurement.* Department of Sanitary and Environmental Engineering, University of Kassel, Germany.

Gendebien, A., Pauwels, M., Constant, M., Ledret-Damanet, M.J, Nyns, E.J., Willumsen, H.C., Butson, J., Fabry, R. and Ferrero, G.L. (1992): *Landfill gas – from environment to energy.* Directorate-General Telecommunications, Information Industries and Innovation, Commission of The European Communities, Luxembourg.

Kim, K.-H. (2006): *Emissions of reduced sulfur compounds (RSC) as a landfill gas (LFG): A comparative study of young and old landfill facilities.* Department of Earth and Environmental Sciences, Sejong University, Seoul, Republic of Korea.

Kjeldsen, P. (1993): Groundwater pollution source characterization of an old landfill. *Journal of Hydrology*, 142, 349–371.

Kjeldsen, P. (1996): Landfill gas migration in soil. In: Christensen, T.H., Cossu, R., Stegmann, R. (eds) *Landfilling of waste: Biogas*, pp. 87–132. E&FN Spon, London, UK.

Kjeldsen, P. and Christensen, T.H. (2001): A simple model for the distribution and fate of organic chemicals in a landfill: MOCLA. *Waste Management and Research*, 19, 201–216.

Kjeldsen, P. and Fischer, E. (1995): Landfill gas migration – Field investigations at the Skellingsted Landfill, Denmark. *Waste Management and Research*, 13, 467–484.

Kjeldsen, P. and Jensen, M.H. (2001): Release of CFC-11 from disposal of polyurethane foam waste. *Environmental Science and Technology*, 35, 3055–3063.

Kromann, A., Ludvigsen, L., Albrechtsen, H.J. and Christensen, T.H. (1998): Degradability of chlorinated aliphatic compounds in methanogenic leachates sampled at eight landfills. *Waste Management and Research*, 16, 54–62.

Lyngkilde, J. and Christensen, T.H. (1992a): Redox zones of a landfill leachate pollution plume (Vejen, Denmark). *Journal of Contaminant Hydrology*, 10, 273–289.

Lyngkilde, J. and Christensen, T.H. (1992b): Fate of organic contaminants in the redox zones of a landfill leachate pollution plume (Vejen, Denmark). *Journal of Contaminant Hydrology*, 10, 291–307.

Manfredi, S., Scharff, H., M., Tonini, D. and Christensen, T.H. (2010): Landfilling of waste: Accounting of greenhouse gases and global warming contributions. *Waste Management and Research*, 27, 825–836.

Misselbrook, T.H., Clarkson, C.R. and Pain, B.F. (1993): Relationship between concentration and intensity of odours for pig slurry and broiler houses. *Journal of Agricultural Engineering Research*,. 55, 163–169.

Neumann, E. and Christensen, T.H. (1996): Effects of landfill gas on vegetation. In: Christensen, T.H., Cossu, R. and Stegmann, R. (eds) *Landfilling of waste: Biogas*, pp. 155–185. E&FN Spon, London, UK.

Nielsen, M.Æ., Carlson, B.B., Bjerg, P.L., Pedersen, J.K. and Christensen, T.H. (1998): Kilder til uorganiske stoffer i grundvandet – herunder gamle lossepladser. (Sources for inorganic substances in the groundwater – including old landfills; in Danish). *Vand og Jord*, 5 (1), 31–35.

Perera, A., Pardo, T., Sundiæ, T., Gutierrez-Osuna, R., Marco, S. and Nicolas, J. (2001): IpNose: Electronic nose for remote bad odour monitoring system in landfill sites. Proceedings of the Third European Congress on Odours, Metrology and Electronic Noses, Paris, France.

Reinhart, D. (2007): *Urban infilling impacts on Florida's solid waste facilities.* State University System of Florida, Florida Center for Solid and Hazardous Waste Management, Gainesville, USA. Located on the internet: 25 July 2007. http://www.hinkleycenter.com/publications/Infilling%20Annual%20Report.pdf.

Ruth, J.H. (1986): Odor thresholds and irritation levels of several chemical substances: a review. *American Industrial Hygiene Association Journal*, 47, A142–A151.

Scheutz, C., Dote, Y., Fredenslund, A.M., Mosbæk, H. and Kjeldsen, P. (2007): Attenuation of fluorocarbons released from foam insulation in landfills. *Environmental Science and Technology*, 41, 7714–7722.

Scheutz, C. and Kjeldsen, P. (2005): Biodegradation of trace gases in simulated landfill soil cover systems. *Journal of Air and Waste Management Association*, 55, 878–885.

Sironi, S., Capelli, L., Céntola, P., Del Rosso, R. and Il grande, M. (2005): *Odour emission factors for assessment and prediction of Italian MSW landfills odour impact.* Politecnico di Milano, Milan, Italy.

US DT (2006): *Construction or establishment of landfills near public airports.* Advisory Circular 150/5200-34A, January 2006. US Department of Transportation. Located on the internet: 25 July 2007. http://rgl.faa.gov/Regulatory_and_Guidance_Library/rgAdvisoryCircular.nsf/9fcfb112c56d8bd885256a620061cf2a/52469b5487918b81862571e10072fb44/$FILE/AC150_5200_34a.pdf.

US EPA (2002): Municipal solid waste landfill location restrictions for airport safety. *Federal Register Environmental Document*, 67 (133), July 2002. Located on the internet: 25 July 2007. http://www.epa.gov/fedrgstr/EPA-WASTE/2002/July/Day-11/f16994.htm.

Williams, G.M. and Aitkenhead, N. (1991): Lessons from Loscoe: The uncontrolled migration of landfill gas. *Quarterly Journal of Engineering Geology*, 24, 191–207.

# 10.3

# Landfilling: Hydrology

**Peter Kjeldsen**

*Technical University of Denmark, Denmark*

**Richard Beaven**

*University of Southampton, UK*

Landfill hydrology deals with the presence and movement of water through a landfill. The main objective in landfill hydrology is usually to predict leachate generation, but the presence and movement of water in a landfill also affect the degradation of the waste, the leaching of pollutants and the geotechnical stability of the fill. Understanding landfill hydrology is thus important for many aspects of landfill, in particular siting, design and operation.

The objective of this chapter is to give a basic understanding of the hydrology of landfills, and to present ways to estimate leachate quantities under specific circumstances. Initially a general water balance equation is defined for a typical landfill, and the different parts of the water balance are discussed. A separate section discusses water flow and the hydrogeology of landfilled wastes and considers the impact of water short-circuiting. In the final section different existing hydrological models for landfills are presented with a special focus on the HELP model. This model is the most widely used tool for the prediction of leachate quantities in landfills, and for the sizing of leachate control and management infrastructure.

## 10.3.1 The Water Balance of a Landfill

As shown in Figure 10.3.1 leachate generation is influenced by many factors including climatic and hydrologic factors, landfill design, operations and management, waste characteristics, and internal landfill processes. Many of these factors are interconnected making leachate generation a rather complicated process that to some degree is controllable by engineering of the landfill.

Figure 10.3.2 shows a finalized section of a landfill covered with a layer of soil, which may be covered with vegetation. The figure also presents the different parts of the water balance. Input of water to the landfill originates

*Solid Waste Technology & Management*   Edited by Thomas Christensen
© 2011 Blackwell Publishing Ltd

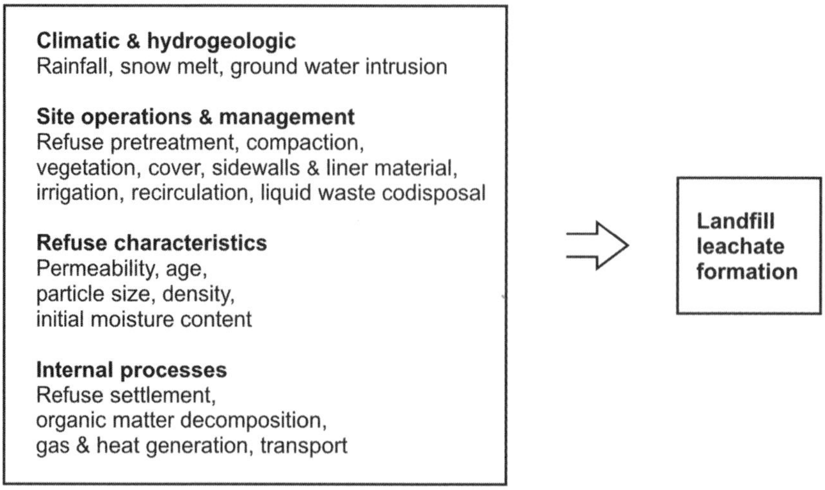

**Climatic & hydrogeologic**
Rainfall, snow melt, ground water intrusion

**Site operations & management**
Refuse pretreatment, compaction,
vegetation, cover, sidewalls & liner material,
irrigation, recirculation, liquid waste codisposal

**Refuse characteristics**
Permeability, age,
particle size, density,
initial moisture content

**Internal processes**
Refuse settlement,
organic matter decomposition,
gas & heat generation, transport

⟹

**Landfill
leachate
formation**

***Figure 10.3.1*** *Factors influencing leachate generation in landfills (after ElFadel, 1997). Reprinted with permission from Waste Management and Research, Modelling Leachate Quality and Quantity in Municipal Solid Waste Landfills by E. D. Yildiz, K. Unlu and R. K. Rowe, 22, 2, 78-92 © (2004) Sage Publications.*

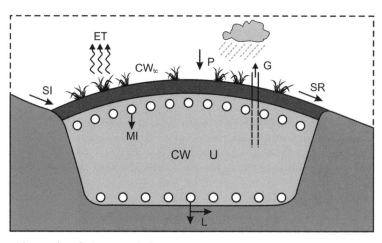

L = Leachate generated
P = Precipitation
$CW_{tc}$ = Change in moisture storage in top cover
ET = Actual evapotranspiration
SR = Surface runoff
CW = Change in moisture content of waste
SI = Surface water inflow
MI = Managed water input
G = Water vapour contained in landfill gas
U = Water consumed by chemical or microbial processes

***Figure 10.3.2*** *The water balance of a landfill.*

from:

- Direct precipitation on the landfill surface (P).
- Surface water inflow of rainwater or melt water (SI).
- Liquid waste or other managed water inputs (MI).

Water is removed or transported away from the landfill by:

- Evapotranspiration from the surface (ET).
- Surface runoff (SR).
- Leachate removed from control systems (e.g. drains and wells) or migration beyond the landfill barriers into underlying strata (L).
- Water vapour contained in landfill gas (G).
- Water consumed by chemical or microbial processes (U).

An important part of the water balance accounts for the change of water content of the landfill (CW). Wastes received at landfills normally have an initial water content that is considerably lower than their field capacity. The wastes are therefore able to take up significant volumes of water inputs.

The water balance equation can be written, using the abbreviations mentioned above:

$$P + SI + MI = ET + SR + L + G + U + CW \qquad (10.3.1)$$

Often it is the leachate generation, which is estimated and the water balance equation is written in the form:

$$L = P + SI + MI - ET - SR - G - U - CW \qquad (10.3.2)$$

The different parts of the water balance equation are discussed in the following sections.

### 10.3.1.1  Leachate Generation

The quantity of leachate (L) that is leaving the lower boundary of the waste includes the volume that is removed from leachate extraction systems and any leachate that may leave the landfill through the bottom liner. This term should not be confused with the volume of freely draining leachate that may build up in the body of a landfill if leachate levels increase. This volume is accounted for in CW.

### 10.3.1.2  Precipitation

The quantity of local precipitation (P) is probably the most important factor governing leachate generation. However, the volume of water actually infiltrating a site will be less than the precipitation because of surface water runoff (SR) and evapotranspiration (ET). Information about the local average precipitation is important, especially in regions with large spatial variation such as mountainous regions. Figure 10.3.3 shows the annual precipitation for the United Kingdom. Within the country a fivefold difference between the highest and the lowest recorded average annual precipitation is observed. In most cases significant seasonal variations are also observed. For instance, in tropical areas rainy seasons give a significant variation. Figure 10.3.4 shows examples of recorded average monthly precipitation from different regions. Besides the variation in average values, large differences can be observed from one year to another.

### 10.3.1.3  Surface Water Inflow

Many landfills are constructed as 'land raise' features on top of the original ground surface. In this case the topography naturally restricts the inflow of surface water (SI) in to the landfill. However, for landfills constructed in old quarries

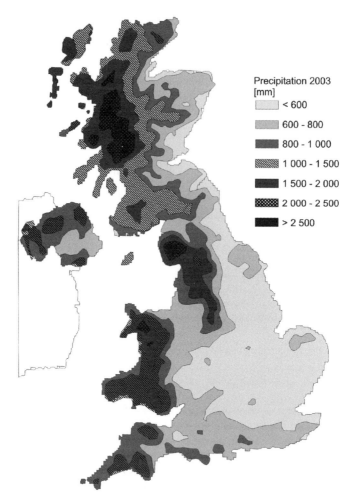

**Figure 10.3.3**    *Average annual precipitation for the United Kingdom based on data obtained from the Met Office, UK. Adapted with permission from Bristich crown © data supplied by the Met Office © (2009) British Crown.*

or in ravines in mountain regions, inflow of water in connection to heavy rain or thaw can be a major problem. Inflow of surface water can be reduced significantly by the construction of ditches around the landfill to redirect surface water around the filling area.

### 10.3.1.4   Surface Water Runoff (SR)

Precipitation that lands on the surface of a landfill may leave the landfill as surface water runoff (SR). This generally takes place after incidents of heavy rain where the intensity of the precipitation is higher than the hydraulic conductivity of the soil cover and the duration of the precipitation is sufficient to fill the suppressions in the topcover. Surface runoff can lead to severe erosion of the soil cover especially in cases where the vegetation is not fully covering the surface. In most temperate areas surface runoff is not a major part of the water balance, but in subtropical and tropical areas surface runoff can constitute a major problem, in particular in landfills with step slopes.

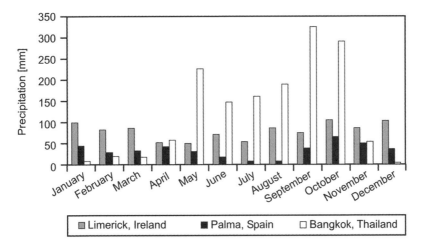

***Figure 10.3.4*** *Examples of recorded monthly average precipitation for different locations (extracted from the HELP database).*

In colder regions some precipitation may fall as snow, and the temperature can be below the freezing point for long periods of time leading to a reduction in water infiltration through the soil cover. In situations where snow overlies frozen ground, a sudden thaw will result in high runoff as water is unable to infiltrate the frozen subbase.

## 10.3.1.5 Evapotranspiration

Water evaporates from land surfaces mainly through the stomata of plants (transpiration) and directly from the soil and leave surfaces (evaporation). Evaporation and transpiration are normally not measured separately and are generally looked at as a single process called evapotranspiration (ET). Potential evapotranspiration is the maximal evapotranspiration that would occur from an area covered by a homogeneous, green and growing crop optimally supplied with water. Potential evapotranspiration is governed by climatic factors as well as ground conditions:

- Solar input.
- Temperature (soil and air).
- Relative humidity.
- Wind speed.
- Type and state of crop (especially depth of root zone and foliage area).

Potential evapotranspiration is seasonally dependent, caused mainly through variations in temperature and solar input. High solar input and high temperatures in the summer season lead to a high potential evapotranspiration resulting in lower water content of the soil layer. Often the soil gets so dry that it affects the state of the vegetation, and the actual evapotranspiration will be lower than the potential evapotranspiration. Figure 10.3.5 shows the relative evapotranspiration (actual ET as a percentage of the potential ET) as a function of the water deficit in the soil layer. The water deficit is defined as the water volume, which should be added to the soil at its actual water content to reach the field capacity of the soil. The field capacity of a soil is the water content when freely draining conditions exist. The figure indicates there is a maximum water deficit at the point where the extrapolation of the data crosses the *x*-axis. At this point, the wilting point, the actual evapotranspiration is zero – there is no more water in the soil, which is accessible to the vegetation.

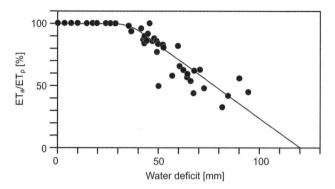

***Figure 10.3.5*** *The relative evapotranspiration (actual ET in percentages of potential ET) as a function of the water deficit in the soil layer (after Aslyng, 1976). Reprinted from Aslyng, H.C. Klima, jord og planter (Climate, soil, and plants; in Danish). Fifth edition. DSR Forlag, Copenhagen © (1976) DSR Forlag Denmark.*

### 10.3.1.6 Use of Water by Reactions

In mineral waste landfills containing fly ashes from waste or coal incineration, the waste takes up the water which is incorporated in the minerals (so-called puzzolanic reactions; U). The extent to which the waste is taking up water is not known, so the importance to the overall water balance is difficult to evaluate.

The anaerobic degradation of organic waste is a water-demanding process. Burton *et al.* (2004) estimated, based on model calculations, that it requires 0.1 l of water to generate 1 m³ of landfill gas. Using these data, the degradation processes in a 20-m deep landfill with a specific gas generation rate of 4 m³ landfill gas/m³ waste/year (which is a relatively high gas generation rate for landfills) would consume water equivalent to an infiltration of approximately 8 mm of water/year (or 0.43 l of water consumption/m³ waste/year). For landfills without top liners this amount of water is in most cases much lower than the actual infiltration into the waste. However, infiltration into some landfills is lower than the potential amount consumed in degradation processes. Examples include landfills in very dry regions of the world or those that have a very tight top cover system intercepting most of the incident precipitation. In these cases the water demand of the degradation process leads to an overall decrease in water content, which again leads to lower degradation rates and slower rates of waste stabilization.

### 10.3.1.7 Loss with Gas Extraction

Landfill gas is normally saturated with water vapour and typically contains 23 g $H_2O$/m³ landfill gas (at a temperature of 20 °C). Using the same assumptions as mentioned above, the loss of water from the landfill by gas extraction (G) is about 2 mm of water/year, which in most cases is insignificant. Although the water as such most be taken care of in the gas extraction system.

### 10.3.1.8 Managed Water Additions

Water, in various forms, may be introduced deliberately into a landfill as a managed input (MI). In some countries, the controlled disposal of compatible liquid wastes into landfills is accepted as good practice. Irrigation of water on the top of a landfill (particularly temporary roads) during dry periods is often undertaken to suppress dust. The introduction and recirculation of water, in most cases in the form of leachate, into the body of a landfill is a technique that is sometimes used to manage processes occurring in the site. Water addition can speed up biodegradation processes, increase rates of landfill settlement, promote landfill gas generation, and through extended leaching, enhance or bring forward landfill stabilization. Finally, the introduction of leachate with high organic strength (from another part of a site) into an area that has methanogenic conditions well established can result in the partial treatment of the leachate. Within the context of the water balance, great care needs to be taken when accounting for managed water inputs. When moderate volumes of water are added (especially if it is done through sub surface percolation pipes) it can be treated as an extra precipitation in the water balance equation. If leachate is being recirculated within the site, then the net balance to the water balance is zero. However, as it can lead to a change in the water content of the landfill, recirculation should be accounted for both as a leachate outflow (L) and as a managed input (MI).

Sprinkling of leachate in the warm season often leads to higher evapotranspiration, since the evapotranspiration is mainly limited by accessible water in this period. This could imply that a smaller volume of leachate needs to be managed overall. However, care should be taken to avoid formation of aerosols of leachate that could spread to surrounding areas.

### 10.3.1.9 Changes in Water Content (CW)

An important part of the water balance relates to changes in the storage of water (CW) in the landfill itself. Within a landfill water can be stored either as water held or retained in the fabric of the waste or in freely draining voids. This is illustrated schematically in Box 10.3.1, where the waste is represented as (dry) solids, liquid and air (gas) phases.

---

**Box 10.3.1   Definitions and Equations in Relation to Water-Unsaturated Waste Materials.**

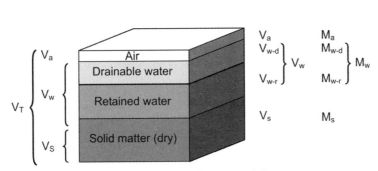

Waste as a three-phase material

In soil mechanics, the water content of a material is defined as the ratio of the mass of water to the mass of dry solids present. It is normally given the symbol w. However, to avoid confusion with an alternative definition of water content generally used in landfill science, the notation $WC_{dry}$ (water content by dry mass) is used here:

$$WC_{dry} = M_W/M_S$$

In landfill science, the water content $WC_{wet}$ is often expressed as a ratio of the mass of water to the total mass of water and solids:

$$WC_{wet} = M_W/(M_S + M_W)$$

The relationship between the two water contents is as follows:

$$WC_{dry} = WC_{wet}/(1 - WC_{wet})$$
$$WC_{wet} = WC_{dry}/(1 - WC_{dry})$$

A further way of expressing the water content of refuse is on a volumetric basis. The volumetric water content $WC_{vol}$ is defined as the ratio of the volume of water to the total volume of air, solids and water:

$$WC_{vol} = \frac{V_W}{V_T}$$

---

Expressing water contents in this form has the advantage that it is possible to relate the water content directly to the drainable porosity. The volumetric water content is related to the wet and dry weight water contents as follows:

$$WC_{vol} = \frac{WC_{dry} \cdot \rho_{dry}}{\rho_w} \tag{1}$$

$$WC_{vol} = \frac{WC_{wet} \cdot \rho_{wet}}{\rho_w} \tag{2}$$

where:

$\rho_{dry} = M_S/V_T$ is the dry density of the waste.
$\rho_{wet} = (M_S + M_W)/V_T$ is the wet density of the waste.

and $\rho_w$ is the density of water.

The definition of terms and the varying ways in which water content can be expressed is potentially confusing. This is illustrated below with an example of a hypothetical waste which has a volumetric water content at field capacity of 40 %. It is further assumed that this water content remains constant with varying waste dry density (see figure below). This example is for illustrative purposes only and is not necessarily meant to relate to actual conditions which may be found in a landfill.

The (dry weight) water content $WC_{dry}$ is calculated according to Equation (1) and is shown in the figure below. Whereas the volumetric water content ($WC_{vol}$) remains constant with changing dry density (a), $WC_{dry}$ plots as a curve (b). As $WC_{dry}$ changes with dry density so does the wet density ($\rho_{wet}$). The relationship is linear (c). Finally the (wet weight) water content $WC_{wet}$ also plots as a curve against both wet density (d) and dry density (not shown).

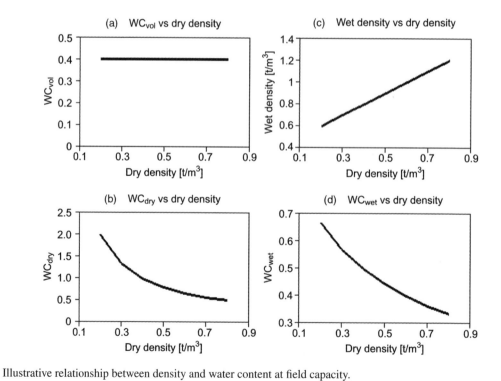

Illustrative relationship between density and water content at field capacity.

## Absorbed Water

In general, fresh waste contains some water but is not saturated. This water, given the notation w-r in Box 10.3.1, is held within the matrix of the refuse and is not freely draining. It is quantified by means of the original moisture or water content which is determined from the loss in weight of a sample of refuse dried at 105 °C. The final weight of the sample after drying gives the mass of dry solids.

After landfilling, the moisture content of wastes may increase through the absorption of water by components such as paper, cardboard, and textiles. Beyond a certain limit the addition of further water leads to the production of free draining pore fluid, which will tend to move downward under the influence of gravity towards a 'water' table. Below this table waste is fully saturated. The overall water content (as opposed to simply the absorbed moisture content) of the drained refuse above the water table may be increased, partly by the trapping of leachate in randomly lined, isolated voids unable to drain under the influence of gravity. In practice it is very difficult to determine whether an increase in the overall water content is due to true absorption or to fluid trapped in non drainable voids. The increases in water content resulting from both processes are therefore usually combined and referred to as the total absorptive capacity of refuse (e.g. Knox, 1991).

Refuse is referred to as being at field capacity when the total absorptive capacity has been fully utilized and free draining conditions exist. Field capacity is often used solely as a qualitative term. However, it is useful to know the water content at field capacity, where it is normally expressed as a volume percentage (i.e. WCvol) but in some cases the unit mass of water per dry mass of waste is used. Most waste types received on landfills have water contents below field capacity. This is the case for municipal solid waste as well as incineration residues. In some cases incineration residues such as fly ashes are watered before transport and landfilling to prevent problems with dust. In such cases the waste can be close to field capacity.

The initial water content of waste at the point of landfilling and the field capacity of the waste are important parameters in relation to landfill hydrology. The difference between the initial water content and the field capacity is referred to as the water deficit of the waste or sometimes as the absorptive capacity of wastes. In addition, to reflect field observations, the total absorptive capacity of refuse has sometimes been split into primary and secondary absorptive capacity. The primary absorptive capacity is taken as the amount of water that can be added to refuse without the creation of any freely draining leachate. Secondary absorptive capacity is taken up more gradually, after leachate production has started, and is probably only fully utilized if the waste becomes completely saturated.

Within the landfill setting, the flow of water within the site will be affected by the heterogeneous nature of the wastes leading to the development of preferential flow paths and the bypassing of waste layers (more details in the next section). Consequently, large volumes of waste may be unaffected by infiltration. This has been observed where waste excavations in landfills have sampled totally dry and undecomposed paper.

In general, most waste types have a high water deficit (absorptive capacity) leading to a potential for water uptake in the order of 100–300 mm precipitation/m waste thickness. Table 10.3.1 gives typical values of initial water content, field capacity, and maximal water deficit for different waste types. The table shows that the values of initial water content exhibit large variation among the different studies. This may be caused by seasonal variations, since higher values are observed in the summer season due to a higher fraction of vegetable content. Values can also depend on the waste composition (degrees of recycling, etc.). Typical values for municipal solid waste are in the order of 35 % and do not vary to a large extent.

A number of researchers have shown that the water content at field capacity varies with waste density (e.g. Campbell, 1982; Beaven, 2000). Stegmann (1982) referred to German experiences showing that, in terms of volumetric absorptive capacity, there is an optimum dry density ($\sim 0.4$ t/m$^3$ for both shredded and undisturbed landfill waste), which leads to the waste having a maximum absorptive capacity.

## Water Held in Drainable Voids

Any change in leachate level within a landfill must be accounted for within a water balance. The volume of water held in freely draining voids (generally as a saturated zone at the base of a site) is related to the effective or drainable porosity of the waste. This is defined as the volume of water released from a unit volume of fully saturated material when the material is allowed to drain freely under the influence of gravity.

**Table 10.3.1** Initial water content, field capacity and resulting water deficit for different waste.

| Test cell size and refuse type | Density (t/m³) | Original WC$_{wet}$ (%) | Original WC$_{vol}$ (%) | Water content at field capacity (WC$_{wet}$; %) | Water content at field capacity (WC$_{vol}$; %) | Primary absorptive capacity (l/t$_{wet}$) | Total absorptive capacity (l/t$_{wet}$) | Total absorptive capacity (%vol) | Reference |
|---|---|---|---|---|---|---|---|---|---|
| MSW | | | 4–20 | | | | | 13–25 | Fungaroli (1971) |
| MSW | | | 16 | | | | | 15–28 | Walsh and Kinman (1982) |
| MSW | | | 8–19 | | 33–38 | | | 18–28 | Wigh (1979) |
| 8 m³ pulverized MSW | 0.5 | 40 | | | | 230 | 225 | | Newton (1976) |
| 8 m³ pulverized MSW | | | | | | | 330 | | Robinson et al. (1981) |
| 300 m³ crude MSW | 0.57 | | | | | | | | Blakey (1982) |
| 0.2 m³ pulverized MSW | 0.76 | 26 | 19.8 | | | 165 | 290 | | Blakey (1982) |
| 4000 m³ crude MSW | 0.66 | 25 | 16.5 | | | 100 | | | Campbell (1982) |
| 4000 m³ crude MSW | 0.95 | 25 | 23.8 | | | 41 | | | |
| 4000 m³ crude MSW | 1.01 | 25 | 25.3 | | | 24 | | | |
| 0.2 m³ drums 17 year old MSW | 0.96 | 31.5 | 30.2 | 38.6 | 41.2 | | 115 | 11 | Holmes (1980) |
| 0.2 m³ drums 17 year old MSW | 0.64 | 31.5 | 20.2 | 47.6 | 39.8 | | 307 | 8.3 | |
| 0.2 m³ drums crude MSW | | 26.5 | | | | | 570 | | Harris (1979) |
| Indoor lysimeter crude MSW | 0.33 | | 5 | | 34 | | 867 | 29 | Fungaroli and Steiner (1979) |
| 6 m³ crude MSW | 0.5 | 35 | 17.5 | 54 | | 345 | 425 | | Kinman et al. (1982) |
| 6 m³ crude MSW | 0.4 | 14.7 | 5.9 | | | | | | Jones and Malone (1982) |
| 1.6 m³ simulated pulverized MSW | 0.4 | | | | | | 1 300 | | Pohland (1975) |
| 9 m³ and 1.8 m³ crude MSW | 0.33 | | 16–21 | | 30–31 | | 372.5 | 10–14 | Rovers and Farquhar (1973) |
| MSW | 17 | | | | | | | 23–32 | Rosquist et al. (1997) |
| Various MSW wastes tested in 2 m diameter compression cell | 0.3–1.4 | 28–40 | 12–24 | 33.6–52.8 | 28.1–48.5 | | | | Beaven (2000) |
| Fly ash (waste incinerator) | | | | | | | | 30–45 | Lundgren and Elander (1986) |
| Fly ash (coal) | | | | | | | | 45 | Lagerkvist and Kylefors (1993) |

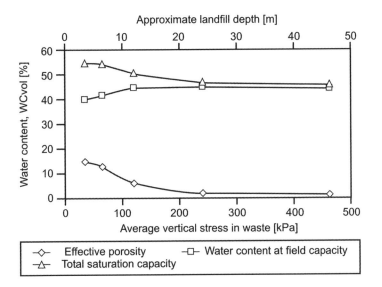

**Figure 10.3.6** *Volumetric water contents versus effective stress for a MSW waste (Beaven, 2000). Beaven, R.P. (2000): The hydrogeological and geotechnical properties of household waste in relation to sustainable landfilling. PhD dissertation. Queen Mary and Westfield College, University of London.*

A number of workers have reported data on drainable porosity from various laboratory experiments and field pumping tests. Values vary from over 40 % in loosely compacted waste in laboratory experiments (Knox, 1991) to less than 10 % in field scale pumping tests (Burrows et al., 1997).

Beaven (2000) investigated the effect of overburden stress on waste density and changes in water content and drainable porosity for a number of different waste types. Figure 10.3.6 illustrates results for a municipal solid waste sample and shows a reduction in drainable porosity from approximately 15 % to less than 2 % as effective stress (and waste density) increases. The figure also illustrates that there is not much change in the water content at field capacity with increasing stress.

## 10.3.2 Water Flow in Landfills

The flow of water in landfills is an exceedingly complex subject that at present is not properly understood. The complexities arise from the following causes:

- Both saturated and unsaturated flow occurs.
- The hydraulic properties of waste may vary over time (through compaction and degradation).
- Waste is a heterogeneous material (on a variety of scales).
- The presence of landfill gas in waste interacts with, and can affect water flow.

### 10.3.2.1 Saturated Flow

Most work on the hydraulic properties of waste has been undertaken on saturated waste materials, with hydraulic conductivity being the focus of several recent investigations. Normally, the traditional Darcy's law for describing water flux has been used (see Box 10.3.2). Measurement of hydraulic conductivity has been carried out either using in situ pump

---

**Box 10.3.2    Darcy's Law for Water Flux Under Saturated Conditions.**

**Darcy's Law**

Hydraulic conductivity is a measure of the ease with which water will flow through a saturated porous media. Darcy (1856) in a series of experiments carried out in Dijon, France derived the following relationship:

$$q = -KiA$$

where:

$q$ is the volumetric flow rate (m$^3$/s).
$K$ is the hydraulic conductivity (m/s).
$i$ is the hydraulic gradient (dimensionless).
$A$ is the cross-sectional area across which flow is taking place (m$^2$).

---

tests or in columns in the laboratory. Results from such investigations are summarized in Table 10.3.2. The table shows that typical values for municipal solid waste are in the range of $10^{-5}$ to $10^{-6}$ m/s, although large variations are observed. Hydraulic conductivities of this range usually do not limit water flow in a landfill on an annual basis but may play a role in short term infiltration events. For incineration residues values are typically lower.

A number of workers (e.g. Bleiker *et al.*, 1995; Powrie and Beaven, 1999; Beaven, 2000) have shown that compaction of waste leads to significantly lower hydraulic conductivities. Powrie and Beaven (1999) produced an empirical relationship between applied stress and hydraulic conductivity, as shown in Figure 10.3.7, and predicted that the hydraulic conductivity would reduce with depth in a landfill. Rowe and Nadarajah (1996) presented data on the hydraulic conductivity of the Fresh Kills landfill (New York, USA) that showed such a reduction in hydraulic conductivity with depth. Despite a considerable amount of scatter their data shows a six orders of magnitude reduction in hydraulic conductivity from $10^{-3}$ to almost $10^{-9}$ m/s with increasing depth from 2 to 40 m within the landfill. An implication of reducing hydraulic conductivities with depth is that waste at the bottom of a landfill may have a lower percolation capacity than waste higher up. If bulk infiltration exceeds this capacity, then some leachate may be prevented from reaching basal leachate drainage layers causing a build up of leachate in the site.

### 10.3.2.2   Unsaturated Flow

The hydraulic conductivity of an unsaturated soil, or waste, is less than that of the same material when saturated. In summary, the main causes for this are:

- Some pores become air filled, reducing the cross sectional area through which flow can occur.
- The larger pores empty first, so that flow is restricted to the smaller pores, which are less conductive.
- The tortuosity of the flow path through the interlinked pores increases.

Water in an unsaturated material is held in the pores by surface tension forces and by the physical attraction of the water to the soil particle interfaces. These forces result in a negative water pressure head, or matrix suction head, $\psi$, in the material. The volumetric water content, $\theta$, is related to this suction head. It has been shown for soils that the relationship between $\psi$ and $\theta$ exhibits hysteresis; it has a different shape when the solids are wetting than when they are drying. At present the significance of this to waste has not been determined.

**Table 10.3.2**  *Collection of reported saturated waste hydraulic conductivities from literature.*

| Method | Number of measurements | Minimum value | Maximum value | Average value | Reference |
|---|---|---|---|---|---|
| **Packed MSW** | | | | | |
| | ? | $2 \cdot 10^{-5}$ | $2 \cdot 10^{-4}$ | — | Fungaroli and Steiner (1979) |
| | 4 | | $1.6 \cdot 10^{-5}$ | — | Oweis and Kehra (1986) |
| | 3 | $1 \cdot 10^{-8}$ | $1.6 \cdot 10^{-6}$ | — | Bleiker et al. (1993) |
| | 13 | $1 \cdot 10^{-9}$ | $2 \cdot 10^{-4}$ | — | Powrie and Beaven (1999) |
| Tests on milled refuse at dry densities between 0.24 and 0.72 t/m$^3$ | | $1 \cdot 10^{-7}$ | $1 \cdot 10^{-4}$ | | Chen et al. (1977) |
| Processed (RDF – Refuse derived fuel) waste packed into columns at different densities | | $5 \cdot 10^{-7}$ | $9.5 \cdot 10^{-4}$ | | Chen and Chynoweth (1995) |
| Wastes excavated from landfill tested in 470 mm diameter odometer | | $6 \cdot 10^{-9}$ | $6.8 \cdot 10^{-5}$ | | Landva et al. (1984) |
| Waste materials at densities between 0.57 and 1.12 t/m$^3$ | | $7 \cdot 10^{-6}$ | $1.5 \cdot 10^{-4}$ | | Oweis and Khera (1986) |
| **MSW in situ** | | | | | |
| | 2 | $1.6 \cdot 10^{-6b}$ | $1 \cdot 10^{-5a}$ | — | Oweis et al. (1990) |
| | ? | $2 \cdot 10^{-7}$ | $2.5 \cdot 10^{-6}$ | — | Hentges et al. (1993)[a] |
| | ? | $1.4 \cdot 10^{-6}$ | $1.8 \cdot 10^{-6}$ | — | Cossu et al. (1997)[a] |
| Site A | 7 | $1.1 \cdot 10^{-6}$ | $2.1 \cdot 10^{-5}$ | $6 \cdot 10^{-6}$ | Burrows et al. (1997)[a] |
| Site B | 2 | $3.5 \cdot 10^{-6}$ | $7.9 \cdot 10^{-6}$ | $5.7 \cdot 10^{-6}$ | |
| Site C | 5 | $3.9 \cdot 10^{-7}$ | $1.5 \cdot 10^{-5}$ | $5.0 \cdot 10^{-6}$ | |
| Site D | 6 | $1.5 \cdot 10^{-5}$ | $6.7 \cdot 10^{-5}$ | $3.0 \cdot 10^{-5}$ | |
| Infiltration from large-scale infiltration ponds | ? | | | $3–4 \cdot 10^{-8}$ | Townsend et al. (1995) |
| Pumping test on 35 m deep landfill with 9 m saturated zone | ? | $9.4 \cdot 10^{-6}$ | $2.4 \cdot 10^{-5}$ | | Oweis and Khera (1986), Oweis et al. (1990) |
| Pumping test on 9 m deep landfill repeated when depth increased to 23 m | ? | $8 \cdot 10^{-6}$ | $1 \cdot 10^{-4}$ | | Beaven (1996) |
| Fly ash | ? | $3 \cdot 10^{-7}$ | $4 \cdot 10^{-6}$ | | Hartlén and Elander (1986) |
| APC residues | ? | $4 \cdot 10^{-9}$ | $2 \cdot 10^{-8}$ | | Hartlén and Elander (1986) |

[a]Pump test
[b]Slug test

Flow in an unsaturated material moves from areas of high pressure head to low pressure head. This implies that, at different points along the flow path, both the water content and the hydraulic conductivity do vary. Therefore, hydraulic conductivity is both a function of water content and suction head (see Box 10.3.3). The application of unsaturated flow modelling to waste requires knowledge of the waste moisture characteristics and the relative permeability characteristics of the waste.

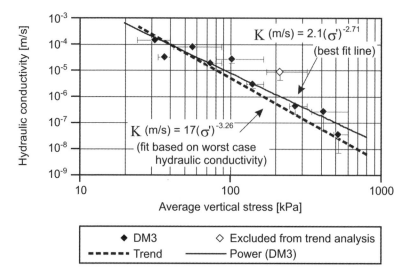

*Figure 10.3.7*   *Variation in saturated hydraulic conductivity with vertical effective stress for crude household waste (from Powrie and Beaven, 1999). Reprinted with permission from Geotechnical Engineering, The hydraulic properties of household waste and implications for landfills by W. Powrie and R.P. Beaven, Proceedings of the Institute of Civil Engineers, 137, 235–247 © (1999) Thomas Telford Limited.*

---

**Box 10.3.3   Equations Describing Water Flow Under Unsaturated Conditions.**

**Soil Moisture Characteristics**

The water flux, $q$ as a function of water content and suction head can be expressed in the following ways:

$$q = -K(\theta)\nabla H$$
$$q = -K(\psi)\nabla H$$

where $\nabla H$ is the hydraulic head gradient, which may include both suctional and gravitational components, and $K(\theta)$ and $K(\psi)$ is, respectively, the hydraulic conductivity ($K$) as a function of the water content, $\theta$, and the suction head, $\psi$.

Two formulations relating suction head to water content are given below following Campbell (1974) and van Genuchten (1980):

**Campbell**

$$\psi = \psi_s \left( \frac{\theta}{\theta_s} \right)^{-b} \tag{1}$$

where:

$\psi_s$ is the suction head at saturation.
$\theta_s$ is the water content at saturation.
$b$ is a dimensionless parameter.

## van Genuchten

$$\theta_e = \left[ \frac{1}{1 + (\alpha \psi)^n} \right]^m \tag{2}$$

where $\theta_e$ is the effective volumetric water content and $\alpha$, $n$ and $m$ are numerical parameters. The effective water content is defined by:

$$\theta_e = \frac{\theta - \theta_r}{\theta_s - \theta_r} \tag{3}$$

where $\theta_r$ is the residual water content (i.e. the lowest achievable water content achievable short of oven-drying the material).

### Relative Permeability Characteristics

Formulations following Campbell (1974) and van Genuchten (1980) for describing the relative permeability, $K_r$, compared to the saturated hydraulic conductivity ($K_s$) are given below:

### Campbell

### van Genuchten

$$K_r = \left[ \frac{\theta}{\theta_s} \right]^{2b+3}$$

$$K_r = \theta_e^{0.5} \left[ 1 - \left( 1 - \theta_e^{1/m} \right)^m \right]^2$$

---

## *Waste Moisture Characteristics*

There are a variety of different mathematical formulations that relate the suction head to the volumetric water content of a soil (see for example Box 10.3.4) but there is limited data regarding waste. McDougall (1996) summarized the work of six researchers (see Box 10.3.4), of which only one (Korfiatis *et al.*, 1984) used direct measurements of the characteristic curve, where the relationship between water content and suction pressure in relatively small samples (15 cm diameter columns) of 6-month-old waste were measured. It was considered that the direct measurement technique generated values for the characteristic curve under static conditions (i.e. no flow) and did not take into account channelling of flow that was observed to take place in column studies. This has implication for the shape of the characteristic curve for waste.

## *Relative Permeability Characteristics*

Direct measurement of the unsaturated hydraulic conductivity of landfilled waste apparently has not been reported. Empirical equations are used to determine the relationship between water content and hydraulic conductivity, usually expressing a relative permeability, where the actual unsaturated hydraulic conductivity is normalized with the saturated hydraulic conductivity ($K_s$). A number of formulations also exist for this relationship as presented in Box 10.3.3.

McDougall (1996) collated results from a number of studies that had derived a relative permeability function for solid waste (Figure 10.3.8). Most of the studies used a solution that had the general form of Campbell's equation as presented in Box 10.3.3. The unique problem associated with applying standard unsaturated flow theory, developed in soil science, to waste materials is that the relative permeability function is based on the saturated hydraulic conductivity, which in waste materials cannot be assumed to be constant. We have already shown how it is strongly influenced by waste density, degree of compaction and effective stress. It may also be affected by the generation of landfill gas (Hudson *et al.*, 2001).

**Box 10.3.4   Soil Moisture Characterization in Waste.**

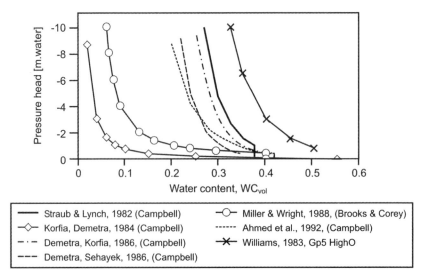

| | |
|---|---|
| —— Straub & Lynch, 1982 (Campbell) | —O— Miller & Wright, 1988, (Brooks & Corey) |
| —◇— Korfia, Demetra, 1984 (Campbell) | ······ Ahmed et al., 1992, (Campbell) |
| — · — Demetra, Korfia, 1986, (Campbell) | —✕— Williams, 1983, Gp5 HighO |
| — — — Demetra, Sehayek, 1986, (Campbell) | |

Soil moisture characterization curves for wastes (McDougall, 1996).

The parameters applicable to Equation (1) in Box 10.3.3 for the four studies presented by McDougall (1996) in the figure above are summarized in the table below. The important point to note about these values, are the high values of *b*. As *b* is the power term in Equation (1), at low water contents very large changes in suctions are generated for small changes in water content.

**Soil moisture characteristic data for modelled waste**

| $\psi_s$ (m) | $\theta_s$ | *b* | Reference |
|---|---|---|---|
| −1.00 | 0.375 | 7.0 | Straub and Lynch (1982) |
| −0.40 | 0.400 | 7.0 | Demetracopoulos *et al.* (1986a) |
| −0.35 | 0.350 | 7.0 | Demetracopoulos *et al.* (1986b) |
| −0.46 | 0.417 | 4.0 | Ahmed *et al.* (1992) |

### 10.3.2.3   Waste Heterogeneity and Structure

The above sections on saturated and unsaturated waste give general solutions to hydrogeological properties of waste. These values will vary over different scales. For example, the material geometry of a landfill facilitates the flow of water in restricted channels and voids (Zeiss and Major, 1993; Bendz *et al.*, 1998) and the existence of rapid flow in favored flow paths in solid waste media has been reported in studies on a field scale (e.g. Ehrig, 1983; Bendz *et al.*, 1997; Rosqvist *et al.*, 1997) and on a laboratory scale (e.g. Uguccioni and Zeiss, 1997; Powrie and Beaven, 1999; Rosqvist and Bendz, 1999; and Beaven *et al.*, 2003).

On a small scale (millimetres to centimetres) the flow of water through waste will be influenced by the composition and distribution of different materials in the waste stream. For example, the localized presence of plastics may introduce a layering effect that will introduce an anisotropy; pieces of metal and glass will create impermeable particles around which flow must take place; and textiles and paper may allow water to pass both through and around them. On a medium

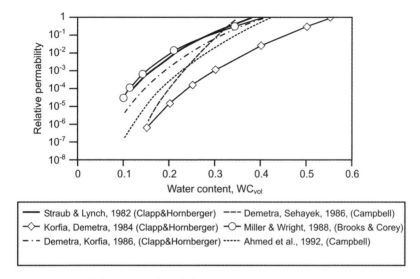

**Figure 10.3.8** *Relative permeability curves for solid wastes (McDougall, 1996). Reprinted with permission from PhD thesis of J. R. McDougall, Application of variably saturated flow theory to the hydraulics of landfilled waste: A finite element solution, University of Manchester/Bolton Instutute (1996).*

scale (decimetres to metres) the juxtaposition of different loads of wastes (with potentially very different compositions) will have an effect on water flow and properties. The presence of daily cover that is not removed prior to a new lift of waste being landfilled can have a very significant effect in terms of layering. As mentioned in the previous section dry waste layers are often observed even in old landfill sites. Leachate generation has been observed from landfilled waste having a large water deficit even for relatively limited water infiltration rates. An explanation for these observations is that water moves in more permeable regions within the heterogeneous waste layers. The use of daily cover soils with low permeability such as clayey or silty loams may either increase water short-circuiting, or alternately impede the downward movement of water. Incineration residues such as bottom ash are of a much more homogeneous nature and in such landfills a more intergranular water flow is expected. Fly ashes and air pollution control residues from incineration may, however, develop into concrete-like boulders, suggesting the predominance of preferential flow in the landfill.

The structure of waste is also important, especially when considering the movement of solutes through waste. In many cases stabilization of landfilled wastes can only be achieved by removing solutes in the leachate by water leaching. It is therefore important (on a medium to large scale) to create conditions in the landfill that lead to as little short-circuiting as possible. This could be achieved in part by avoiding low permeable cover soils and by not introducing high spatial variability in waste hydraulic properties. Finally, on a much larger scale factors such as the presence of site haul roads (which are very often made of relatively permeable 'hard core' material) can introduce large zones that can influence the flow of water.

Waste is now generally considered to be a dual porosity media, containing a macro porosity (equivalent to drainable porosity) through which water moves and a matrix porosity containing water that is relatively immobile. Solutes (or contaminants) in the mobile water are relatively easy to remove by advective flow (flushing). However, the removal of solutes from the matrix porosity is reliant to a large extent on diffusive processes. When the concentration of a pollutant is higher in water held in the matrix pores than in macropores, the fissure porosity causes the movement of the pollutant from the matrix to the mobile water by diffusive processes. The development and application of dual porosity models to flow in waste materials using tracers has been undertaken on both unsaturated (e.g. Bendz *et al.*, 1998; Rosqvist and Destouni, 2000) and on saturated wastes (e.g. Beaven *et al.*, 2003).

On a basic level, tracer experiments can also be used to evaluate the retention time distribution in a landfill, and from that the importance of water short-circuiting. Figure 10.3.9 shows the results of such tracer experiments carried out on a full scale landfill and a waste lysimeter. The full-scale experiment was carried out on a Danish landfill in the 1980s (Christiansen *et al.*, 1985). Tritium was momentarily added to the pump sump and leachate was recirculated to the landfill

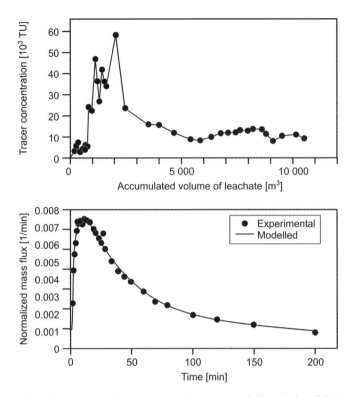

***Figure 10.3.9*** *The results of tracer experiments carried out on a full-scale landfill (top graph; Christiansen et al., 1985) and a waste lysimeter (lower graph; Rosqvist and Destouni, 2000). Reproduced with permission from Journal of Contaminant Hydrology. © (2000) Elsevier Ltd.*

through established infiltration pipes. The resulting tritium concentration in the leachate collected from the leachate drainage system is shown in Figure 10.3.9. A two-peak pattern is observed. This indicates that the flow in the 7–11 m landfill was governed by a fast part, presumably taking place through high permeability regions, and a slower part, where a larger part of the voids in the waste was participating in the flow process. The result of the pilot-scale experiment in Figure 10.3.9 shows the same pattern.

### 10.3.2.4   Effect of Landfill Gas on Hydraulic Properties of Waste

There is growing evidence that the generation of landfill gas within a waste will affect its hydrogeological properties. Hudson *et al.* (2001, 2004) used a large-scale compression cell to investigate the hydrogeological properties of a gas-generating household waste under different applied loads. It was found that large volumes of gas would accumulate in saturated wastes, displacing water from the voids between waste particles. This results in a reduction of the drainable porosity, water content and the hydraulic conductivity of the waste. However, accumulated gas in the waste may be compressed at higher pore water pressure, resulting in significant increases in water content and hydraulic conductivity.

### 10.3.3   Hydrological Models of Landfills

Estimates of leachate generation can be obtained with differing levels of detail. A basic method could use a regional annual average value for the nett precipitation; a sophisticated method could use a dynamic hydrological landfill model for time-dependent leachate generation rate estimates. The first method can be used if almost no local data exists. However,

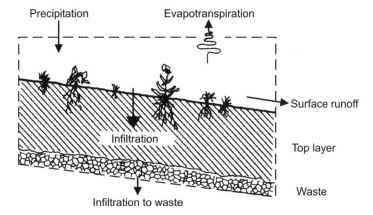

**Figure 10.3.10** *The hydrological elements of a top cover model.*

the precision of the estimate is not very good, especially if engineered top cover systems have been installed at the landfill. Also, using a sophisticated dynamic model demands a high level of good quality local data to be justified. Another option is to set up a hydrological model of the top cover system in order to estimate the water recharge to the waste layers.

### 10.3.3.1 Top Cover Models

Figure 10.3.10 shows the elements of a top cover model. One important assumption of most top cover models for estimating leachate generation is that there is no water transport from the waste layer up in to the top cover. This is probably a valid assumption for mixed waste landfills due to the very coarse nature of the waste. However, for fly ashes and other fine-grained waste types some back flow of water through capillary action may occur into the top covers, especially in dry seasons.

A water balance for the top cover can be defined, assuming that there is no input of managed infiltration:

$$RW = P + SI - ET - SR - CW_{top\ cover}$$

RW is recharge to waste and the other terms correspond to those introduced earlier. If the estimation of RW is made for a whole year, it may be assumed that the $CW_{top\ cover}$ can be neglected in the water balance. The water content of the top cover affects the actual amount lost by actual evapotranspiration, especially in warm seasons (as explained earlier). This means that a top cover model needs to keep track of its actual water content in order to estimate the water lost by evapotranspiration. To justify the use of a top cover model, local values of precipitation and potential evapotranspiration should be used, and the annual SI and SR should be estimated based on local experience. The top cover model can be set up as a spreadsheet model using monthly time steps.

### 10.3.3.2 The HELP Model

The HELP model (HELP = *H*ydrological *E*valuation of *L*andfill *P*erformance) is a computerized water balance model able to calculate water fluxes through the whole vertical profile of the landfill. The HELP model was developed by the United States Army Engineer Waterways Experiment Station for the Environmental Protection Agency in the 1980s, and the first version was issued in 1984 (Berger, 2000). The model has been updated several times. The use of HELP is now required in the United States for obtaining landfilling operation permits. The code is shareware and can be downloaded from Waterways Experiment Stations web site (www. wes.army.mil/el/elmodels/index.html). A Windows-based version of HELP (VisualHelp) is now commercial available from the company 'Waterloo Hydrologic'. HELP can be used for the comparison and optimization of different landfill concepts, for the long-term evaluation of liner effectiveness, and for the evaluation of worse case scenarios (e.g. clogging of drainage systems, etc). The model is based on programme modules that take into account the following processes (Schroeder *et al.*, 1994):

- Water storage on landfill surface.
- Snow melt.
- Surface runoff.
- Evapotranspiration.
- Vegetational growth.
- Water storage in soil layers.
- Lateral subsurface drainage.
- Leachate recirculation.
- Vertical unsaturated water flow.
- Leakage through soil layers and liner systems.

The model calculates daily values of surface runoff, evapotranspiration, vertical, and lateral flow and drainage in different layers, and the leakage through the bottom liner system. HELP uses specific data for climate, vegetation, soil and landfill design. Climate data consists of daily values of precipitation, temperature and solar input, which can be provided by the user, or in the United States be generated by the model through interpolation from existing data originating from 139 American cities. This option has been expanded in the commercial VisualHELP version to cover the whole world. For soils, the porosity, field capacity, wilting point, saturated hydraulic conductivity, and a code number used for calculating the surface runoff using the SCS runoff submodel (Schroeder *et al.*, 1994) is needed. In the HELP model, data for 42 different soils and materials (geotextiles, geomembranes) is included.

HELP is a 'quasi two-dimensional layer model' (Schroeder *et al.*, 1994). Figure 10.3.11 shows an example of a landfill design giving the different horizontal layers. The layers in the landfill is characterized by the hydraulic function, and three types of layers are defined: (1) layers with vertical percolation, (2) drainage layers (having lateral drainage), (3) barrier layers. The top cover and the waste are typical percolation layers, while sand or gravel layers on top of barrier layers are drainage layers. Clay liners and geomembrane liners are typical barrier layers. Design data (such as thickness, slope, etc.) is allocated to each layer. For geomembranes the number and sizes of defects can be given (for further details, see Chapter 10.8).

HELP is used widely on a routine basis, especially in the United States. The use of HELP has been presented in more detail on several occasions. However, in most instances the presented case studies (Dho *et al.*, 2002; Jang *et al.*, 2002; Yalcin and Demirer, 2002) have not included detailed measurements, which would have made a true test of the validity of the HELP model possible. In some cases the HELP model has been compared with other more detailed models (Khire *et al.*, 1997; Yuen *et al.*, 2001). The most extensive evaluation of HELP has been performed on full-scale test data from a German landfill (Berger *et al.*, 1996; Berger, 2000). This has led to changes to the computer code and a new version of HELP (version 3.55; Berger, 2003; Institut für Bodenkunde, 2003).

Based especially on the detailed studies by Khire and coworkers and Berger and coworkers, several limitations of the HELP model can be listed. The most important limitations are found in relation to snow cover and frozen soil, surface runoff, migration through liner systems, and long-term changes. Overall, even though the HELP model has its limitations, its ease of use makes it preferable to more sophisticated computer models for the initial evaluation of different technical alternatives, and for the optimization of different layers in the design of landfills (Berger, 2000). However, care should be taken with its use for areas of more extreme climate.

### 10.3.3.3  Saturated and Unsaturated Flow Models

Although many researchers have developed there own numerical models, there is commercially available software that can be used to simulate basic saturated and unsaturated flow. A commonly used programme (but certainly not the only one available) that models both saturated and unsaturated flow is SUTRA. SUTRA is the United States Geological Survey's

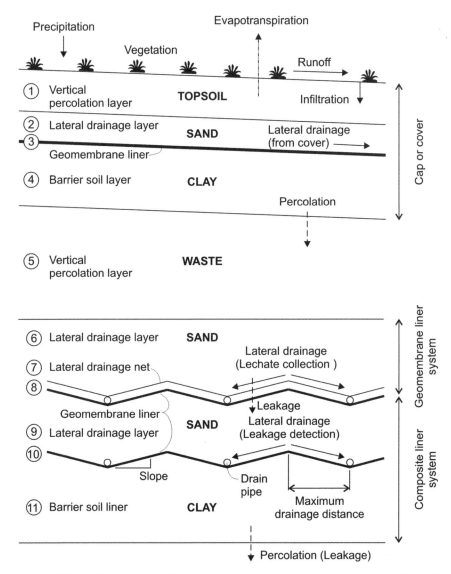

***Figure 10.3.11*** *Landfill profile showing different layer, as defined in the HELP model (after Schroeder* et al., *1994). Reproduced with permission from Dr. J. McDougall, UK. © (1996) McDougall.*

saturated–unsaturated flow and transport model, which can model unsaturated flow in two dimensions. McCreanor and Reinhart (1999, 2000) used SUTRA to model leachate movement in a leachate recirculating landfill.

MODFLOW is an example of a saturated groundwater flow model that has been applied to landfill. Oweis and Biswas (1993) used it to model the build up of leachate in the base of a landfill. Beaven (2000) introduced an effective stress dependent hydraulic conductivity module into MODFLOW to account for the anticipated change in permeability with depth in a landfill due to compaction for self weight. Al-Thani *et al.* (2004) used MODFLOW to model the vertical and spatial distribution of leachate heads around leachate pumping wells in landfills.

## 10.3.4   Leachate Generation

As evident from the previous sections many factors are affecting leachate generation. The leachate generation may vary from close to nothing in very dry climates only occasionally subject to rainstorms to more than 1000 mm/year in wet climates. Also the properties of the landfill cover are important aspects. For compacted German landfills with a vegetated soil cover, Ehrig (1983) report on leachate generation in the range 100–300 mm or 12–30 % of total precipitation. In contrast, for American landfills provided with tight top covers (consisting of vegetated layer, a drainage layer, a geomembrane, and a clay layer) Barlaz *et al.* (2002) report leachate generation in the range of 1–14 mm/year. On a European scale Hjelmar *et al.* (1995) mention that leachate generation for landfills without tight top covers may vary between 50 and 400 mm/year, depending mainly on local climatic conditions.

## References

Ahmed, S., Khanbilvardi, R.M., Fillos, J. and Gleason P.J (1992): 2-Dimensional leachate estimation through landfills. *Journal of Hydraulic Engineering – ASCE*, 118 (2), 306–322.

Al-Thani, A.A., Beaven, R.P. and White J.K. (2004): Modelling flow to leachate wells in landfills. *Waste Management*, 24 (3), 271–276.

Aslyng, H.C. (1976): *Klima, jord og planter* (Climate, soil, and plants; in Danish). Fifth edition. DSR Forlag, Copenhagen, Denmark.

Barlaz, M.A., Rooker, A.P., Kjeldsen, P., Gabr, M.A. and Borden, R.C. (2002): A critical evaluation of factors required to terminate the post-closure monitoring period at solid waste landfills. *Environmental Science and Technology*, 36 (16), 3457–3464.

Beaven, R.P. (1996): Geotechnical and hydrogeological properties of wastes. In: Bentley, S.P. (ed.) *Engineering geology of waste storage and disposal*. Geological Society Engineering Geology Special Publication No. 11, pp. 57–65.

Beaven, R.P. (2000): *The hydrogeological and geotechnical properties of household waste in relation to sustainable landfilling*. PhD dissertation. Queen Mary and Westfield College, University of London.

Beaven R.P, Barker, J. and Hudson, A.P. (2003): Description of a tracer test through waste and application of a double porosity model. In: Christensen T.H., Cossu R. and Stegmann R. (eds) *Proceedings of the Ninth International Waste Management and Landfill Symposium*. CD-ROM. CISA – Environmental Sanitary Engineering Centre, Cagliari, Italy.

Bendz, D., Singh, V.P. and Åkesson, M. (1997): Accumulation of water and generation of leachate in a young landfill. *Journal of Hydrology*, 203, 1–10.

Bendz, D., Singh, V.P., Rosqvist, H. and Bengtsson L. (1998): Kinematic wave model for water movement in municipal solid waste. *Water Resource Research*, 34, 2963–2970.

Berger, K., Melchior, S. and Miehlich, G. (1996): Suitability of hydrologic evaluation of landfill performance (HELP) model of the US Environmental Protection Agency for the simulation of the water balance of landfill cover systems. *Environmental Geology*, 28, 181–189.

Berger, K. (2000): Validation of the hydrologic evaluation of landfill performance (HELP) model for simulating the water balance of cover systems. *Environmental Geology*, 39, 1261–1274.

Berger, K. (2003): Validation and enhancement of the HELP model to simulate the water balance of surface covers. In: Christensen T.H., Cossu R. and Stegmann R. (eds) *Proceedings of the Ninth International Waste Management and Landfill Symposium, vol. 2*. CD-ROM. CISA – Environmental Sanitary Engineering Centre, Cagliari, Italy.

Blakey, N. (1982): Absorptive capacity of refuse – WRC Research. In: Harwell Research (ed.) *Proceedings of a Landfill Leachate Symposium*. Harwell Research, Harwell, UK.

Bleiker, D.E., McBean, E. and Farquhar, G. (1993): Refuse sampling and permeability testing at the Brock West and Keele Valley landfills. In: University of Wisconsin (ed.) *Proceedings of the 16th International Madison Waste Conference*. CD-ROM. Department of Engineering Professional Development, University of Wisconsin, Madison/Extension, Madison, USA.

Bleiker, D.E., Farquhar, G. and McBean, E. (1995): Landfill settlement and the impact on the site capacity and refuse hydraulic conductivity. *Waste Management and Research*, 13, 533–554.

Burrows, M.R., Joseph, J.B. and Mather, J.D. (1997): The hydraulic properties on in-situ landfilled waste. In: Christensen, T.H., Cossu, R., Stegmann, R. (eds) *Proceedings of the Sixth International Landfill Symposium, Sardinia, vol. 2.* CISA – Environmental Sanitary Engineering Centre, Cagliari, Italy, pp. 73–83.

Burton, S.A.Q., Beaven, R.P. and White J.K. (2004): The effect of moisture content in controlling landfill gas production and its application to a model for landfill refuse decomposition. In: *Proceedings of the Waste 2004 Conference*, pp. 333–342. Stratford upon Avon, UK.

Campbell, G.S (1974): A simple method for determining unsaturated conductivity from moisture retention data. *Soil Science*, 117, 311–314.

Campbell, D. (1982): Absorptive capacity of refuse – Harwell Research. In: Harwell Research (ed.) *Proceedings of a Landfill leachate symposium*. Harwell Research, Harwell, UK.

Chen, W.W.H., Zimmerman, R.E. and Franklin, A.G. (1977): Time settlement characteristics of milled urban refuse. In: ASCE (ed.) *Proceedings of the Conference on Geotechnology: Practice for Disposal of Solid Waste Materials*. ASCE, Washington, D.C., USA.

Chen, Ten-Hong and Chynoweth, D.P. (1995): Hydraulic conductivity of compacted municipal solid waste. *Bioresource Technology*, 51, 205–212.

Christiansen, K., Prisum, M. and Skov, C. (1985): *Undersøgelse af lossepladsers selvrensende effekt ved recirkulation af perkolat* (Investigation of landfill stabilization by leachate recirculation; in Norwegian). Enviroplan A/S, Lyngen, Norway.

Cossu, R., Frongia, G., Muntoni, A., Nobile, A. and Raga, R. (1997): *Use of pumping test for the assessment of leachate flow regimes, waste hydraulic parameters and well efficiency*. In: Christensen, T.H., Cossu, R. and Stegmann, R. (eds) Proceeding of the sixth international landfill symposium, Sardinia, vol. 2, pp. 53–61. CISA – Environmental Sanitary Engineering Centre, Cagliari, Italy.

Darcy, H. (1856): *Les fontaines publiques de la ville de Dijon*. Dalmont, Paris, France.

Demetracopoulos, A.C, Korfiatis, G.P, Bourodimos, E.L. and Nawy, E.G. (1986a): Unsaturated flow through solid waste landfills: Model and sensitivity analysis. *Water Resources Bulletin*, 22 (4), 601–609.

Demetracopoulos, A.C., Sehayek, L. and Erdogan, H. (1986b): Modelling leachate production from municipal landfills. *ASCE Journal of Environmental Engineering*, 112 (5), 849–866.

Dho, N.Y., Koo, J.K. and Lee, S.R. (2002): Prediction of leachate level in Kimpo metropolitan landfill site by total water balance. *Environmental Monitoring and Assessment*, 73, 207–219.

Ehrig, H-J. (1983): Quality and quantity of sanitary landfill leachate. *Waste Management Research*, 1, 53–68.

ElFadel, M., Findikakis, A.N. and Leckie, J.O. (1997): Modeling leachate generation and transport in solid waste landfills. *Environmental Technology*, 18, 669–686.

Fungaroli, A.A. (1971): *Pollution of subsurface water from sanitary landfills*. Interim Report SW-12. US Environmental Protection Agency, Washington, D.C., USA.

Fungaroli, A.A. and Steiner, R. (1979): *Investigation of sanitary landfill behaviour – volume 1*. Final report. EPA-600-2-79-053a US Environmental Protection Agency, Washington, D.C., USA.

Hartlén, J. and Elander, P. (1986): *Residues from waste incineration – chemical and physical properties*. SGI Varia 172. Statens Geotekniska Institut, Linköping, Sweden.

Harris, M.R.R. (1979): Geotechnical characteristics of landfilled domestic refuse. In: Midland Geotechnical Society (ed.) *Proceedings of the Engineering Behavior of Industrial and Urban Fill*. Midland Geotechnical Society, Bolton, UK.

Hentges, G.T., Theis, F. and Lemar, T.S. (1993): Leachate extraction well assessment Des Moines, Iowa Metropolitan Park East sanitary landfill, Hamilton County, Iowa sanitary landfill. In: University of Wisconsin (ed.) *Proceedings of the 16th International Madison Waste Conference*. CD-ROM. Deparment of Engineering Professional Development, University of Wisconsin, Madison/Extension, Madison, USA.

Hjelmar, O., Johannessen, L.M., Knox, K., Ehrig, H.J., Flyvbjerg, J., Winther, P. and Christensen, T.H. (1995): Composition and management of leachate from landfills within the EU. In: Christensen, T.H., Cossu, R., Stegmann, R. (eds) *Proceedings of the fifth international landfill symposium, Sardinia, vol. 1*, pp. 5243–262. CISA – Environmental Sanitary Engineering Centre, Cagliari, Italy.

Holmes, R. (1980): The water balance method of estimating leachate production from landfill sites. *Solid Wastes*, 1.

Hudson, A.P, Beaven, R.P and Powrie (2001): Interaction of water and gas in saturated household waste. In: Christensen, T.H., Cossu, R. and Stegmann, R. (eds) *Proceedings of the Eighth International Waste Management and Landfill Symposium,, vol. 3*, pp. 585–594. CISA – Environmental Sanitary Engineering Centre, Cagliari, Italy.

Hudson, A.P., White J.K., Beaven R.P. and Powrie, W. (2004): Modelling the compression behaviour of landfilled domestic waste. *Waste Management*, 24 (3), 259–269.

Institut für Bodenkunde (2003): *Validation and adaptation of the simulation model HELP for the computation of the water balance of German landfills*. Institut für Bodenkunde, Universität Hamburg, Germany. Located on the internet: 19 July 2007. www.geowiss.uni-hamburg.de/i-boden/fhelpeng.htm.

Jang, Y.S., Kim, Y.W. and Lee, S.I. (2002): Hydraulic properties and leachate level analysis of Kimpo metropolitan landfill, Korea. *Waste Management*, 22, 261–267.

Jones, R.J.A. and Malone, P.G. (1982): *Disposal of treated and untreated electroplating waste in a simulated municipal landfill*. EPA (ed.) Proceedings of the EPA eighth annual research symposium: Land disposal of hazardous wastes, pp. 294–314. EPA-600/9-82-002. EPA, Kentucky, USA.

Khire, M.V., Benson, C.H. and Bosscher, P.J. (1997): Water balance modeling of earthen final covers. *Journal of Geotechnical and Geoenvironmental Engineering*, 123, 744–754.

Kinman, R.N., Rickabaugh, J.I., Walsch, J.J. and Vogt, W.G. (1982): Leachate from co-disposal of municipal and hazardous waste in landfill simulators. Proceedings of the EPA eighth annual research symposium: Land disposal of hazardous wastes, pp. 274–293. EPA-600/9-82-002. EPA, Kentucky, USA.

Knox, K. (1991): *A review of water balance methods and their application to landfills in the UK*. Report CWM 031/91. The technical aspects of controlled waste management. Waste Technical Division, Department of the Environment, London, UK.

Korfiatis, G.P., Demetracopoulos, A.C., Bourodimos, E.L., and Nawy, E.G. (1984): Moisture transport in a solid-waste column. *Journal of Environmental Engineering – ASCE*, 110 (4), 780–796.

Lagerkvist, A. and Kylefors, K. (1993): *Composition and treatment of leachates from different wastes*. In: Christensen, T.H., Cossu, R. and Stegmann, R. (eds) Proceedings of the fourth international landfill symposium, Sardinia, vol. 1, pp. 811–819. CISA – Environmental Sanitary Engineering Centre, Cagliari, Italy.

Landva, A.O., Clark, J.I., Weisner, W.R. and Burwash, W.J. (1984): *Geotechnical engineering and refuse landfills*. Proceedings of the sixth national conference on waste management, Vancouver, Canada, pp. 1–37.

Lundgren, T. and Elander, P. (1986): *Deponering av avfall från kol och torveldning, handledning*. Rapport 3144. Naturvårdsverket, Stockholm, Sweden.

McCreanor, P.T. and Reinhart, D.R. (1999): Hydrodynamic modeling of leachate recirculating landfills. *Waste Management and Research*, 17 (6), 465–469.

McCreanor, P.T. and Reinhart, D.R. (2000): Mathematical modeling of leachate routing in a leachate recirculating landfill. *Water Research*, 34 (4), 1285–1295.

McDougall, J.R. (1996): *Application of variably saturated flow theory to the hydraulics of landfilled waste: A finite element solution*. PhD thesis, University of Manchester/Bolton Institute, Manchester, UK.

Miller, C.J. and Wright, S.J. (1988): Application of variably saturated flow theory to clay cover liners. *Journal of Hydraulic Engineering – ASCE* 114 (10), 1283–1300.

Newton, J.R. (1976): *Pilot-scale experiments on leaching from landfills. II. Experimental procedures and comparison of overall performance of all experimental units, November 1973 – August 1975*. WLR Technical Note No 17. Department of the Environment, London, UK.

Oweis, I.S. and Kehra, R. (1986): Criteria for geotechnical construction of sanitary landfills. In: Fang, H.-Y. (ed.) *Proceedings of the International Symposium on Environmental Geotechnology, vol. 1*. CD-ROM. Envo Publishing Company, Inc., Allentown, USA.

Oweis, I.S., Smith, D.A., Ellwood, R.B. and Greene, D.S. (1990): Hydraulic characteristics of municipal refuse. *Journal of Geotechnical Engineering (ASCE)*, 116 (4), 539–553.

Oweis I.S. and Biswas G.C. (1993): Leachate mound changes in landfills due to changes in percolation by a cap. *Ground Water*, 31 (4), 664–674.

Pohland, F.G. (1975): *Sanitary landfill stabilization with leachate recycle and residual treatment*. EPA-600/2-75-043. US EPA, Washington, D.C., USA.

Powrie, W. and Beaven, R.P. (1999): The hydraulic properties of household waste and implications for landfills. *Proceedings of the Institution of Civil Engineers, Geotechnical Engineering*, 137, 235–247.

Robinson, H.D, Barber, C. and Maris, P.J. (1981): Generation and treatment of leachate from domestic wastes in landfills. *Water Pollution Control*, 81 (4), 465–478.

Rosqvist, H., Bendz, D., Öman, C. and Meijer, J.-E. (1997): *Water flow in a pilot-scale landfill.* In: Christensen, T.H., Cossu, R., Stegmann, R. (eds) Proceedings of the sixth international landfill symposium, Sardinia, vol. 2, pp. 85–96. CISA – Environmental Sanitary Engineering Centre, Cagliari, Italy.

Rosqvist, H. and Bendz, D. (1999): An experimental evaluation of the solute transport volume in biodegraded municipal solid waste. *HESS – Hydrology and Earth Systems Science* 3, 429–438.

Rosqvist, H. and Destouni, G. (2000): Solute transport through preferential pathways in municipal solid waste. *Journal of Contaminant Hydrology*, 46, 39–60.

Rovers, F.A. and Farquhar, G.J. (1973): Infiltration and landfill behaviour. *ASCE Journal of Environmental Engineering*, 99, 671–690.

Rowe, R.K. and Nadarajah, P. (1996): Estimating leachate drawdown due to pumping wells in landfills. *Canadian Geotechnical Journal*, 33 (1), 1–10.

Schroeder, P.R., Lloyd, C.M. and Zappi, P.A. (1994): *The hydrologic evaluation of landfill performance (HELP) model. Users guide for version 3.* EPA 600/R-94/168a. US Environmental Protection Agency, Cincinnati, USA.

Straub, W.A. and Lynch, D.R. (1982): Models of landfill leaching: Moisture flow and inorganic strength. *Environmental Engineering Division* 108, 231–250.

Stegmann, R. (1982): Absorptive capacity of refuse – West German research. In: Harwell Research (ed.) *Proceedings of the Landfill Leachate Symposium.* Harwell Research, Harwell, UK.

Townsend, T.G., Miller, W.L., and Earle, J.F.K. (1995): Leachate – recycle infiltration ponds. *Journal of Environmental Engineering – ASCE*, 121 (6), 465–471.

Uguccioni, M. and Zeiss, C. (1997): Comparison of two approaches to modelling moisture movement through municipal solid waste. *Journal of Environments and Systems* 25 (1), 41–63.

Walsh, J.J and Kinman, R.N. (1982): Leachate and gas production under controlled moisture conditions. In: *Proceedings of the Eighth Annual Research Symposium on Land Disposal of Solid Waste.* EPA-600-82-002. US Environmental Protection Agency, Cincinatti, USA.

Wigh, R.J. (1979): *Boone county field study interin report.* EPA-600-2-79-058. US Environmental Protection Agency, Cincinatti, USA.

Williams, J., Prebble, R.E., Williams, W.T and Hignett, C.T. (1983): The influence of texture, structure and clay mineralogy on the soil moisture characteristic. *Australian Journal of Soil Research*, 21, 15–32.

van Genuchten, M.T. (1980): A closed form equation for predicting the hydraulic conductivity of unsaturated soils. *Soil Science Association of America Journal*, 44, 892–898.

Yalcin, F. and Demirer, G.N. (2002) Performance evaluation of landfills with the HELP (hydrologic evaluation of landfill performance) model: Izmit case study. *Environmental Geology*, 42, 793–799.

Yuen, S.T.S., Wang, Q.J., Styles, J.R. and McMahon, T.A. (2001) Water balance comparison between a dry and a wet landfill – a full-scale experiment. *Journal of Hydrology*, 251, 29–48.

Zeiss, C. and Major, W. (1993): Moisture flow through municipal solid waste: pattern and characteristics. *Journal of Environments and Systems*, 22, 211–231.

# 10.4

# Landfilling: Geotechnology

**R. Kerry Rowe**

*Queen's University, Kingston, Ontario, Canada*

**Jamie F. VanGulck**

*Arktis Solutions Inc., Yellowknife, Northwest Territories, Canada*

The geotechnical properties of waste are strongly influenced by its type. From a geotechnical perspective, waste can be classified into two main groups. The first is soil-like waste, which is defined as particulate material for which the principles of soil mechanics are applicable. Examples of soil-like waste include excavated soil, road construction debris, and incineration residue (i.e. slag, ash, dust). The second is nonsoil-like waste, which includes material for which the principles of soil mechanics may have limited applicability, but have been used in describing waste behaviour. Nonsoil-like waste may include municipal solid waste (MSW), some industrial waste, bulky waste, 'green' waste, and residues from mechanical-biological treated waste. It is desirable during the design of landfill facilities to have details of the mechanical, hydraulic, and geotechnical properties of the disposed waste. Such information is necessary for proper investigations into issues such as landfill stability, settlement behaviour, and the design of landfill engineered components (i.e. barrier system, collection system).

This chapter provides a summary of reported nonsoil-like waste parameters, specifically municipal solid waste, such as unit weight, shear strength, hydraulic conductivity, settlement parameters, and dynamic stability values taken from literature. For each parameter, the chapter discusses factors influencing their value, methods of determination, values reported in the literature, and relevance to engineering design. Additionally, a summary of landfill failures and lessons learned from these case histories is provided.

## 10.4.1 Geotechnical Classification of Waste

Classification and analysis of soil-like waste for geotechnical properties should be performed in accordance with conventional soil mechanics principles (König and Jessberger, 1997). Caution must be exercised in the interpretation of the results and their implications for design because certain soil-like waste may produce geotechnical properties that cannot

*Solid Waste Technology & Management*   Edited by Thomas Christensen
© 2011 Blackwell Publishing Ltd

***Table 10.4.1*** *Classification proposed by Konig and Jessberger (1997) for material class analysis of waste materials. Reproduced with permission from Waste mechanics. Report of subcommittee 3, ISMMFE Technical Committee TC5 on Environmental Geotechnics, by Konig, D. and Jessberger, H.L, pp. 35–76. © (1997) Ruhr-Universitat, Bochum, Germany.*

| Analytical steps | Material classes |
|---|---|
| Step 1: Sorting the material classes from total waste sample | • Bulky components, paper, cardboard<br>• Soft synthetics<br>• Hard synthetics<br>• Metals<br>• Minerals<br>• Wood<br>• Organics<br>• Sludge<br>Remainder ≤ 120 mm |
| Step 2: Particle size distribution for remainder ≤ 120 mm. Distinguish at a minimum between three broad size fractions | • 8 mm<br>• 8–40 mm<br>• 40–120 mm |
| Step 3: Identification and description of each material class | • Geometry of waste components<br>  ○ Grain<br>  ○ Stick, fibre<br>  ○ Foil-like<br>  ○ Box<br>• Water content<br>• Organic content |

be fully explained based on the theory of soil mechanics. Information on the composition of waste can be obtained by an audit of incoming waste to a landfill or by analysing already landfilled waste in test pits or borings.

Fassett *et al.* (1994) stated that the determination of engineering properties of nonsoil-like waste (specifically MSW) is problematic due to:

- Widely variable properties resulting from the inconsistent and heterogeneous composition of the waste.
- Difficulty of obtaining representative samples of a sufficient size.
- Erratic nature of the waste particles which makes sampling and testing difficult.
- Change in waste properties with time and location (due to biodegradation and physicochemical decay processes).

A waste audit can be made on a large quantity of waste that enters a landfill facility. Konig and Jessberger (1997) proposed the following description of the waste (Table 10.4.1) including the following three steps: (1) sorting of material classes from total waste sample (bulky components like paper and cardboard, soft synthetics, hard synthetics, metals, minerals, wood, organics, sludge), (2) determining the particle size distribution (at least three size fractions) for the remainder from the sorting procedure ≤ 120 mm, and (3) identification and description of each material class with respect to particle geometry, water content, and content of organics. The organic content may be classified according to the relative ease of breakdown over time (Landva and Clark, 1990), as summarized in Table 10.4.2.

For existing landfills, samples are obtained by borings and excavation of pits. But it is difficult or expensive to obtain a large number of samples. Landva and Clark (1990) summarized their experiences in drilling and sampling within landfills as follows:

- The use of split spoon samplers of various diameters in municipal waste recovered very little material.
- Auger drilling is most suitable for sampling all types of municipal waste above and below the water table. The best overall continuous auger with respect to production rate and quality and size for the sample was a solid-stem 130-mm auger (140-mm bit). The range of auger diameters used varied between 100 and 230 mm in diameter. A reasonably

***Table 10.4.2***   *Classification of the biodegradability of waste (adapted from Landva and Clark, 1990).*

| Class | Examples |
|---|---|
| 1. Organics | |
|    (a) Putrescible (monomers and low-resistance polymers, readily biodegradable) | Food waste<br>Garden waste<br>Animal waste<br>Material contaminated by such waste |
|    (b) Nonputrescible (highly resistant polymers, slowly biodegradable) | Paper<br>Wood[a]<br>Textiles<br>Leather[a]<br>Plastic[a], rubber[a]<br>Paint, oil, grease, chemicals, organic sludge |
| 2. Inorganic | |
|    (a) Degradable | Metals (corrodible to varying degrees)[a] |
|    (b) Nondegradable | Glass[a], ceramics[a]<br>Mineral soil, rubble<br>Tailings, slimes<br>Ash<br>Concrete, masonry (construction debris) |

[a]May contain numerous void-forming constituents that will affect the geotechnical behaviour of the waste fill.

   heavy drill rig is required to penetrate or displace problematic materials such as wood, tires, rock, concrete blocks, and steel objects.
- Test pits can be used in most landfills but are limited to depths of about 4 m and generally it is not feasible to excavate below the ground water level.

## 10.4.2   Density and Unit Weight of Waste

### 10.4.2.1   Definitions: Density and Unit Weight of Waste

There are two densities commonly used in landfill engineering, they are the in-place density and total average landfill density.

   The in-place density ($\rho$) is defined as the mass of waste ($M_w$, includes mass of refuse and moisture) per unit volume of waste ($V_w$, includes mass of refuse, moisture, and voids).

$$\rho = \frac{M_w}{V_w}$$

The in-place density has units [M/L$^3$], typically reported as kg/m$^3$ or t/m$^3$. The in-place density will depend on the composition of the waste and can be used as an assessment of the degree of compaction.

   The total average landfill density ($\rho_{av}$) is equal to the mass of waste plus cover material (including the moisture contained in both) per unit volume of landfill and has the same units as in-place density.

$$\rho_{av} = \frac{M_w + M_c}{V_w + V_c}$$

This is the density used in stability calculations for loads on collection pipes, foundation, etc.

The bulk unit weight of the waste ($\gamma$) can be calculated by multiplying the total average landfill density by gravitational acceleration constant ($g$):

$$\gamma = \rho_{av} g$$

The bulk unit weight has units [Force/L$^3$], typically reported as kN/m$^3$.

The bulk unit weight of waste has a reported range of 1 to 19 kN/m$^3$. The average landfill density can be calculated by dividing the unit weight by gravitational acceleration constant (9.81 m/s$^2$), which results in a density range of 102 to 1936 kg/m$^3$ or 0.10 to 1.93 t/m$^3$ (note: 1 kN = 1000 kg/m/s$^2$; 1000 kg = 1 t).

### 10.4.2.2 Influencing Factors: Density and Unit Weight

The density or unit weight of landfill waste is highly variable and is dependent on many factors such as:

- *Waste composition:* MSW containing relatively greater proportions of dense material such as metal, glass or rock, would be expected to have a higher unit weight than MSW comprised primarily of less dense material such as organic matter or textiles (Landva and Clark 1990; Manassero *et al.* 1996). Similarly, the frequency of placement and the thickness of relatively dense daily cover soil will influence the bulk unit weight of the waste mass. Additionally, waste composition changes due to decomposition. Thus, the unit weight may change depending on the age of the waste and therefore may have different values with time and location within the landfill.
- *Degree of compaction:* An increase in the compacted effort applied to the waste increases the unit weight of the waste (e.g. Fassett *et al.* 1994; Van Impe 1994).
- *Moisture content:* The larger the percentage of fluid (e.g. water or leachate) in the waste, the greater the bulk unit weight. The moisture content of MSW may be highly variable, with reported values between 10 and 100 %, depend on factors such as the organic content and age of the waste (Knochenmus *et al.* 1998). Other factors that may influence the moisture content include: (1) landfill operations, such as leachate recirculation which inject leachate back into the waste to promote rapid waste stabilization, (2) composition of the waste, (3) local climatic conditions, (4) percolation through the landfill cap/cover, and (5) effectiveness of the leachate collection and removal systems.
- *Applied stress:* The unit weight of waste generally increases with depth due to the compression and consolidation under the overburden stress (Knochenmus *et al.*, 1998).

### 10.4.2.3 Methods of Determination: Density and Unit Weight of Waste

The most common and reliable method of evaluating the in-situ unit weight of landfilled MSW is through the excavation of test pits (Gachet *et al.* 1998). This method involves the excavation of waste using a mechanical shovel (backhoe). The waste is then weighed, and the excavated volume is deduced by either (1) lining the excavation with an impermeable material (PVC, HDPE) and filling with a known volume of water, or (2) surveying the volume of the cut. Detailed specifications and analysis of the advantages and disadvantages of this and other density determination methods may be found in Gachet *et al.* (1998). Other techniques that have been employed include indirect methods such as core and auger drilling, and direct methods, such as $\gamma$-adsorption, microgravimeter, and penetrodensitometer (Gachet *et al.* 1998). Some studies have used test fills (e.g. Marques *et al.* 1998) and large-scale laboratory methods (e.g. Powrie and Beaven 1999) to obtain the unit weight of waste.

Other indirect methods for density determination involve surveying landfill volume and weight of incoming waste and cover material, and measuring the unit weight of individual components of the waste and making an estimate of the overall unit weight by using percentages of each component (Landva and Clark 1990; Qian *et al.* 2002). Oweis and Khera (1990) provide values for unit weight of individual waste components in municipal landfills. Landva and Clark (1990) also provide a summary of typical refuse composition and corresponding dry and saturated unit weights. Fassett *et al.* (1994) suggested the least reliable values are computed using indirect methods of determination.

**Table 10.4.3**   *Summary of density/unit weight of landfilled waste.*

| | Bulk unit weight (kN/m³) | Density (tonne/m³) | Source |
|---|---|---|---|
| Refuse landfill: | 9.0–13.2 | 0.91–1.34 | Landva and Clark (1986)[a] |
| Refuse to soil over ratio: 2:1 to 10:1 | | | |
| For 6:1 refuse to daily cover soil | 7.2 | 0.73 | EMCON Associates (1989)[a] |
| Municipal refuse: | | | Oweis and Khera (1990)[b] |
| Poor compaction | 2.8–4.7 | 0.28–0.47 | |
| Moderate to good compaction | 4.7–7.1 | 0.47–0.72 | |
| Good to excellent compaction | 7.1–9.4 | 0.72–0.96 | |
| Baled waste | 5.5–10.5 | 0.56–1.07 | |
| Shredded and compacted | 6.4–10.5 | 0.65–1.07 | |
| Incinerator residue | | | Oweis and Khera (1990)[b] |
| Poorly burnt | 7.2 | 0.73 | |
| Intermediately burnt | 11.8 | 1.20 | |
| Well burnt | 12.6 | 1.28 | |
| Ashes | 6.4–8.2 | 0.65–0.83 | |
| Densified MSW (heavy tamping) | 10 | 1.02 | Van Impe (1994) |
| MSW | | | Fassett et al. (1994) |
| Poor compaction | 3–9 | 0.31–0.91 | |
| Moderate compaction | 5–8 | 0.51–0.81 | |
| Good compaction | 9.0–10.5 | 0.91–1.07 | |
| MSW | | | Gachet et al. (1998)[c] |
| Zero compaction | 1–8 | 0.10–0.81 | |
| Slight compaction | 2.4–6.5 | 0.24–0.66 | |
| Mean compaction | 3.5–7.7 | 0.36–0.78 | |
| Extreme compaction | 2.5–12 | 0.25–1.22 | |
| 'Not important' | 1.0–17.3 | 0.10–1.76 | |
| MSW | | | Machado Santos et al. (1998) |
| Compacted waste, heavy cover | 14–19 | 1.43–1.94 | |
| MSW | | | Marques et al. (1998) |
| Compacted to varying degrees | 7.0–9.4 | 0.71–0.96 | |
| Loose (precompaction) | 3.6–6.6 | 0.37–0.67 | |

[a]Compiled based on summaries by Sharma et al. (1990), modified from Qian et al. (2002).
[b]Modified from a summary compiled by Oweis and Khera (1990).
[c]Summary of various reported studies (approximately 200 values).

### 10.4.2.4   Reported Values: Density and Unit Weight of Waste

A summary of literature reported waste unit weight is provided in Table 10.4.3. The range of reported values is indicative of the heterogeneous nature of waste. The great degree of scatter indicates the importance of site-specific unit weight determinations for any landfill design analyses. Note the importance of compactive effort on waste unit weight in the reported values. It should also be noted that these values do not reflect the increase in unit weight that can occur with the uptake of additional moisture due to heavy rain, leachate recirculation, or leachate mounding.

### 10.4.2.5   Relevance to Design: Density and Unit Weight of Waste

The weight of the waste material within a landfill typically provides the main driving force in slope stability analyses, and a reliable estimate of the unit weight is necessary for an accurate assessment of stability (discussed in Section 10.4.3). Likewise, the weight of waste provides the main component of overburden pressure giving rise to settlement (discussed in Section 10.4.5). Due to the heterogeneity of MSW and the change in parameters with time (due to physicochemical processes), an accurate assessment of unit weight is often difficult, time-consuming and costly, and therefore the unit weight is often estimated for stability and settlement analyses. Acknowledging the difficulty in measuring this parameter, Manassero *et al.* (1996) recommended that testing be undertaken to assess the unit weight prior to undertaking stability or settlement analyses. Additionally, a sensitivity analysis should be performed over a range of likely unit weights to assess the potential effect of uncertainty regarding this parameter on the design. Careful consideration should be given to potential changes in unit weight due to uptake of moisture after placement and compaction.

## 10.4.3   Shear Strength

### 10.4.3.1   Definition: Shear Strength

The shear strength of municipal solid waste is typically reported in terms of the traditional Mohr–Coulomb failure criterion:

$$\tau_f = c' + \sigma_n' \tan \phi'$$

where $\tau_f$ is the shear strength (kPa), $c'$ is the apparent cohesion (kPa), $\sigma_n'$ is the applied normal effective stress (kPa), and $\phi'$ is the mobilized friction angle (degree). The shear strength of MSW has been shown to increase with increasing normal stress due to increased particle interlocking, and hence the Mohr–Coulomb criterion seems appropriate (Knochenmus *et al.*, 1998). Classical soil mechanics principles relate $c'$ and $\phi'$ to failure conditions. If it is not possible to identify failure conditions up to a predetermined deformation level, a deformation criterion has to be defined (Konig and Jessberger 1997) as discussed below.

Kockel and Jessberger (1995) idealized MSW as: (1) a basic matrix comprised of fine and medium grained soil-like particles that contribute to the frictional behaviour, and (2) a reinforcing matrix comprised of fibrous waste components like plastics, textiles, paper, and wood that contribute to the apparent cohesion (Konig and Jessberger, 1997; Knochenmus *et al.*, 1998; Powrie *et al.*, 1999). Despite its heterogeneity, it is generally assumed in analyses that MSW is a homogeneous isotropic material over a certain volume of the waste body (Konig and Jessberger 1997). Kockel and Jessberger (1995) showed that the shear strength of the basic matrix (frictional) is only activated at large strains (about 20 % although this value may not be universally applicable) while only slightly influenced by the reinforcing components at this strain. The reinforcing matrix (apparent cohesion) is also activated at very large strain but after the frictional component is almost fully mobilized. Jesseberger *et al.* (1995) indicated that axial strains of up to 50 % were necessary to achieve peak shear strength.

Powrie *et al.* (1999) raised questions about the applicability of the Mohr–Coulomb failure criterion to MSW: Specifically, concerns regarding the physical interpretation of the shear strength parameters, the high strains required for mobilization of MSW shear strength, the extrapolation of high normal stress data to low normal stress conditions, and the applicability of effective stress concepts to relatively compressible MSW. These are justifiable concerns but they do not preclude the application of traditional soil mechanics concepts to problems involving MSW. Although the mechanism leading to apparent cohesion may be different for soils (surface interaction between clay mineral particles) than for waste (reinforcing effect of sheet-like material), there is evidence to support the development of shearing resistance even at low confining stresses (e.g. Richardson and Reynolds 1991; Koda 1998), which corroborates data taken from high normal stress studies. The mechanisms that lead to the development of apparent cohesion for waste may be different than for soils but, at least for the time being, these soil mechanics techniques are the best available and when used with sound engineering judgement and experience appear to result in reasonable designs.

### 10.4.3.2   Influencing Factors: Shear Strength

The shear strength of MSW and its mobilization are influenced by many of the same factors previously identified for unit weight, including:

- **Waste composition:** An increase in materials that contribute to the reinforcing effect, such as plastics or wood, will have the effect of increasing the apparent cohesion. As well, increasing grain size generally increases the friction angle for a given strain level).
- **Degree of compaction:** The friction angle for compacted waste is significantly higher than for uncompacted waste, especially at small axial strains.
- **Moisture content:** Mobilized cohesion tends to decrease with increasing moisture content (Gabr and Valero, 1995).
- **Age of waste:** Shear strength tends to increase over time due to densification of the waste. However, this effect may be at least partially offset by biodegradation processes that may decrease waste strength (Knochenmus *et al.*, 1998).

### 10.4.3.3  Methods of Determination: Shear Strength

The heterogeneity of waste and the tendency of many of the above parameters to change with time make accurate determinations of waste shear strength difficult. Methods of evaluating MSW shear strength parameters include laboratory tests, *in situ* tests, and back calculations; each are discussed below:

- **Laboratory:** Laboratory tests used for estimating MSW shear strength parameters are generally the same as those used for soils testing, with modifications occasionally made to increase the sample size. The most commonly performed tests are direct shear and triaxial tests. Although these tests are useful for identifying the effects of changing variables under controlled testing conditions, they may be limited in their applicability to design. The very large range of particle size and the heterogeneity of MSW make it difficult to obtain truly representative shear strength parameters from laboratory tests (Eid *et al.* 2000).
- **In situ testing:** In-situ shear strength measurement techniques that have been applied to MSW include: Vane shear tests, standard penetration tests, cone penetration tests, pressuremeter tests, dynamic penetration tests, spectral analysis of surface waves (SASW) technique. Heterogeneity results in a high degree of scatter in the results from these tests and make interpretation difficult. No correlation has yet been identified between the results of SPT and CPT tests and the shear strength parameters of MSW (Knochenmus *et al.*, 1998).
- **Back calculations:** Back calculation of the shear strength of MSW has been made using results from: Plate load tests, test fills or embankments (e.g. Oweis *et al.*, 1985), unfailed waste slopes (e.g. Kavazanjian *et al.*, 1995), failed waste slopes (e.g. Oweis *et al.*, 1985; Reynolds, 1991; Eid *et al.*, 2000). In general, back-calculations made from plate load tests, test fills, and unfailed waste slopes are comparatively unreliable, since assumptions of failure plane location or factor of safety necessary for such calculations may result in inaccuracies (Knochenmus *et al.*, 1998; Eid *et al.*, 2000). The most reliable estimates of MSW shear strength parameters come from large-scale direct shear measurements and from back-calculations made on failed waste slopes (Eid *et al.*, 2000).

### 10.4.3.4  Reported Values: Shear Strength

A comprehensive analysis of investigations into the shear strength behaviour of MSW, taken from numerous investigations employing large-scale direct shear tests or back-analyses of failed waste slopes, was presented by Eid *et al.* (2000). From this summary they concluded that MSW may reasonably be assumed to have shear strength parameters of $c' = 25\,\mathrm{kPa}$ and $\phi' = 35°$. They found that the apparent cohesion term may vary between 0 and $50\,\mathrm{kPa}$, depending on whether the waste contains more soil (low $c'$) or plastic material (high $c'$). As previously noted, these values could (and probably will) decrease with time as waste degrades. It should also be re-emphasized that this strength may only be mobilized at large strain.

Knochenmus *et al.* (1998) summarized shear strength parameter results from various sources; these along with other studies are listed in Table 10.4.4. The significant scatter in these results is indicative of both the heterogeneous nature of MSW and the limitations of smaller-scale laboratory tests and back-calculations from unfailed slopes. The reported values from larger-scale tests and back-calculations of failed slopes fall into a much narrower range and suggest greater reliability of these parameters.

**Table 10.4.4**  *Shear strength characteristics of MSW.*

| c' (kPa) | φ' (°) | Comments | Test method | Source |
|---|---|---|---|---|
| 15.7 | 21 | Low density bales | Direct shear | Del Greco and Oggeri |
| 23.5 | 22 | Higher density bales | | (1994) |
| 16–19 | 38–42 | Old waste | Direct shear | Landva and Clark (1990) |
| 16 | 33 | Old waste +1 year | | |
| 23 | 24 | Fresh, shredded waste | | |
| 10 | 18–43 | 14 kPa $< \sigma <$ 38 kPa | *In situ*, direct shear | Richardson and Reynolds (1991) |
| — | 25–34 | 45 kPa $< \sigma <$ 180 kPa | Back-calculations | Kavazanjian *et al.* (1995) |
| 25 | 20 | Noncomposted, $\gamma = 9$ kN/m$^3$, $\sigma = 35$ kPa | Slope failure, CPT, WST | Koda (1998) |
| 23 | 25 | Noncomposted with sand, $\gamma = 12$ kN/m$^3$, $\sigma = 50$ kPa | Slope failure, CPT, WST | |
| 20 | 26 | Old MSW, $\gamma = 14$ kN/m$^3$, $\sigma = 65$ kPa | Back-calculations, CPT, WST | |
| 15 | 21 | Fresh MSW, $\gamma = 11$ kN/m$^3$, $\sigma = 125$ kPa | Back-calculations, CPT, WST | |
| 25 | 35 | Design recommendation based on summary of current literature | Back-calculations, large-scale tests | Eid *et al.* (2000) |
| 20 | 0 | Design recommendation based on summary of reported results | Various laboratory and field tests | Manassero *et al.* (1996)[a] |
| 0 | 38 | | | |
| 20 | 30 | | | |
| 43 | 31 | Reconstituted MSW | Direct shear (large-scale) | Kavazanjian *et al.* (1999) |

[a]Design recommendation based on much of the same data as analysed by Eid *et al.* (2000), therefore it is not surprising that the recommended strength parameters are similar.

### 10.4.3.5  Relevance to Design: Shear Strength

Some researchers (e.g. Knochenmus *et al.*, 1998) have advocated the use of a limiting strain value in assessing suitable MSW shear strength parameters for stability analysis. Eid *et al.* (2000) argued that parameters taken from back analyses of failed waste slopes represent the shear strength mobilized at failure and are thus appropriate for use in design. They suggested that peak strengths measured in laboratory tests are lower than peak *in situ* values due to the removal of some particle interlocking during sample sorting and reconstitution processes. Evidence to this effect is the observed behaviour of vertical waste cuts remaining stable for months to years.

In general, for case histories involving failed landfill slopes, failure was not attributed to low values of the shear strength of the waste body itself but, rather, due to low interface friction angles between geosynthetic lining layers (e.g. Kettleman Hills slide, described in Section 10.4.3) or to other extenuating circumstances such as low foundation soil strength (e.g. Beirolas slide, described in Section 10.4.3) and toe excavations (e.g. Maine landfill; Reynolds, 1991). It may be concluded that although the shear strength of MSW is important in assessing stability, failures have been largely due to an inadequate assessment of more conventional geotechnical parameters. Nevertheless, careful assessment of MSW shear strength using site-specific field and large-scale laboratory testing is required when uncertainty regarding waste shear strength has a significant impact on estimates of stability.

## 10.4.4  Hydraulic Conductivity

The hydraulic conductivity of waste is important when assessing the rate and pattern of moisture movement within MSW, particularly for landfills where leachate recirculation is being contemplated. Darcy's Law has generally been used and considered to be valid (e.g. Powrie and Beaven, 1999) for assessing the flow of fluids in saturated municipal solid waste.

### 10.4.4.1   Definition: Hydraulic Conductivity

Hydraulic conductivity is an empirical derived value that represents the ease at which water can pass through a porous media. Hydraulic conductivity is a function of the porous media and the property of the fluid. Commonly hydraulic conductivity is reported in units of m/s and reported for water as the fluid. Hydraulic conductivity is related to intrinsic permeability, which represents the ease at which any fluid can pass through a porous media and therefore is a function of the porous media only. Intrinsic permeability, or permeability, has units [$L^2$], typically reported as $cm^2$.

### 10.4.4.2   Influencing Factors: Hydraulic Conductivity

Like other waste properties, the hydraulic conductivity of waste may be expected to vary considerably within the waste mass due to its high dependency on many factors, including:

- *Composition of waste:* The relative grain size and porosity of the waste material will greatly influence the hydraulic conductivity.
- *Degree of compaction:* More heavily compacted waste shows lower hydraulic conductivity, due primarily to a reduction in void space (e.g. Landva and Clark, 1990; Manassero *et al.*, 1996).
- *Overburden pressure:* Increasing the stress level decreases the hydraulic conductivity, again due primarily to a reduction in void space at higher overburden pressures (Powrie and Beaven, 1999).
- *Waste ageing:* The degradation process tends to densify the waste, causing a decrease in hydraulic conductivity as the waste ages (Powrie and Beaven, 1999).

### 10.4.4.3   Methods of Determination: Hydraulic Conductivity

The methods of assessing the hydraulic conductivity of waste follows the same techniques traditionally used for soils. However, laboratory testing usually employs a larger apparatus to reduce the effects of waste heterogeneity in attempts to obtain more representative results. For example, Landva *et al.* (1998) used a consolidometer of 440 mm diameter to evaluate vertical hydraulic conductivity and a 760 mm diameter by 450 mm deep consolidometer to investigate horizontal hydraulic conductivity. Beaven and Powrie (1995) constructed the large scale (2 m diameter, 3 m high) Pittsea compression cell that uses hydraulic pistons to simulate the stress levels at different depts. in the waste mass. This cell is being used to evaluate the behaviour of waste on a realistic scale for both new and aged waste and a wide range of applied loadings. Laboratory tests have the advantage of controlling the influencing factors, such as overburden pressure and degree of compaction, to assess their impact on the hydraulic conductivity of the waste. However, the main disadvantage is the difficulty in accounting for large-scale waste heterogeneity and variations and hydraulic conductivity.

Field investigations of hydraulic conductivity have included: Test pits (Oweis *et al.*, 1990; Landva *et al.*, 1998), falling head tests (Oweis *et al.*, 1990; Blengino *et al.*, 1996; Jang, 1998), and pumping tests (Manassero, 1990; Oweis *et al.*, 1990; Jang, 1998). Field tests have the advantage of utilizing a much larger waste sample than laboratory methods, and generally are considered to give results more representative of field conditions. However, one cannot control the other factors influencing the hydraulic response; and in most cases these are unknown. This makes field tests difficult to analyse and also to compare with data from other sources.

### 10.4.4.4   Reported Values: Hydraulic Conductivity

Manassero *et al.* (1996) presented a summary of laboratory and field hydraulic conductivity measurements, and the results are presented in Table 10.4.5 along with other published values. Based on the summary by Manassero *et al.* (1996), the hydraulic conductivity ranges from $10^{-4}$ to $10^{-6}$ m/s. These authors proposed that when no other data is available a hydraulic conductivity of $10^{-5}$ m/s could be used as an initial approximation.

Large-scale tests completed by Powrie and Beaven (1999), within the Pittsea compression cell, studied the effect of applied vertical stress on waste hydraulic conductivity. The hydraulic conductivity of unprocessed household waste at 40 kPa was $1.5 \times 10^{-4}$ m/s and this decreased to $3.7 \times 10^{-8}$ m/s at an applied stress of 600 kPa. These authors presented

**Table 10.4.5**  *Hydraulic conductivity of MSW (modified from Manassero et al., 1996).*

| Hydraulic conductivity (m/s) | Unit weight (kN/m$^3$) | Method | Source |
|---|---|---|---|
| $10^{-5}$ to $2 \times 10^{-4}$ | 1.1–4.0 | Lysimeters | Fungaroli *et al.* (1979) |
| $10^{-5}$ | 6.45 | Estimation from field data | Oweis and Khera (1986) |
| $10^{-5}$ | 6.45 | Pumping head | Oweis *et al.* (1990) |
| $1.5 \times 10^{-6}$ | 9.4–14.0 | Falling head | |
| $1.1 \times 10^{-5}$ | 6.3–9.4 | Test pit | |
| $1 \times 10^{-5}$ to $4 \times 10^{-4}$ | 10.1–14.4 | Test pit | Landva and Clark (1990) |
| $10^{-7} \times 10^{-5}$ | — | Laboratory tests | Gabr and Valero (1995) |
| $3 \times 10^{-7}$ to $3 \times 10^{-6}$ | 9–11 | Deep borehole (30–40 m), falling head | Blengino *et al.* (1996) |
| $1.5 \times 10^{-5}$ to $2.6 \times 10^{-4}$ | 8–10 | Pumping test | Manassero (1990) |
| $10^{-7}$ to $10^{-4}$ | 5–13 | Laboratory tests (Pittsea compression cell) | Beaven and Powrie (1995) |
| $1 \times 10^{-5}$ to $3.4 \times 10^{-4}$ | — | Test pit | Landva *et al.* (1998) |
| $5 \times 10^{-4}$ to $1 \times 10^{-8}$ | — | Large-scale laboratory tests, $\sigma =$ 20–400 kPa | |
| $2.2 \times 10^{-5}$ | — | Pumping test | Jang (1998) |
| $3.4 \times 10^{-5}$ | — | Slug tests | |
| $1.5 \times 10^{-4}$ to $3.7 \times 10^{-8}$ | — | Large-scale laboratory tests, $\sigma =$ 40–600 kPa | Powrie and Beaven (1999) |

the following empirical correlation relating vertical hydraulic conductivity ($k_V$) and effective stress ($\sigma'$):

$$k_V(\text{m/s}) = 17[\sigma' \, (\text{kPa})]^{-3.26}$$

Landva *et al.* (1998) in their large-scale hydraulic conductivity test also observed a decrease in hydraulic conductivity with applied stress that ranged between $5 \times 10^{-4}$ m/s at 20 kPa and $1 \times 10^{-8}$ m/s at 400 kPa. These authors also reported that the ratio between horizontal and vertical hydraulic conductivity was relatively constant at $k_H = 8\,k_V$ over the applied vertical stress range (150–500 kPa). Hudson *et al.* (1999) found that the ratio of $k_H$ to $k_V$ increased from approximately 2 at an applied vertical stress of 40 kPa to approximately 5 at an applied stress of 600 kPa.

## 10.4.4.5  Relevance to Design: Hydraulic Conductivity

A decrease in hydraulic conductivity of the waste with depth may have significant implications on the design of some leachate collection systems. This is particularly true for landfills that do not have an underlying drainage blanket, for those where leachate extraction is by horizontal or vertical wells, or landfills where leachate recirculation is being considered.

The hydraulic conductivity of the waste will have a significant effect on the achievable infiltration rate through the waste (Powrie and Beaven, 1999). High infiltration rates are key to leachate recirculation practices which inject fluid (i.e. leachate or water) back into the waste for rapid stabilization. The decrease in waste hydraulic conductivity with depth may limit the maximum infiltration rate for leachate recirculation.

Powrie and Beaven (1999) showed that even for a fully functioning leachate collection system (i.e. zero pore water pressure at the base), a decrease in waste hydraulic conductivity with depth can lead to the development of significant excess pore water pressures within the waste for a given infiltration rate. The development of excess pore water pressures within a landfill was identified as the principal factor in the Dona Juana Landfill failure (described in Section 10.4.3) which employed a leachate recirculation strategy. Since the horizontal hydraulic conductivity of the waste is higher than the vertical hydraulic conductivity, this also may lead to leachate seepage from the sides of the landfill. The effect of the development of excess pore pressures on landfill stability must be carefully considered in the design and operations of landfills.

Vertical drainage wells have been employed as a means to control the leachate mound in landfill with either no leachate collection system or failed leachate collection system. The effectiveness of the well to control the leachate mound is affected in part by waste anisotropy and the decrease in waste hydraulic conductivity with depth (Rowe and Nadarajah, 1996; Jang, 1998; Powrie and Beaven, 1999). Generally, the specific yield of the well will decrease as drawdown is increased (drawdown increases with decreases in hydraulic conductivity), thus limiting the useful lifespan of such systems (Rowe and Nadarajah, 1996).

## 10.4.5   Settlement

Estimation of the amount of waste settlement is essential for storage volume calculations, the design of final cover, leachate recirculation systems, and gas extraction systems. Waste settlement is characteristically irregular and has been reported to range from 5 to 30 % of the original waste thickness.

### 10.4.5.1   Definition: Settlement

Settlement is the vertical deformation, or movement, that occurs within a waste layer due to a change in effective stress resulting from a change in total stress and/or a change in pore-water pressure. Settlement has units [L] and is typically reported in m. Commonly, settlement is divided by the original thickness of the waste layer and reported as a percentage of reduction in layer thickness.

### 10.4.5.2   Influencing Factors: Settlements

The settlement of MSW depends on many factors, including the initial unit weight and organic content of the waste, the landfill and overburden thickness, stress history, leachate levels, and various environmental factors (Edil *et al.*, 1990). Manassero *et al.* (1996) summarized the following settlement behaviour mechanisms:

- *Physical compression* due to mechanical distortion, bending, crushing and reorientation.
- *Raveling settlement* due to migration of small particles into voids among large particles.
- *Viscous behavior and consolidation* phenomena involving both solid skeleton and single particles or components.
- *Decomposition settlement* due to the biodegradation of the organic components.
- *Collapse of components* due to physicochemical changes, such as corrosion, oxidation and degradation of inorganic components.

Due to the heterogeneity of the waste and the interrelated and complex mechanisms involved in settlement, the settlement of waste is typically irregular and difficult to predict. For estimation purposes it is typical to separate settlement into two components (Fassett *et al.*, 1994; Knochemnus *et al.*, 1998), as follows:

- *Primary consolidation:* relatively uniform, load-dependent settlement due primarily to the mechanical compression of the waste. This component of settlement is completed within the first several months following load application.
- *Secondary compression:* time dependent, biologically and chemically induced compression. This component of settlement may occur over several years and is relatively unaffected by variation in applied stress.

### 10.4.5.3   Methods of Determination: Settlement

Settlements can be assessed theoretically as well as empirically.

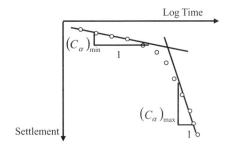

**Figure 10.4.1** *Idealized plot of landfill settlement versus log time (modified from Edgers et al., 1992).*

## Theoretical Techniques

One dimensional consolidation theory has been used to model settlement behaviour of MSW (Sowers, 1973; Gabr and Valero, 1995). Waste material properties needed for analyses are: (1) the compression index ($C_c$) or modified compression index ($C_c'$) to estimate the primary consolidation, and (2) the secondary compression index ($C_\alpha$) or modified secondary compression index ($C_\alpha'$) to estimate secondary compression. Traditional oedometer testing on large specimens is typically used to determine these parameters. Unlike soils, the compaction or secondary compression of MSW includes contribution from the decomposition of the waste as well as creep (Konig and Jessberger, 1997). Powrie *et al.* (1999) stated that the use of traditional secondary compression (creep) models are invalid due to the fundamentally different processes that cause this behaviour in soils (i.e. creep) compared to MSW (i.e. creep and waste decomposition). Knochenmus *et al.* (1998) noted several factors that make application of this methodology difficult, including:

- Saturation is not typical in the majority of the waste body; however it is a prerequisite in the application of consolidation theory.
- The initial void ratio or initial waste height is difficult to estimate; however the compression and secondary compression indices are a function of these parameters.
- The e-log($\sigma'$) or e-log($t$) relationships are frequently nonlinear and consequently the compression indices $C_c$ and $C_\alpha$ can significantly vary with the stresses generated in the landfill.
- Primary settlements are a function of the effective stresses, which depend on the unit weight of the waste and leachate levels both of which are difficult to evaluate.
- In a study on 22 landfills, Edgers *et al.* (1992) showed that the rate of settlement changed with time; $C_\alpha$ was relatively small at the beginning and this value increased at larger times (schematically shown in Figure 10.4.1). The change in settlement rate may be attributed to degradation (biological and physicochemical decay processes) of the waste.

## Empirical Techniques

Due to the difficulties in applying consolidation theory to estimate landfill settlement, various empirical methods have been proposed, as follows:

- Edil *et al.* (1990) proposed a linear viscoelastic model which requires estimates of the final settlement and a reference time for 63 % of the final settlement to occur.
- Fassett *et al.* (1994) both proposed an empirical model which requires estimations of the compression index and the secondary compression index (therefore may have the limitations associated with classical consolidation theory).
- van Meerten *et al.* (1997) and Ling *et al.* (1998) presented empirical formulations based on parameters determined from settlement observations.

### 10.4.5.4   Reported Values: Settlement

Literature reported values of settlement or compression indexes are limited, and they generally do not consider the component of total settlement dependent on waste decomposition, due to the long testing times this would require. Thus, the value for $C_\alpha$ may be significantly higher than those reported due to the influence of biodegradation (Knochenmus *et al.* 1998). Sowers (1973) reported upper and lower bounds of $C_c$ and $C_\alpha$ based on void ratio, organic content, and environmental conditions of $0.15e_0$ (low organic content) and $0.55e_0$ (high organic content) for $C_c$, and $0.03e_0$ (anaerobic conditions) and $0.09e_0$ (aerobic conditions) for $C_\alpha$. Landva and Clark (1990) reported a $C_c'$ range of 0.2 to 0.5 based on five different Canadian landfills measured in a 450 mm diameter consolidometer, and a $C_\alpha$ range of 0.002 to 0.03. In a summary provided by Oweis and Khera (1990), $C_c$ ranged from $0.1e_0$ to $0.55e_0$ based on four different studies and $C_\alpha'$ ranged from 0.02 to 0.03 based on nine studies. Gabr and Valero (1995) conducted small-scale compression tests using conventional equipment on waste samples that were 15–30 years old. They found that $C_c$ was 0.4 and 0.8 for a void ratio of 1.0 and 3.0, respectively. Kavazanjian *et al.* (1999) performed large-scale oedometer tests on reconstituted MSW and found an average $C_c$ of 0.185 (range 0.121 to 0.247) and $C_\alpha$ of 0.0035 (range $\sim$0 to 0.01). Qian *et al.* (2002) reported values, based on experience, for $C_c'$ to vary from 0.17 to 0.36, and $C_\alpha'$ to vary from 0.03 to 0.1.

Several researchers have attempted to assess the compression behaviour of MSW through various parameters such as stiffness modulus, modulus of subgrade reaction, and compressibility from field tests such as plate load test (e.g. Machado Santos *et al.*, 1998), test fills and pressuremeter tests. In a summary by Manassero *et al.* (1996), the stiffness modulus generally ranged from 0.5 to 3.0 MPa, depending on the vertical stress level. Machado Santos *et al.* (1998) reported a modulus of subgrade reaction of 1.72 MPa/m and a Young's modulus of $< 2$ MPa under an applied stress of 50–250 kPa based on plate load tests. Powrie *et al.* (1999) reported compressibility for raw MSW to range from 7.45 to 1.07/MPa under an applied stress of 34–463 kPa, and for aged waste a compressibility of 7.38 to 0.66/MPa over the same stress range.

## 10.4.6   Stability

The stability of the landfill relates to dynamic (earthquakes) as well as stationary conditions.

### 10.4.6.1   Dynamic Stability

The dynamic strength properties of refuse material are required for the analysis of landfill slope stability under earthquake loading. Singh and Murphy (1990) provide a summary of the limited number of shear wave velocity test completed on waste fill; values are given in Table 10.4.6.

Singh and Murphy (1990) conclude that static stress–strain data or the downhole shear wave velocity data to estimate dynamic shear modulus of refuse should be used with caution due to the highly compressive nature of the refuse and its nonsoil-like strength deformation characteristics. The authors suggest that the excellent performance of sanitary landfills when subjected to strong earthquake motions indicate that the refuse has inherently strong energy absorption characteristics. For a more comprehensive examination of dynamic stability of MSW, see Singh and Murphy (1990) and Konig and Jessberger (1997).

*Table 10.4.6*   *Summary of shear wave velocity measurement on waste fill (compiled by Singh and Murphy 1990).*

| Shear wave velocity (m/s) | Measurement technique | Source |
| --- | --- | --- |
| 26 | Estimated shear wave velocity based on static load settlement results | Volpe (1985) |
| 274 | Average value based on geophysicial cross-hole and downhole shear wave velocity tests | Earth Technology, Inc. (1998) |
| 91 and 213 | Seismic survey by downhole shear wave velocity tests | EMCON (1989) |

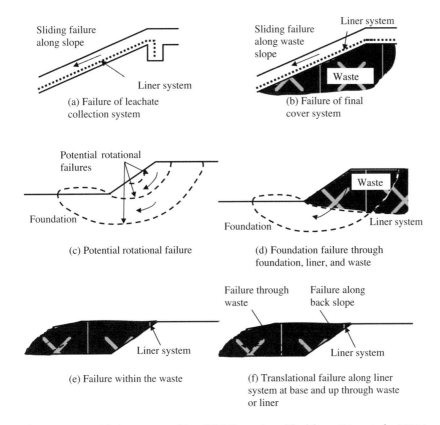

**Figure 10.4.2** *Various types of landfill failures (modified from Qian et al., 2002).*

### 10.4.6.2 Landfill Stability

There are numerous failure modes that need to be considered in the design of landfills during all stages of construction, for example, during cell excavation, liner system construction, waste filling, and after landfill closure (Qian *et al.*, 2002), as schematically summarized in Figure 10.4.2. All classical failure modes are possible, depending upon site-specific conditions and the placement and geometry of the waste mass. The stability of liners and drainage material (see Figure 10.4.2a–e) on the side slopes of the excavation (a) and waste material (b) needs to be carefully examined both during leachate collection system placement and during the placement of waste. The stability of the capping systems also needs careful examination. Additionally, conventional slope stability analyses are required for the excavation prior to landfill liner placement (c).

The analysis of the above mentioned failure modes require strength properties for both the soil at the landfill site and the construction materials. There are also potential failure modes that require the geotechnical properties of the waste to be specified. These include general stability of the foundation soil, liner, and waste, failure within the waste mass, and transitional failure of the waste along the base liner system (d, e).

Stable waste slope angles will be highly dependant on the specific conditions but a common value would be 1 vertical : 3 horizontal (1v:3h; Figure 10.4.3). Steeper slopes may be stable in some circumstances while there are also instances where 1v:3h is not stable. The safe slope angle can only be properly assessed by doing a stability calculation for the particular conditions being considered. Stability calculations, which may involve considering failure in the refuse, liner components, and/or foundation soil, are performed using normal geotechnical methods. Based on findings from centrifuge model tests, shear planes develop in MSW only after very large deformations have occurred (Konig

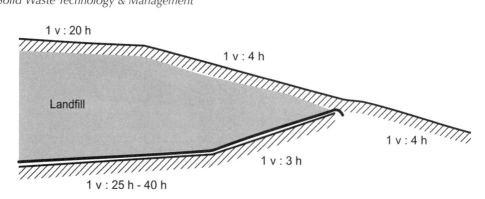

**Figure 10.4.3**   Typical slopes used in landfill design (the stability of the slopes depend on the material and site specific assessment should always be made).

and Jessberger, 1997). Konig and Jessberger (1997) concluded that this demonstrates the applicability of using limit equilibrium methods for slope stability analysis. However, it also highlights the importance of considering the strain at which strength is mobilized in assessing the parameters to be used in the stability calculations.

In addition to ensuring stability of the landfill, consideration must also be given to the expected deformations of the waste mass (even if waste slope is stable). Excessive deformation may negatively affect the performance of the liner and drainage systems as well as structures within the landfill (Konig and Jessberger, 1997).

Table 10.4.7 summarizes a number of case histories of landfill failures. In addition to the lessons learned in Table 10.4.7, these case histories illustrate a number of design issues including the importance of:

- Specifying site- and material-specific testing procedures in order to obtain truly representative shear strength parameters, permitting reasonable assumptions related to the slope geometry, suitable slope stability analyses.
- Good construction quality control and assurance in order to ensure the facility is constructed to the specification required by the design.
- Planning landfill operations consistent with what was envisaged in design, particularly with respect to stability.
- Having appropriate monitoring and contingency plans.

Expansion of existing landfills may involve integrating an old landfill with no barrier system with the new landfill that includes a barrier system. Careful consideration must be given to the effects the new construction will have on existing facilities to avoid stability problems when excavating at the toe of the existing landfill. Table 10.4.8 summarizes four landfill failures associated with new construction at an existing facility. Based on these case histories, Rowe (1999) concluded that the likelihood of failure in future landfill expansions can be minimized by:

- Ensuring a proper geotechnical investigation of the subsoil properties.
- Carefully considering all potential failure mechanisms.
- Avoiding optimism regarding geotechnical properties when faced with data that is inconsistent with that optimism.
- Taking account of the effect of potential increase in leachate in terms of both increasing the unit weight of the waste and decreasing the shear strength of components of the geotechnical system (e.g. covers, waste, liners, and foundation materials).
- Appropriate design and materials selection (including appropriate laboratory tests and stability analysis).
- Good construction quality control and assurance to ensure that the barrier system is installed as designed.
- Taking account of the effect of excavation on stability (e.g. at the toe of existing waste).
- Having development plans for expanded landfills that limit toe excavation and overfilling and define allowable conditions for construction of the expansion area and a means of monitoring adherence to the development plans.

**Table 10.4.7** *Case histories of landfill failures.*

| Source and landfill description | Cause of failure and outcome | Lessons learned |
|---|---|---|
| Kettleman Hills Landfill, Kettleman City, California (Mitchell *et al.* 1990; Seed *et al.* 1990; Stark and Poeppel, 1994)<br>– Hazardous waste landfill.<br>– Design included a liner system employing many geosynthetics. | – During filling, and after placement of approximately 490 000 m$^3$ of waste, a slope stability failure occurred.<br>– The waste mass sliding approximately 11 m laterally with a vertical slump of more than 5 m along the side slopes.<br>– Failure primarily occurred along the interface between the secondary clay liner and the secondary geomembrane, and the low shear strength and undrained conditions at this interface were deemed to be the primary cause of failure. | – Geomembrane-compacted clay interfaces may exhibit essentially undrained behaviour during placement of waste, and are quite sensitive to the moisture content of the clay. Care must be taken during the placement of such layers.<br>– Residual shear strength parameters may be considerably lower than peak parameters, and the decrease from peak to residual strength may be rapid during deformation.<br>– There may be significant zones where the peak shear strength is not mobilized. Thus analyses utilizing peak parameters over the entire failure surface are inappropriate.<br>– 3-D analyses may aid in the interpretation of slope stability, but 2-D analyses along representative cross-sections are able to indicate potential problems.<br>– Laboratory testing under conditions representative of anticipated field conditions should give results appropriate for use in design. |
| Mahoning Landfill, Youngstown, Ohio (Stark *et al.* 1998)<br>– Municipal solid waste landfill.<br>– Located at a former strip mining site. | – Precise failure mechanism was not determined.<br>– Observations indicated that a translational failure occurred primarily at the interface between the bentonite components of a geomembrane-backed GCL and the underlying compacted clay liner. | – The stability of interim slopes should be investigated to ensure that overstressing and damage of underlying geosynthetic materials does not occur.<br>– Unreinforced GCLs should not be placed in contact with CCLs compacted wet of optimum due to the low shear strength of the hydrated bentonite component.<br>– The critical failure interface may change along the length of the failure surface. Thus, composite failure envelopes should be considered in design. |
| Dona Juana Landfill, Bogata, Columbia (Hendron *et al.* 1999)<br>– Located in a valley and incorporates a PVC geomembrane overlying compacted clay liner. | – Sudden waste slope failure of approximately 1.5 million tonnes of waste.<br>– Failure surface occurred along horizontal planes defied by intermediate cover layers.<br>– Recirculation of large volumes of leachate resulted in excess pore pressures and likely led to the failure. | – A critical assessment of the potential for excess pore pressure development and the associated impacts on the design of leachate collection systems and slope stability is required.<br>– The importance of obtaining accurate assessment of waste properties for slope stability design.<br>– When employing novel technologies (i.e. leachate recirculation) operational guidelines are required to ensure that assumptions used in design are translated to the construction and operation of the facility. |

*(continued)*

**Table 10.4.7**   (continued)

| Source and landfill description | Cause of failure and outcome | Lessons learned |
| --- | --- | --- |
| Location not given (Thiel 1999)<br>– Landfill cover included a 15 cm surface layer, 30 cm protective layer, 30 cm drainage layer, PVC geomembrane, GCL (needlepunched), 30 cm gas collection layer and a foundation layer.<br>– Cover included at 4H:1V slope. | – Sliding failure of the slope occurred after placement of the geomembrane and 3.2 ha were covered with the drainage layer.<br>– The geomembrane tore at the top of the slope and slide down, with the sand, along the GM/GCL interface.<br>– As the failure progressed, gas bubbles were seen in the GM and gas pressure uplifted the GCL.<br>– The factor of safety for 2.5 kPa of gas pressure for the slope was 1. Post monitoring of the gas pressures report values up to 4 kPa. | – Localized failure occurred due to a build up of gas pressure that stemmed from poor gas permeability of the gas drainage layer that consisted of fine sand.<br>– The fine sand demonstrated good hydraulic conductivity, but lost its gas permeability due to the presence of field moisture.<br>– It was recommended that strip drains be installed to relieve gas pressure. |

**Table 10.4.8**   *Case histories of landfill failures due to new construction at existing facilities.*

| Source and landfill description | Cause of failure and outcome |
| --- | --- |
| Rumpke Landfill, Cincinnati (Stark and Evans, 1997; Kenter *et al.*, 1997; Rowe, 1998; Eid *et al.*, 2000; Stark *et al.*, 2000)<br>– In operation since 1955, the landfill consisted of ravines that have been filled with waste.<br>– Landfill expansion granted in 1994, with provision to include an engineered barrier system consisting of a leachate collection system and a composite liner. | – A new landfill cell was excavated to a depth of approximately 40 m at the toe of the slope.<br>– At near completion of the excavation, cracks were noted at ht crest of the landfill, but were largely disregarded.<br>– Block translational slide of approximately 1 300 000 m$^3$ of waste along a weak foundation soil. |
| Bulbul Drive Landfill, South Africa (Brink *et al.*, 1999; Rowe, 1998)<br>– Valley fill type landfill with a longitudinal slope of approximately 10 % and slide slopes of approximately 36 %.<br>– Waste placement procedures involved filling each phase against the previous phase in a diagonal layered configuration. | – When the waste height reached 45 m, failure occurred and approximately 160 000 m$^3$ of waste flowed into the valley below.<br>– Failure surface occurred along the interface between the two phases of waste, and along the interface between the geomembrane and compacted clay liner at the base of the landfill.<br>– Analyses of the failure led to the use of textured geomembrances and benching the waste phase interface in subsequent phases of waste placement. |
| Beirolas Landfill, Lisbon, Portugal (Santayana and Pinto, 1998)<br>– Barrier system consisted of a composite geomembrane/geosynthetic clay liner/compacted clay liner on the landfill base, with a geomembrane above a granular drainage layer and clayey structural fill along the side slopes over the existing waste.<br>– Foundation soil consisted of 4–5 m of silty clay fill overlying a 20–35 m thick estuarine and alluvial soft clay deposit overlying an irregular, thin basal sand and gravel overlying bedrock. | – Slope movement was detected in February of 1995 and attributed to the excavation at the toe of the slope in conjunction with the development of positive pore water pressures within the waste due to the placement of the relatively impermeable clayey fill along the side slopes.<br>– Liner construction continued and waste was placed (110 000 m$^3$ of contaminated soil) until late June of 1995.<br>– A major landslide occurred, with a nearly vertical failure scarp associated with 4 m of vertical and several metres of horizontal movement. |

(continued)

***Table 10.4.8***   (continued)

| Source and landfill description | Cause of failure and outcome |
| --- | --- |
| | – Postfailure analyses revealed that the shear strength of the soft clay stratum underlying the landfill was over-estimated. |
| | – It was assumed that placement of the fill in the 1970s and 1980s led to an increase in strength of this stratum, it was reported that very little dissipation of excess pore pressures had occurred. Additionally, the failure extended to an area where no fill had been placed (thus, no significant strength gain would be expected); this failure mechanism was not considered in design. |
| | – This case highlights the importance of an appropriate detailed geotechnical investigation before expansions are approved. |
| Maine Landfill (Reynolds, 1991; Rowe, 1998) | – Waste slide originating with a rotational failure passing under the original landfill slope, progressed into the movement of large blocks of waste over a soft, remolded clay underlying the landfill. |
| | – Approximately $500\,000\,m^3$ of soil and waste was involved in the failure, and horizontal movements of up to 120 m were observed. |
| | – Failure occurred due to a combination of excavation at the toe of the slope to allow for expansion, stockpiling of cover soil at the crest of the slope, and heavy rainfall leading to excess pore pressure development. |

- Having operation plans that include consideration of stability as the waste is placed and means of monitoring adherence to the operation plans.
- Avoiding co-disposal of liquid waste or increasing the amount of liquid waste without fully assessing the potential impact on both stability and geoenvironmental protection.
- Having contingency plans for changed conditions that may occur during construction (e.g. excessive rain, unexpected foundation conditions).
- Having disposal alternatives so that waste can be diverted if expansion schedules are not met.

## 10.4.7   Conclusion

The heterogeneous nature of municipal solid waste, which has characteristics that may change with time due to physico-chemical processes, makes selection of geotechnical parameters a challenge for landfill designers. This chapter summarizes estimates of values for waste unit weight, hydraulic conductivity, and shear strength based on values reported in the literature. Selection of waste geotechnical parameters is required in order to estimate landfill settlement and waste stability, which are generally assessed according to conventional soil mechanics procedures. An examination of case histories of landfill failures shows that it is critical to both use representative waste geotechnical parameters in design, and ensure that the construction and operation procedures employed at a landfill facility conform with design assumptions. Care must be given to the potential effect of changes in waste characteristics with time (and therefore geotechnical parameters) in assessing long-term performance.

## References

Beaven, R.P. and Powrie, W. (1995): Determination of the hydrogeological and geotechnical properties of refuse using a large compression cell. *Proceedings of the International Landfill Symposium*, 5 (2), 745–760.

Blengino, A.M., Manassero, M., Rossello, A., Spanna, C. and Veggi, S. (1996): Investigation, monitoring system and stability analyses of an old canyon landfill. *Proceedings of the Second International Congress on Environmental Geotechnics*, Osaka, Japan.

Brink, D., Day, P.W. and Du Preez, L. (1999): Failure and remediation of Bulbul Drive landfill: KwaZulu-Natal, South Africa. *Proceeding of the International Landfill Symposium*, 7 (3), 555–562.

Del Greco, O. and Oggeri, C. (1994): Shear resistance tests on solid municipal wastes. *Proceedings of the First International Congress on Environmental Geotechnics*, Edmonton, Canada.

Earth Technology Corporation (1988): *Instability of landfill slope, Puente Hills Landfill, Los Angeles County, California*. Report to Los Angeles County Sanitation District. Earth Technology Corporation, Los Angeles, USA.

Edil, T.B., Ranguette, V.J. and Wuellner, W.W. (1990): Settlement on municipal refuse. In: ASTM (ed.) *Goetechnics of waste fills – theory and practice*. ASTM STP 1070, pp. 225–239. ASTM, Washington, D.C., USA.

Edgers, L., Noble, J.J. and Williams E. (1992): A biological model for longterm settlement in landfills. *Environmental Geotechnology*. Balkema, Rotterdam, The Netherlands.

Eid, H.T., Stark, T.D., Evans, W.D. and Sherry, P.E. (2000): Municipal solid waste slope failure. I: Waste and foundation soil properties. *Journal of Geotechnology and Geoenvironmental Engineering, ASCE*, 126 (5), 397–407.

EMCON (1989): *Analysis of deformations under static and seismic loading – West Contra Costa Sanitary Landfill, Richmond, CA*. Richmond Sanitary Service, Richmond, USA.

Fassett, J.B., Leonards, G.A., and Repetto, P.C. (1994): Geotechnical properties of municipal solid wastes and their use in landfill design. *Landfill Technology, Technical Proceedings*, 94, 13–14.

Fungaroli, A.A. and Steiner, R.L. (1979) *Investigation of sanitary landfill behavior, vol. 1, p. 331*. Final report. EPA-600/2-79/053a. US Environmental Protection Agency, Cincinnati, USA.

Gachet, C., Gotteland, P., Lemarechal, D. and Prudhomme, B. (1998): An in-situ household refuse density measurement protocol. *Proceedings of the International Congress on Environmental Geotechnics*, 3, 849–854.

Gabr, M.A. and Valero, S.N. (1995): Geotechnical properties of municipal solid waste. *Geotechnical Testing Journal*, 18, 241–251.

Hendron, D.M., Fernandez, G., Prommer, P.J., Giroud, J.P. and Orozco, L.F. (1999): Investigation of the cause of the 27 September 1997 slope failure at the Dona Juana landfill. *Proceedings of the International Landfill Symposium*, 7 (3), 545–554.

Hudson, A.P., Beaven, R. and Powrie, W. (1999): Measurement of the horizontal hydraulic conductivity of household waste in a large scale compression cell. *Proceedings of the International Landfill Symposium*, 7 (3), 461–468.

Jang, Y.-S. (1998): Analysis of flow behavior in a landfill with cover soil of low hydraulic conductivity. *Environmental Geology*, 39 (3/4), 292–298.

Kavazanjian, E. Jr, Matasovic, N., Bonaparte, R. and Schmetrmann, G.R. (1995): Evaluation of MSW properties for seismic analysis. Proceedings, specialty conference geoenvironment 2000, ASCE. *Geotechnical Special Publication*, 46 (2), 1126–1141.

Kavazanjian, E. Jr, Matosovic, N. and Bachus, R. (1999): Large-diameter static and cyclic laboratory testing of municipal solid waste. *Proceedings of the International Landfill Symposium*, 7 (3), 437–444.

Kenter, R.J., Schmucker, B.O. and Miller, K.R. (1997): The day the earth didn't stand still: The Rumpke landslide. *Waste Age* 1997.

Knochenmus, G., Wojnarowicz, M. and Van Impe, W.F. (1998): Stability of municipal solid wastes. *Proceedings of the International Congress on Environmental Geotechnics*, 3, 977–1000.

Koda, E. (1998): Stability conditions improvement of the old sanitary landfills. *Proceedings of the International Congress on Environmental Geotechnics*, 3, 223–228.

Kockel, R. and Jessberger, H.L. (1995): Stability evaluation of municipal solid waste slopes. Proceedings, 11th ECSMFE, Copenhagen, Denmark. *Danish Geotechnical Society Bulletin* 11, 2.

Konig, D. and Jessberger, H.L. (1997): *Waste mechanics*. Report of subcommittee 3, ISMMFE Technical Committee TC5 on Environmental Geotechnics, pp. 35–76. Ruhr-Universität, Bochum, Germany.

Landva, A.O. and Clark, J.I. (1986). Geotechnical testing of waste fill. *Proceedings of the Canadian Geotechnical Conference*, 1986, 371–385.

Landva, A.O. and Clark, J.I. (1990): *Geotechnics of waste fills – Theory and practice*. ASTM special technical publication 1070. ASTM, Wasington, D.C., USA.

Landva, A.O., Pelkey, S.G. and Valsangkar, A.J. (1998): Coefficient of permeability of municipal refuse. *Proceedings of the International Congress on Environmental Geotechnics*, 3, 63–68.

Ling, H.I., Leshchinsky, Y., Mohri, Y. and Kawabata, T. (1998): Estimation of municipal solid waste landfill settlement. *Journal of Geotechnology and Geoenvironmental Engineering, ASCE*, 124 (1), 21–28.

Machado Santos, S., Jucá, J.F.T. and Aragão, J.M.S. (1998): Geotechnical properties of a solid waste landfill: Muribeca's case. *Proceedings of the International Congress on Environmental Geotechnics*, 3, 181–184.

Manassero, M. (1990): *Pumping tests in a municipal solid waste landfill*. Geotechnics Seminar, ENEL-CRIS, Milan, Italy.

Manassero, M., Van Impe, W.F. and Bouazza, A. (1996): Waste disposal and containment. *Proceedings of the International Congress on Environmental Geotechnics*, 2, 1425–1474.

Marques, A.C.M., Vilar, O.M. and Kaimoto, L.S.A. (1998): Urban solid waste – Conception and design of a test fill. *Proceedings of the International Congress on Environmental Geotechnics*, 3, 127–132.

Mitchell, J.K., Seed, R.B. and Seed, H.B. (1990): Kettleman Hills waste landfill slope failure 1: Liner system properties. *Journal of the Soil Mechanics and Foundations Division, ASCE*, 116 (4), 647–668.

Oweis, I. and Khera, R. (1986): *Criteria for geotechnical construction of sanitary landfills*. In: Fang, H.Y. (ed.) International symposium on environmental geotechnology, vol. 1, pp. 205–222. Lehigh University Press, Bethlehem, USA.

Oweis, I. and Khera, R. (1990): *Geotechnology of waste management*. Butterworths, London, UK.

Oweis, I.S., Mills, W., Leung, A. and Scarino, J. (1985): *Stability of sanitary landfills. Geotechnical aspects of waste managements*. Foundation and Soil Mechanics Group, Metropolitan Section, ASCE, New York, USA.

Oweis, I.S., Smith, D.A., Enwood, R.B. and Greene, D.S. (1990): Hydraulic characteristics of municipal refuse. *Journal of Geotechnological Engineering, ASCE*, 116 (4), 539–553.

Powrie, W. and Beaven, R.P. (1999): Hydraulic properties of household waste and implications for landfills. *Proceedings of the Institution of Civil Engineers, Geotechnical Engineering*, 137, 235–247.

Powrie, W., Beaven, R. and Harkness, R.M. (1999): Applicability of soil mechanics principles to household wastes. *Proceedings of the International Landfill Symposium*, 7 (3), 429–436.

Qian, X., Koerner, R.M. and Gray, D.H. (2002): *Geotechnical aspects of landfill design and construction*. Prentice-Hall, New Jersey, USA.

Reynolds, R.T. (1991): Geotechnical field techniques used in monitoring slope stability at a landfill. In: Sorum, G. (ed.) *Field measurements in geomechanics*. Balkema, Rotterdam, The Netherlands, pp. 883–891.

Richardson, G.N. and Reynolds, R.D. (1991): *Geosynthetic considerations in a landfill on compressible clays*. Proceedings, Geosynthetics '91, vol. 2. Industrial Fabrics Association International, St. Paul, USA.

Rowe, R.K. (1998): From the past to the future of landfill engineering through case histories. Keynote lecture. *Proceedings of the International Conference on Case Histories in Geotechnical Engineering*, 4, 145–166.

Rowe, R.K. (1999): Soiled waste disposal facilities for urban environments. Keynote lecture. *Proceedings of the XI Pan American Conference on Soil Mechanics and Geotechnical Engineering*, Foz do Iguassu, Brazil.

Rowe, R.K. and Nadarajah, P. (1996): Estimating leachate drawdown due to pumping wells in landfills. *Canadian Geotechnical Journal*, 33 (1), 1–10.

Santayana, P.D. and Pinto, A.A.V. (1998): The Beirolas landfill eastern expansion landslide. In: Seco e Pinto, P. (ed) *Environmental Geotechnics*. Balkema, Rotterdam, pp. 905–910.

Seed, R.B., Mitchell, J.K. and Seed, H.B. (1990): Kettleman Hills waste landfill slope failure. II: Stability analysis. *Journal of Soil Mechanics and Foundation Division, ASCE*, 116 (4), 669–690.

Singh, S. and Murphy, B.J. (1990): Evaluation of the stability of sanitary landfills. In: Landva, A. and Knowles, D. (eds) *Geotechnics of waste fills – Theory and practice*. ASTM STP 1070. ASTM, West Conshohocken, USA, pp. 240–258.

Sowers, G.F. (1973): Settlement of waste disposal fills. *Proceedings of the ICSMFE*, 8 (2), 207–210.

Stark, T.D, and Evans, W.D. (1997): Stability of grandfathered landfills. *ASCE Civil Engineering Magazine* 1997.

Stark, T.D. and Poeppel, A.R. (1994): Landfill liner interface strengths from torsional–ring shear tests. *Journal of Geotechnology and Geoenvironmental Engineering, ASCE*, 120 (3), 597–615.

Stark, T.D., Arellano, D., Evans, W.D., Wilson, V.L., and Gonda, J.M. (1998): Unreinforced geosynthetic clay liner case history. *Geosynthetics International*, 5 (5), 521–544.

Stark, T.D., Eid, H.T., Evans, W.D. and Sherry, P.E. (2000): Municipal solid waste slope failure. II: Stability analyses. *Journal of Geotechnology and Geoenvironmental Engineering, ASCE*, 126 (5), 408–419.

Thiel, R. (1999): Design of a gas pressure relief layer below a geomembrane cover to improve slope stability. *Proceedings, Geosynthetics*, 1999, 235–251.

Van Impe, W.F. (1994): Municipal and industrial waste improvement by heavy tamping. *Proceedings of the Meeting of Geotechnical Engineering, Geotechnics in the Design and Construction of Controlled Waste Landfills*. Associazione Poligeotecnici Riuniti, Milan, Italy.

van Meerten, J.J., Sellmeijer, J.B. and Pereboom, D. (1997): Prediction of landfill settlement. *Proceedings of the International Landfill Symposium*, 6 (3), 535–544.

Volpe, R.L. and Associates (1985): *Slope stability investigation, intermediate and final slopes static and dynamic loading conditions, Kirby Canyon sanitary landfill, San Jose, California*. Report for EMCON. Volpe and Associates, San Jose, USA.

# 10.5

# Landfilling: Mineral Waste Landfills

**Ole Hjelmar**

*DHI - Water, Environment & Health, Hørsholm, Denmark*

**Hans A. van der Sloot**

*ECN, Petten, the Netherlands*

Mineral waste landfills are characterised by containing predominantly inorganic (mineral) waste. The stabilisation of a mineral waste landfill occurs primarily by leaching and migration of inorganic contaminants and/or by inorganic reactions in the waste (e.g. the dissolution of existing and formation of new minerals). Mineral waste landfills may contain a limited amount of organic material, but the chemical environment within the mineral waste landfill is generally inorganic and not dominated by the degradation of organic materials and substances. The major environmental problem associated with mineral waste landfills is the leachate containing primarily inorganic contaminants.

Several countries in the European Union (EU; e.g. Denmark, The Netherlands, Sweden and Germany) have implemented landfilling strategies that basically ban the landfilling of organic, biologically degradable waste. Landfill legislation within the EU requires a gradual, substantial reduction of the landfilling of biodegradable waste within the 27 member states to reach a level of no more than 35 % of the amount of biodegradable waste produced in 1995 by 2016 (CEC, 1999, 2003). The motivation for this has been a desire to reduce the greenhouse effects of methane gas escaping from landfills with mixed waste.

Many of the existing mineral waste landfills are so-called monofills dedicated to specific types of industrial waste or groups of waste with comparable properties and leaching behaviour. However, landfills containing only MSWI residues (bottom ash and/or treated/untreated fly ash and acid gas cleaning residues) and similar predominantly inorganic waste types are also examples of mineral waste landfills.

This chapter defines mineral waste and describes the stabilisation process and the leaching from mineral waste landfills.

## 10.5.1  Mineral Waste Types

In this context the term 'mineral waste' refers to the inorganic, mineral-like nature of the waste regardless of its origin and is not necessarily associated with wastes produced by the extraction and processing of minerals (although such waste

*Solid Waste Technology & Management*   Edited by Thomas Christensen
© 2011 Blackwell Publishing Ltd

**Table 10.5.1**   *EU leaching criteria for landfilling of waste (CEC, 2003). L/S = 2 l/kg and L/S = 10 l/kg refer to the conditions under which the column or batch leaching tests are carried out. $C_0$ refers to the initial leachate from the column leaching test (the fraction from L/S = 0.0 to 0.1 l/kg).*

| Parameter | Landfill for inert waste | | | Landfill for nonhazardous waste receiving stable, nonreactive hazardous waste | | | Landfill for hazardous waste | | |
|---|---|---|---|---|---|---|---|---|---|
| | L/S = 2 l/kg mg/kg | L/S = 10 l/kg mg/kg | $C_0$ mg/l | L/S = 2 l/kg mg/kg | L/S = 10 l/kg mg/kg | $C_0$ mg/l | L/S = 2 l/kg mg/kg | L/S = 10 l/kg mg/kg | $C_0$ mg/l |
| As | 0.1 | 0.5 | 0.06 | 0.4 | 2 | 0.3 | 6 | 25 | 3 |
| Ba | 7 | 20 | 4 | 30 | 100 | 20 | 100 | 300 | 60 |
| Cd | 0.03 | 0.04 | 0.02 | 0.6 | 1 | 0.3 | 3 | 5 | 1,7 |
| Cr (total) | 0.2 | 0.5 | 0.1 | 4 | 10 | 2.5 | 25 | 70 | 15 |
| Cu | 0.9 | 2 | 0.6 | 25 | 50 | 30 | 50 | 100 | 60 |
| Hg | 0.003 | 0.01 | 0.002 | 0.05 | 0.2 | 0.03 | 0.5 | 2 | 0.3 |
| Mo | 0.3 | 0.5 | 0.2 | 5 | 10 | 3.5 | 20 | 30 | 10 |
| Ni | 0.2 | 0.4 | 0.12 | 5 | 10 | 3 | 20 | 40 | 12 |
| Pb | 0.2 | 0.5 | 0.15 | 5 | 10 | 3 | 25 | 50 | 15 |
| Sb | 0.02 | 0.06 | 0.1[c] | 0.2 | 0.7 | 0.15 | 2 | 5 | 1 |
| Se | 0.06 | 0.1 | 0.04 | 0.3 | 0.5 | 0.2 | 4 | 7 | 3 |
| Zn | 2 | 4 | 1.2 | 25 | 50 | 15 | 90 | 200 | 60 |
| Chloride | 550 | 800 | 460 | 10 000 | 15 000 | 8 500 | 17 000 | 25 000 | 15 000 |
| Fluoride | 4 | 10 | 2.5 | 60 | 150 | 40 | 200 | 500 | 120 |
| Sulfate | 564 | 1000 | 1500 | 10 000 | 20 000 | 7 000 | 25 000 | 50 000 | 17 000 |
| Phenol ind. | 0.5 | 1 | 0.3 | – | – | – | – | – | – |
| DOC[a] | 240 | 500 | 160 | 380 | 800 | 250 | 480 | 1 000 | 320 |
| TDS[b] | 2500 | 4000 | – | 40 000 | 60 000 | – | 70 000 | 100 000 | – |

[a]If the waste does not meet these values for DOC at its own pH, it may alternatively be tested at L/S = 10 l/kg and a pH of 7.5 – 8.0. The waste may be considered as complying with the acceptance criteria for DOC, if the result of this determination does not exceed 800 mg/kg.
[b]The values for TDS can be used alternatively to the values for sulfate and chloride.
[c]The $C_0$ limit value for Sb for inert waste landfills should actually be 0.01 mg/kg (error in the official version of the regulations).

would almost certainly fall within the category of mineral waste). There is no universal technical or legal definition of mineral waste. In generic terms it could be defined as predominantly inorganic waste with a relatively low content of biodegradable organic matter.

In EU landfill legislation two types of waste are defined which could be classified as mineral waste types (CEC, 1999, 2003): Inert waste and nonhazardous or nonreactive hazardous waste.

Inert waste is defined in the EU Landfill Directive (CEC, 1999) as: 'waste that does not undergo any significant physical, chemical or biological transformations. Inert waste will not dissolve, burn or otherwise physically or chemically react, biodegrade or adversely affect other matter with which it comes into contact in a way likely to give rise to environmental pollution or harm human health. The total leachability and pollutant content of the waste and the ecotoxicity of the leachate must be insignificant, and in particular not endanger the quality of surface water and/or groundwater.' In reality, inert waste as defined in CEC (1999) is a special type of nonhazardous waste since the definition excludes all properties that could render the waste hazardous. In more practical terms, the EU criteria for acceptance of waste at landfills for inert waste (CEC, 2003) include limit values for leaching (see Table 10.5.1) and content of various specific organics (BTEX = 6 mg/kg, PCBs = 1 mg/kg, mineral oil = 500 mg/kg) and total organic carbon, TOC (3 % w/w). The leaching criteria for inert waste include limit values for dissolved organic carbon (DOC) of 500 mg/kg and for total dissolved solids (TDS) of 4 g/kg or 0.4 % w/w (as an alternative to the criteria for chloride and sulfate), both determined at *L/S* = 10 l/kg.

The mineral nonhazardous/hazardous waste is defined in the EU landfill legislation (CEC, 2003) by a maximum TOC content of 5 % w/w and for the hazardous waste a minimum pH of 6 as well as a set of leaching limit values (see Table 10.5.1). The leaching criteria include limit values for DOC of 800 mg/kg and for TDS of 60 g/kg or 6 % w/w, also as an alternative to criteria for chloride and sulfate, determined at L/S = 10 l/kg.

**Table 10.5.2** *Various examples of waste types that may be regarded as mainly mineral waste. For some waste types, pretreatment may be necessary prior to acceptance at a mineral waste landfill.*

| | |
|---|---|
| Bottom ash from coal-fired power plants | Phosphorous slag |
| Fly ash and flue gas desulfurisation waste from coal fired power plants | Phosphogypsum |
| | Construction and demolition (C&D) waste |
| Bottom ash from MSW and hazardous waste incinerators | Glass and waste glass-based fibrous materials without organic binders |
| Fly ash and acid gas cleaning residues from MSW and hazardous waste incinerators | Slightly contaminated (sub)soil |
| | Gypsum-based construction materials |
| Waste rock and tailings from mining | Foundry sand |
| Blast furnace slag | Sandblasting waste |
| Steel slag | Inorganic sludges |
| Vitrified waste | Metal waste |
| Other industrial thermal waste | Road sweepings |

Table 10.5.2 lists a number of waste types that are likely to fit the generic definition of mineral waste although they may not necessarily fulfil the EU criteria for waste to be accepted at landfills for inert waste or landfills for nonhazardous waste accepting stable, nonreactive hazardous waste. A waste type such as MSW incinerator fly ash or acid gas cleaning residues may be almost entirely inorganic or mineral in nature, but due to a high content of readily soluble constituents (salts) they may not be acceptable at a mineral waste landfill without pretreatment because of their potential environmental impact and adverse influence on sustainability goals and physical stability. Some types of waste that are classified in one regulatory system or another as hazardous waste due to their total content of certain (inorganic) substances also fit the generic definition of mineral waste and behave as such when landfilled, e.g. in a monofill for a specific type of hazardous waste.

## 10.5.2 Leaching Processes within the Mineral Waste Landfill

The most important stabilising processes in a mineral waste landfill are leaching processes and chemical/physicochemical reactions. In the early part of the leaching, biological processes may also play a role for waste with a certain residual content of organic, biodegradable material. Below, leaching and chemical reactions are discussed separately, although the two types of stabilisation processes are often strongly interrelated. More thorough discussions of leaching from inorganic waste materials may be found in Chapter 8.4.

### 10.5.2.1 Kinetic Versus Equilibrium Approach

The release or transfer of a given substance from a solid (waste) phase to a contacting liquid phase involves both liquid/solid phase reactions and interactions and transport within the solid phase as well as in the liquid phase. The release to and presence in the liquid (aqueous) phase of dissolved substances may be controlled by (local) equilibrium conditions or reaction kinetics. Both slow reaction kinetics and physical resistance to mass transfer may delay or prevent the achievement of chemical equilibrium between a given substance in the solid phase and the liquid phase.

The internal porosity and the complicated (tortuous) transport through the internal pores within a solid waste particle together with possible chemical reactions cause an inner resistance to diffusion of dissolved substances or liquid into or out of the particle. In this context the size of the particle and hence the distance of transport will be crucial for the rate of transfer between the solid and liquid phases of a substance or for the achievement of equilibrium between the solid and liquid phases for that substance: the smaller the particle size, the faster equilibrium is attained. For monolithic waste materials, i.e. large coherent units of waste (e.g. cement-stabilised waste), the achievement of equilibrium will evidently take a long time or, if the aqueous phase is renewed, it may never happen.

The boundary layer of a liquid adjacent to a particle may also, due to an elevated concentration level and hence a reduced driving force, constitute an external resistance to the diffusion of dissolved substances between the solid/particle surface and the bulk liquid phase. If the diffusion through this boundary layer is rate controlling for a substance, then the rate of transfer of that substance between a particle and the bulk aqueous phase depends on the hydraulic flow conditions around the particle.

The relationship between equilibrium control and kinetic control in a percolation system may e.g. be evaluated by means of the so-called Damkoehler number. The Damkoehler number, Damk, for intraparticular diffusion may be defined as follows (after Bold, 2004; Christ and Hofmann, 2004):

$$\text{Damk} = \tau/\lambda, \tag{10.5.1}$$

where $\tau$ is the average retention time for the percolating water and $\lambda$ is the reaction rate for the transport of a substance between the particle and the percolating water body (moving by advection). The Damkoehler number increases with increasing effective diffusion coefficient of the substance and with decreasing particle size. If the liquid/solid partitioning of the substance can be described by a linear sorption model, it may be assumed that Damkoehler numbers above 100 describe systems in equilibrium. For further discussion of this, the reader is e.g. referred to Christ and Hofmann (2004).

In the speciation model LeachXS-Orchestra, a dual porosity model is applied to account for the different modes of release in a percolation test. By varying the diffusion distance, the diffusion from smaller and larger particles can be assessed. Dijkstra *et al.* (2008) demonstrated the suitability of that approach.

In a leaching system where the diffusion through the boundary layer is rate controlling and where the bulk fluid is not saturated with respect to a given dissolved substance, the rate of transfer for that substance from the surface of a particle to the bulk fluid (all other things being equal) increases when the liquid flow around the particle increases.

### 10.5.2.2   Leachate Characteristics as a Function of *L/S* and Time

The flux of contaminants released or discharged from a mineral waste landfill depends on the one hand on the flow of water through the landfill (e.g. infiltrating precipitation percolating through the landfill) and on the other hand on the composition of the discharged water (leachate).

For granular waste, i.e. waste present as powder or particles of limited size, a (local) equilibrium-like condition between many of the contaminants in the percolating water phase and the solid phase of the landfilled waste will often exist for normally occurring flow rates. This local equilibrium controls the composition of the water phase/the leachate, which generally changes with time or with the amount of percolating water since contaminants are removed from the system. It is therefore often useful to describe the composition of leachate from leaching of mineral waste as a function of the liquid/solid ratio (*L/S*), where *L* is the cumulated amount of water that at any given time has percolated through the system, while *S* is the dry weight of the waste being percolated. At lower *L/S* values the leaching system may be totally dominated by the properties of the waste, and for some substances saturation may occur in the water/leachate phase. Such a system is often referred to as being solubility controlled. For higher values of *L/S* the system may be more dominated by the properties of the liquid which, in the extreme, may act as an 'infinite' sink for some of the contaminants released from the waste. If the progression of the leaching of contaminants is described as a function of *L/S*, it will under certain conditions (including the existence of local equilibrium) be possible to ignore specific leaching conditions and, to a certain extent, use the results of certain accelerated laboratory leaching tests to estimate and predict the progression of leachate quality for a specific mineral waste landfill.

When the water percolation conditions for a given landfill containing mineral waste are known, the description of the progression of the leaching of various contaminants as a function of *L/S* may be converted to description of leaching as a function of time, where the relationship between time, *t* (years), and *L/S* (l/kg or m³/t) can be described by the following equation (Hjelmar, 1990):

$$t = t_0 + (L/S)\, d\, H/I \tag{10.5.2}$$

where $t_0$ (years) is the period passing until the first leachate appears at the base of the landfill, $d$ (t/m³) is the dry bulk density of the landfilled waste, $H$ (m) is the height of the landfill and $I$ (m/year) is the net rate of infiltration into the landfill. Using this equation, it can e.g. be estimated that for a 10 m landfill containing waste with a dry bulk density of 1.5 t/m³ and a rate of infiltration on 200 mm/year (= 0.2 m/year), it will take 75 years from the occurrence of the initial leachate to reach an average degree of leaching corresponding to *L/S* = 1 l/kg. For a 20 m landfill it would take 150 years under similar conditions. This is obviously a crude, average estimate which does not account for local differences in flow conditions within the landfill. In practice, preferential flow often occurs and causes some parts of the landfill to be leached faster than others. The achievement of an effective leaching of the entire landfill may require substantial efforts to enhance an even distribution of the flow within the waste in the landfill. It is not unusual to find 'dry pockets' as well as 'wet pockets' within a mineral waste landfill. Both are indicators of uneven flow distribution and/or hydraulically isolated spots within the landfill.

For monolithic mineral waste, e.g. waste which has been solidified and made coherent or block-like, the release of contaminants will occur by diffusion through the surface of the material, provided that the convective flow through the material is negligible due low permeability and/or a small hydraulic gradient. Whether or not the diffusion through the surface is the rate-controlling step depends on the rate of removal of the contaminants from the surface, i.e. the flow of water on the surface of the material and the build-up of concentrations in the boundary layer and the bulk water phase. If the diffusion through the surface is rate-controlling and there is no direct dissolution of the surface of the monolithic waste material, the rate of release or the flux, $J$ (mg/m$^2$/s), of a contaminant decreases as a function of the square root of the time. Assuming that the diffusion depth is small compared to the size of the surface and using a simplified one-dimensional model, the flux $J$ may be described by the following equation:

$$J = S_0 D_e^{0.5} \pi^{-0.5} t^{-0.5} \tag{10.5.3}$$

where $S_0$ (mg/kg) is the content of the contaminant in the waste material that is available for leaching, $D_e$ (m$^2$/s) is the effective diffusion coefficient for the contaminant (ion) in the material and $t$ is the contact time (s; van der Sloot *et al.*, 1989). The effective diffusion coefficient, $D_e$ (m$^2$/s), must be determined experimentally for the actual monolithic material and the relevant substances. Cai *et al.* (2003) and van der Sloot (1989) offer $D_e$ values for treated ashes in cement-based monoliths. Recently, a semimechanistic approach for modelling the release of some 25 major, minor and trace elements from cement-stabilised MSWI fly ash was presented (van der Sloot *et al.*, 2007).

The flux from a given amount of solidified/monolithic waste in a landfill normally decreases with decreasing surface: the larger the units or blocks and the fewer cracks, the lower is the flux in general. Normally the surface to mass or surface to volume ratio decreases with increasing size. This also means that the release of contaminants from solidified waste may increase if the integrity of the waste material deteriorates with time. For further discussion of the leaching conditions within monolithic landfills, the reader may for example refer to van Zomeren *et al.* (2003) and Hjelmar *et al.* (2005a).

For many granular waste types the concentration in the leachate of most contaminants decrease with time and/or increasing *L/S*. This is not necessarily true for contaminants that initially are solubility controlled. In a landfill containing MSWI bottom ash an initial increase in the concentration of sulfate may e.g. occur over a certain *L/S* range (due to a simultaneous decrease in the Ca concentration), of Ba (as the sulfate concentration decreases) and of As (possibly due a decrease in the concentration of carbonate or due to changes in the adsorption conditions). The release and hence the composition of the leachate is influenced by several factors, including pH conditions, redox potential, ionic strength, complexation with inorganic and organic contaminants as well as the adsorption conditions within the landfill.

### 10.5.2.3 Preferential Flow

Mobile non-interacting constituents (e.g. Cl) can provide insight in the role of preferential flow aspects at different scales of testing. This has been evaluated in a pilot project on sustainable landfill concepts (van der Sloot *et al.*, 1999, 2001). The study integrates testing at laboratory scale, lysimeter scale (1.0–1.5 m$^3$) and pilot field scale (12 000 m$^3$) with modelling of long-term release. The mass of all waste materials delivered to the pilot cell was recorded. A comparison between the release of mobile constituents (Cl, Na, K) in lysimeters and field leachate with the column leaching test data obtained with CEN/TS 14405 with upflow (minimal channelling) allowed conclusions to be drawn on the possible role of preferential flow. Figure 10.5.1 shows the cumulative release of K as a function of L/S (l/kg). The release from lysimeters and from the pilot field is lower than from the laboratory test and can be explained by preferential flow. From the comparison of release at the corresponding L/S a factor can be calculated. This has been done for Cl, K and Na as these parameters do not interact appreciably with the matrix. In Table 10.5.3 the release as derived from lysimeter and field data is given relative to release obtained for column leaching tests. Figure 10.5.1 and Table 10.5.3 indicate that about 25 % of the mobile species is released at a given L/S from lysimeter and full-scale systems relative to the release obtained for column leaching tests. This implies that more than 70 % of the waste is not affected by infiltrating rainwater. In full-scale monitoring of leachate production at Lostorf MSWI bottom ash landfill, Johnson *et al.* (1998, 1999) demonstrated from the observed fluctuations

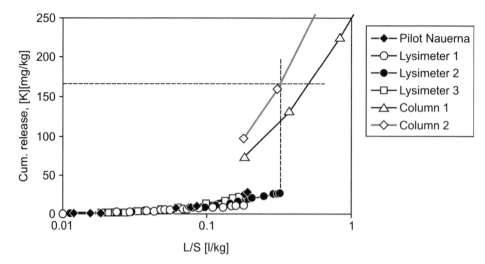

**Figure 10.5.1**   *Comparison of release of K from lysimeters (lysimeter 1 – regular fill; lysimeter 2 – enhanced organic matter content; lysimeter 3 – encapsulation of high leaching waste with less reactive waste; all lysimeters: 1 m³) and field leachate data (pilot Nauerna; 12 000 m³) with results from upflow column leaching tests (column 1: same mix as lysimeter 1 and column 2 same mix as lysimeter 2; each 0.0005 m³).*

in leachate concentrations between dry weather runoff and after rain events that the micropore leachate of the landfill is diluted by macropore leachate. This is a clear indication of preferential flow.

## 10.5.3   Chemical Reactions

The chemical reactions that are part of the stabilising processes that mineral waste placed in a landfill undergoes will vary in importance and depend on the character and properties of the waste in question as well as on the design and mode of operation of the landfill. Some of the most commonly occurring reactions are:

- Hydration.
- Dissolution/precipitation and formation of new minerals.
- Carbonation.
- Acid/base reactions and changes in alkalinity.
- Redox reactions.
- Complexation reactions and ionic strength (common ion) effects.
- Sorption reactions.

**Table 10.5.3**   *Relative release for lysimeter and field experiments to column data for mobile constituents. Calculations based on corresponding L/S values in the L/S range = 0.05 and 0.7 l/kg.*

| Experiment[a] | % Na leached | % K leached | % Cl leached |
|---|---|---|---|
| Field experiment | 39 | 31 | 24 |
| Lysimeter 1[b] | 20 | 14 | 19 |
| Lysimeter 2 | 28 | 20 | 18 |
| Lysimeter 3 | 32 | 28 | 31 |

[a]The data have been corrected for the preferential flow in an upflow column of around 85 % (mobile zone)
[b]Lysimeter 1 – regular fill; lysimeter 2 – enhanced organic matter content; lysimeter 3 – encapsulation of high leaching waste with less reactive waste; field data from pilot Nauerna, 12 000 m³.

### 10.5.3.1 Hydration

Several mineral waste types will undergo hydration reactions when they are brought into contact with water, before or after landfilling. Various types of thermal waste (e.g. bottom ash, fly ash, acid gas cleaning residues from incineration and energy production processes) often contain significant amounts of oxides of alkali metals (Na, K) and earth alkali metals (Ca, Mg). Upon contact with water these oxides are transformed into the corresponding hydroxides which eventually will be released into the leachate, which then become alkaline, unless a similar amount of acidic substances is released or absorbed. Also iron oxides and aluminium oxides are hydrolysed, and hygroscopic compounds such as calcium chloride first take up water and then dissolve. Hydration reactions are often strongly exothermic, and temperatures above 90 °C have been observed during the hydration of bottom ash and scrap iron.

### 10.5.3.2 Dissolution/Precipitation

Simple dissolution and precipitation reactions are among the most commonly occurring stabilisation processes in landfills for mineral waste. Depending on the type of mineral waste, the leachate may have a high content of salts, particularly at lower *L/S* values. Most chlorides are strongly soluble and their release is generally availability-controlled, whereas sulfate often is solubility-controlled by gypsum in the solid phase. The release of many trace elements is also controlled by solubility reactions. Many types of waste formed at high temperatures, such as bottom ashes and fly ashes from incineration, energy production and industrial processes, are thermodynamically unstable at ambient temperatures and pressures, and some of the minerals of which they consist (when they come into contact with water after landfilling) dissolve and recrystallise to form new, more stable minerals. Several trace elements partially precipitate, e.g. as carbonates or, if conditions are reducing, as sulfides (which may redissolve if conditions eventually become oxidising).

### 10.5.3.3 Carbonation

Landfilled mineral waste types which after hydration have become alkaline, and the leachate they produce, may subsequently react with carbon dioxide and form carbonates and hydrogen carbonates. The carbon dioxide may be atmospheric and may enter the landfill as part of the pore gas or dissolved in infiltrating rainwater. In some cases, if the mineral waste contains residual degradable organic matter, the carbon dioxide may be produced by biological processes within the landfill. Carbonation may typically reduce the pH of the leachate from e.g. MSWI bottom ash from pH 11–12 to a level of pH 7–8. Carbonation starts at the surface of granular mineral waste particles and slowly moves inwards over a long period of time. After several years such particles may still have a strongly alkaline core, which may be exposed if the landfill is disturbed and the particles are crushed. Due to preferential flow, the pH lowering effect of carbonation on alkaline waste types may be surprisingly quick. This may have a mitigating effect on the leaching of some contaminants which are soluble and hence released at high pH values (e.g. Pb and Zn) but may increase the leaching of other contaminants (e.g. Sb).

### 10.5.3.4 Acid/Base Reactions/Changes in Alkalinity

The pH is the factor that has the strongest influence on the solubility and release of many trace elements in the waste. As already indicated above, the introduction of acidic substances such as e.g. carbon dioxide may cause a substantial decrease in the leachate pH and hence a significant change in the solubility conditions for many trace elements and metals. A possible formation and dissolution of ammonia will have the opposite effect and increase pH in the leachate. Many trace elements and heavy metals exhibit minimum solubilities in the pH range 8–9, while the release of nearly all contaminants increases at low pH values (some of the trace elements that form oxyanions (e.g. Cr, Mo, Sb) often exhibit local maxima in the weakly acidic to weakly alkaline range. The carbonate system may be expected to stabilise the pH of the leachate within a landfill containing originally alkaline mineral waste at pH 7–8 over a longer period of time. Van der Sloot *et al.* (1997) point out that both mineral waste landfills containing MSWI bottom ash and biological reactor landfills with time move towards a state where pH in the leachate is approximately 8 and the solubility of Ca is controlled by calcite (calcium carbonate). While this development may take less than 20 years for MSWI bottom ash landfills, it can be expected to take 500–1700 years for biological reactor landfills.

### 10.5.3.5  Redox Reactions

Oxidation/reduction reactions are common and important reactions occurring in mineral waste landfills. Degradation of even small residual amounts of biodegradable organic material may during a period give rise to reducing conditions within a mineral waste landfill. This may lead to the formation of sulfide and ammonia. As mentioned above, several trace elements form sulfides of low solubility. If the content of sulfide is limited or the pH is low, iron is mobilised as ferrous ions under reducing conditions.

A mineral waste landfill where conditions in the earlier stages are reducing may eventually (possibly over a very long time) through external influences such as ingress of atmospheric oxygen as part of the pore gas or dissolved in infiltrating rainwater, become oxidised to the ambient level of redox potential. Oxidation changes sulfide to sulfate, pH decreases and trace elements/heavy metals that have been bound as sulfides may be mobilised. Later, ferrous iron again becomes immobilised as ferrous oxides unless pH has become permanently low ($< 4$).

For certain types of mineral waste, namely wastes containing substantial amounts of iron sulfides (e.g. pyrite), acid/base reactions are particularly important. Such wastes are produced in large quantities within the mining industry (as waste rock and tailings). They are stable as long as reducing conditions are maintained, but as soon as oxygen and moisture are present, iron sulfides begin to oxidise and form sulfuric acid. This may lead to the production of so-called acid rock drainage (ARD) if the acid production potential exceeds the neutralisation potential of the mineral waste or if the kinetics of the acid production reactions are faster than those of the neutralisation reactions. ARD production is one of the main problems associated with landfilling of pyrite-containing mining waste.

### 10.5.3.6  Complexation, Common Ion Effects and Organic Matter

High concentration of salts in the (initial) leachate from a mineral waste landfill may, because of the common ion effect alone, contribute to a strongly increased solubility and release of several trace elements/heavy metals. High concentrations of chloride will, for instance, form complexes with and hence increase the solubility and release of e.g. Cd, Pb and Hg. Organic matter, in either dissolved (DOC) or particulate (POM) form will affect the leachability of many inorganic as well as organic (trace) substances. Subfractionation of either DOC or POM may prove necessary and useful in properly describing the mobilisation and retention by organic matter (van Zomeren and Comans, 2007).

### 10.5.3.7  Sorption

The release of some trace elements, e.g. As, from mineral waste in landfills may be controlled both by solubility and precipitation processes and sorption processes. Under oxidising conditions at neutral pH, As occurs as arsenate, which may sorb to minerals containing Fe, Mn and Ca. Both sorption and solubility conditions may change with changes in pH. As described in Chapter 8.4 (on ageing of MSWI residues), the release of several trace elements from landfilled MSWI bottom ash may be expected to remain at a relatively low level, even in the long term, due to adsorption on the clay-like minerals that may form upon ageing of the bottom ash.

### 10.5.4  Description and Prediction of the Leaching Behaviour of Landfilled Mineral Waste

The expected leaching behaviour of waste materials is best characterised by a combination of a pH dependence leaching test and a percolation leaching test. This combination covers a wide range of relevant conditions from which different exposure conditions from those actually tested can be derived. The pH dependence leaching test provides information on chemical speciation aspects, whereas the percolation test through the *L/S* (liquid to solid ratio, l/kg) provides a measure for the time scale (Kosson *et al.*, 2002).

Laboratory characterisation of individual wastes is important to understand the behaviour and potential modification of leaching properties of individual wastes. However, when it comes to assessing the release from a landfill, where several waste types may be mixed in, it is the release of contaminants from the integral waste mixture that should be considered. This implies that more attention should be given to the release behaviour of waste mixtures to enable a better assessment of mixed minerel landfills in the future, as ultimately an entire landfill will need to meet criteria to safeguard against

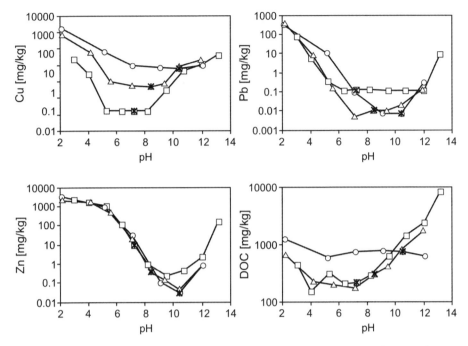

***Figure 10.5.2*** *pH-dependent leaching behaviour of Cu, Pb, Zn and DOC from predominantly inorganic waste (□), MSWI bottom ash (○) and a mixture of predominantly inorganic waste (80 %) and MSWI bottom ash (20 %; △). The star (\*) indicates the material's own pH.*

unacceptable impacts on soil and groundwater. This is illustrated in Figure 10.5.2, which shows the effect of mixing two fundamentally different waste materials on the leaching behaviour of Cu, Pb, Zn and DOC. Depending on the pH of the material, the effect of mixing predominantly inorganic waste with MSWI bottom ash creates an effect much larger than expected based on dilution. The leaching behaviour of Cu in the waste mix is drastically influenced by the addition of 20 % MSWI bottom ash. Cu leaching around neutral pH increases by almost two orders of magnitude. The leaching behaviour of Pb in the waste mix is appreciably lower than the original waste mix, which is attributed to a substantially larger mineral component in the new mix. The leaching of Zn in the waste MSWI bottom ash mix is also lower than the original waste mix. This like in the case of Pb can only be understood when the partitioning in the solid phase is considered. Figure 10.5.3 shows this change in partitioning for Cu. The leaching of DOC in MSWI bottom ash is lower than the waste mix at acidic to neutral pH. When bottom ash is added to the waste mixture, the DOC leaching is not significantly influenced in that pH range. The resulting concentration of DOC is however enough to complex additional Cu from the MSWI bottom ash and thus the increased availability of Cu results in enhanced Cu leaching through organic carbon complexation. These results indicate the different conclusions that follow from a study of individual wastes versus the study of waste mixtures.

Geochemical speciation modelling is a useful approach in understanding leaching behaviour and for predicting long-term leaching behaviour. The ORCHESTRA modelling framework embedded in the LeachXS expert system is briefly introduced in Box 10.5.1.

## 10.5.5 Examples of Properties of Leachate from Mineral Waste Landfills

Leachate from mineral waste landfills is generally characterised by having a varying content of inorganic salts and trace elements/heavy metals and a low content of organic constituents. The concentration level of most components is highest

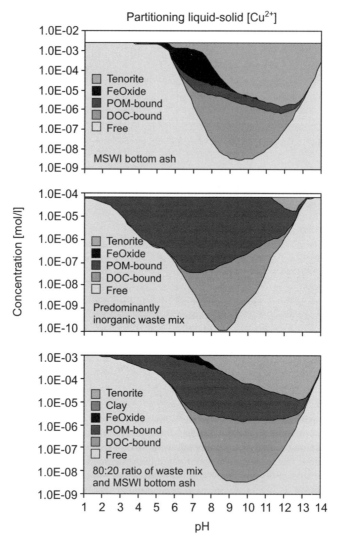

**Figure 10.5.3** *Partitioning of Cu in solution and in the solid for MSWI bottom ash, the Nauerna waste mix and the 80:20 ratio of waste mix and MSWI bottom ash. The boundary between light and dark colouring is the boundary between solid and liquid. Both solution and solid composition are expressed in mol/l (unit volume containing both solids and liquid).*

in the beginning and decreases with time, except for contaminants for which the release is sensitive to changes in e.g. pH and redox potential, when such changes actually occur.

Relatively few detailed field observations exist of mineral waste landfills that account for both the leachate quality and the water balance, allowing a description of the composition of the leachate as a function not only of time but also of *L/S* for the landfill.

One such dataset that is available describes the development of leachate quality at a Danish mineral landfill containing MSW incinerator residues over a period of 33 years (Hjelmar, 1996; Hjelmar and Hansen, 2005). The landfill, which was established and filled up in 1973 to 1976, contains approximately 10 000 m³ of MSWI bottom ash, including scrap

**Box 10.5.1   Geochemical Speciation and Release Modelling with ORCHESTRA/LeachXS.**

Chemical speciation of the contaminants in eluates from leaching tests and leachate from landfills may be determined with the ORCHESTRA modelling framework (Meeussen, 2003) embedded in LeachXS - a database/expert system (LeachXS, 2007). Aqueous speciation reactions and selected mineral precipitates are taken from the MINTEQA2 database and amended with additional thermodynamic data. Ion adsorption onto organic matter is calculated with the NICA-Donnan model (Kinniburgh *et al.*, 1999), with the generic adsorption reactions as published by Milne *et al.* (2003). Adsorption of ions onto iron and aluminium oxides is modelled according to the generalized two layer model of Dzombak and Morel (1990).

The database/expert system LeachXS is used for data management, e.g. pH dependent leaching data, percolation test data, lysimeter and field leachate data and for visualization of the calculated and measured results. The coupled LEACHXS - ORCHESTRA combination allows for very quick data retrieval, automatic input generation for modelling, processing of calculated results and graphical and tabular data presentation (see figure for graphical output).

The input to the different sub-models consists of metal availabilities, selected possible solubility controlling minerals, active Fe-and Al-oxide sites (Fe- and Al-oxides were summed and used as input for hydrated ferric oxides (HFO) as described by Meima and Comans (1998)), particulate organic matter, and a description of the dissolved organic carbon (DOC) concentration as a function of pH (polynomial curve fitting procedure). Basically, the speciation of all elements as obtained in the pH dependence test is calculated in one problem definition using the same parameter settings. This provides a chemical speciation fingerprint. This approach limits the degrees of freedom in selecting parameter settings considerable, as improvement of the model description for one element may deteriorate the outcome for other elements. The figure below shows, as an example, for Al, Pb and Zn model prediction of the total dissolved concentrations as well as the fraction of dissolved species and partitioning in the solid phase (limited to the fraction corresponding with the metal availability).

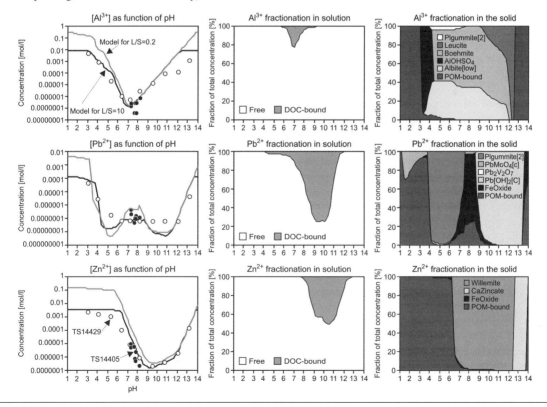

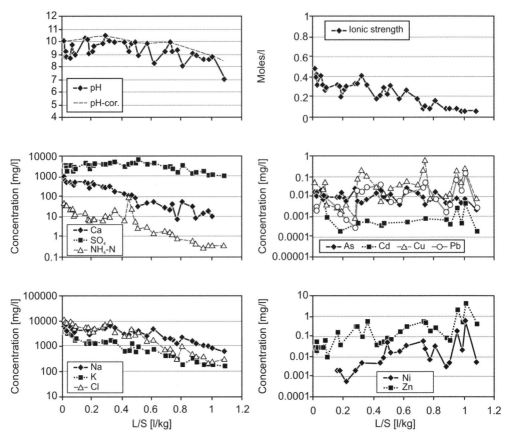

**Figure 10.5.4**  *Development in the composition of leachate from a Danish MSWI residue landfill as a function of L/S over a period of 33 years.*

metal and an estimated content of MSWI fly ash of 15 % (w/w). The landfill, which is approximately 9 m thick at the highest point, has a 1 m top cover of soil which allows infiltration of approximately 200 mm/year. It is equipped with a PVC bottom liner and a leachate collection tank. Leachate is removed at regular intervals via a central pumping well and transported to a wastewater treatment plant after adjustment of pH (with acid) and redox potential (with hydrogen peroxide). The amounts of leachate removed have been registered throughout the period. In 2006, a total of 13 000 m³ of leachate had been removed from the landfill. Once a year, a filtered sample taken from the leachate before it is treated is subjected to chemical analysis. The dry bulk density of the waste is estimated at 1.2 t/m³. It is therefore possible to describe the leachate quality as a function of *L/S*. This is done in Figure 10.5.4 for pH and selected salts and trace elements. It is clear both that substantial changes with time occur for some of the components and that the nature of the changes varies from one component to another. A slight, but significant decrease is seen in pH due to hydration and subsequent carbonation of the alkaline oxides of Ca, Na and K. The fluctuations of pH seen in the figure are likely to have been caused by unintended uptake of carbon dioxide before or during sampling. The dotted curve through the highest pH values is probably more representative of the true pH development. Due to the substantial leaching and removal of salt ions (Ca, Na, K, chloride, sulfate), the ionic strength of the leachate is drastically reduced during the period. The same is true for the concentrations of Ca, Na, K and chloride. The concentration of sulfate originally increases with time, but then starts to drop. As indicated e.g. by an observed decrease in the concentration of sulfide (and by the redox potential, which has been measured sporadically), the leachate becomes less reduced with time. This, together with the general flushing effect, may also have influenced the concentration of ammonia, which has decreased by a factor of 100 over the 33 years.

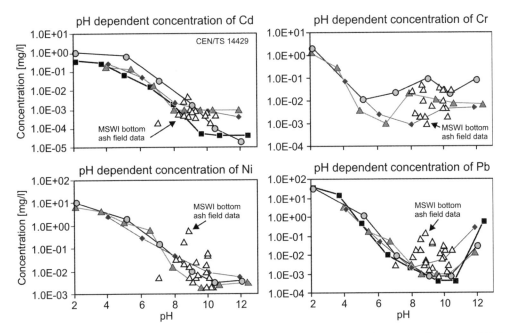

***Figure 10.5.5*** *Field leachate data in comparison with laboratory pH dependence test data on MSWI bottom ash. Triangles are field data from Danish MSWI residue landfill. The pH dependence test data stem from Dutch MSWI bottom ashes (L/S = 10). The field data cover an L/S range up to about 1.*

The concentrations of some of the trace elements (Ni, Zn) appear to exhibit a slightly increasing trend with time, whereas the variations in concentration of most of the others (e.g. As, Cd, Cr, Cu, Pb) exhibit no definite trends and seem to depend more on short-term variations in pH and redox potential than on *L/S* or time. Most of the trace element concentrations are fluctuating at relatively low levels. In Figure 10.5.5 the field data from Figure 10.5.4 have been placed in context with data from the pH dependence leaching test (CEN/TS14429). The agreement between the pH dependence test data and the field data for Cd, Cr, Ni and Pb at significantly different L/S values points towards solubility control by minerals and/or sorptive phases that apparently are the same in the laboratory as they are in the field situation. If this condition can be proven independently, then prediction of long-term release behaviour is a substantial step closer to realisation.

In many cases, monitoring programmes for leachate at landfills only include a limited number of analytical parameters, and often leachate from different landfill cells that may contain different types of waste are mixed in the leachate collection system before sampling. In addition, sampling conditions, waste types and water balances may not be well documented, and data series such as the one discussed above are therefore not generally available for the various types of mineral waste landfills. Table 10.5.4 presents some ranges of leachate composition that have been observed at landfills for different types of MSW incinerator residues, namely bottom ash (with a limited content of fly ash) and fly ash containing acid gas cleaning residues (dry/semidry process, wet scrubbing process). Note the high (initial) salt concentrations caused by the fly ash and the acid gas cleaning residues. Table 10.5.5 shows some examples of composition of leachate from 'inert' waste/C&D waste landfills in three European countries as well as ranges of composition of leachate from 'mineral waste' landfills in Denmark and The Netherlands.

## 10.5.6 Environmental Impact: Technical Control Options

The release and subsequent migration of contaminants with the leachate is considered by far the most important potential long-term impact from mineral waste landfills. The leachate management strategy for a mineral waste (or any) landfill should be tailored to ensure that the impact on the environment during operation and aftercare does not at any time exceed

**Table 10.5.4**   *Ranges of leachate composition observed at three different landfills containing MSWI residues (after AV-Miljø, 1996; Flyvbjerg and Hjelmar, 1997).*

| Parameter | Unit | MSWI bottom ash including approx. scrap metal and approximately 15 % fly ash | Dry/semidry MSWI acid gas cleaning residues (including fly ash) | MSWI wet scrubber sludge mixed with fly ash |
|---|---|---|---|---|
| Observation period[a] | years | 33 | 6 | 2 |
| pH | – | 7.1 – 10.5 | 6.6 – 10.9 | 7.1 – 7.7 |
| Electric conductivity | mS/m | 380 – 3900 | 780 – 17000 | 4900 – 16000 |
| Chloride | mg/l | 220 – 11400 | 2300 – 110000 | 3300 – 82000 |
| Sulfate | mg/l | 1100 – 7200 | 160 – 1300 | 980 – 1900 |
| Ammonia-N | mg/l | 0.27 – 87 | – | – |
| Na | mg/l | 600 – 7300 | – | – |
| K | mg/l | 170 – 4300 | 65 – 18000 | 7100 – 35000 |
| Ca | mg/l | 7.0 – 1000 | 250 – 45000 | 1400 – 4100 |
| As | mg/l | 0.0021 – 0.0026 | – | – |
| Cd | mg/l | 0.0002 – 0.0048 | 0.005 – 7.3 | 0.23 2.3 |
| Cr | mg/l | 0.001 – 0.05 | 0.003 – 0.98 | 0.026 – 0.3 |
| Cu | mg/l | 0.0009 – 0.66 | 0.004 – 0.53 | 0.0006 – 0.017 |
| Fe | mg/l | 0.005 – 1.6 | 0.15 – 3 | 0.03 – 0.75 |
| Hg | mg/l | < 0.00005 – 0.003 | < 0.0005 – 0.003 | 0.0005 – 0.002 |
| Ni | mg/l | 0.0006 – 0.6 | < 0.01 – 0.15 | 0.01 – 0.05 |
| Pb | mg/l | 0.0007 – 0.14 | 0.04 – 1600 | 0.12 – 1.3 |
| Zn | mg/l | 0.01 – 4.5 | < 0.01 – 7.6 | 0.09 – 0.43 |
| Accumulated L/S | l/kg | 1.1 | – | – |

[a] Period starting at the commencement of landfilling.

an acceptable limit, and that the highest possible degree of sustainability is achieved, i.e. the necessary aftercare period should be minimised, and environmental safety should, at least in the longer term, not rely on active protection systems.

The potential impact or the flux of contaminants reaching the base of the landfill from where it may be collected or leak into the environment, depends at any time on the quality and quantity of the leachate. The quality of the leachate may be influenced by the leaching properties of the waste, the physical and chemical conditions within the landfill (flow pattern, redox potential, pH, etc.) and the leaching history (time, $L/S$). The quantity of the leachate may be influenced by the amount and intensity of rainfalls (the rate of infiltration of precipitation), the properties of the top cover, and by the geometry, design and operation of the landfill.

The leaching properties of the waste may be changed by treatment prior to landfilling, e.g. by removal of soluble salts by washing, possibly integrated in the waste production process, or by temporary storage to reduce pH (carbonation) of alkaline waste or to reduce the content of organic matter by degradation of biodegradable material. The leaching properties will also change after landfilling as a function of $L/S$ and other conditions. In order to optimise this process, it is important to know (from the waste characterisation) how the leaching of various contaminants may be expected to change under various conditions (for many contaminants, the level of leaching decreases with increasing time and $L/S$). However, developments e.g. in pH and/or redox potential may influence this, so they must be taken into consideration when planning the strategy. The influence of the leachate itself and various conditions (e.g. atmospheric $CO_2$) on the environmental protection systems (e.g. clogging of drainage systems) must be taken into account.

It is likely that a two-stage strategy will be optimal for many mineral waste landfills (unless the leachate quality is fully compatible with the surrounding environment from the outset – in which case no liners are needed). In the initial phase, including the operation and possibly the initial part of the aftercare, active environmental protection systems may be employed along with other measures. In order to remove and collect the leachable contaminants as extensively as possible, leachate production could be maximised during a period not exceeding the expected lifetime of the bottom liner (which may be estimated to be up to 100 years). This means that a top cover of high permeability, possibly combined with

**Table 10.5.5** *Examples of composition of leachate from inert waste landfills/C&D waste landfills and a landfill for mixed waste with low organic content/mineral waste (Hjelmar et al., 1994; Roskilde County, 2004; Avfalzorg, 2006).*

| Parameter | Unit | Landfills for inert waste/C&D waste | | | Landfill for predominantly inorganic waste | Landfill for mixed waste with low organic content |
|---|---|---|---|---|---|---|
| Country | | Denmark | Germany | UK | The Netherlands | Denmark |
| Observation period[a] | Years | — | — | — | 11 | 13 |
| pH | | 7 | 7.5 | 8.1 | 6.9–7.4 | 6.6–8.0 |
| Electric conductivity | mS/m | 120 | 250 | — | | 280–1200 |
| BOD | mg/l | | 20 | — | 20–300 | |
| COD | mg/l | 220 | 130 | 240 | | 27–2900 |
| DOC/TOC | mg/l | | 40 | 93 | max. 300 | |
| Chloride | mg/l | | 100–600 | 370 | 500–2000 | 97–2200 |
| Sulfate | mg/l | < 5–170 | 450 | 210 | 30–500 | — |
| Ammonia-N | mg/l | | 13 | 28 | 50–300 | 1–140 |
| Na | mg/l | 170 | 270 | 100 | 2000–5000 | 12–1400 |
| K | mg/l | | 50 | 50 | 400–600 | 2.7–1300 |
| Ca | mg/l | | 200 | 340 | max. 500 | — |
| As | mg/l | | 0.009–0.037 | | 0.012–0.14 | — |
| Cd | mg/l | bd | bd | | 0.001–0.049 | < 0.00009–0.55 |
| Cr | mg/l | bd | 0.003–0.008 | | 0.0035–0.12 | — |
| Cu | mg/l | | 0.001–0.011 | | 0.0013–0.088 | — |
| Fe | mg/l | 4–260 | 0.0035 | 0.07 | < 0.05–5.0 | < 0.02–39.0 |
| Hg | mg/l | | bd | – | 0.00005–0.00046 | — |
| Ni | mg/l | | 0.007 | | 0.013–0.081 | — |
| Pb | mg/l | 0.2 | 0.003–0.006 | bd | 0.001–0.19 | < 0.003–0.13 |
| Zn | mg/l | 500 | 0.1–0.2 | 0.6 | 0.014–1.1 | < 0.05–0.2 |
| Number of landfills included | | 1 | 3 | 6 | 1 | 1 |

[a]Period starting at the commencement of landfilling.

recirculation or flushing, should be applied during that period or until the concentration level has reached a predetermined, acceptable level. Then the active environmental protection systems may be abandoned (not necessarily a trivial exercise), and the leachate allowed to leak into the surroundings. Some passive hydraulic control may possibly, if necessary, still be maintained at this stage by establishing a geologically stable top cover that may reduce the rate of infiltration and thus the flux of contaminants to a desired level.

The leachate collected during the active stage will need to be managed properly, i.e. to be discharged with or without treatment (as needed). To achieve the best possible degree of sustainability, care should be taken to choose a treatment that corresponds to the properties of the leachate. It is not uncommon that, for convenience, leachate from mineral waste landfills containing basically salts and trace elements and very low levels of degradable organic matter is discharged to a biological wastewater treatment plant. Since the only effect is dilution of the salts and possibly contamination of the sludge from the plant with trace elements/heavy metals, it might be advantageous to pursue other discharge options (e.g. direct discharge to the sea after specific treatment to reduce the concentration of trace elements to an acceptable level).

# References

Avfalzorg (2006): *Personal communication*, The Netherlands.

AV Miljø (1996): *Annual report 1995*. AV Miljø, Avedøre Holme, Hvidovre, Denmark.

Bold, S. (2004): *Process-based prediction of the long-term risk of groundwater pollution by organic non-volatile contaminants*. Tübinger Geowissenschaftliche Arbeiten, Reihe C Nr. 72.

Cai, Z.; Jensen, D.L.; Christensen, T.H. and Bager, D. (2003): Reuse of stabilized flue gas ashes from solid waste incineration in cement-treated base layers for pavement. *Waste Management and Research*, 21, 42–53.

CEC (1999): *Council Directive 1999/31/EC of 26 April 1999 on the landfill of waste*. European Commission, Brussels, Belgium.

CEC (2003): *Council Decision 2003/33/EC of 19 December 2002 establishing criteria and procedures for the acceptance of waste at landfills pursuant to Article 16 of and Annex II to Directive 199/31/EC*. European Commission, Brussels, Belgium.

CEN (2005): *CEN TS 14405 Characterisation of waste – Leaching behaviour tests – Upflow percolation test*. CEN, Brussels, Belgium.

CEN (2005): *CEN TS 14429, Characterisation of waste – Leaching behaviour tests – Influence of pH on leaching with initial acid/base addition*. CEN, Brussels, Belgium.

Christ, A. and Hofmann, T. (2004): *Bewertung der Mobilisierung, des Transports und der Rückhaltung von kolloidalen und suspendierten Partikeln in der Bodenzone* (Evaluation of the mobilisation, transport and retention of colloidal and suspended particles in the soil zone; in German). 02WP0199. Statusseminar Sickerwasserprognose, Berlin, Germany.

Dijkstra, J.J., Meeussen, J.C.L., van der Sloot, H.A. and Comans, R.N.J. (2008): A consistent geochemical modelling approach for the leaching and reactive transport of major and trace elements in MSWI bottom ash. *Applied Geochemistry*, 23, 1544–1562.

Dzombak, D.A. and Morel, F.M.M. (1990): *Surface complexation modeling: hydrous ferric oxide*, John Wiley & Sons, Ltd, New York, USA.

Flyvbjerg, J. and Hjelmar, O. (1997): *Restprodukter fra røggasrensning ved affaldsforbrænding*. (Residues from incinerator flue gas cleaning; in Danish), Arbejdsrapport fra Miljøstyrelsen nr. 92. Miljøstyrelsen, Copenhagen, Denmark.

Hjelmar, O. (1990): Leachate from land disposal of coal fly ash. *Waste Management and Research*, 8, 429–449.

Hjelmar, O. (1996): Disposal strategies for municipal solid waste incineration residues. *Journal of Hazardous Materials*, 47, 345–368.

Hjelmar, O., Johannessen, L.M., Knox, K., Ehrig, H.-J., Flyvbjerg, J., Winther, P. and Christensen, T.H. (1994): *Management and composition of leachate from landfills*. Final Report for the Commission of the European Communities, DGX A.4. Waste 92. Water Quality Institute, Carl Bro Environment/Knox Associates, University of Wuppertal, Technical University of Denmark, VKI, Hørsholm, Denmark.

Hjelmar O. and J.B. Hansen (2005): Sustainable landfill: The role of final storage quality. *Proceedings of the tenth international landfill symposium*. Environmental Sanitary Engineering Centre, Cagliari, Italy.

Johnson, A.C., Richner, G.A., Vitvar, T., Schittli, N. and Eberhard, M. (1998): Hydrological and geochemical factors affecting leachate composition in municipal solid waste incinerator bottom ash. Part I: The hydrology of landfill Lostorf, Switzerland. *Journal of Contaminant Hydrology*, 33, 361–376.

Johnson, A.C., Kaeppeli, M., Brandenberger, S., Ulrich, A. and Baumann W. (1999): Hydrological and geochemical factors affecting leachate composition in municipal solid waste incinerator bottom ash. Part II: The geochemistry of leachate from landfill Lostorf, Switzerland. *Journal of Contaminant Hydrology*, 40, 239–259.

Kinniburgh, D.G., van Riemsdijk, W.H., Koopal, L.K., Borkovec, M., Benedetti, M.F. and Avena, M.J. (1999): Ion binding to natural organic matter: competition, heterogeneity, stoichiometry and thermodynamic consistency. *Journal of Colloids and Surfactants, Part A*, 151, 147–166.

Kosson, D.S., van der Sloot, H.A., Sanchez, F. and Garrabrants, A.C. (2002): An integrated framework for evaluating leaching in waste management and utilization of secondary materials. *Environmental Engineering Science*, 19 (3), 159–204.

LeachXS (2004): *Homepage for LeachXS©*. Located on the internet: 19 July 2007. www.leachxs.org.

Meeussen, J.C.L. (2003): ORCHESTRA: An object-oriented framework for implementing chemical equilibrium models. *Environmental Science and Technology*, 37, 1175–1182.

Meima, J.A.and Comans, R.N.J. (1998): Application of surface complexation/precipitation modeling to contaminant leaching from weathered municipal solid waste incinerator bottom ash. *Environmental Science and Technology*, 32, 688–693.

Milne, C.J., Kinniburgh, D.G., van Riemsdijk, W.H. and Tipping, E. (2003): Generic NICA-Donnan model parameters for metal-ion binding by humic substances, *Environmental Science and Technology*, 37, 958–971.

Roskilde County (2004): *Landfill leachate monitoring data*. Roskilde County, Roskilde, Denmark.

van der Sloot, H.A., de Groot, G.J. and Wijkstra, J. (1989): *Leaching characteristics of construction materials and stabilization products containing waste materials*. In: Côté, P. and Gilliam, M. (eds) Environmental aspects of stabilization and solidification of Hazardous and radioactive wastes. STP 1033. ASTM, Philadelphia, USA.

van der Sloot, H.A., Kosson, D.S., Cnubben, P.A.J.P., Hoede, D. and Hjelmar, O. (1997): Waste characterization to modify waste quality prior to disposal. In: CISA (ed.) *Proceedings of the sixth international landfill symposium*, pp. 315–328. CISA, Environmental Sanitary Engineering Centre, Cagliari, Italy.

van der Sloot, H.A., Cnubben, P.A.J.P. and Scharff, H. (1999): Predominantly inorganic equilibrium disposal – part of the total concept sustainable recycling and storage of solid waste. *Proceedings of the International Landfill Symposium*, 7, 103–110.

van der Sloot, H.A., van Zomeren, A., Rietra, R.P.J.J., Hoede, D., and Scharff, H. (2001): Integration of lab-scale testing, lysimeter studies and pilot scale monitoring of a predominantly inorganic waste landfill to reach sustainable landfill conditions. *Proceedings of the International Landfill Symposium*, 8 (1), 255–264.

van der Sloot, H.A., van Zomeren A., Meeuwsen J.C.L., Seignette, P. and Bleyerveld, B. (2007): Interpretation of test method selection, validation against field data, and predictive modelling for impact evaluation of stabilised waste disposal. *Journal of Hazardous Materials*, 141, 354–369.

van Zomeren, A. and Comans R.N.J (2007): Measurement of humic and fulvic acid concentrations and dissolution properties by a rapid batch procedure. *Environmental Science and Technology*, 41, 6755–6761.

van Zomeren, A., van Wetten, H., Dijkstra, J.J., van der Sloot, H.A. and Bleijerveld, R. (2003): Long term prediction of release from a stabilised waste monofill and identification of controlling factors. In: CISA (ed.) *Proceedings of the ninth international waste management and landfill symposium*. CD-ROM. CISA, Environmental Sanitary Engineering Centre, Cagliari, Italy.

# 10.6

# Landfilling: Reactor Landfills

**Thomas H. Christensen and Simone Manfredi**

*Technical University of Denmark, Lyngby, Denmark*

**Keith Knox**

*Knox Associates, Nottingham, UK*

Reactor landfills are engineered landfills that receive waste containing untreated or partly treated organic matter and thus produce gas as well as a contaminated leachate. The term reflects the fact that the presence of organic matter in the landfill and the degradation reactions that it undergoes make the landfill a reactor that must be engineered in order to control the gas and leachate that it produces. The reactor landfill is subject to leaching processes, as is the mineral landfill described in Chapter 10.5, but the degradation of the organic waste in the reactor landfill is such a dominant process that it controls the chemical environment in the landfill and thus also controls the leaching. Emphasis with regard to reactor landfills is on the degradation processes and their consequent effect on the gas generation although leaching is also very important.

The reactor landfill may receive a variety of waste categories and types, including demolition waste, industrial/commercial waste, and contaminated soil, but municipal solid waste with a high content of organic waste always is a significant fraction. The mix of received waste may vary over a wide range within and between reactor landfills and this should be kept in mind in the later discussion on gas and leachate quality. Reactor landfills may be categorized as:

- Conventional reactor landfills.
- Bioreactor landfills (enhancing gas generation primarily by leachate recirculation).
- Flushing bioreactors (enhancing gas generation and ammonium wash out).
- Semiaerobic bioreactors (introducing a partially aerobic environment to control gas and leachate).

The conventional landfill is described with respect to the time-dependent development in gas and leachate quality as well as quantity. Typical values for gas and leachate are provided for the conventional landfill. The bioreactor, the flushing bioreactor, and the semiaerobic landfills are described relative to the conventional landfill in terms of main objectives, technical approaches and consequences for gas and leachate. Figure 10.6.1 illustrates the technical differences of the

*Solid Waste Technology & Management*   Edited by Thomas Christensen
© 2011 Blackwell Publishing Ltd

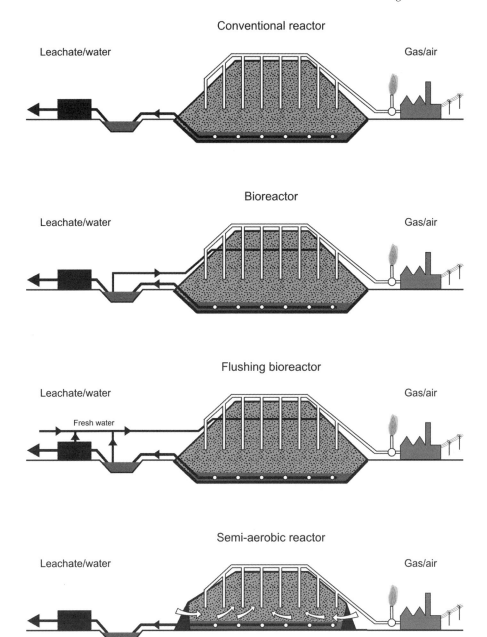

**Figure 10.6.1** *Main technical differences of the various reactor landfills. The activities to the left of the landfill body illustrate leachate collection, storage pond, treatment facility and discharge to surface water, while the activities to the right illustrates gas collection, flaring and utilization with electricity production.*

various reactor landfills. The latter landfill technologies are relatively new technologies and the database on full-scale performance is still very limited, in particular with respect to the development of characteristics over time.

Landfills receiving pretreated municipal solid waste (by mechanical biological treatment) are described in Chapter 10.7. They share many common characteristics with reactor landfills.

## 10.6.1   Conventional Reactor Landfill

The conventional reactor landfill is a landfill which receives a mix of waste, except hazardous waste, and has a conventional leachate collection system (Chapter 10.8) and gas collection system (Chapter 10.10). Landfilling is undertaken as a disposal operation with little focus on optimizing the degradation processes within the waste. The waste is typically compacted to a wet density of $0.7–1.0\,t/m^3$ and the waste is regularly covered with soil. The cells and sections (group of cells with joint leachate collection system) are filled as the waste is received.

### 10.6.1.1   Degradation Processes

The degradation of organic matter in the conventional landfill is predominantly anaerobic. The initial aerobic phase (see later) may last only few days until the oxygen contained in the air entrapped with the landfilled waste is used up. The dominant microbiological processes in the anaerobic landfill are the same as those taking place in the anaerobic digester as described in detail in Chapter 9.4. However, the conditions in the landfill are far from optimal because of lack of pretreatment, heterogeneity of the waste, lack of mixing and very often low moisture content. In addition, landfills usually contain cells of different age and degree of degradation. Consequently, depending on the physical structure of the landfill, leachate may move from younger cells to older cells. The conventional landfill hence constitutes a 'reactor' with characteristics somewhat different from the anaerobic digester. A further difference is that in a landfill cell it may take from several months to several years before the balance is established between the acidogenic microbial community and the $CH_4$ formers.

The time scale of organic matter degradation in landfills may be several decades or even centuries. Whilst the majority of gas generation may have occurred within 10–20 years, complete degradation of the organic matter may then proceed only very slowly. Considering the anaerobic degradation of the organic waste and the losses by gas and leachate Manfredi and Christensen (2009) estimated for conventional reactor landfills that, typically, about half of the biogenic carbon entering the landfill through the waste remains in the landfill after 100 years. The main microbial anaerobic degradation processes in a landfill are summarized in Table 10.6.1.

### 10.6.1.2   Phases in the Life of a Conventional Landfill

The degradation of organic waste in a landfill cell determines the environment within the landfill and this is reflected in the composition of the gas and the leachate generated. The development of these characteristics over time may be explained by a multiphase degradation sequence before the landfilled waste is comparable to the surrounding soil. Christensen and Kjeldsen (1995), explained the development in gas and leachate composition over time with an eight-phase model (Figure 10.6.2), while Rettenberger (2005) used a nine-phase model for the development in gas composition over time. The phases have highly varying durations, ranging from days to decades. The figures are illustrative and do not reflect a particular set of data. A lot of data from a variety of conventional landfills are available for the first four phases representing the first 20–35 years of the landfill, but the characteristics of the later phases are of a more speculative nature. Data from excavation of old dumps provide some support for the later phases.

With respect to gas and leachate composition the phases of the conventional landfill are:

- *Initial aerobic phase:* Oxygen in air voids and air penetrating into thin layers of landfilled waste may sustain a short aerobic composting process (few days), but for compacted waste in lifts of 2–3 m this is an unimportant phase in relation to the overall degradation of the wastes. It may however, be of importance in inactivating pathogens, as the temperature can rise to 60–70 °C for a period of several weeks as result of the oxidation reactions.

**Table 10.6.1** *Major gas generating processes involved in the anaerobic degradation of waste.*

Fermentative processes

$C_6H_{12}O_6 + 2H_2O \quad\rightarrow\quad 2CH_3COOH + H_2 + 2CO_2$

$C_6H_{12}O_6 \quad\rightarrow\quad CH_3C_2H_4COOH + 2H_2 + 2CO_2$

$C_6H_{12}O_6 \quad\rightarrow\quad 2CH_3CH_2OH + 2CO_2$

Acetogenic processes

$CH_3CH_2COOH + 2H_2O \quad\rightarrow\quad CH_3COOH + 3H_2 + CO_2$

$CH_3C_2H_4COOH + 2H_2O \quad\rightarrow\quad 2CH_3COOH + 2H_2$

$CH_3CH_2OH + H_2O \quad\rightarrow\quad CH_3COOH + 2H_2$

$C_6H_5COOH + 6H_2O \quad\rightarrow\quad 3CH_3COOH + 3H_2 + CO_2$

Methanogenic processes

$4H_2 + CO_2 \quad\rightarrow\quad CH_4 + 2H_2O$

$CH_3COOH \quad\rightarrow\quad CH_4 + CO_2$

$CH_3OH + H_2 \quad\rightarrow\quad CH_4 + 2H_2O$

Sulfate-reducing processes

$4H_2 + SO_4^{-2} + H^+ \quad\rightarrow\quad HS^- + 4H_2O$

$CH_3COOH + SO_4^{-2} \quad\rightarrow\quad CO_2 + HS^- + HCO_3^- + H_2O$

$2CH_3C_2H_4COOH + SO_4^{-2} + H^+ \quad\rightarrow\quad 4CH_3COOH + HS^-$

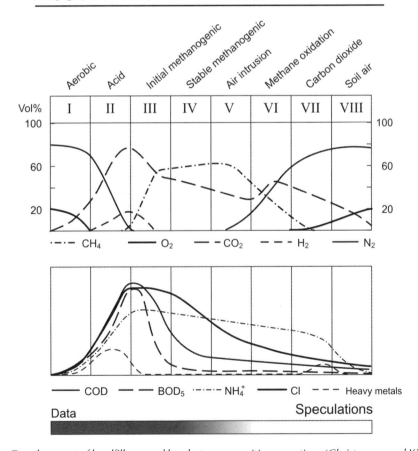

**Figure 10.6.2** *Development of landfill gas and leachate composition over time (Christensen and Kjeldsen, 1995).*

- *Acidic phase:* As oxygen is depleted, hydrolysis and fermentation of carbohydrates, fats and proteins by anaerobic bacteria start producing primarily volatile fatty acids (VFA) and $CO_2$, as well as ammoniacal nitrogen. The generation of $CO_2$ displaces remaining $N_2$ from the voids. Fermentative bacteria produce VFAs, $CO_2$, and $H_2$. Subsequently, acetogenic bacteria convert the previously formed VFAs into simpler acids (for instance acetic acid), which are utilized later by the methanogens. The generation of VFAs and $CO_2$ contributes to an acidic pH ($< 6$). The leachate is low in pH, high in VFAs (high BOD/COD ratio), high in soluble inorganics such as $NH_4$-N, Cl, Na and high in metals such as Fe, Ca, and some heavy metals. The elevated metals are due to the low pH and to complexation by dissolved organic carbon and salts. This phase may last several months or occasionally even years.

- *Initial methanogenic phase:* Locally, where niches of near-neutral pH value exist, a balance between the acid formers and the methanogens evolves and $CH_4$ starts appearing. Table 10.6.1 shows that $CH_4$ is formed by several reactions. The overall result is that $CH_4$ concentration in the landfill gas increases while $CO_2$ concentration decreases. $H_2$ practically disappears from the gas. This phase may last from several months to, typically, not more than $\sim$2 years. The conversion of VFAs increases the pH, which further improves the conditions for the methane formers.

- *Stable methanogenic phase:* Over time the balance between the acid formers and the methane formers becomes stable, generating a gas that contains slightly more $CH_4$ than $CO_2$. The ratio between $CH_4$ and $CO_2$ reflects the nature of the organic waste being degraded and possibly the fact that some of the $CO_2$ is dissolved in the leachate. Since pH is at or above neutral the leachate will be strongly buffered by bicarbonate, $HCO_3^-$. In large landfills, the landfill temperature may reach more than $40\,°C$ as a consequence of the energy released by the anaerobic processes and the insulation provided by the thick waste layers. The rate of landfill gas generation peaks in this phase which may last 10–30 years. The leachate has a pH slightly above neutral, a stable content of COD (2000–6000 mg COD/l) with a low degradability (low BOD/COD ratio since most of the VFAs have been degraded inside the landfill body), and a low content of heavy metals. The content of salts (Cl, Na, $NH_4$-N, etc.) in the leachate may still be high, since it typically takes more than an L/S ratio of 2 or 3 (l/kg) to wash out the salts from waste. Most of the settlement takes place during the methanogenic phase, because this is when the main degradation occurs.

- *Air intrusion phase:* When gas generation becomes slow, air will begin to intrude the outskirts of the landfill and $N_2$ will appear in the gas. The oxygen is used up by bacteria, for the oxidation of $CH_4$ in the gas or of carbon in the partially degraded waste. The length of this period is hardly known. The leachate will probably have changed little from the previous period except that the salt concentrations may start decreasing as a consequence of flushing.

- *Methane oxidation phase:* As methane generation slows even further, air intrusion is dominating the outskirts of the landfill and migrating methane generated in the lower part of the landfill will be oxidized in the outer parts of the landfill ($CH_4 + 2O_2 \rightarrow CO_2 + 2H_2O$). The oxidation reaction may locally lower the pressure (the reaction consumes 3 mol gas but produces only 1 mol gas; water condenses), potentially increasing the mass flux to the active methane oxidizing niches. Methane is present in only low concentrations, while $N_2$ and $CO_2$ concentrations are increasing. Concentrations of soluble leachate components (e.g. salts, $NH_4$-N, DOC) are likely to have further decreased due to flushing.

- *Carbon dioxide phase:* After methanogenesis has ceased, intruding oxygen will partially oxidize solid organic carbon left in the landfill body. $CO_2$ is still produced but in less quantity, and $N_2$ is now the dominant component in the gas.

- *Soil air phase:* After decades, or perhaps centuries, the organic matter in the landfill is so stable that, with respect to oxidation, it is approaching organic matter in soil and the landfill gas will approach the composition of soil air with high concentration of nitrogen, 10–15 % $O_2$ and 5–10 % of $CO_2$. The leachate may, depending on the effective *L/S* ratio reached, still contain salts, DOC and some ammonium. Depending on the buffer capacity of the waste, a slight decrease in pH could occur in the later phases of the landfill, which potentially could cause a slight release of heavy metals from the landfill. However, there is no hard evidence of this. Solid organic matter and, in the long term, iron oxides, may provide sufficient binding sites to keep the metal concentrations low in the leachate even after stabilization of the waste.

It should be remembered that the characteristics of the later phases are primarily based on theoretical speculations combined with observations from laboratory scale experiments and old dumps, since no data are currently available from actual landfills.

### 10.6.1.3 Conventional Landfill: Controlling Factors

The anaerobic microbial degradation of organic matter depends on several environmental factors as also described for the anaerobic digester in Chapter 9.4. In the landfill the most important ones are:

- *Water content* affects methanogenesis (refer to Christensen *et al.*, 1996). At low water content the mass transfer of substrate and nutrients is limited between niches in the waste and across microbial membranes. Low water content is the most common factor limiting gas generation rates at landfills in most European and North American landfills. Water content close to field capacity seems the optimal to promote degradation (EPA, 2002). Addition of water or recirculation of leachate may be useful in some cases but a fully saturated landfill body is probably not a requirement for good methanogenesis. Moreover, water addition programs have to be carefully planned as too high a water content during the premethanogenic degradation phases may lead to excessive acid formation, thus to inhibition of methanogenesis. In addition, high water contents can adversely affect the efficiency of gas collection systems, due to flooding.
- *Temperature* is a key factor in landfills like in any other microbial degradation process. An increase in temperature from 20 to 30 °C approximately doubles the rate of methanogenesis (refer to review by Christensen *et al.*, 1996), suggesting that landfills with elevated temperature convert the organic waste faster.
- *Oxygen* inhibits methanogenesis. Oxygen ingress into the landfill is easily avoided if the pumping of landfill gas is controlled, reducing the intrusion of air to a minimum. Oxygen may however infiltrate in the upper waste layer (1–2 m) but it is here rapidly consumed through respiration of aerobic bacteria. The thickness of this upper partially aerobic layer can be reduced by effective waste compaction and by a soil cover.
- *Hydrogen* is produced by fermentative and acetogenic bacteria: the more vigorous the activity of these two families of bacteria the higher hydrogen pressure in the gas. Some VFAs, such as butyric and propionic acids, cannot be degraded into simpler acids when the hydrogen pressure is above a certain value. This may lead to accumulation of VFAs and to inhibition of methanogenesis. In practice this is rare.
- *Sulfate,* for example released from demolition waste (plasterboard) and bottom ashes, may be reduced to sulfides ($HS^-$) during the oxidation of organic matter or even of $CH_4$. This reduces the amount of methane, increases the generation of $CO_2$ and potentially releases toxic gases in the form of $H_2S$, at low pH (Fairweather and Barlaz, 1998). In most cases, a moderate amount of sulfate-containing waste does not lead to significant amounts of $H_2S$, because most of the reduced S precipitates in the waste as metal sulfides (for example as iron sulfides).
- *Nutrients* are usually not considered to be limiting the degradation processes in landfills for mixed municipal waste. Salts like Ca and Mg as well as ammonium ($NH_4^+$) may inhibit anaerobic processes, but usually they are not present at sufficiently high concentrations in leachates.
- *Inhibitors* may appear in the form of high concentrations of heavy metals and organic chemicals in certain niches in the waste, but at large scale this is usually not considered as a problem in mixed municipal waste. $H_2$ as an end product may be an inhibitor for fermentation of certain VFAs and is some times considered an indicator for an unbalanced anaerobic degradation (hydrogenotrophic conversion of $H_2$ to $CH_4$ is slower than the acidotrophic degradation of certain higher VFAs generating $H_2$). $CO_2$ may also exhibit inhibitory effects on methanogenesis. Hansson and Molin (1981) detected that the conversion of acetic acid to methane is inhibited at $CO_2$ partial pressures in the range 0.2–1.0 atm. This encompasses the typical range of $CO_2$ partial pressure during anaerobic degradation in landfills and most other digesters (0.5–0.9 atm), and is unavoidable.

In practice it is difficult to control most of the above mentioned factors for a landfill. The practical options available often affect several of the basic factors. The following management options can be considered for the conventional landfill (e.g. see Christensen *et al.*, 1992):

- *Daily soil covers* with a permeable but buffered soil may also provide buffering capacity and thus help in establishing the right pH conditions in the waste to support methanogenesis. Clays and other heavy soil should preferably be avoided because they are hard to handle, difficult to drive on during rain and they may sustain local perched leachate water tables within the landfill body, and interfere with gas collection.

- *Section geometry* in a landfill section affects the leachate generation per tonne of waste and the possibility of obtaining enhanced temperatures in the landfill body. A deep landfill provides better insulation than a shallow landfill and thus more likely obtains an enhanced temperature in the landfill body and faster degradation. In contrast, the infiltrating precipitation (*L*iquid) percolates through more waste (*S*olid) in a deep landfill, resulting in a lower *L/S* ratio for a fixed time frame. This results in higher leachate concentration and longer time frame before leachable substances are depleted from the waste.
- *Incoming waste* may be mixed during the disposal process. For example by distributing soil and contaminated demolition waste as layers within the sections providing buffer capacity, or sludge and wet waste as layers providing moisture.
- *Shredding* may be introduced in order to increase the surface area of the waste and its compactability. For easily degradable waste, this may however increase the activity of the acid phase to a level that could make the establishing of stable methanogenesis more difficult.
- *Compaction* of waste ensures a better utilization of the landfill capacity, improves the geotechnical stability of waste and reduces the potential for subsequent waste settlement. Compaction may by itself increase the available moisture in the waste and thus increase degradation. However, leaving the first shallow layer of waste uncompacted for some months may provide some composting, degrading all the easily degradable organic matter, increasing the temperature of the waste, and establishing an inoculum of methanogens in anaerobic microenvironments. A firm compaction just before the next lift is installed may help in maintaining elevated temperatures in the bottom layer and the anaerobic acid phase starting immediately after the compaction will not be too vigorous, since most of the easily degradable waste is already decomposed. This gives methanogenesis a better chance to establish quickly. The active methanogenic bottom layer will later function as a treatment layer for acid phase leachate from the lifts above.

The addition of water and leachate recirculation are discussed later as part of the bioreactor and flushing-bioreactor technology.

### 10.6.1.4   Conventional Landfill: Gas

The amount of gas and its main constituents depend on the composition of the waste degraded and can be calculated as specified in Chapter 9.4. The majority of the gas is generated during the stable methanogenic phase. The gas generation rate for a real landfill can be estimated from the equations provided in Chapter 10.10. It should be emphasized that there are widespread concerns that the accuracy of these equations is poor. The overall amount of landfill gas generated is typically in the range of 100–200 m$^3$/t depending on type of waste and time frame considered. The actual amount of gas collected is typically lower, since it is neither practical nor economical to collect all gas.

An average landfill gas composition during methanogenesis is shown in Table 10.6.2. The trace components found in landfill gas originate from the waste landfilled, both directly and indirectly. Some (e.g. vinyl chloride) are formed *in situ* by microbial processes. Their concentrations depend on the release rate from the waste and the physicochemical characteristics of the substance. Kjeldsen and Christensen (2001) calculated for a range of organic substances that, when released from the waste, most of the chemical substances would leave the landfill with the landfill gas, rather than the leachate, because of their relative high volatility. The degradability of the organic trace substances was the main uncertainty in estimating their fate, partly because relatively few data are available and partly because the rate of degradation is thought to vary over a wide range among landfills. Kromann *et al.* (1998) showed that PCE (tetrachloroethylene) added to leachate from different landfills degraded at highly varying rates supposedly reflecting differences among the landfills. Figure 10.6.3 shows the fate of PCE over time after addition to four different leachates. The experiment also illustrates that volatile substances in the gas may originate from the degradation of other substances. This applies in this case of dichloroethylene (DCE) and vinyl chloride (VC). Of the trace substances in landfill gas, vinyl chloride and benzene are often considered the most critical from a health perspective because they are very volatile and highly toxic. Others are of concern for their deleterious impact on gas to energy systems, including siloxanes, hydrogen sulfide, and chlorinated hydrocarbons generally.

The development of gas composition over time is, for the main components, presented in Figure 10.6.2. Definitive data on the time dependence of trace component are not available, but most decrease significantly during the first few years

**Table 10.6.2** *Average landfill gas composition during methanogenesis (based on Deipser et al., 1996; Rettenberger and Stegman, 1996; NSCA, 2002; Scheutz et al., 2004; Scheutz and Kjeldsen, 2005).*

| Components | Concentration (mg/m$^3$) |
|---|---|
| Methane | 50–60% |
| Carbon dioxide | 40–50% |
| Arsenic | 0.03 |
| Benzene | 7 |
| Cadmium | 0.08 |
| Carbon monoxide | 0.01 |
| Carbon tetrachloride | 0.03 |
| CFC 11 | 10 |
| CFC 12 | 50 |
| CFC 114 | 4 |
| Chlorobenzene | 2 |
| Chloroform | 5 |
| Dichloroethylene | 5 |
| Dichloromethane | 50 |
| Ethyl benzene | 50 |
| HCFC 21 | 12 |
| HCFC 22 | 13 |
| Hydrogen chloride | 6 |
| Hydrogen fluoride | 2 |
| Hydrogen sulfide | 10 |
| Mercury | 0.004 |
| Nickel | 0.8 |
| Propylbenzene | 50 |
| Tetrachloroethylene | 27 |
| Toluene | 150 |
| Trichloroethylene | 15 |
| Vinyl chloride | 20 |
| Xylenes | 60 |
| Total VOCs | 200 |

unless new releases happen inside the landfill body because of corrosion of containers holding organic substances. Some highly volatile substances such as CFCs tend to decline more rapidly, whilst others such as BTEX compounds tend to be released over a much longer period.

### 10.6.1.5 Conventional Landfill: Leachate

Leachate is the drainage collected at the bottom of the landfill. Water percolating through the waste layers dissolves a range of substances reflecting the environment within the landfill body. Leachate usually contains a significant concentration of particulate matter, depending on how it is collected, and filtration is sometimes advisable before chemical analysis. Constituents of conventional landfill leachates may conveniently be classified into five main groups of substances:

- **Bulk organic matter:** expressed as chemical oxygen demand (COD, mg $O_2$/l), biochemical oxygen demand (BOD, mg $O_2$/l over 5 or 21 days), total organic carbon (TOC, mg C/l), or dissolved organic carbon (DOC, mg C/l on a filtered sample). The main components are VFAs and partly humified substances (fulvic-like and relatively short humic-like substances).

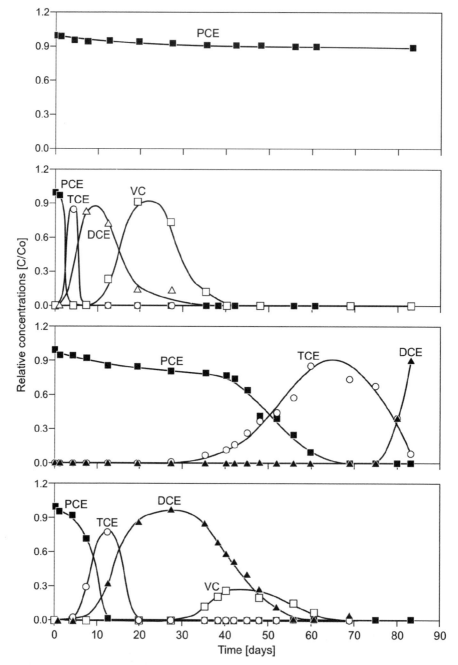

**Figure 10.6.3**  *Fate of PCE over time after addition to landfill leachate from four different landfills (Kromann et al., 1998).*

**Table 10.6.3**   *Typical conventional landfill leachate concentrations levels (based on Robinson, 2005; Kjeldsen and Christophersen, 2001; Reinhart and Grosh, 1998; Jones-Lee and Lee, 1993).*

| Concentration level | Acidic phase leachate | Methanogenic phase leachate | Old leachate (L/S > 3 l/kg) |
|---|---|---|---|
| > 10 g/l | BOD, COD | | |
| 1–10 g/l | $Cl^-$, $NH_4$-N | $Cl^-$, BOD, COD | |
| 100–1000 mg/l | Na, Ca, K | Na, Ca, (BOD), $NH_4$-N, K | $Cl^-$, Na, Ca, COD, $NH_4$-N |
| 10–100 mg/l | | | BOD, K |
| 1–10 mg/l | Zn, Si | Zn, Si | Si |
| 100–1000 µg/l | Cu, Mn, Mo | Mn, Mo | Zn, Mn |
| 10–100 µg/l | As, Pb, Cd, Ni, Se, Cr | As, Cu, Pb, Ni, Se, Cr | As, Cu, Ni, Cr, Mo |
| 1–10 µg/l | Hg, Sn | Cd, Sn | Pb, Cd, Se |
| < 1 µg/l | | Hg | Hg, Sn |

- **Nitrogenous compounds:** Organic nitrogen and ammonium ($NH_4^+$), both primarily originating from the degradation of proteins and to a lesser extent from detergents. These are together expressed as total Kjeldahl nitrogen (TKN).
- **Inorganic macrosubstances:** calcium (Ca), magnesium (Mg), sodium (Na), potassium (K), iron ($Fe^{2+}$), manganese ($Mn^{2+}$), chloride ($Cl^-$), sulfate ($SO_4^{2-}$), and bicarbonate ($HCO_3^-$).
- **Heavy metals and alike:** cadmium (Cd), lead (Pb), copper (Cu), nickel (Ni), zinc (Zn), chromium (Cr), mercury (Hg), arsenic (As), antimony (Sb), and tin (Sn), including their inorganic and organic complexes. Zinc and nickel are usually the most prevalent. All are usually at levels below mg/l and many are at levels of µg/l or lower.
- **Organic trace components:** originating from household hazardous waste and industrial waste. Aromatic hydrocarbons (BTEX, PAH), phenols, chlorinated solvents (e.g. PCE, TCE) and pesticides are often present in low concentrations, typically at µg/l or below µg/l levels.

Leachate composition varies according to the main phases of the life of the landfill and data are often provided separately for the acidic phase and the stable methanogenic phase. Kjeldsen *et al.* (2003) reviewed the short and long term composition of conventional landfill leachate. Table 10.6.3 presents typical levels of substances in conventional landfill leachate. Table 10.6.4 presents observations regarding trace organics in conventional landfill leachate.

**Table 10.6.4**   *Examples of trace organic concentrations in conventional landfill leachate during methanogenesis (based on Kjeldsen and Christophersen, 2001; Reinhart and Grosh, 1998; Jones-Lee and Lee, 1993).*

| Components | Concentration (µg/l) |
|---|---|
| 1,2-Dichlorobenzene | 2 |
| Benzene | 6 |
| Chlorobenzene | 3 |
| Chloroform | 0.3 |
| Ethylbenzene | 20 |
| Ethylene dichloride | 50 |
| Methylene Chloride | 20 |
| Napthalene | 20 |
| Phenolic compounds | 100 |
| Tetrachloroethylene | 10 |
| Toluene | 150 |
| Trichloroethylene | 6 |
| Vinyl chloride | 50 |
| Xylenes | 50 |

The difference in biochemistry between the acid phase and the methanogenic phase is clearly reflected in the composition of landfill leachate (Figure 10.6.2). Leachate from the acid phase is, relative to the methanogenic phase:

- Lower in pH.
- Higher in BOD and COD.
- Higher BOD/COD ratio.
- Higher in Fe and Ca.
- Higher in heavy metals.

Sulfate, magnesium, and manganese are often also higher during the acid phase. Sulfate-reduction does not take place unless the redox potential is very low. This reduction typically occurs during and after the transition to methanogenesis.

## 10.6.2   Bioreactor Landfill

The bioreactor landfill is a conventional landfill where leachate and/or other sources of moisture are introduced to accelerate the degradation of the organic waste. The aim is to obtain as rapid waste degradation and gas development as possible. The concept is based on the knowledge that waste has to be kept sufficiently wet to achieve a fast degradation. Waste moisture contents close to field capacity are considered by some to be optimal for promoting degradation (EPA, 2002). This is most often achieved through the recirculation of the collected leachate to the waste body. Leachate recirculation also provides better interaction between microorganisms, soluble nutrients and insoluble substrates, leading to an optimization of the degradation process (Barlaz *et al.*, 1990). Other sources of water may also be added to bring the landfill to optimum water content. Flushing has not so far been part of the design of bioreactor landfills (Benson *et al.*, 2007). The benefits of the bioreactor landfill are potentially (Benson *et al.*, 2007):

- Reduce the period needed for long term maintenance and monitoring.
- Increase rate of settlement providing more capacity or a more stable surface for final use of the site.
- Increase rate of gas generation improving the viability of gas utilization.
- Reduce leachate treatment cost because some treatment takes place within the landfill body.

Bioreactor landfills require a more sophisticated degree of management and monitoring than conventional landfills. Potential problems from recirculation or moisture addition, if sufficient care is not taken, include: increased odour and gas emissions, surface leachate breakouts, slope instability, and clogging of recirculation infrastructure (Knox *et al.*, 2007).

The implementation of bioreactor landfill systems in cold climates may be more challenging, as low temperatures drastically decrease the rate of waste degradation. Some operators have injected heated leachate to stimulate degradation rates in cold landfills. Dry climates may also be problematic for bioreactor landfills because with low precipitation, the leachate generation is not sufficient to sustain the recirculation operation and therefore huge volumes of supplementary liquids may be needed to keep the waste sufficiently wet.

Implementation of bioreactor technology can be started at the design stage or at a later stage by installing a leachate recirculation system in a conventional landfill section. Leachate is usually collected in a lagoon or tank from where it is injected either at the surface or via preinstalled infrastructure at depth.

The environmental performance achieved by bioreactor landfills practising leachate recirculation was assessed by means of a life cycle assessment (LCA) approach by Manfredi and Christensen (2009). In a 100-year perspective the environmental benefits obtained by leachate recirculation may not be evident since leachate recirculation does not directly reduce the cumulative amount of emissions to the environment from the landfilled waste. However, the emissions are concentrated during the time frame of leachate recirculation (8–10 years), during which the level of control over the emissions to the environment is the highest. Thereafter, the potential for emissions is low because the waste is already largely stabilized. The main benefit of the bioreactor landfill is that a major part of the degradation and the associated emissions take place early in the life of the landfill when the technical systems (leachate collection and treatment; gas collection and utilization) are at their most efficient. However, it must be borne in mind that gas collection efficiency may be low in recently deposited wastes until they are capped and abstraction infrastructure installed. It may be advisable in such circumstances to delay raising moisture contents until a suitable infrastructure is in place.

## 10.6.3  Flushing Bioreactor Landfill

The flushing bioreactor landfill is a bioreactor landfill where supplementary liquids are added to the waste mass and recirculated together with the leachate in order to flush out the soluble waste constituents (primarily ammonium, but also salts and hard COD) and, as for the bioreactor, to accelerate the waste degradation process. It should be noted that while there has been much theoretical discussion and some lysimeter-scale study of flushing bioreactors, there have been no full scale applications published.

A major factor in the operation of flushing landfills is the flushing rate. The cumulative amount of liquid passed through and removed from the waste mass per tonne of waste landfilled is known as the liquid/solid ($L/S$) ratio. The required magnitude of flushing rate is usually decided based on the average porosity of the waste landfilled, and the dilution in leachate concentrations that is needed. One typical estimate suggests that flushing by approximately five times the hydraulic bed volume of the wastes is required for a 100-fold reduction in concentration, corresponding to a $L/S$ ratio of $\sim 3\,m^3/t$ (Hupe *et al.*, 2003). As an example, a flushing rate of at least $2\,m^3/t$ waste is needed if the waste average porosity is 20 % (v/v). Another approach to determine the flushing rate is to consider single leachate constituents. For instance, COD concentrations may be significantly reduced with a flushing of only $1-2\,m^3/t$ waste, while an effective removal of $NH_3/NH_4^+$ may require $3-5\,m^3/t$ waste (Blakey *et al.*, 1997; Hupe *et al.*, 2003). These differences depend on the specific mechanism governing the release through leachate of a certain compound. As a general rule, for a given flushing rate, the efficiency of the flushing operation is high for those substances in leachate controlled by the availability of the substance. Their removal through leaching is proportional to the $L/S$ ratio ($m^3$ liquids/t waste) accounting for the amount of water that has percolated through the waste. This is the case for e.g. $NH_4$-N, COD, Na, Cl, Br, and K. However, the release through leachate of substances such as Ca, Mg, Al, $SO_4^{2-}$, and most heavy metals is mainly controlled by their solubility and therefore the flushing does not necessarily lead to proportionately decreasing concentrations of these substances. The leachate concentration of most heavy metals is largely controlled by their solubility which is in turn highly dependent on the pH in leachate, with little influence of the $L/S$ ratio and thus of the flushing rate.

The flushing landfill may employ nitrification combined with recirculation of the leachate as an active method to reduce $NH_3/NH_4^+$ concentrations. The nitrified leachate, when recirculated, denitrifies and the nitrogen is released as $N_2$ (see e.g. Knox and Gronow, 1995). In this way the need for addition of external sources of water can be reduced. A potential concern in flushing landfills would be if excessive flushing rates caused too vigorous an acid generation and therefore caused inhibition of methanogenesis. There has also been a concern that inhibition of methanogenesis may also come from the leachate denitrification operation due to the increased concentration of $NO_3^-$ in leachate causing a change of the redox condition suitable for methanogenesis. However, El-Marouki and Watson-Craik (2004) showed that the inhibitory effect of $NO_3^-$ on methanogenesis usually lasts for no longer than 45 days. Knox and Gronow (1995) showed that denitrification and methanogenesis both proceeded efficiently in the same $2\,m^3$ MSW reactor, possibly in separate zones or microenvironments.

Monitoring of moisture content and moisture distribution in the waste mass may be desirable during the flushing operation. Moisture monitoring may be achieved by means of techniques such as geoelectric sounding (Wens *et al.*, 2001) and measurements of waste resistivity (Grellier *et al.*, 2003; Moreau *et al.*, 2003).

## 10.6.4  Semiaerobic Bioreactor Landfill

The semiaerobic bioreactor landfill (or simply 'semiaerobic landfill'), developed in Japan, combines anaerobic and aerobic degradation processes (Matsufuji *et al.*, 2005). The technology has been applied primarily to relatively shallow landfills. The operational period is divided into two distinct phases symbolizing the two dominating degradation regimes. In the first phase, waste degradation is anaerobic and enhanced through leachate recirculation. The goal in this phase is to generate, collect and utilize as much gas as possible. This phase is kept active as long as the methane yield is high enough to justify gas utilization (typically 6–10 years), which mainly depends on quantity and composition of the landfilled waste. The second phase is initiated by injecting air from the bottom of the waste mass through the leachate collection pipes. This drastically stops the methane generation and leads to the formation of an aerobic environment that, once established, warms the waste to 60–70 °C. The upward moving air flow may then proceed autonomously thanks to the temperature gradient between the waste (warm) and the external environment (cold), by convection (Hanashima, 1999). The purpose of the aerobic phase is to rapidly stabilize the waste fractions that the anaerobic phase did not degrade. The air flow will naturally cease when the waste is fully stabilized. Anaerobic conditions may eventually reestablish but the waste does not then produce methane in quantities that can sustain any gas recovery scheme. During the aerobic phase leachate generation is significantly reduced due to the drying effect of the air flow. Matsufuji *et al.* (2005) reported halved leachate generation in a semiaerobic lysimeter experiment compared to the anaerobic case.

European application of aerobization of old landfills has developed to a significant extent and has found that active rather than passive introduction of air may be needed, the aeration rates being reduced as the rate of oxygen uptake by the remaining degradable wastes declines (Ritzkowski and Stegmann, 2007).

## 10.6.5   Comparison of Reactor Landfill Technologies

While full-scale experience is available for several conventional reactor landfills covering up to 35 years of the life of landfills, only limited time series data are available for the more advanced technologies. None at all are available for the flushing bioreactor. This makes comparison difficult. Manfredi and Christensen (2009) estimated typical values for the generation of gas and leachate from the various reactor landfill technologies and calculated the overall mass flow out of the landfills in a 100-year time period. For reasons of simplicity it was assumed that the landfills all contained household waste. Some of the data used were subject to major uncertainty because the data were extrapolations of laboratory and pilot-scale results. The results are shown in Table 10.6.5 for selected substances. The table also shows estimates of how

**Table 10.6.5**   *Estimated mass balances after 100 years for selected substances in 1 t of waste landfilled in different reactor landfills (based on Manfredi and Christensen, 2009).*

| | $C_{biogenic}$ | N | Cl | Pb | Zn | As | Cd | Cr | Hg |
|---|---|---|---|---|---|---|---|---|---|
| Unit (per tonne of wet waste) | | kg | | | g | | | mg | |
| Input through waste | 194.1 | 5.8 | 4.9 | 135.0 | 390.0 | 1610 | 6690 | 72 300 | 1090 |
| **Open dump** | | | | | | | | | |
| Output through gas | 74.0 | 0.0 | 0.0 | 0.0 | 0.0 | 5.1 | 13.6 | 0.0 | 0.6 |
| Output through leachate | 15.7 | 1.7 | 2.9 | 0.1 | 5.0 | 88.0 | 30.4 | 194.5 | 0.4 |
| Remaining (%) | 53.8 | 71.2 | 40.9 | 99.9 | 98.7 | 94.2 | 99.3 | 99.7 | 99.9 |
| **Conventional (with flares and with energy recovery)** | | | | | | | | | |
| Output through gas | 74.0 | 0.0 | 0.0 | 0.0 | 0.0 | 5.1 | 13.6 | 0.0 | 0.6 |
| Output through leachate | 8.4 | 0.8 | 2.6 | 0.1 | 2.5 | 41.6 | 14.5 | 92.0 | 0.4 |
| Remaining (%) | 57.5 | 86.3 | 46.4 | 99.9 | 99.4 | 97.1 | 99.6 | 99.9 | 99.9 |
| **Standard Bioreactor** | | | | | | | | | |
| Output through gas | 73.8 | 0.0 | 0.0 | 0.0 | 0.0 | 5.0 | 13.4 | 0.0 | 0.6 |
| Output through leachate | 8.4 | 0.8 | 2.5 | 0.04 | 2.4 | 40.6 | 14.1 | 90.2 | 0.4 |
| Remaining (%) | 57.6 | 86.7 | 48.5 | 99.8 | 99.4 | 97.2 | 99.6 | 99.9 | 99.9 |
| **Flushing Bioreactor** | | | | | | | | | |
| Output through gas | 73.8 | 0.0 | 0.0 | 0.0 | 0.0 | 5.0 | 13.4 | 0.0 | 0.6 |
| Output through leachate | 22.6 | 1.1 | 3.7 | 0.1 | 7.4 | 111.8 | 41.8 | 264.6 | 1.1 |
| Output through nitrification | 0.0 | 1.1 | 0.0 | 0.0 | 0.0 | 0.0 | 0.0 | 0.0 | 0.0 |
| Remaining (%) | 50.3 | 61.8 | 23.2 | 99.9 | 98.1 | 92.7 | 99.2 | 99.6 | 99.8 |
| **Semiaerobic** | | | | | | | | | |
| Output through gas | 75.9 | 0.0 | 0.0 | 0.0 | 0.0 | 5.4 | 14.4 | 0.0 | 0.6 |
| Output through leachate | 6.5 | 0.8 | 2.7 | 0.1 | 2.5 | 45.0 | 15.4 | 99.0 | 0.5 |
| Remaining (%) | 57.5 | 85.4 | 44.1 | 99.9 | 99.4 | 96.9 | 99.6 | 99.9 | 99.9 |

much of a substance remains in the landfill after 100 years. Typically, less than 1 % of the heavy metals have been leached and of the most soluble substances (e.g. Cl) nearly half still remains in the landfill. Only 15 % of the expected content of N in the waste has been removed, except in the flushing bioreactor which is expected to remove about 40 % of N during 100 years. The flushing bioreactor seems to remove substantially more N and salts from the landfill than the other technologies, as would be expected, but besides this the technologies show little difference in overall behaviour during 100 years. However, Manfredi and Christensen (2009) also assessed the distribution of the environmental emissions over the period, and here the reactor landfill technologies showed some differences. The benefit of the bioreactor landfill and the flushing bioreactor landfill is that the majority of the emissions take place early, when the technical measures installed are likely to be better performing than later in the life of the landfill.

An aspect of concern with all types of reactor landfills is that even after application of acceleration methods, evidence emerging from pilot and field studies suggests that degradation rates may fall inevitably to very low levels even while a considerable portion (30–50 %) of the degradable matter still remains in the waste (Knox *et al.*, 2005). This may be due to cellulose being biologically inaccessible to bacteria, due to, for example, a lignin coating. The implication is that there could be a very long 'tail' during which degradation continues at rates high enough to require continued active control of leachate and gas but too low for the gas to be of economic value. This 'tail' is likely to delay for many decades the achievement of final storage quality, i.e. the point at which all active environmental controls can be stopped.

# References

Barlaz, M.A., Ham, R. and Schaefer, D. (1990): Methane production from municipal refuse: a review of enhancement techniques and microbial dynamics. *Critical Reviews in Environmental Control*, 19 (6), 557–584.

Benson, C.H., Barlaz, M.A., Lane, D.T. and Rawe J.M. (2007): Practice review of five bioreactor/recirculation landfills. *Waste Management*, 27, 13–29.

Blakey, N.C., Bradshaw, K., Reynolds, P. and Knox, K. (1997): Bio-reactor landfill – A field trial of accelerated waste stabilisation. *Proceedings of the International Landfill Symposium*, 6 (1), 375–386.

Christensen, T.H. and Kjeldsen, P. (1995): Landfill emissions and environmental impact: An introduction. *Proceedings of the International Landfill Symposium*, 5 (3), 3–12.

Christensen, T.H., Kjeldsen, P. and Stegmann, R. (1992): *Effects of landfill management procedures on landfill stabilization and leachate and gas quality.* Chapter 2.7. In: Christensen, T.H., Cossu, R. and Stegmann, R. (eds) Landfilling of waste: Leachate. Elsevier Applied Science, London, UK, pp. 119–137.

Christensen, T.H., Kjeldsen, P. and Lindhardt, B. (1996): *Gas-generating processes in landfills.* Chapter 2.1. In: Christensen, T.H., Cossu, R. and Stegmann, R. (eds) *Landfilling of waste: Biogas.* E&FN Spon, London, UK, pp. 270–50.

Deipser, A., Poller, T. and Stegmann, R. (1996): *Emissions of volatile halogenated hydrocarbons from landfills.* In: Christensen, T.H., Cossu, R. and Stegmann, R. (eds) Landfilling of waste: biogas. Elsevier, London, UK, pp 59–71.

El-Mahrouki, I.M. and Watson-Craik, I.A. (2004): The effects of nitrate and nitrate-supplemented leachate addition on methanogenesis from municipal solid waste. *Journal of Chemical Technology and Biotechnology*, 8, 842–850.

EPA (2002): *State of practice for bioreactor landfills.* Workshop on bioreactor landfills. National Risk Management Research Laboratory Office of Research and Development, US Environmental Protection Agency, Cincinnati, USA.

Fairweather, R.J. and Barlaz, M.A. (1998): Hydrogen sulfide production during decomposition of landfill inputs. *Journal of Environmental Engineering, ASCE*, 124, 353–361.

Grellier, S., Duquennoi, C., Guerin, R., Munoz, M.C. and Ramon, M.C. (2003): *Leachate recirculation study of two techniques by geophysical survey.* In: Christensen, T. H., Cossu, R. & Stegmann, R. (eds.): Sardinia 2003, Ninth International Waste Management and Landfill Symposium, 6–10 October, Sardinia, Italy. Proceedings. CD-ROM, CISA – Environmental Sanitary Engineering Centre, Cagliari, Italy.

Hanashima, M. (1999): Pollution control and stabilization process by semiaerobic landfill type: the Fukuoka method. *Proceedings of the International Waste Management and Landfill Symposium*, 7 (1), 313–325.

Hansson, G. and Molin, N. (1981): End product inhibition in methane fermentation: effects of carbon dioxide and methane on methanogenic bacteria utilizing acetate. *European Journal of Applied Microbiology and Biotechnology.* 13, 136–241.

Hupe, K., Heyer, K. and Stegmann, R. (2003): *Water infiltration for enhanced in situ stabilization.* In: Christensen, T. H., Cossu, R. & Stegmann, R. (eds.): Sardinia 2003, Ninth International Waste Management and Landfill Symposium, 6–10 October, Sardinia, Italy. Proceedings. CD-ROM, CISA – Environmental Sanitary Engineering Centre, Cagliari, Italy.

Jones-Lee, A. and Lee, F. (1993): *Groundwater pollution by MSW landfills: leachate composition, detection and water quality significance.* In: Christensen, T. H., Cossu, R. & Stegmann, R. (eds.): Sardinia '93, Fourth International Landfill Symposium, S. Margherita di Pula, Italy, 11–15 October, Proceedings, Vol. II, pp. 1093–1103. CISA – Environmental Sanitary Engineering Centre, Cagliari, Italy.

Kjeldsen, P. and Christensen, T.H. (2001): A simple model for the distribution and fate of organic chemicals in a landfill: MOCLA. *Waste Management and Research*, 19, 201–216.

Kjeldsen, P. and Christophersen, M. (2001): Composition of leachate from old landfills in Denmark. *Waste Management and Research*, 19, 249–256.

Kjeldsen, P., Barlaz, M.A., Rooker, A., Baun, A., Ledin, A. and Christensen, T.H. (2003): Present and long term composition of MSW landfill leachate – a review. *Critical Reviews in Environmental Science and Technology*, 32, 297–336.

Knox, K. and Gronow, G. (1995): A pilot scale study of denitrification and contaminant flushing during prolonged leachate recirculation. *Proceedings of the Waste Management and Landfill Symposium*, 5 (1), 419–436.

Knox, K., Braithwaite, P., Caine, M. and Croft, B. (2005): *Brogborough landfill test cells: the final chapter – a study of landfill completion in relation to final storage quality (FSQ) criteria.* In: Cossu, R. & Stegmann, R. (eds.): Sardinia 2005, Tenth International Waste Management and Landfill Symposium, 3–7 October, Sardinia, Italy. Proceedings. CD-ROM, CISA – Environmental Sanitary Engineering Centre, Cagliari, Italy.

Knox, K., Beaven, R.P., Rosevear, A. and Braithwaite, P. (2007): *A technical review of leachate recirculation.* In: Cossu, R., Diaz, L. F. & Stegmann, R. (eds.): Sardinia 2007, Eleventh International Waste Management and Landfill Symposium, 1–5 October, Sardinia, Italy. Proceedings. CD-ROM, CISA – Environmental Sanitary Engineering Centre, Cagliari, Italy.

Kromann, A., Ludvigsen, L., Albrechtsen, H.-J., Christensen, T.H., Ejlertsson, J. and Svensson, B.H. (1998): Degradability of chlorinated aliphatic compounds in methanogenic leachates sampled at eight landfills. *Waste Management and Research*, 16 (1), 54–62.

Manfredi, S. and Christensen, T.H. (2009): Environmental assessment of solid waste landfilling technologies by means of LCA-modeling. *Waste Management*, 29, 32–43.

Matsufuji, Y., Tachifuji, A. and Matsugu, H. (2005): *Biodegradation process of municipal solid waste by semiaerobic landfill type.* In: Cossu, R. & Stegmann, R. (eds.): Sardinia 2005, Tenth International Waste Management and Landfill Symposium, 3–7 October, Sardinia, Italy. Proceedings. CD-ROM, CISA – Environmental Sanitary Engineering Centre, Cagliari, Italy.

Moreau, S., Bouye, J.M., Barina, G. and Oberti, O. (2003): *Electrical resistivity survey to investigate the influence of leachate recirculation in a MSW landfill.* In: Christensen, T. H., Cossu, R. & Stegmann, R. (eds.): Sardinia 2003, Ninth International Waste Management and Landfill Symposium, 6–10 October, Sardinia, Italy. Proceedings. CD-ROM, CISA – Environmental Sanitary Engineering Centre, Cagliari, Italy.

NSCA (2002): *Comparison of emissions from waste management options.* National Society for Clean Air and Environmental Protection, Brighton, UK.

Reinhart, D.R. and Grosh C.J. (1998): Analysis of Florida MSW landfill leachate quality. Report No. 97-3. Civil and Environmental Engineering Department, University of Central Florida/Florida Center for Solid and Hazardous Waste Management, Miami, USA.

Rettenberger, G. (2005): Landfill gas characterization over time – the 9-phase model. In: Cossu, R. and Stegmann, R. (eds) *Proceedings of the Tenth International Landfill Symposium.* CD-ROM. CISA – Environmental Sanitary Engineering Centre, Cagliari, Italy.

Rettenberger, G. and Stegmann, R. (1996): *Landfill gas components.* In: Christensen T. H., Cossu R. and Stegmann R. (eds) Landfilling of waste: biogas. Elsevier, London, UK.

Ritzkowski, R. and Stegmann, S. (2007): *Biostabilizing a MSW landfill by means of in situ aeration – results of a 8 year project*. In: Cossu, R., Diaz, L. F. & Stegmann, R. (eds.): Sardinia 2007, Eleventh International Waste Management and Landfill Symposium, 1–5 October, Sardinia, Italy. Proceedings. CD-ROM, CISA – Environmental Sanitary Engineering Centre, Cagliari, Italy.

Robinson, H. (2005): The composition of leachates from very large landfills. In: Cossu, R. and Stegmann, R. (eds) *Proceedings of the Tenth International Landfill Symposium*. CD-ROM. CISA – Environmental Sanitary Engineering Centre, Cagliari, Italy.

Scheutz, C. and Kjeldsen, P. (2005): Biodegradation of trace gases in simulated landfill soil covers. *Journal of Environmental Quality*, 37, 5143–5149.

Scheutz, C., Mosbæk, H. and Kjeldsen, P. (2004): Attenuation of methane and volatile organic compounds in landfill soil covers. *Environmental Science and Technology*, 33, 61–77.

Wens, P., Vercauteren, T. and Verstraete W. (2001): *Factors inhibiting anaerobic degradation in a landfill*. In: Christensen, T. H., Cossu, R. & Stegmann, R. (eds.): Sardinia 2001, Eight International Waste Management and Landfill Symposium, 1–5 October, Sardinia, Italy. Proceedings. Vol. I. The Sustainable Landfill, pp. 13-20. CISA – Environmental Sanitary Engineering Centre, Cagliari, Italy.

# 10.7

# Landfilling: MBP Waste Landfills

**Rainer Stegmann**

*Technical University of Hamburg-Harburg, Germany*

Emissions from MSW landfills may last, as discussed in Chapter 10.1, for decades, maybe even centuries. It is expected that landfill gas will be produced in significant amounts for more than 30 years, while leachate probably must be treated for 100 to 200 years to meet, for example, the German target values for discharge to surface water (Heyer and Stegmann, 1997). These emissions mean that treatment systems and maintenance must be active long after closure of the landfill. For this reason closed landfills are a major problem since high aftercare costs evolve and environmental risks exist. In many countries, the law requires that a significant part of the disposal fee must be saved to cover the aftercare cost.

Leachate and biogas emissions from a MSW landfill are to a very high degree influenced by the degradation of the organic fraction of the landfilled waste. In addition, the degradation of the organic matter causes the landfill surface to settle (maybe as much as 20–25 % of the depth of the landfill). This may damage technical installations such as surface liners, gas extraction and leachate collection systems. By partially degrading the organic matter prior to landfilling, as done in a mechanical biological pretreatment plant, many of the problems known to MSW landfills can be reduced significantly: emissions are reduced, settlements are reduced and the aftercare period shortened.

The landfilling of mechanical biological pretreatment (MBP) waste is a relatively new technology and reported experiences are few. The introduction of the European Union (EU) landfill directive (Council Directive 99/31/EC), reducing the amount of organic waste going to landfills (See Box 10.7.1) has been a driving force in developing the MBP concept, and as the technology is being implemented in several countries in Europe, the database on experiences with BMP waste landfills will develop in the years to come.

This chapter describes briefly MBP and provides information on how leachate emissions and gas production are reduced when the MBP waste afterwards is landfilled. The chapter also addresses how design and operation of MBP landfills differ from ordinary MSW landfills.

## 10.7.1 Mechanical Biological Pretreatment

The pretreatment of waste prior to landfilling usually includes both a mechanical part, which has the purpose to remove valuable fractions and prepare the waste for further treatment, and a biological part, which has the purpose to degrade

*Solid Waste Technology & Management*   Edited by Thomas Christensen
© 2011 Blackwell Publishing Ltd

---

**Box 10.7.1    EU Regulation Fostered the MBP Landfill.**

The EU introduced in 1999 political targets for the reduction in organic waste being landfilled (Anonymous, 1999). In three steps the amount of organic waste being landfilled should be reduced by 25 % (at the latest in 2006), 50 % (at the latest by 2009) and 65 % (at the latest by 2016). The basis for these reductions is the MSW composition in the year 1995. Those countries that in the year 1995 had landfilled more than 80 % of their MSW were granted another 4 years to fulfill these targets (Anonymous, 1999).

The reduction of 25–65 % of the biologically degradable MSW fraction as prescribed in the EU Directive is not easy to achieve. This goal can in general be reached by means of MSW incineration, by separate collection, treatment and utilisation of biowaste and/or paper as well as by mechanical biological pretreatment of MSW or a combination of these possibilities. The introduction of these reduction targets was in EU the start of landfills receiving only mechanically biologically pretreated waste.

---

the easily degradable organic fraction. The biological processes may be composting, anaerobic digestion or anaerobic digestion followed by composting. Both the mechanical and the biological pretreatment release odorous and potentially toxic components to the air, and modern mechanical biological pretreatment plants most often require off gas treatment.

### 10.7.1.1    Mechanical Pretreatment

Mechanical pretreatment usually involves shredding, size screening, magnetic iron removal and some kind of man-operated sorting. Mechanical pretreatment provides an opportunity, prior to landfilling, to remove waste fractions, which have a value high enough to support their removal. The fractions to consider are primarily a refuse-derived fuel (RDF) fraction and magnetic iron.

The separation of a highly calorific, light fraction, to be used as RDF, can be achieved by sieving the waste using screen sizes between 40 and 100 mm. In many cases rotating trommel screens are used. As an example, the grain size distribution of unshredded MSW shown in Figure 10.7.1 suggests that, with a mesh size of 80 mm, over 90 % of the biodegradable fraction will report to the screen underflow and nearly 70 % of the materials with a high calorific value (like paper and plastics) will report to the screen overflow. The RDF fraction may in Germany account for about 15–20 % of the MSW input.

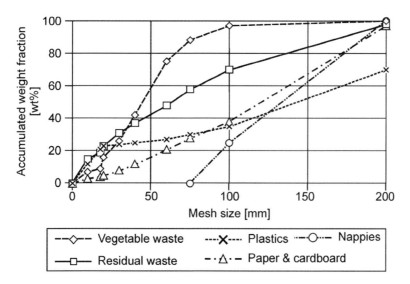

**Figure 10.7.1**    *Grain size distribution of MSW (Leikam and Stegmann, 1996).*

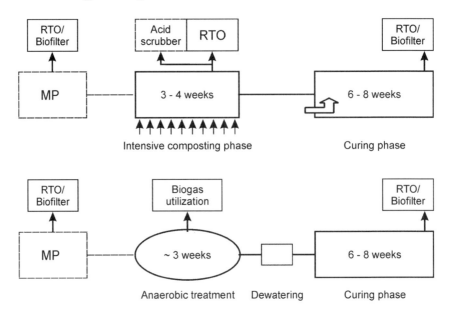

**Figure 10.7.2**   *Mechanical biological treatment concepts for municipal solid waste: aerobic and anaerobic treatment including off gas emission control.*

If the metal content of the waste is significant, a magnet can be used to remove iron metal. Potentially also aluminium may be removed mechanically.

Mechanical pretreatment may also involve the removal of foreign items in the waste. This is most often a man-operated activity (manually or by man-operated picking equipment).

Shredding may be involved depending on the waste heterogeneity and the subsequent biological process performed.

### 10.7.1.2   Biological Pretreatment

The technological approaches to biological pretreatment are illustrated in Figure 10.7.2.

After mechanical biological pretreatment the volatile solid content (VS is an expression of the organic matter content) of the waste to be landfilled amounts typically in Germany to 20–40 % by weight depending on the waste composition and the mechanical separation procedure. However, the MBP waste to be landfilled must meet a set of criteria, as shown in Table 10.7.1, which relates to total content, biological activity and availability for leaching. The criteria of a 96-h respiration rate less than 5 mg $O_2$/g dry matter is usually the most demanding quality criterion.

**Table 10.7.1**   *Selection of target values for the landfilling of mechanically biologically pretreated municipal solid waste in Germany (Anonymous, 2001).*

| Parameter | Target value |
|---|---|
| Respiration activity during 4 days | $\leq$ 5 mg $O_2$/g dry mass |
| Gas formation potential during 21 days | $\leq$ 20 N ml/g dry mass |
| $TOC_{eluat}$ at L/S 10 l/kg for 24 h | $\leq$ 250 mg C/l |
| $TOC_{solid}$ | $\leq$ 18 wt% |
| Gross calorific value | $\leq$ 6000 kJ/kg |

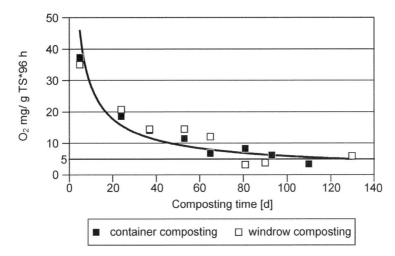

**Figure 10.7.3**   *Respiration rates of MSW samples after different periods of composting (Leikam and Stegmann, 1996). d: Days.*

***Composting:*** In the composting process, the organic waste components are degraded under the release of heat to carbon dioxide, water and biomass. High odour concentrations are emitted during the initial intensive degradation phase. In addition, during the first weeks of composting volatile componens, including hazardous chemicals like benzene, toluene and xylene are emitted into the gas phase. The intensive phase of the composting process (approximately 3–4 weeks) must be practised in a closed system with a forced aeration system and off gas treatment. The curing phase (about 6–8 weeks) has to be operated in-house or under a roof depending on the degree of maturation reached. In total a composting period of about 16 weeks or longer is necessary. More details about the biological pretreatment technology are presented elsewhere (Fricke *et al.*, 2001).

The progression of the composting process with time is expressed in Figure 10.7.3 in terms of the respiration rate of mechanically pretreated waste (Leikam and Stegmann, 1996). The respiration rate is the oxygen consumption per mass unit of dry matter at optimal conditions during 96 h (see Chapter 2.1). The actual waste sample was German MSW that was sieved in a rotating trommel screen (80 mm screen) and thereafter the screen undersize was aerobically treated in a composting windrow which was turned over weekly. For the respiration test the waste was sieved again and a particle size less than 20 mm was used in the respiration test. The samples were adjusted to a water content of 50 %.

***Anaerobic digestion:*** After an intensive mechanical pretreatment (removal or RDF, metals, sand, etc.) including shredding and in many cases water addition, also anaerobic biological treatment is effective. During anaerobic pretreatment the organic waste residues are converted into biogas and a digestion residue. The anaerobic pretreatment has several advantages compared to aerobic treatment, e.g. saving energy due to avoidance of forced aeration in the intensive phase of composting, net gain of energy from biogas production as well as a less odorous off gas. As a result the removal of odours and other compounds to meet the local off gas criteria is less costly (Table 10.7.2 shows the off gas criteria in force in Germany). Anaerobic digestion should always be combined with a postcomposting step, since not all organic substances (e.g. lignin-containing components) can be degraded under anaerobic conditions to the degree required to meet the national regulations (see Table 10.7.1 for the German regulation). In addition, the digest is in a reduced stage and should be converted into the oxidised form. The anaerobic treatment step often takes about 3 weeks, followed by a postcomposting step lasting about 4–6 weeks.

### 10.7.1.3   Off Gas Treatment

In order to reduce external odour problems it is often required that the unloading of the waste as well as the mechanical and biological pretreatment are practised inhouse. This is, for instance, the case of Germany. Consequently, all buildings

**Table 10.7.2**   *German target values for off gases from MBP plants (Anonymous, 2001).*

| Parameter | Target value |
|---|---|
| TOC mg/Nm$^3$ | 20/40 |
| TOC g/t | 55 |
| Dust mg/Nm$^3$ | 10/30 |
| Odour (odour units)/m$^2$ | 500 |
| PCDD/F ng/Nm$^3$ | 0.1 |
| N$_2$O g/t | 100 |

have to be ventilated and the off gases have to be captured and treated. Off gas from the unloading and mechanical pretreatment area in general can be treated in biofilters, which may be combined with a wet acid scrubber for NH$_3$ removal. The off gas from the composting, especially from the intensive first degradation phase, must often be treated thermally (e.g. noncatalytic incineration) for the removal of the organic compounds and probably also by acid scrubbing for the reduction of NH$_3$ (see also Doedens, 2001). The off gas treatment contributes significantly to the costs of the entire MBP plant. The strict German off gas target values, according to the authorities, are argued by the fact that off gas criteria for MBP plants should be equally strict as for incineration plants, since both treatment processes are seen as a pretreatment before landfilling.

## 10.7.2   Landfilling of MBP Waste

### 10.7.2.1   Landfill Emissions

To describe the landfill behaviour of MBP waste, landfill simulation experiments have been carried out in the laboratory. The test system ensures that typical landfill phases, in particular the acid phase and the stable methane phase, take place in the reactors. By choosing appropriate experimental conditions, enhanced biological degradation is achieved, and within a reasonable period of time the emissions and the emission potentials of the waste sample can be assessed in terms of the gas quality and quantity as well as the leachate concentrations and amount leached. As an example, cumulative leachate emissions (Figure 10.7.4) and gas production (Figure 10.7.5) determined in laboratory experiments with untreated as well as mechanically biologically treated MSW (high calorific value fraction was first removed, followed by four months of composting) are discussed below.

MBP waste typically has a significantly lower permeability than untreated waste. This results in a lower leachate generation in MBP landfill. Typical leachate generations in untreated waste landfills range from 25 to 50 % of precipitation (Plinke *et al.*, 2000; Wallmann, 1999), whereas range from 9 to 13 % of precipitation in MBP landfill (Rettenberger and Fricke, 1998). In addition, the acidogenic phase is eliminated or is very weak when MBP waste is landfilled. As a result, the generated leachate has a pH of at least 7.5–8.0, even in the initial degradation phases (Binner, 2002). Comparing the leachate emissions at an *L/S* value of 1.0 l/kg (i.e. the amount of dry landfilled waste was brought into contact with the same amount of water which in full scale corresponds to a landfill 'life' of about 50 years respecting northern European climatic conditions), the total emissions were with respect to COD and nitrogen up to 90 % lower for MBP waste than for untreated MSW. As a result, the period of aftercare for a landfill where MBP waste has been landfilled, is considerably shorter, especially due to the reduction of the nitrogen load.

Comparing the landfill gas production of untreated and pretreated residual municipal solid waste in the landfill simulation tests, Figure 10.7.5 clearly shows that the biological pretreatment reduces the landfill gas emission potential significantly. The duration of the pretreatment phase has a strong influence on the potential for gas generation. The untreated residual waste showed a landfill gas production around 200 l gas/kg dry matter, while after 4 months of pretreatment the gas generation potential was reduced by 75–90 %.

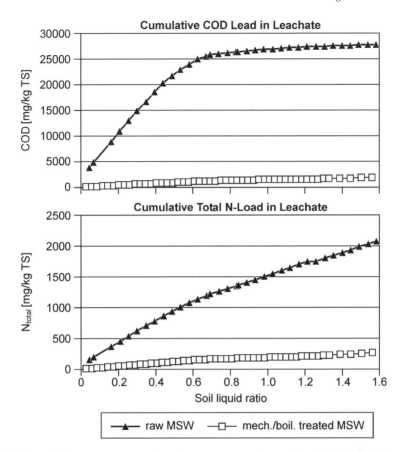

**Figure 10.7.4** *COD and nitrogen concentrations for an untreated as well as biologically (4 months) pretreated RMSW results from landfill simulation tests.*

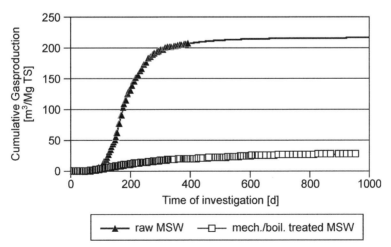

**Figure 10.7.5** *Cumulative gas production of untreated and biologically pretreated (4 months) MSW results from landfill simulation tests.*

**Table 10.7.3**    *Effects of mechanical biological pretreatment (MBP) on landfill emissions based on laboratory scale tests (Stegmann and Heyer, 2001).*

| Emission | Raw waste | MBP waste | Reduction (%) |
|---|---|---|---|
| COD (mg/kg TS) | 25 000–40 000 | 1000–3000 | 90 |
| Total Nitrogen (mg/kg TS) | 1500–3000 | 150–300 | 90 |
| Gas production, 21 days (l/kg TS) | 150–200 | 0–20 | 90 |

A summary of the differences of the emissions from landfills without or with mechanical biological pretreatment is presented in Tables 10.7.3 to 10.7.5.

## 10.7.2.2    Landfilling Characteristics of MBP Waste

The landfilling characteristics of MBP waste are in many ways different from ordinary MSW. The main differences are shortly described below:

- The MBP waste has a low content in structural material, which reduces the mechanical stability of the landfilled waste. This is due to the separation of the RDF fraction (see also Ziehmann *et al.*, 2002).
- The emission potential of the MBP-treated waste is reduced by approximately 90 % (Leikam and Stegmann, 1996).
- The gas formation potential is low. According to current knowledge an active gas extraction will not be necessary; however, the residual amount of produced gas needs to be biologically oxidised before leaving the landfill.

**Table 10.7.4**    *Average leachate composition (mg/l) in MBP landfills and in conventional MSW landfills (Binner, 2003; Bone et al., 2003; Ehrig, 1983; Kjeldsen and Christophersen, 2001; Lee and Jones, 1993; Oberstainer et al., 2007; Reinhart and Grosh, 1998).*

| Substance | Young leachate MBP (years 1–5) | Young leachate Conventional (acidogenic) | Mature leachate MBP (years 6–30) | Mature leachate Conventional (methanogenic) | Old leachate MBP (years 31–100) | Old leachate Conventional (postmethanogenic) |
|---|---|---|---|---|---|---|
| pH | 7.5–8.5 | 6 | 8.0–8.8 | 7 | 7.7 | 8 |
| TOC | 20 | 5000 | 10–14 | 600 | 500 | 500 |
| $BOD_5$ | < 1000 | 12 000 | 10–350 | 1000 | 5 | 100 |
| COD | < 6000 | 15 000 | 200–2000 | 3000 | 1200 | 500 |
| $NH_4$-N | 30–290 | 1 000 | 60–140 | 600 | 30 | 400 |
| $NO_2$-N | 1–10 | 0.1 | 1–8 | 0.7 | 0.84 | 0.8 |
| $NO_3$-N | 30–40 | 4 | 30–40 | 10 | 10 | 80 |
| $SO_4$ | 20–900 | 98 | 200–950 | 170 | 80 | 80 |
| Fe | 1–10 | 50 | 1–7 | 22 | 2 | 13 |
| Zn | 0.5–2.5 | 4 | 0.1–0.5 | 2.5 | 0.54 | 1 |
| Na | 300–600 | 700 | 300–580 | 500 | 440 | 200 |
| K | 200–500 | 1220 | 150–400 | 800 | 275 | 600 |
| Ca | 80–350 | 1 000 | 75–85 | 1000 | 95 | 1000 |
| Mn | 0.3–2.0 | 4 | 0.1–0.5 | 2 | 0.4 | 1 |
| Pb | 0.05–0.2 | 0.1 | 0.02–0.1 | 0.05 | 0.03 | 0.01 |
| Cu | 0.04–0.15 | 0.12 | 0.05–0.5 | 0.1 | 0.04 | 0.07 |
| Ni | 0.01–0.4 | 0.07 | 0.08–0.3 | 0.05 | 0.12 | 0.04 |
| Cr | 0.05–0.5 | 0.07 | 0.2–0.5 | 0.05 | 0.18 | 0.04 |
| Hg | < 0.01 | 0.0004 | < 0.01 | 0.0003 | < 0.01 | 0.0002 |

**Table 10.7.5** *Average gas emission per tonne wet waste in MBP landfills and in conventional MSW landfills (Deipser et al., 1996; NSCA, 2002; Oberstainer et al., 2007; Rettenberger and Stegmann, 1996).*

| Substance | Young biogas | | Mature biogas | | Old biogas | |
|---|---|---|---|---|---|---|
| | MBP (years 1–5) | Conventional (acidogenic) | MBP (years 6–30) | Conventional (methanogenic) | MBP (years 31–100) | Conventional (postmethanogenic) |
| $CH_4$ ($m^3$) | 2.8 (60%) | 9 (30%) | 5.6 (60%) | 54 (60%) | 0.9 (60%) | 1 (10%) |
| $CO_2$ ($m^3$) | 1.9 (40%) | 21 (70%) | 3.7 (40%) | 36 (40%) | 0.6 (40%) | 3 (30%) |
| Benzene (mg) | 2 | 980 | 4 | 980 | 0.8 | 980 |
| Toluene (mg) | 4 | 28000 | 8 | 28000 | 2 | 28000 |
| Dichloromethane (mg) | 0.1 | 7000 | 0.2 | 7000 | 0.05 | 7000 |
| Trichloromethane (mg) | 0.004 | 6600 | 0.008 | 200 | 0.002 | 7 |
| Tetracholomethane (mg) | 0.002 | 16 | 0.003 | 10 | 0.001 | 2 |
| Trichloroethene (mg) | 0.03 | 2800 | 0.07 | 2800 | 0.01 | 2800 |
| Tetrachloroehtene (mg) | 0.02 | 4000 | 0.04 | 4000 | 0.006 | 4000 |

- The MBP waste can be landfilled with high density (about $1.5\,t/m^3$ w/w), which may increase with an increasing waste load on top. This results in a low permeability (around $10^{-8}$ m/s). For this reason a low leachate production and a high amount of surface water runoff can be expected. This is also dependent on the kind of compaction (for example, compacted by a compactor or a roller) and the characteristics of the landfill surface and its slope (Maak, 2001).
- The water content of the waste during landfilling and the kind of landfill operation can affect the stability of the landfill. If for instance the pore gas pressure and/or the pore water pressure are too high, landfill stability may be reduced (Stegmann and Heyer, 2001).

### 10.7.2.3 Design and Operation of a MBP Landfill

Today there are only limited experiences regarding the operation of MBP waste landfills. The conditions are not yet optimised especially regarding the installation of optimum water content of the MBP waste and adequate emplacement and compaction technique including the achievement of an even final surface structure. The results from investigations from operational MBP landfills may therefore not yet be representative.

First results on a MBP waste landfill were presented by Doedens (2004). In this specific case the MBP waste was not landfilled at optimum water content and the waste surface was not flattened. Water was soaked up in the upper part of the landfilled waste and the main compaction took place due to the load of the landfilled waste. As already mentioned, in this case high water pore pressures in the landfill can be expected. As a result, mechanical stability of the landfill may not be achieved and uncontrolled leachate migration out of the landfill slopes can be expected. Results from a controlled test field, where MBP waste at optimum water content had been landfilled, were different (Maak, 2001). In this case a low permeability of the landfilled MBP waste and a plane landfill surface with a significant slope had been achieved. The rainwater infiltrated only to a very low extent into the landfill. Therefore, more than 90% of the rainwater was collected as surface water. It should be aimed at that the surface water runoff does not get in contact with the waste in order to prevent surface water contamination.

Based on the present experiences the following procedures for the minimisation of leachate and gas production from a MBP landfill may be effective:

- In order to achieve a low permeability the waste must be carefully compacted in thin layers, i.e. by means of compactor and a subsequent compaction by a roller (Maak, 2001). Using a roller as a final compaction step, a plane surface with adequate slopes can be achieved, resulting in high surface water runoff. Experiences with this technology have been gained only from test fields.

- The operational face should be covered, for example, with plastic membranes at the end of the day (see also Stegmann and Heyer, 2002). In a similar way the areas that are not in operation and not yet recultivated should be intermittently covered. In this way the production of polluted surface water can be minimised.

- Construction and operation of a surface water collection system are necessary, where up to 100 % of the precipitation can be expected as surface runoff. The plastic membranes used for temporary cover should be reused as far as possible after resuming operation of those areas.

- In addition to other top covers and lining systems, a final top cover consisting of at least 1.5 m soil with a high water storage capacity and a capillary barrier must be installed with a slope > 8° (see also Stegmann and Heyer, 2002). It is necessary to collect the surface water as well as the water from the capillary barrier and to divert them from the landfill body. This water has to be collected in ditches outside the landfill or its quality has to be controlled. Treatment of the surface water that had contact with the waste surface may be necessary. A significant pollution reduction can be achieved by settling of the soil particles (Ziehmann *et al.*, 2002).

- Dependent on the kind of landfill operation, low amounts of leachate may still be observed. For example in Germany around 2–10 % of the yearly precipitation rate can be expected. Using common waste compactors without water content optimisation, the leachate production may even be higher. The leachate is collected at the base and diverted through the drainage system from the landfill body (Stegmann and Heyer, 2002). The leachate has concentrations approximately corresponding to leachate known from landfills in the methanogenic stage, but with significantly lower concentrations of ammonia (only about 10 %; Leikam and Stegmann, 1996). First results from full scale MBP landfills are presented in Table 10.7.5 (Doedens, 2002).

- The leachate should be treated, if possible, together with municipal wastewater in sewage treatment plants or treated separately. Due to a low content of biological degradable substances in the leachate, activated carbon absorption and membrane separation processes are suitable for a treatment in order to achieve the target values for the discharge of treated leachate (Anonymous, 1996).

- Aerobic conditions should be ensured permanently within the landfill body by means of suitable constructional and operational measures. Should anaerobic conditions restore, a rise in leachate concentrations and methane formation – however low – can be expected.

- As already mentioned, over decades still very low gas formation rates are to be expected with a total remaining residual gas potential of about 30–40 m³/t dry matter MBP waste. The produced gas must be able to escape from the landfill body so that the build-up of gas pore pressures is prevented. The gas should be oxidised in soil layers in order to avoid emissions of methane.

***Above-ground MBP waste landfills:*** An above-ground MBP waste landfill could be an attractive design in Germany although not yet constructed (Figure 10.7.6). The main feature of such a landfill, compared to traditional MSW landfills, is the emplacement of horizontal layers for the reduction of pore water and gas pressures in the waste. Every 1.5–2.0 m sloped horizontal layers of coarse material should be installed. It should be investigated if coarse bottom ash from MSW incinerators among other materials could be used as coarse material. It is expected that, due to the reduction of gas and pore water pressures, the landfill will have better geotechnical stability. Also the higher friction angle of the introduced layers filled with coarse materials will improve the stability of the MBP waste. This concept has to be further developed on the basis of soil-mechanical investigations. By means of suitable constructional and operational measures the aforementioned layers of coarse material should enable also a passive aeration. Thus semi aerobic conditions in the landfill can be expected (Stegmann and Heyer, 2001). Only low or moderate settling rates of the landfill surface are to be expected due to the pretreatment and degree of compaction of the waste.

***Baled MBP waste landfills:*** Baling of MBP waste could also be an attractive design (Stegmann and Heyer, 2002). Due to the often difficult placement of the MBP waste in a landfill and the operational limitations during rainy weather conditions, the operation of a bale landfill is expected to be easier. With regard to the residual emissions, no significant differences are expected. The elevated costs of baling could be compensated for by savings in the size of the operational face of the landfill, installation of intermediate layers and no or considerably reduced interim roofed storage area to compensate for bad weather conditions. Bales with a cube length of 1 m are appropriate. The features of this type of landfill are:

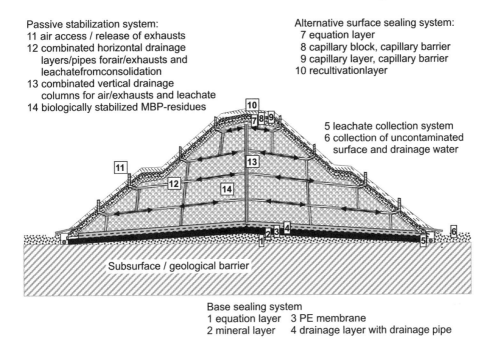

**Passive stabilization system:**
11 air access / release of exhausts
12 combined horizontal drainage
   layers/pipes forair/exhausts and
   leachatefromconsolidation
13 combined vertical drainage
   columns for air/exhausts and leachate
14 biologically stabilized MBP-residues

**Alternative surface sealing system:**
7 equation layer
8 capillary block, capillary barrier
9 capillary layer, capillary barrier
10 recultivationlayer

5 leachate collection system
6 collection of uncontaminated
   surface and drainage water

Subsurface / geological barrier

**Base sealing system**
1 equation layer   3 PE membrane
2 mineral layer    4 drainage layer with drainage pipe

**Figure 10.7.6**   *Concept of a landfill for MBP municipal solid waste (Stegmann and Heyer, 2001).*

- Due to the considerable pressure necessary for baling, water is pressed out of the waste and no elevated load-induced pressure on the bales is to be expected. As a result, no pore water pressure built up is to be expected when the bales are landfilled.
- Bales with relatively even edges and high density can be produced due to the high content of fine-grained material and the possibility to adjust to optimum water content. The regular bales allow exact placement of the bales in the landfill.
- The baling process could be integrated in the same building where the mechanical biological treatment takes place.
- The placement of the bales could be carried out also during moderate rainfall if the stored bales were immediately covered with plastic membranes.
- The space in between the bales could be used for passive aeration, which at the same time allows for preferential flow of residual biogas and probably also for biological oxidation of the methane.
- Intermediate layers made from coarse-grained material are probably not necessary.
- The final layer on the final bale surface on which the surface sealing/cover is placed could be made of approximately 0.5–1.0 m layer of MBP waste put in place in thin layers.

## References

Anonymous (1996): *Allgemeine Rahmen-Verwaltungsvorschrift über Mindestanforderungen an das Einleiten von Abwässer in Gewässer.* Anhang 51: Ablagerung von Siedlungsabfällen. Rahmen-Abwasser VwV-GMBI. Located on the internet: 19 July 2007. http://www.bmu.de.

Anonymous (1999): Richtlinie 1999/E6 des Rates vom 26.04.1999 über Abfalldeponien. *Amtsblatt der Europäischen Gemeinschaften*, 182.

Anonymous (2000): *Verordnung über die umweltverträgliche Ablagerung von Siedlungsabfällen und über biologische Behandlungsanlagen.* Located on the internet: 19 July 2007. http://www.bmu.de.

Anonymous (2001): *30. Verordnung zur Durchführung des Bundesemissionsschutzgesetzes* (Verordnung über Anlagen zur biologischen Behandlung von Abfällen – 30. BImSchV). Located on the internet: 19 July 2007. http://www.bmu.de.

Binner, E. (2002): *The impact of mechanical-biological pretreatment on the landfill behavior of solid wastes.* Paper presented at the Workshop: Biological treatment of biodegradable wastes – technical aspects. Brussels, Belgium.

Bone, B.D., Knox, K., Picken, A. and Robinson, H.D. (2003): The effect of mechanical and biological pretreatment on landfill leachate quality. In: CISA (ed.) *Proceedings of the Ninth International Waste Management and Landfill Symposium.* CD-ROM. CISA – Environmental Sanitary Engineering Centre, Cagliari, Italy.

Deipser, A., Poller, T. and Stegmann, R. (1996): *Emissions of volatile halogenated hydrocarbons from landfills.* In: Christensen T.H., Cossu R. and Stegmann R. (eds) Landfilling of waste: Biogas. Elsevier, London, UK.

Doedens, H. (2002): *Emissionen aus MBV-Deponien-Praxiserfahrungen.* In: Stegmann, Rettenberger, Bidlingmaier, Ehrig (eds) Deponietechnik 2002, Hamburger Berichte 18. Verlag Abfall Aktuell, Stuttgart, Germany.

Doedens, H. (2004): *Erfahrungen mit der Ablagerung von MBV-Material.* In: Stegmann, Rettenberger, Bidlingmaier, Ehrig, Fricke (eds) Deponietechnik 2004, Hamburger Berichte 22. Verlag Abfall Aktuell, Stuttgart, Germany.

Doedens, H., (2001): *Abgas aus Mechanisch-Biologischer Abfallbehandlung – Einleitung.* In: Stegmann, Doedens, Heusel (eds) Abluft 2001, Hamburger Berichte 17. Verlag Abfall aktuell, Stuttgart, Germany.

Ehrig, H.J. (1983): Quality and quantity of sanitary landfill leachate. *Waste Management and Research*, 1, 53–68.

Fricke, K., Münnich, K., Ziehmann, G. and Wallmann, R. (2001): Auswirkungen der neuen Abfallverordnungen auf die MBA- und Deponietechnik (Korrespondenz Abwasser). *KA-Wasser-Wirtschaft Abwasser, Abfall*, 48, 10.

Heyer, K.-U. and Stegmann, R. (1997): The long-term behaviour of landfills: Results from the joint research project Landfill Body. In: CISA (ed.) *Proceedings of the Sixth International Landfill Symposium.* CD-ROM. CISA – Environmental Sanitary Engineering Centre, Cagliari, Italy.

Heyer, K.-U., Stegmann, R. and Ehrig, H.-J. (1998): Leachate treatment principles and options. In: CISA (ed.) *Proceedings of an International Training Seminar: Management and Treatment of MSW Landfill Leachate*, pp. X1–X18. CISA – Environmental Sanitary Engineering Centre, Cagliari, Italy.

Kjeldsen, P. and Christophersen, M. (2001): Composition of leachate from old landfills in Denmark. *Waste Management and Research*, 19, 249–256.

Lee, G.F. and Jones, R.A. (1993): Groundwater pollution by municipal landfills: leachate composition, detection and water quality significance. In: CISA (ed.) *Proceedings of the Fourth International Landfill Symposium*, pp. 1093–1103. CISA – Environmental Sanitary Engineering Centre, Cagliari, Italy.

Leikam, K. and Stegmann, R., (1996): Stellenwert der mechanisch-bioloigschen Restabfallvorbehandlung. *Abfallwirtschaftsjournal*, 9/96, 39–44.

Maak (2001): *Reduzierung der Infiltration in Deponien für mechanisch-biologisch behandelte Abfälle – Einflüsse der Einbautechnik.* Dissertation, Fachbereich Bauingenieurwesen, Technical University Braunschweig, Germany.

NSCA (2002): *Comparison of emissions from waste management options.* National Society for Clean Air and Environmental Protection, Brighton, UK.

Oberstainer, G., Binner, E., Mostbauer, P. and Salhofer, S. (2007): Landfill modelling in LCA – A contribution based on empirical data. *Waste Management*, 27, S58–S74.

Plinke, E., Schonert, M., Meckel, H., Detzel, A., Giegrich, J., Fehrenbach, H., Ostermayer, A., Schorb, A., Heinisch, J., Luxenhofer, K. and Schmitz, S. (2000): *Ökobilanz für Getränkeverpackungen II.* Umweltforschungsplan des Bundesministeriums für Umwelt, Naturschutz und Reaktorsicherheit im Auftrag des Umweltbundesamtes, Bonn, Germany.

Reinhart, D.R. and Grosh, C.J. (1998): *Analysis of Florida MSW landfill leachate quality.* Report No 97-3. Civil and Environmental Engineering Department, University of Central Florida/Florida Center for Solid and Hazardous Waste Management. Miami, USA.

Rettenberger, G. and Fricke, K. (1998): *Anforderung an die Deponierung von Restabfällen aus der mechanisch-biologischen Abfallbehandlung.* In: Wiemer, K, Kern, M. (eds) Bio- und Restabfällen II. M.I.C. Baeza-Verlag, Witzenhausen, Germany.

Rettenberger, G. and Stegmann, R. (1996): *Landfill gas components*. In: Christensen, T.H., Cossu, R. and Stegmann, R. (eds) Landfilling of waste: Biogas. Elsevier, London, UK.

Stegmann, R. (1993): Design and management of a dry landfill system (DLS). In: CISA (ed.) *Proceedings of the Fourth International Landfill Symposium*, pp. 1797–1805. CISA – Environmental Sanitary Engineering Centre, Cagliari, Italy.

Stegmann, R. (1997): Description of a laboratory scale method to investigate anaerobic degradation processes taking place in landfills. In: Christensen, T.H., Cossu, R. and Stegmann, R. (eds) *Proceedings of the Third International Landfill Symposium*. CD-ROM. CISA – Environmental Sanitary Engineering Centre, Cagliari, Italy.

Stegmann, R. and Heyer, K.-U. (2001): Landfill concept for mechanical-biologically treated residual waste. CISA (ed.) *Proceedings of the Eighth International Landfill Symposium*, pp. 381–388. CISA – Environmental Sanitary Engineering Centre, Cagliari, Italy.

Stegmann, R. and Heyer, K.-U., (2002): Konzept für eine nachsorgearme MBV-Deponie. In: Stegmann, R., Rettenberger, E., Bidlingmaier, B. and Ehrig, H. (eds) *Deponietechnik 2002, Hamburger Berichte*, 18, 145–157. Verlag Abfall aktuell, Stuttgart, Germany.

von Felde, D. and Doedens, H., (1997): Mechanical-biological pretreatment: Results of full scale plants. In: CISA (ed.) *Proceedings of the Sixth International Landfill Symposium*, pp. 531542. CISA – Environmental Sanitary Engineering Centre, Cagliari, Italy.

Wallmann R. (1999): *Ökologische Bewertung der mechanisch-biologischen Restabfallbehandlung und der Müllverbrennung auf Basis von Energie- und Schadgasbilanzen*. Schriftenreihe d. Arbeitskreises für die Nutzbarmachung von Siedlungsabfällen.

Ziehmann, G., Münnicke, K. and Fricke, K. (2002): *Einbau von MBV-Materialien*. In: Stegmann, Rettenberger, Bidlingmaier, Ehrig (eds) Deponietechnik 2002, Hamburger Berichte 18. Verlag Abfall aktuell, Stuttgart, Germany.

# 10.8

# Landfilling: Bottom Lining and Leachate Collection

**Thomas H. Christensen, Simone Manfredi and Peter Kjeldsen**

*Technical University of Denmark, Denmark*

**Robert B. Wallace**

*Solid Waste Engineering Consultant, Laguna Niguel, California, USA*

The critical element of a landfill, which is essential for the protection of the environment in general, and prevention of contamination of the underlying soils and groundwater in particular, is the bottom lining system. The major focus of the bottom lining system development is to prevent leachate from entering the groundwater or surface water.

The bottom lining system should cover the full footprint area of the landfill, including both the relatively flat bottom and the sideslopes in the case of an excavated configuration. This prevents the lateral migration of leachate from within the landfill, as well as the migration of landfill gas, preventing contact between gas and groundwater.

The bottom lining system is composed of a relatively impermeable liner or lining system. This very low hydraulic conductivity system controls the movement of the leachate out of the landfill. The bottom lining system works together with the overlying leachate management system, also referred to as the leachate collection and removal system (LCRS), which consists of a drainage layer that provides easy horizontal drainage of the leachate to a point of gravitational collection or pumping.

Although individual liners, whether composed of soils or geosynthetic barriers, are able to prevent leachate emission to the environment for a relatively long time (50 years or longer), it should be realized that no liner is 100 % efficient. However, modern lining systems, which include composite liners and multiple (double, or even triple) liners, are extremely effective in preventing leachate from entering into the environment. In addition, the risk of polluting the groundwater at a landfill by any leakage of leachate depends on several factors related to siting of the landfill: distance to the water table, distance to surface water bodies, and the properties of the soil beneath the landfill.

In addition to the lining and drainage systems described in this chapter, the siting and hydrogeology of the landfill site (Chapter 10.12) and the top cover (Chapter 10.9) are also part of the barrier system, contributing to reducing the environmental risk associated with landfills. This chapter provides information about the materials used in the construction

*Solid Waste Technology & Management*   Edited by Thomas Christensen
© 2011 Blackwell Publishing Ltd

of liners and drainage systems, tools to calculate migrations through liners, as well as information about requirements for lining systems in the European Union (EU) and the United States (USA).

## 10.8.1    Landfill Bottom Lining System: Materials

The selection of a lining system suitable for a landfill depends on the type of waste landfilled, the quantity and quality of the leachate produced, and the hydraulic resistance that the liner has to provide. These factors (along with regulatory requirements) influence the choice of materials for the lining system, as well as for the leachate collection and removal system (LCRS), since the latter aims at limiting the buildup of leachate head above the bottom liner. By regulation, in the USA, as promulgated by the RCRA Subtitle D regulation for municipal solid waste (MSW) landfills, these systems shall be designed to ensure that the leachate head acting on the bottom liner shall be no greater than 30 cm. Different kinds of materials, either natural or synthetic, are available for the bottom liner barrier layer. These can be utilized individually or as a component of a composite liner, composed of two or more low-permeability layers in intimate contact.

### 10.8.1.1    Natural Barrier Materials

*Low Permeability Soil*

The most common natural lining material is low permeability soil, usually clay. Clay liners are generally constructed in lifts of 15–20 cm compacted thickness, and developed to a total layer thickness of, typically, 0.5–1.5 m. EU regulations specify that the clay used for hazardous and nonhazardous waste landfills can only allow leachate to penetrate at a rate of less than 3 cm/year, which corresponds to an infiltration rate of approximately $10^{-9}$ m/s (CEC, 1999).

The major factors controlling the performance of clay liners are porosity and permeability, which in turn depend on many other factors: its constituents, expressed in terms of clay composition and mineralogy; and its soil properties, such as particle size distribution, plasticity, density (controlled by compaction during placement), moisture content, and by the field placement technique and liner thickness (Christensen *et al.*, 1994; Williams, 1998).

Clay moisture content is a key parameter that must be controlled during placement and compaction, as it can define the range of achievable permeability. For most clays, in order to attain the prescribed maximum permeability, moisture contents typically on the order of 4–6 % wet of the optimum moisture content for compaction are required, so that selection of the moisture content to obtain suitable permeability as a barrier layer may be quite different than the criteria based solely on density and compaction for an engineered fill. The design moisture content can be estimated through a dedicated laboratory testing program. However, as noted above, moisture content slightly higher than the optimum moisture content associated with the maximum soil density should ensure suitably low permeability of the clay liner (Benson and Daniel, 1990; Williams, 1998).

Theoretically, the permeability of the liner is not related to its thickness, but the flow through the liner is a function of the head gradient, so that flow rates are higher at smaller thicknesses due to higher pressure gradients. The efficiency of the clay liner is improved by spreading and compacting the clay in lifts (e.g., 20 cm thick), then compacting each lift with a heavy roller. *In situ* compaction increases the homogeneity of the layer by breaking up clumps of clay, thereby reducing the void space between the pieces and increasing density. In addition, homogeneous liners that are made of a single type of clay normally perform better than liners that may be constructed using different types of mineral soils (Hughes *et al.*, 2003).

Nevertheless, in certain circumstances, the as-constructed condition of a clay liner can deteriorate over time. The effectiveness of clay liners can be compromised by a secondary structure (cracks) induced by freeze–thaw or wet–dry cycles, drying out during placement, and insufficient geotechnical sub-base support. Once formed, these preferential paths through the clay layer do not self-heal with reversal of their causative influence. For this reason, clay liners should always be protected or otherwise covered to prevent deterioration, which can commence immediately after placement (e.g., due to rain, freezing, or drying). Much is known about the placement and compaction of soils and the controls that are required to achieve a successful effective liner (Daniel, 1987, 1993).

### *Bentonite and Bentonite–Sand Mixtures*

'Bentonite' represents clay minerals (primarily montmorillonite) that are able to swell, in wet conditions, up to 15–18 times their dry volume, without confinement. Thanks to this property, bentonite and sandy soil can be used to construct low permeability liners that can be used instead of clay. When the naturally occurring clay soil does not have a sufficient level of clay minerals to create a properly low permeability, bentonite clay can be added (up to 5–10 % by volume) to form a bentonite-enhanced soil. Moreover, bentonite-enhanced soil further swells under pressure and as a result, lower permeability is obtained, as the mass of waste increases in the landfill body (Williams, 1998). Nevertheless, due to the significant field working of the mixture and the ready availability of geosynthetic liner materials, bentonite-amended soil liners are seldom selected.

## 10.8.1.2  Geosynthetic Materials

Geosynthetics are utilized as liner materials, and for separation, drainage, filtration, and reinforcement. Due to their ready availability, small volume consumption, and good performance, geosynthetics have, to a large degree, progressively replaced natural soil materials in landfill construction (Christensen *et al.*, 1994; Giroud, 2005). For bottom liners, the geosynthetics utilized in landfills include geomembranes and geosynthetic clay liners (GCLs). They are used extensively both individually and together to form composite liners that exhibit much better performance than soil-only clay liners (or, for that matter, either of them alone). Other geosynthetics used in conjunction with lining systems include: geotextiles, geogrids, geonets, and geocomposites, but geomembranes and GCLs are the key materials comprising the barrier layer.

### *Geomembranes (GM)*

Geomembranes (GM), also known as flexible membrane liners (FML), are made from a variety of polymers and elastomers with extremely low permeabilities including, most commonly, polyvinyl chloride (PVC), polypropylene (PP), and high-density polyethylene (HDPE). HDPE is often the final choice due to its very low permeability (on the order of $10^{-14}$ m/s), strength, biological resistance, and cost. Even so, however, the mechanism of leachate diffusion through the geomembrane cannot be ignored.

In addition, the incidence of small holes due to poor seaming practices, handling damage, or construction-related damage has been shown to typically occur at a frequency of five to 10 holes per hectare. Even seemingly minuscule holes, as small as 0.1 cm², can yield leakage flows far in excess of the diffusive process. Hence, diffusion is generally not of considerable concern with regard to overall leakage.

Geomembranes are available in sheets or rolls typically 5–10 m wide and up to 100 m in length. The thickness of the geomembrane used in a landfill liner is regulated by country-specific legislation, but it is usually in the range 1.0–2.5 mm. A geomembrane for a landfill project is characterized by density, tensile strength, puncture resistance, tear resistance, resistance to ultraviolet light and ozone, and chemical resistance to aggressive leachate constituents (Sharma and Lewis, 1994). In addition, however, design properties must be addressed during analysis. These include elements such as the interface friction between two adjacent layers, necessary for veneer stability of the layered lining system. Those calculations should be carried out for every design, and should include laboratory testing employing the specific geosynthetics and soils to be utilized at the site.

### *Geosynthetic Clay Liners (GCL)*

Generically, geocomposites are geosynthetics that are composed of two or more different geosynthetics bonded together. In practice, however, the term geocomposite is generally reserved for bonded geosynthetics that are used for drainage purposes (e.g., a drainage geocomposite formed by a geonet bonded on each side with a geotextile filter). By definition, geosynthetic clay liners (GCL) are also geocomposites, but they are different in that they are a barrier layer, consist of a combination of geosynthetics and mineral soil (bentonite), and are not typically included in the geocomposite category.

A GCL can be used as a stand-alone barrier layer, or as an alternative to a natural compacted clay liner in a GM/GCL composite liner. They are principally composed of a layer of bentonite soil sandwiched between two geotextiles, and are bonded together by needle-punching to join the geotextiles through the bentonite. Variations on this fabrication approach include bonding the geotextiles together by stitching. Another variation includes chemical bonding the bentonite to a thin geomembrane, although this is a specialty product that falls outside the conventional definition of a GCL.

***Figure 10.8.1*** *Installation of a geomemebrane on top of a geotextile (Giroud, 2005). Reprinted with permission from The Mercer Lecture 2005–2006, Contribution of geosynthetic to the geotechnical aspects of waste containment by J.P. Giroud © (2005) J.P. Giroud.*

Geosynthetic clay liners are typically supplied in rolls that are 5 m wide by 30 m long. The thickness of the bentonite layer is typically 1.5 cm. A GCL can be installed much more quickly than traditional compacted clay liners and, additionally, it exhibits an ability to self-heal after influenced by wet–dry or freeze–thaw cycles. Because of these advantages, the use of a GCL as the bottom component of a composite liner has become common practice for landfill design.

### *Geotextiles (GT)*

Geotextiles (GT) are principally woven or nonwoven fabrics that are used for filtration, separation, drainage, reinforcement, and cushioning. The nature of the geotextile spans a large variety of materials having properties important to its particular function. Properties of geotextiles that relate to different applications vary: (1) for filtration, the opening or pore size of the geotextile is paramount, (2) for separation, its continuity and some strength are important, (3) for drainage, its transmissivity governs its capacity; for reinforcement, (4) its strength is critical, and (5) for cushioning, its mass provides the function.

In a landfill, filtration is one of the most important functions: these materials allow the movement of water but restrict the movement of soil particles, thereby reducing clogging in the LCRS. Geotextile cushions can be installed above a geomembrane layer to protect the geomembrane from damage from the LCRS granular soils, as shown in Figure 10.8.1. Joining of the geotextiles may be achieved simply by overlapping, by sewing, or by heat bonding (Christensen *et al.*, 1994; McBean *et al.*, 1995; Sharma and Lewis, 1994).

### *Geonets (GN)*

Geonets (GN) are porous sheets of plastic netting (Figure 10.8.2) which may be used in landfills as drainage layers in combination with sand or gravel to carry leachate or landfill gas (Christensen *et al.*, 1994; Giroud, 2005; Williams, 1998). Geonets are mostly made of polyethylene.

## 10.8.2 Landfill Bottom Lining System: Functions of Barrier System Components

A barrier system for a landfill is expected to provide, as its principal function, the prevention of migration of liquids (and solids) from the containment system into the environment. It also serves as a platform for the LCRS, which allows flow over the liner to facilitate removal and eliminate the buildup of leachate head on the liner.

Nevertheless, the main function of a landfill lining system is to provide the barrier layer. This function can be supplied either by natural materials (such as clay and bentonite mixtures) or synthetic materials (geomembranes) or a combination of natural and synthetic materials, such as GCLs.

As veneer stability problems can arise along the bottom sideslopes of a landfill, the barrier system should also provide mechanical support and reinforcement to soil. However, during design, the strength of the geomembrane or GCL should never be included in the stability analysis (these materials should be designed as zero-tension elements, as they are prone

**Figure 10.8.2** *Geonet rolls utilized at landfill sites (Giroud, 2005). Reprinted with permission from The Mercer Lecture 2005–2006, Contribution of geosynthetic to the geotechnical aspects of waste containment by J.P. Giroud © (2005) J.P. Giroud.*

to elongation when stressed, which is undesirable for the integrity of the entire system). Hence, when the steepness of the sideslopes requires enhancement to ensure stability of the veneer (layered) barrier system, geogrids or high strength woven geotextiles may be employed for this purpose.

Because of their relatively thin cross-section, geosynthetic liners are somewhat susceptible to puncture, and therefore some form of cushioning protection must be provided. Puncturing of the liner may occur due to sharp objects contained in the waste; the pressure contact of coarse-grained, angular granular drainage materials, and the action of vehicles and mobile equipment in the landfill. A simple and effective way to achieve this mechanical protection is to cover the liner with layers of sand or small, rounded gravel eventually coupled with geotextiles and geocomposites. When combining granular material of diverse size, a *separation layer* should be provided in order to avoid penetration of one material into the other. This function can be successfully achieved by using nonwoven geotextiles (Christensen *et al.*, 1994). This protection can be provided by appropriate design of the overlying LCRS, described later in this chapter. Other measures can be included in the project specifications for construction and operations, including: prohibition of rubber-tired vehicles and tracked vehicles of certain contact pressures to travel above the lining system with less than a prescribed thickness of suitable cushioning materials; providing a thick protection layer (usually satisfied by the LCRS) that inhibits the intrusion of sharp waste materials to the liner – or by making specific prohibition of angular, sharp, or hard waste materials in the bottom lift of the landfill, providing protection by imposition of a physical separation of these materials.

Some of the components described for the bottom lining system are also regularly used in the construction of landfill top covers and side barriers. In this case, they provide functions such as *erosion control* (protection against water and wind erosion), water diversion (to direct surface water and precipitation from infiltrating into the landfill), and biogas migration control (drainage layer for biogas to be connected to the pumping system) (Christensen *et al.*, 1994). Table 10.8.1 gives functions and levels of applicability of the landfill liner components.

### 10.8.3 Landfill Bottom Lining System: Design

Lining systems are composed of either single or double liners, with each liner consisting of either a single element (clay, geomembrane, or GCL), or of composite liners, typically consisting of a geomembrane directly over either a clay layer or a GCL. Figure 10.8.3 (Benson *et al.*, 2004) gives examples of bottom liners adopted in six full-scale North American bioreactor landfills. Figure 10.8.4 gives schematic examples of typical single, composite, and double liners.

- *A single component liner* (Figure 10.8.4a) consists of a clay liner, a GCL, or a geomembrane. Single liners can be utilized in containment structures where the subgrade consists of relatively impermeable soil or rock, where the groundwater is extremely deep, or where the consequences of leakage (e.g., a noncontaminating liquid, such as raw water) are minimized. In most countries, however, a single liner is not allowed for landfills, except under exceptional

**Table 10.8.1**   *Level of applicability of different landfill materials, with regard to the required function. The evaluation is given on the base of material property, performance and cost (Christensen et al., 1994). Christensen, T.H., Cossu, R. and Stegmann, R. (1994): Principles of landfill barrier systems. In: Christensen, T.H., Cossu, Reprinted from Christensen, T.H., Cossu, R. and Stegmann, R. (1994): Principles of landfill barrier systems. In: Christensen, T.H., Cossu, R. and Stegmann, R. (eds) Landfilling of waste: Barriers, pp. 3–10. E&FN Spon, London, UK. © (1994) E&FN Spon.*

| APPLICABILITY LEVEL | | FUNCTIONS | | | | | | | | | |
|---|---|---|---|---|---|---|---|---|---|---|---|
| High (dark) / Medium (grey) / Low (light) | | Lining | Leachate percolation | Leachate drainage | Soil reinforcement | Mechanical protection | Separation | Erosion control | Water filtration | Water drainage | Gas migration control |
| **TYPE OF MATERIAL** | **COMPONENTS** | | | | | | | | | | |
| Natural | Bentonite soil mixture | ■ | | | | ■ | | | | | |
| | Clayey soil | ■ | | | | | | | | | |
| | Sand | | | ▨ | | ▨ | | | | ■ | ■ |
| | Gravel | | | ■ | | | | | ▨ | ■ | ■ |
| Synthetic | Geomembrane | ■ | | | | | | | | | |
| | Geotextile | | | ▨ | ■ | ▨ | ■ | | ■ | ■ | |
| | Geonet | | | ■ | | | | | | ■ | |
| | Geogrid | | | | ■ | | | ■ | | | |
| | Geocomposite bentonite | ■ | ■ | | | | | | | | ■ |
| | Geocomposite drain | | | ▨ | | | ▨ | | ■ | ■ | |

circumstances. A landfill holding relatively inert construction and demolition waste might be an example of a landfill where the leaching process is not of great concern (Hughes *et al.*, 2003).

- *A composite liner* (Figure 10.8.4b) typically consists of a geomembrane on top of a clay liner or a GCL. Thanks to this structure, a composite lining system shows a much higher capability to limit leachate migration into the subsoil than a single component liner of any type. Giroud (2005) noted that: 'if there is intimate contact between geomembrane and clay, leakage through a composite liner is typically two to four orders of magnitude less than leakage through a geomembrane alone or through a clay liner alone.' Composite liners are therefore suitable for landfills where leaching is a major process, such as municipal solid waste landfills (Hughes *et al.*, 2003) and in the USA, are mandated for those facilities.

- *A double lining system* (Figure 10.8.4c) consists of a combination of either two single liners, two composite liners, or a single and composite liner. The primary (or top) liner usually functions as the barrier to collect the leachate, whereas the secondary (or bottom) liner acts to collect leakage through the primary liner. The two liners of a double lining system are separated by the leakage detection, collection, and removal system (LDCRS), which collects any leakage that may penetrate the primary liner. Double liners are used in some municipal solid waste landfills and in all hazardous waste landfills. In most cases, when double liners are required, at least the bottom liner will be a composite system. For hazardous waste landfills, both liners are composite systems. To prevent lateral movement of leachate, it is of chief importance that primary and secondary liners are in tight contact (Giroud, 2005).

The first and probably most important issue related to the application of any composite liner is to ensure tight and long-lasting contact between the geomembrane and clay/GCL interfaces. This is a vital condition for the liner to be

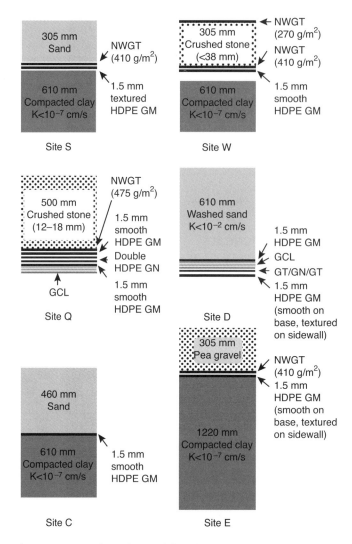

***Figure 10.8.3*** *Bottom lining systems adopted in six full-scale American bioreactor landfills. NWGT = nonwoven geotextile; GM = geomembrane; GN = geonet; GT/GN/GT = geocomposite of geonet with geotextiles bounded to each side; GCL = geosynthetic clay liner; HDPE = high-density polyethylene; K = saturated hydraulic conductivity (Benson et al., 2004). Reprinted with permisison from Geo engineering report No. 04-03. Benson, C.H., Barlaz, M.A., Lane, D.T. and Rawe, J.M. (2004): State-of-practice review of bioreactors landfills.Geoengineering Program, University of Wisconsin–Madison. Madison, USA.*

fully effective in minimizing leachate infiltration over time (Giroud, 2005). Field observations may detect that intimate contact is not achieved, due to the formation of wrinkles on the surface of the geomembrane, caused by the relatively high coefficient of thermal expansion and contraction of the geomembrane, which is particularly problematic with stiff HDPE geomembranes. Wrinkles may result in cracking of the liner under load, which can result in holes and decreased efficiency of the composite liner containment (Figure 10.8.5). Knowledge of the phenomenon, and care during construction, can decrease the probability of this occurring. For example, covering of the geomembrane with the next layer of soil in the coldest part of the day can allow covering when the geomembrane is at its maximum degree of contraction in the day, and the insulating effect of the soil layer will mitigate the temperature variation thereafter.

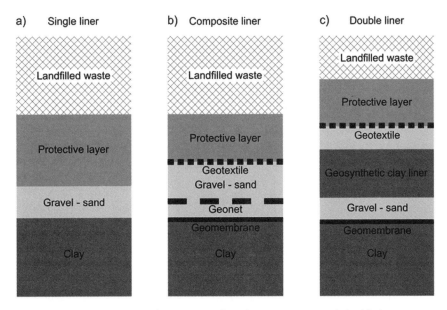

**Figure 10.8.4** *Typical structures of single, composite, and double liners.*

As noted above, wrinkles have a more pronounced effect on HDPE geomembranes than, for example, PVC geomembranes. The manner in which wrinkles form is a function of the stiffness of the material and the action of two types of forces acting on the surface of the geomembrane: bending forces (that tend to extend the wrinkle) and friction forces (that tend to lift up the wrinkle). The strength of friction forces is proportional to the spacing between wrinkles. Since HDPE is a stiff material, wrinkles may be formed only quite far from each others (2–3 m or more); with such a spacing, friction forces become strong enough to overwhelm bending forces. This leads to the formation of large ($\geq 10\,\text{cm}$ in height), visible wrinkles. Conversely, PVC geomembranes are much more flexible and can develop wrinkles that are typically very close to each other. Friction forces are therefore weaker than bending forces, which leads to the formation of a large number of small, almost invisible wrinkles (Giroud, 2005).

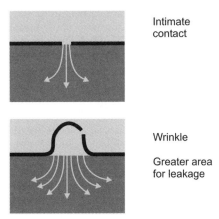

**Figure 10.8.5** *Leaking of leachate due to holes in the geomembrane in case of intimate contact or in case of wrinkle (Giroud, 2005). Reprinted with permission from The Mercer Lecture 2005–2006, Contribution of geosynthetic to the geotechnical aspects of waste containment by J.P. Giroud © (2005) J.P. Giroud.*

The detailed design of bottom liner systems is covered in the literature, with description of methods for evaluating leakage through liners (Giroud, 2000), veneer stability of lining systems (Koerner and Soong, 2005), and removal of leachate head above liners (Giroud, 2000). Considerable literature in the conference papers from international conferences on landfills and geosynthetics are available and should be consulted to develop a comprehensive understanding of bottom liner system design.

## 10.8.4   Landfill Bottom Lining Systems: Construction

The development of a bottom liner is an in-step process (EPA, 2001) that includes:

- *Soil investigation*. The soil layer beneath the liner should be carefully analyzed prior to liner construction. Soil investigations should at least comprehend measurements of parameters such as the moisture–density relationship (both *in situ* and laboratory measurements), hydraulic conductivity, Atterberg limits (soil-specific, limit water content values, identifying the four physical states the soil may appear: solid, semisolid, plastic, and liquid) and grain size distribution. Visual control during construction of soil layers, in order to remove unwanted materials, is also recommended.
- *Detailed design*. Comprehensive evaluation of the stability, integrity, and configuration of the lining system must be completed, in order to meet environmental regulations and provide adequate protection of the soils, groundwater, and surface waters at the site.
- *Construction of excavations, berms, and/or embankments*. Excavations or embankments should clearly define the physical borders of the landfill section and allow for liners to be installed along the sides of the landfill. Berms are also useful for temporary and permanent subdivision of the landfill area into operational units.
- *Subbase preparation*. Subbase preparation includes fine grading intended to develop the subgrade surface ready for the liner installation. Careful grading is also required to provide bottom slopes that will facilitate the drainage of leachate above the liner, to sumps for removal.
- *Construction of the bottom liner system*. The bottom lining system for a landfill typically consists, at least, of a single composite liner, composed of a geomembrane directly over a compacted clay layer or a GCL. If a geomembrane/GCL composite section is selected, the materials can simply be deployed over the prepared subgrade. Generally speaking, only as much clay/GCL surface should be final-prepared as can be covered by geomembrane in the same day. This protects the GCL (or clay, if used) from damage due to excessive drying in the sun and wind, or by inundation by rain. GCLs, in particular, should never be deployed in the rain or onto a very wet surface, as the bentonite layer will hydrate, making it virtually impossible to deploy the overlying geomembrane without thorough drying of the GCL.

A. For clay components of composite liners, a careful selection of the clay soil is essential to meet the required hydraulic conductivity of, for instance, $10^{-9}$ m/s. Monitoring and optimization of the moisture content is also of high importance. In particular, these general steps should be followed:
  - Clay shall be placed in lifts, with a maximum compacted thickness of about 15–20 cm. Lifts are then compacted until a density based on tests such as the Standard Proctor density is achieved that corresponds to the moisture content determined by laboratory testing to be representative of the required density conditions, usually at a moisture content of 4–6 % wet of the optimum moisture content corresponding to maximum density (see Box 10.8.1 regarding compaction of clay liners and Box 10.8.2 regarding the measurement of hydraulic conductivity). Specialized compaction vehicles and proper methods are required to ensure the integrity of the liner.
  - In the case of construction in cold weather conditions, freezing of the clay liner must be avoided. In view of that, temporary soil covers may be placed above the liner, eventually in combination with insulation of the clay. Frozen soil should be removed. In practice, clay liners should not be constructed when day or night temperatures fall below freezing. If lining systems must be constructed under such circumstances, then GCLs should be considered rather than a compacted clay layer.
  - Construction of clay liners on steep sideslopes (typically, three horizontal to one vertical, 3:1) requires special care, as it is extremely difficult to apply an appropriate level of compactive effort – again, the use of a GCL for that purpose results in a more easily constructed barrier layer.

**Box 10.8.1  Compaction of Clay Liners. Reprinted from Daniel, D.E. Clay liners. In: Daniel, D.E. (ed.) Geotechnical practice for waste disposal, pp. 137–163. Chapman and Hall, New York, USA. © (1993) Taylor and Francis.**

Compaction is the process of mechanically densifying a soil. Compaction is the major technique to decrease the hydraulic conductivity and thus to improve the performance of the clay liner. Compaction is accomplished by pressing the soil particles together into a close state of contact as air is expelled from the soil mass. When soil particles are forced together by compaction, both the number of voids contained in the soil mass and the size of the individual void spaces are reduced. At the same time, the structure of the clay particles in the soil is changed. Higher densities are achieved when soil particles pack closer together.

The maximum density occurs at the so-called optimum moisture content (OMC), or optimum water content (OWC), which varies with the type of soil and compaction effort (see figure, adapted from Daniel, 1993). At optimum, the lubrication effect of the water allows soil particles to become more easily realigned during the compaction procedure, resulting in closer packing and higher density. At higher moisture contents, the lubrication effect is offset by dilution. Therefore, for any given textural composition of clay and compaction effort, there is a maximum dry density that can be achieved at the optimal moisture level. Maximum density does not represent a soil with no void space remaining but rather one where the tightest possible packing arrangement is achieved for the given compaction conditions. This condition cannot be reached unless the soil is completely saturated to begin with and is seldom achieved during conventional compaction operations (Gray, 2002). Benson and Daniel (1990) reported that, in the case of clayey soil, the lowest hydraulic conductivity is achieved when the soil is compacted at moisture content slightly higher than OMC. Increasing water content generally results in an increased ability to break down clay aggregates and to eliminate interaggregate pores. Too high water contents should be avoided because the soil becomes less workable, although the hydraulic conductivity may further decrease.

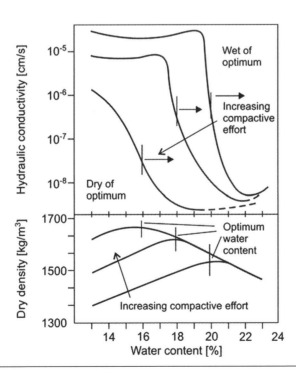

**Box 10.8.2   Measurement of Hydraulic Conductivity.**

The most important property of a compacted clay liner is its hydraulic conductivity and this should be documented for any clay liner installed. Unfortunately, the methods available for measuring hydraulic conductivity do not lead, for a given clay, to similar results. Differences up to two orders of magnitude depending on the experimental procedure are not uncommon (Farquhar, 1994). It should therefore be stressed that statistical analyses are highly needed when interpreting and reporting data about liner hydraulic conductivity (Korfiatis *et al.*, 1987). The only correct hydraulic conductivity is the one that the liner exhibits in place and the only way to determine this value is through several seepage measurements in the field. However, several approaches are being used including laboratory tests on clay samples.

## Laboratory Tests

Laboratory experiments make use of permeameters to measure the hydraulic conductivity of soil samples. There exist two kinds of permeameters: ***constant head*** (left figure below) and ***falling head permeameters*** (right figure below). They provide only a coarse estimation of field values because only small core samples are normally collected and analyzed from a limited number of locations at a site. A flow is maintained through a small column of material, while flow rate and head loss are measured. For a constant head permeameter, Darcy's law can be directly applied to determine the hydraulic conductivity $K$ (Bedient *et al.*, 1999):

$$K = \frac{V \cdot L}{A \cdot t \cdot \Delta H} \quad \text{(Constant head permeameter)}$$

where $V$ (m$^3$) is the volume flowing, $A$ (m$^2$) and $L$ (m) are, respectively, the cross-section and length of the sample, $t$ (s) is the time needed for the volume $V$ to flow throughout the sample, and $\Delta H$ (m) is the constant head.

When estimating the hydraulic conductivity by mean of a falling head permeameter, the rate of fall of the water level in the column has to be measured. The following expression, based on Darcy's law, can be utilized (Bedient *et al.*, 1999):

$$K = \left(\frac{r}{r_s}\right)^2 \cdot \frac{L}{t} \cdot \ln\left(\frac{H_0}{H_1}\right) \quad \text{(Falling head permeameter)}$$

where $r$ (m) and $r_s$ (m) are, respectively, the radius of the column and soil sample, and $t$ (s) is the time needed for the level in the column to fall from $H_0$ (m) to $H_1$ (m).

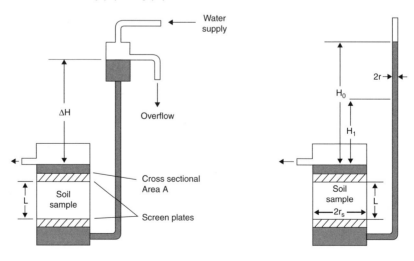

The validity of laboratory tests is often compromised by:

- The hydraulic gradients utilized. Hydraulic gradients up to several hundred times greater than those existing in the field are utilized. This leads to unnatural flow conditions which adversely affect hydraulic conductivity and therefore lead to errors (Quigley *et al.*, 1988).
- The duration of the tests. The test length is too short to be representative of biomass build-up on the liner surface and long-term interactions between the liner materials and some organic leachate constituents.
- The size of the samples. Too small a sample is often utilized. Small samples are not representative of field heterogeneities, which leads to bigger hydraulic conductivities (Farquhar, 1994; Quigley *et al.*, 1988).
- Stresses induced to the clay during the sampling phase. Hydraulic conductivity decreases with an increase in overburden stress for a constant sample size (Korfiatis *et al.*, 1987).

### Field Measurements

Field conditions cannot be reproduced in the laboratory or are not practical in term of scale and time needed (e.g. Daniel, 1987). Field measurements, in general, lead to hydraulic conductivities up to two orders of magnitude bigger than laboratory estimations (Farquhar, 1994). The most common field tests utilized for determining hydraulic conductivity in clay liners are **infiltrometers**. Large-scale infiltrometers are likely to produce quite realistic values because they do not require sample removal or elevated applied pressure and they can include the effect of field heterogeneities as their size increases. A major drawback is the long time needed to carry out these trials, which finally results in high expenses. Double-ring infiltrometers are being increasingly utilized to perform this kind of field tests (see below; adapted from Qian *et al.*, 2002). The experimental apparatus uses two open cylinders driven into the ground, one inside the other. The test consists of maintaining a constant water elevation in the rings. The volume of water added to the inner ring over time is a measure of the soil infiltration rate. The water infiltrating through the outer ring confines the flow from the inner ring to approximately one-dimensional flow in the vertical direction, so that the Darcy's law can be utilized to estimate the hydraulic conductivity.

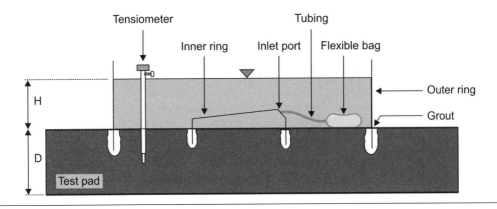

**B.** Construction of GCL liners is very straightforward, although it is necessary to take care not to allow the rolls to become wet during handling, because they become hydrated quite easily, which generally renders them unmanageable and cannot be used. During the installation of GCLs, care should be taken at all times to avoid hydration. Hydration results in unconfined swelling of the bentonite and can impede the subsequent covering with the geomembrane components of the composite liner. For GCL-only liners (generally not allowed for bottom liners of landfills), once the installer covers the GCL with soil, the GCL could hydrate by drawing moisture from the soil, however this does not necessarily impact the performance of the GCL.

**C.** For geomembrane components of composite liners, the following measures should be carefully followed:
- ○ Prior to installation, the manufacturer must provide a certification of the geomembrane that includes quality control (QC) documentation.
- ○ The installation of geomembrane liners requires specially qualified personnel. Specification of installer qualifications must be met before any materials can be deployed.
- ○ Geomembranes are normally transported to the landfill in rolls or on pallets, which should be carefully inspected upon arrival to ensure that there has not been handling or transit damage, and to determine whether the material fulfills the project specifications.
- ○ Trial welds are completed prior to any production welding of the geomembrane. These trial welds are prepared on pieces of the geomembrane, using the same equipment and personnel as will be doing the production welding (seaming) for the installation. Samples are then cut from the trial welds and tested. No welding may proceed before a passing result is obtained for the trial welds. Upon passing, welding of the geomembrane can then be started.
- ○ During production welding, samples to be used for laboratory destructive testing are taken at the frequency of, typically, one for every 150–200 m of field seam length. Figure 10.8.6 illustrates construction and placement of landfill liners.
- ○ All of these specialized requirements for installation are typically presented in a construction quality assurance (CQA) plan, which is a document used in conjunction with project specifications and should outline all requirements to be met for installation of soils and geosynthetic systems. It is also a required document for most permit approvals.

- **Protection and repair measures**. The major factors resulting in damage to both clay and synthetic liners include drying, freezing, exposure to solar (UV) radiation, rain and wind, and the traffic of vehicles above the liners at the site. The most common protection measure is to prohibit any equipment or vehicles from traveling on these systems, but it is also feasible to cover the liner with soil until the drainage layer is installed. Even so, regardless of the care and testing taken during installation, the vast majority of liner damage occurs after the completion of installation, during subsequent covering operations. It is imperative that CQA oversight be provided, both during installation, as well as during subsequent work, until the lining system is covered with a protective layer of soil.

### 10.8.5   Landfill Bottom Liner: Transport of Leachate Constituents Through Liners

No individual liner can be considered absolutely impermeable, especially on a long-term basis (Giroud, 2005). The transport of leachate constituents through liners due to diffusion and advection affects any liner, including new, intact ones. Besides, if the liner becomes damaged in any way, additional leakage of leachate would occur because additional advective fluxes would develop.

#### 10.8.5.1   Diffusion Through Liners

Independent of the type of liner (natural or geosynthetic), diffusive transport is governed by Fick's laws for diffusion:

$$J_D = -D_e \cdot \frac{\partial C}{\partial z} \qquad \text{Fick's first law} \tag{10.8.1}$$

$$\frac{\partial C}{\partial t} = D_e^* \cdot \frac{\partial^2 C}{\partial z^2} \qquad \text{Fick's second law} \tag{10.8.2}$$

where

$J_D$ (kg/m$^2$s) is the diffusive mass flow of pollutant
$C$ (kg/m$^3$) is the concentration of pollutant
$z$ (m) is the flow direction (perpendicular to the liner)
$t$ (s) is time
$D_e^*$ (m$^2$/s) is the effective diffusion coefficient:

$$D_e^* = \frac{D_e}{R}$$

where

$D_e$ is the diffusion coefficient in the liner material (m²/s)
$R$ is the retardation factor reflecting the sorption properties of the liner (dimensionless; see later).

For a clay liner, the diffusion coefficient $D_e$ is estimated approximately by:

$$D_e = \varepsilon \, \tau \, D_w \qquad (10.8.3)$$

where

$\varepsilon$ (–) is the porosity of the liner
$D_w$ (m²/s) is the diffusion coefficient in water
$\tau$ (–) is the tortuosity.

The tortuosity is estimated by the empirical equation:

$$\tau = \varepsilon^{1.33}$$

In a geomembrane, the diffusion coefficient $D_e$ is approximately:

$$D_e = D_{gmb} \qquad (10.8.4)$$

where

$D_{gmb}$ (m²/s) is the diffusion coefficient of the considered pollutant through the geomembrane.

Fick's second law, which describes the diffusion process in term of concentration, can be integrated and solved when assuming that:

1. The thickness of the liner is infinite: $z = \infty$.
2. $C_i$ is the initial (at $t = 0$) concentration of the pollutant throughout all the liner ($z > 0$):

$$C(z, t) = C_i \quad \text{for } z > 0; \; t = 0.$$

3. $C_i$ is also the concentration of pollutant at an infinite distance and at any time:

$$C(z, t) = C_i \quad \text{for } z = \infty; \; t > 0.$$

4. $C_w$ is the concentration of the pollutant at the top of the liner ($z = 0$) at any time:

$$C(z, t) = C_w \quad \text{for } z = 0; \; t > 0.$$

***Figure 10.8.6***   *Example of bottom liner construction (Carbofol – Geomembranes for Environmental Protection, Hüls Troisdorf AG, Kempen/Tönisberg, Germany).*

The resulting solution is:

$$C(z, t) = C_i + (C_w - C_i) \cdot \text{erfc} \left( \frac{z}{2 \cdot \sqrt{D_e^* \cdot t}} \right) \tag{10.8.5}$$

Fick's first law, which describes the diffusion process in terms of flux of pollutant, can be solved under the following assumptions:

1. Liner of finite thickness ($L$).
2. Steady state.
3. $C_w$ is the concentration of the pollutant at the top of the liner ($z = 0$) at any time:

$$C(z, t) = C_w \quad \text{for } z = 0, \ t > 0.$$

4. Rapid removal of pollutant at the outside of the liner leading to a zero concentration:

$$C(z, t) = 0 \quad \text{for } z > L, \ t > 0.$$

The resulting solution for a *clay liner* is:

$$J_w = (\varepsilon \cdot D_w \cdot \tau) \cdot \frac{C_w}{L} \tag{10.8.6}$$

where

$J_w$ (kg/m$^2$s) is the diffusive mass flow of pollutant;
$\varepsilon$ (–) is the porosity of the liner;
$D_w$ (m$^2$/s) is the diffusion coefficient of water in the clay liner;
$\tau$ (–) is the tortuosity;
$L$ (m) is the thickness of the liner;
$C_w$ (kg/m$^3$) is the concentration in leachate.

The resulting solution for a *geomembrane liner* is:

$$J_D = D_{gmb} \cdot \frac{C_w}{L} \tag{10.8.7}$$

where

$J_D$ (kg/m$^2$s) is the diffusive mass flow of pollutant;
$D_{gmb}$ (m$^2$/s) is the diffusion coefficient of the considered pollutant in the geomembrane;
$C_w$ (kg/m$^3$) is the concentration of pollutant in leachate;
$L$ (m) is the thickness of the geomembrane.

Box 10.8.3 gives examples of the application of Equation (10.8.6).

---

**Box 10.8.3    Calculating Chloride Migration Through a Clay Liner and A GCL by Diffusion and Advection.**

## Clay Liner

A clay liner is installed as bottom liner in a landfill. The liner is $L = 60$ cm thick, has hydraulic conductivity $K = 8^*10^{-10}$ m/s and porosity $\varepsilon = 0.45$. The chloride concentration in the leachate is 2500 g/m³ and the diffusion coefficient of chloride in water is $D_w = 0.064$ m²/year. It is estimated that the depth of leachate on the top of the liner is $Z = 20$ cm.

The diffusive flux is given by:

$$J_{w,\text{diff}} = (\varepsilon \cdot D_w \cdot \tau) \cdot \frac{C_w}{L} = (0.45 \cdot 0.064 \cdot 0.45^{1.33}) \cdot \frac{2500}{0.6} = 41.5 \text{g/m}^2/\text{year}$$

The advective flux is given by:

$$J_{w,\text{adv}} = q \cdot C_w = (K \cdot i) \cdot C_w = \left( K \cdot \frac{Z+L}{L} \right) \cdot C_w = 8 \cdot 10^{-10} \cdot \frac{0.2 + 0.6}{0.6} \cdot 2500$$

$$= 2.67 \cdot 10^{-6} \text{ g/m}^2/\text{s} = 84.0 \text{ g/m}^2/\text{year}$$

The advective flux is thus the bigger. The theoretical hydraulic conductivity $K^*$ at which the two fluxes are equally important can be estimated by:

$$K = 8 \cdot 10^{-10} \text{ m/s} = 0.0252 \text{ m/year}$$

$$J_{w,\text{diff}} = J_{x,\text{adve}} \Leftrightarrow (\varepsilon \cdot D_w \cdot \tau) \cdot \frac{C_w}{L} = K \cdot \frac{Z^* + L}{L} \cdot C_w \Rightarrow\Rightarrow K^* = \frac{\varepsilon \cdot D_w \cdot \tau}{(Z + L)}$$

$$= 0.0124 \text{ m/year} = 3.9 \cdot 10^{-10} \text{ m/s}$$

A hypothetical hydraulic conductivity of $3.9 \cdot 10^{-10}$ m/s is needed in order for advective and diffusive fluxes to be equally important.

## GCL

Using the same equations as above the migration through a geosynthetic clay liner (GCL) can be evaluated. With the same data as above except for the liner thickness and the hydraulic conductivity of the GCL (a thickness of $L = 10$ mm and a hydraulic conductivity $K = 1 \cdot 10^{-11}$ m/s is used), the diffusive and advective fluxes are:

$$J_{w,\text{diff}} = (0.45 \cdot 0.064 \cdot 0.45^{1.33}) \cdot \frac{2500}{0.01} = 2500 \text{g/m}^2/\text{year}$$

$$J_{w,\text{adv}} = 1 \cdot 10^{-11} \cdot \frac{0.2 + 0.01}{0.01} \cdot 2500 = 16 \text{g/m}^2/\text{year}$$

The calculation shows that a stand-alone liner solution consisting of a single GCL is not acceptable due to a very high diffusive migration.

---

**Box 10.8.4  Compatibility of Liners and Organics Solvents in Leachate.**

Bottom liners are exposed to a variety of pollutants present in the leachate. Short- and long-term effects of this exposure on the hydraulic performance of the liner have to be assessed to prevent liner failures. In particular, the effect of organic solvents on the hydraulic conductivity of clay and synthetic liners should de addressed.

### Clay Liners

Insoluble, hydrophobic (high $K_{OW}$) organic liquids (NAPLs) are generally harmless to water-compacted clays. This is due to the strong interparticle surface tension of water at the clay-organic interface and to the small pore size of the clay. Quigley and Fernandez (1994) showed that the hydraulic conductivity of water-compacted clays does not appreciably increase when exposed to cyclohexane during a 30-day laboratory test.

    Several leachate constituents are polar in nature, which makes them highly soluble in water (low $K_{OW}$). The effect of these organic compounds on clay liner is therefore different from that described for NAPLs. Quigley and Fernandez (1994) carried out experiments with leachate progressively enriched with ethanol and observed no effect on the hydraulic conductivity until the ethanol concentration in the leachate reached about 70 % in volume. Such high ethanol concentrations are unlikely and any real effects are not expected.

### Synthetic Liners

When synthetic liners are exposed to organic chemicals in leachate several problems can potentially occur, ranging from minor effects such as discoloration to serious issues such as swelling and solvent permeation. The latter often has been disregarded by constructors, probably because it does not lead to any visible deterioration of the liner. Several organic chemicals may permeate a synthetic liner including chlorinated solvents, benzene, TCE, and vinyl chloride. The process consists in the diffusion of these organics into the plastic sheeting and subsequently to their emission to the environment beneath the liner. It occurs not only at very high solvent concentrations, but also at concentrations that are likely detectable in MSW landfill leachates. All plastic-like liners may be affected by solvent permeation, including unfractured ones (Buss *et al.*, 1995; Haxo and Lahey, 1988).

---

## 10.8.5.2  Advection Through Liners

Advection contributes to the migration of leachate constituents through a clay liner by advection through the porous media and through cracks in the clay liners. Advection contributes to the migration of leachate constituents through geomembrane liners by advection through pinholes and cracks.

### *Clay Liners*

Mathematically, the advective transport through an intact porous medium is governed by Darcy's equation:

$$J_w = q \cdot C_w = (K \cdot i) \cdot C_w \tag{10.8.8}$$

$$i = \frac{Z + L}{L} \tag{10.8.9}$$

where

$J_w$ (kg/m$^2$s) is the mass flow of pollutant.
$q$ (m$^3$/m$^2$s) is the volumetric flow per unit area. It can also be defined as percolation of leachate and in this case a convenient unit is mm/year.
$C_w$ is the aqueous concentration of the pollutant.

**Table 10.8.2**   *Hydraulic conductivities of selected materials and times required for water to pass through a layer, 1 m thick, of each material at a hydraulic gradient of 1 (adapted from Abramson et al., 1995). Reprinted with permission from The Mercer Lecture 2005–2006, Contribution of geosynthetic to the geotechnical aspects of waste containment by J.P. Giroud © (2005) J.P. Giroud.*

|  | Hydraulic conductivity $K$ (m/s) | Time to pass through 1 m (at a gradient of 1 m) |
|---|---|---|
| Natural materials |  |  |
| Well graded, clean gravels, gravel sand mixture | $2 \times 10^{-4}$ | 1.39 h |
| Poorly graded, clean sands, gravely sands | $5 \times 10^{-4}$ | 0.56 h |
| Silty sands, poorly graded, sand silt mixture | $5 \times 10^{-5}$ | 5.6 h |
| Inorganic silts and clayey silts | $2 \times 10^{-8}$ | 1.6 years |
| Mixture of inorganic silts and clay | $2 \times 10^{-9}$ | 15 years |
| Inorganic clays of high plasticity | $5 \times 10^{-10}$ | 63 years |
| Synthetic or processed materials |  |  |
| Compacted clay liner | $1 \times 10^{-8}$–$10^{-10}$ | 3–317 years |
| Bentonite-enhanced soil | $5 \times 10^{-10}$ | 63 years |
| Geosynthetic clay liner | $1 \times 10^{-10}$–$10^{-12}$ | 317–31 710 years |
| Flexible membranes FML | $1 \times 10^{-13}$ | 317 010 years |
| Geotextile | $1 \times 10^{-4}$–$10^{-5}$ | 2.8–28 h |
| Geonet | $2 \times 10^{-1}$ | 5 s |
| Waste |  |  |
| Municipal solid waste as placed | $1 \times 10^{-5}$ | 28 h |
| Shredded municipal solid waste | $1 \times 10^{-4}$–$10^{-6}$ | 2.8 h to 12 days |
| Baled municipal solid waste | $7 \times 10^{-6}$ | 1.7 days |

$k$ (m/s) is the hydraulic conductivity of the clay liner; typical values of k for clay liners are in the range of $10^{-8}$ to $10^{-10}$ m/s (for instance, with $k = 10^{-10}$ m/s and $Z = 1$ m, the time needed for leachate to pass through the liner is $10^{10}$ s = 317.1 years). Table 10.8.2 gives examples of permeability for various materials.

$i$ (–) is the hydraulic gradient on the clay liner.

$Z$ (m) is the leachate depth on the top of the liner.

$L$ (m) is the thickness of the clay liner.

Box 10.8.3 gives an example of the application of Equations (10.8.6) and (10.8.8).

Clay liners can become fractured due to problems in the construction of the liner or due to stresses from the weight of the waste layers on the top of the barrier system. Advective transport of leachate through fractures can be modeled as a laminar flow if the fractures are sufficiently thin. The discharge of leachate through the fracture, $Q$ (m$^3$/s), is then mathematically described by:

$$Q = \frac{W \rho g b^3}{12 \eta} \cdot i \tag{10.8.10}$$

where

$b$ (m) and $W$ (m) are parameters relative to the fracture: $b$ is the width of the aperture, and $W$ is the width perpendicular to the flow direction.

$\rho$ (kg/m$^3$) is the density of water (typical values are $1.01 * 10^3$ to $1.05 * 10^3$ kg/m$^3$).

$\eta$ (kg/ms) is the dynamic viscosity of water (typical value for leachate: $10^{-5}$ kg/ms).

$g$ is the gravity (9.806 m/s$^2$).

$i$ (–) is the hydraulic gradient.

## Geomembrane Liners

Geomembrane liners may have pinholes or larger holes. Pinholes are defined as holes having a dimension significantly smaller than the thickness of the liner. The flow of leachate in pinholes can be considered to be laminar and is described by Poiseuille's equation:

$$Q = \frac{\pi \rho g d^3}{128 \eta L} \tag{10.8.11}$$

where

$Q$ (m³/s) is the flow of leachate.
$\rho$ (kg/m³) is the density of water.
$\eta$ (kg/ms) is the dynamic viscosity of water.
$g$ is the acceleration due to gravity (9.806 m/s²).
$i$ (–) is the hydraulic gradient.
$d$ (m) is the pinhole diameter.
$L$ (m) is the thickness of the geomembrane.

Geomembrane-only liners may also have larger holes, usually induced by mechanical damage. The flow of leachate, $Q$ (m³/s) through these holes is governed by Bernoulli's equation:

$$Q = c \cdot a \cdot \sqrt{2gZ} \tag{10.8.12}$$

where

$Q$ (m³/s) is the flow of leachate.
$c$ (–) is a dimensional coefficient (typical values in the range 0.4 to 0.8).
$a$ (m²) is the hole surface area.
$g$ is the acceleration due to gravity (9.806 m/s²).
$Z$ (m) is the leachate depth on the top of the liner and assumed constant in spite of the leachate flow through the hole.

Equation (10.8.12) is a conservative estimate of the flow of leachate through the hole. If the hole is large, the leachate flow could quickly empty out the standing leachate on the liner reducing the leachate depth, Z. In practice, this does not occur because there is a material (soil) of finite permeability beneath the geomembrane. For cases of composite liners, alternative analysis has been defined by Bonaparte *et al.* (1989).

### 10.8.5.3 Sorption in Liners

Sorption is herein defined as the association of a dissolved pollutant (a substance in the leachate) with a solid material (the liner material). For organic pollutants, sorption is correlated to the octanol-water partitioning coefficient of the actual compound (the more hydrophobic, the more sorption) and the organic content in the liner material. For inorganic pollutants, sorption is particularly significant for positively charged species, since clay minerals are dominated by negative surface charges and thus attract positively charged pollutants in the leachate.

The presence of sorption processes means that the migration of pollutants may be retarded relative to the water movement through a porous media (advection) or diffusion in a nonsorbing material. In practice, it is expected that heavy metal migration in clay liners will be significantly affected by sorption, while the migration of organic pollutants is much less affected. The effect of sorption is temporary, which means that the migration of a pollutant is retarded in time until the capacity of the sorbent media is used up. This means that migration requires a longer time, for example, for a heavy metal to diffuse through a clay layer, but after full breakthrough, the flux will be the same as if there was no sorption.

The delaying effect of sorption is usually described by the retardation factor $R$ (dimensionless), depending on both pollutant and soil characteristics:

$$R = \frac{T_p}{T_s} = 1 + K_d \cdot \frac{\rho_b}{\varepsilon} \tag{10.8.13}$$

where

$T_p$ (years) is the time it takes to migrate through a liner with sorptive properties.

$T_s$ (years) is the time it takes to migrate through a liner by physical means (advection in clay liner, diffusion in clay liner, diffusion in synthetic liner) with no sorption.

$K_d$ (cm$^3$/g, or l/kg) is the distribution coefficient and expresses the pollutant's distribution between solid matrix and water at equilibrium or, in other worlds, the ratio between the concentrations of the pollutant in clay ($C_s$) and in water ($C_w$): $K_d = C_s/C_w$ (Chiou et al., 1990).

$\rho_b$ (g/cm$^3$) is the bulk density of the liner (e.g., bulky density of clay).

$\varepsilon$ (–) is the porosity of the transport media.

This shows that the retardation, or delay, increases as the distribution coefficient ($K_d$) increases. In fact, the higher the value of $K_d$, the higher the tendency of the pollutant to bind to the solid. $K_d$ is therefore the key factor in the sorption process and estimating its numerical evaluation is a crucial task. $K_d$ values can be either measured or estimated. For organic pollutants, estimation of $K_d$ is often preferred because the list of organic pollutants is very long, while for heavy metals, experimental determination of $K_d$ is usually the only approach. Traditionally, $K_d$ for organic pollutants has been estimated based on the content of solid organic matter, $f_{oc}$, in the clay liner material, and a physical–chemical parameter, the octanol–water partition coefficient, $K_{ow}$, specific for each organic pollutant (see Bedient et al., 1999). However, newer research has shown that for materials with high clay contents, estimation of $K_d$ in the traditional way is uncertain, and will lead to underestimation of $K_d$. Due to the microporous nature of the clay minerals, physical sorption phenomena of hydrophobic compounds are likely to occur. This may lead to strong attenuation of the compounds in clay liners (Cheng and Reinhard, 2006; Allen-King et al., 1996).

For heavy metals, $K_d$ may typically vary between 1 and 1000 l/kg, depending on the pH of the liner–leachate system, the fraction of clay in the liner, and the degree of complexation present in the leachate. Organic matter and inorganic salt ions may complex heavy metals in solution, which means that the heavy metals are present in the leachate with a different charge and size (see Baun and Christensen, 2004). High $K_d$ values are observed at high pH, high clay contents and in the absence of complexation. If $K_d$ is on the order of 200 l/kg, then $R$ typically is on the order of 1000, which means that it takes about 1000× longer to migrate through a clay liner than if no sorption had taken place.

## 10.8.6 Leachate Collection and Drainage System: Materials

Accumulation of leachate should be avoided because it could inhibit the waste degradation process and may lead to instability of the waste mass. In addition, by regulation and in order for bottom liners to function properly, the leachate head on the bottom liner should be minimized, and kept below about 30 cm in any case. This is achieved through leachate drainage. Natural granular materials, drainage geocomposites, and drain pipes are utilized to make the horizontal transfer of leachate faster (drainage). Clogging and damage of the drain pipes are frequent concerns and they may considerably decrease the effectiveness of the drainage and collection system (LCRS). Key parameters are the size of the granular particles and the geometry of the drainage bed. This can, however, be mitigated by proper design (Rowe, 2005).

### 10.8.6.1  Natural Drainage Materials

Sand is used for cushioning of liners, in particular geomembranes, from punctures caused by sharp items in the waste compacted as the first layer in the landfill, or from the concentrated loads from equipment before there is a suitable protective waste thickness above the liner. This layer is typically a part of the LCRS, usually 15–30 cm deep, depending on the overall drainage layer implemented, which may also include a geocomposite for effective transmissive flow to remove leachate from above the liner. When combining granular materials of diverse size, a separation layer should be provided in order to avoid penetration of one material into the other. This function can be successfully achieved by using geotextiles (Christensen *et al.*, 1994).

Gravel is often used to construct the LCRS drainage layer for leachate, typically about 20–30 cm thick. Gravel particle sizes can vary depending on the design, from small pea gravel (about 6–8 mm diameter) to coarse angular crushed rock, up to >50 mm.

Similarly, gravel and/or crushed rock may be used for construction of the leachate removal system, directing leachate to sumps for extraction from the landfill where small gradients exist. Stone drains are often preferred in deep landfills because pipes (concrete, plastic) may collapse under the heavy load of the waste, but with large drainage distances (> 30 m) a combination of stone drains and pipes may be needed.

### 10.8.6.2  Geosynthetic Drainage Materials

Geosynthetic drainage materials are principally composed of high transmissivity materials such as geonets, with a built-in filter and cushioning layer, usually composed of geotextiles bonded to each side of the geonet. These drainage geocomposites typically rest directly on top of the bottom liner and effectively remove leachate along the liner surface gradient to pipes for more efficient flow and removal, and/or sumps where the leachate can accumulate and be removed by pumping. Sumps are typically provided with double composite liners to relieve the potential impacts of leachate head on the liner.

### 10.8.6.3  Pipes and Pumps

Leachate collection and removal systems typically include perforated pipes within the drainage layer, so that flows within the geocomposite and drainage soils can be collected and transmitted directly and efficiently to the sumps for removal. Clogging of the pipes is prevented by the proper design of geotextile filters to prevent the migration of soil particles as the flow progresses from areas of lower permeability to higher permeability (e.g., from the sand to the pipe), representing an increasingly higher gradient as the flow moves downgradient.

Temporary and/or permanent removable (in order to allow maintenance and repair) pumps are emplaced within the sumps for pumping accumulated leachate to a treatment system or other management alternative (such as an evaporation pond).

## 10.8.7  Leachate Collection and Drainage System: Design

As the landfill bottom liner prevents the vertical movement of the leachate out of the landfill, a leachate head starts to build up on the liner. The purpose of the leachate drainage and removal system (LCRS) is to effectively remove this leachate from the landfill. If leachate is not removed, three potentially unwanted consequences may result:

- A free leachate mound will build up within the waste, reducing the friction forces in the waste and thus threatening the geotechnical stability of the waste. This is in particular critical where the landfill bottom is sloped.
- An increasing leachate mound increases the hydraulic head, and as a result, the hydraulic gradient acting on the liner, thus increasing advective flow of leachate through clay liners, fractures in clay liners, as well as pinholes and cracks in geomembrane liners.
- Leachate may be spilled over the sidewalls of the bottom liner system if the leachate level rises above the liner on the landfill sides.

The principal components of an LCRS (adapted from Reinhart and Manoj, 2000) are:

- Filter layer: sand or geotextile.
- Drainage layer: coarse sand, gravel or geocomposite.
- Perforated or slotted pipe, used to collect leachate from the LCRS and direct it towards a sump.
- Collection sumps (or trenches): depressions filled with gravel in low points at the bottom of the landfill, where the leachate is collected and subsequently removed by pumping from vertical manholes or riser pipes.
- Cleanout port pipes: pipes used for removing and cleaning fragments or precipitate that might accumulate inside the leachate pipes. These pipes usually extend from the leachate pipe at the bottom of the landfill to the outside of the landfill. The cleaning is achieved by pumping highly pressurized water injected into the leachate pipe using a hose.
- Leachate storage lagoon: leachate is typically stored in an at-surface lagoon where it can be eliminated by evaporation or by transmission to an on-site or nearby wastewater treatment facility. In some cases, storage can be achieved in a tank with suitable protection against corrosive attack by the leachate.
- Leak detection system.

For effective removal of the generated leachate to be achieved, these components should ensure free and rapid leachate flow, thus a condition where leachate is able to exit from a landfill following a natural gradient. In order for free leachate flows to establish, it is essential to prevent the formation of solid deposits in the drainage systems. Solid deposits may in fact result in clogging and cause excessive accumulation of leachate on the top of the liner, and in fact, failure of the LCRS. The USA EPA (1993) requires a leachate depth of 30 cm as a maximum.

In order to enhance the rate of the waste degradation process at landfills, waste must be kept sufficiently moist and compacted. Operations such as leachate recirculation and water addition are commonly practiced to keep the waste wet. This may however lead to development of water tables, so appropriate design from normal operation must accommodate the additional moisture within the cell and the eventual saturation of the waste. Based on this, the landfill may be operated as a bioreactor to accelerate decomposition and accelerate landfill gas generation, both of which can be used for the ultimate benefit of better landfill management. Special permitting issues apply before most jurisdictions will allow a landfill to be operated as a bioreactor.

In this condition, the drainage of leachate can become problematic, ultimately worsening the performance of the leachate collection system, unless the transmissivity and filtration of the leachate collection system are designed to address this condition. A solution to this problem includes the provision of higher-gradient inclined drainage layers (at least 5 %) connected with the base drainage by means of vertical drainage slits (Lechner, 1994). Strong gas pressure in the area of the base drainage may eventually be generated and should be monitored.

To design a leachate drainage and collection system a number of parameters should be taken into account, including the amount of leachate generated, collection method, local topography and long-term functioning of the system (Sharma and Lewis, 1994). The first design parameter to be estimated is the maximum leachate head over the liner, $y_{max}$ (m), which represents the highest saturated depth of leachate that the drainage and collection system induces. The United States EPA (1989) recommends the following equation:

$$y_{max} = L \cdot \left(\frac{r}{K}\right)^{1/2} \cdot \left[\frac{K S^2}{r} + 1 - \frac{K S}{r} \cdot \left(S^2 + \frac{r}{K}\right)^{1/2}\right] \qquad (10.8.14)$$

where

$y_{max}$ (m) is maximum leachate head in the drainage system.
$L$ (m) is the maximum horizontal (or nearly horizontal) distance of leachate flow, which corresponds to the half of the distance between two adjacent pipes measured along the liner slope.
$r$ (m/s) is the rate of vertical leachate inflow towards the drainage layer.
$S$ (–) is the slope of the liner: $S = \tan\beta$, where $\beta$ is the angle that the liner forms with the horizontal direction.
$K$ (m/s) is the hydraulic conductivity of the drainage layer.

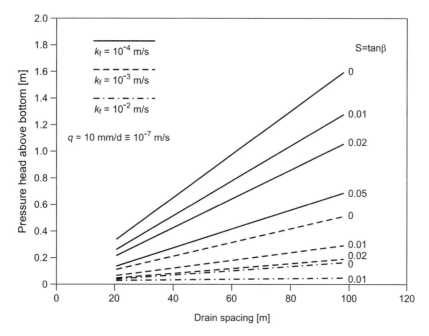

**Figure 10.8.7** *The infiltration q is an intensive short-term event. The $k_f$ corresponds to the K in the equations in the text (Ramke 1989). Reprinted with permission from Sanitary Landfilling: Process, technology and environmental impact by by T. H. Christensen, R. Cossu and R. Stegmann, Leachate Collection systems pp 343–364 by H.G. Ramke © (1989) Elsevier Ltd.*

For instance, with $L = 15\,\text{m}$, $r = 500\,\text{mm/year} = 1.6 \times 10^{-5}\,\text{m/s}$, $S = \tan\ 1.2° = 0.02$, $K = 10^{-3}\,\text{m/s}$, the result is $y_{max} = 0.03\,\text{m}$.

Alternatively, Giroud's equation, which often is used in practice, can be applied:

$$y_{max} = \frac{L}{2\cos^2\beta} \cdot \left( \left[ 4 \cdot \left( \frac{r}{K} \right) + \tan^2\beta \right]^{1/2} - \tan\beta \right) \tag{10.8.15}$$

With the same data as entered in Equation (10.8.14), Giroud's equation gives a depth of leachate of 0.01 m. Ramke (1989) provided a graphical mean (Figure 10.8.7) for relating drain spacing and pressure head build up for intensive short term infiltration events.

In the case of utilization of a geonet between the liner and the gravel, the distance between the collection pipes, $D$ (m), can be calculated from this equation (EPA, 1989):

$$\theta = \left( \frac{r \cdot D^2}{4 y_{max} + 2D \cdot \sin\beta} \right) \tag{10.8.16}$$

where

$D$ (m) is the distance between leachate collection pipes;
$\theta$ (m/s) is the transmissivity of the geonet;
$\beta$ (–) is the slope of the drainage system;
$r$ (m/s) is the rate of vertical inflow towards the drainage layer;
$y_{max}$ (m) is the head of leachate over the liner.

Manning's equation expresses the relationship between the flow of leachate and the physical dimensions of the collection pipes:

$$Q = \frac{1}{n} \cdot R_h^{2/3} \cdot S^{1/2} \qquad (10.8.17)$$

where

$Q$ (m$^3$/s) is the discharge of leachate.

$n$ (–) is the Manning's coefficient of roughness. It depends on the material the collection pipe is made of; for instance, for a pipe made of PVC, $n = 0.010$.

$R_h$ is the hydraulic radius of the pipe. It corresponds to one-quarter of the diameter ($\Phi$) of the collection pipe. $R_h = A/P$, where $A$ (m$^2$) and $P$ (m) are the cross-sectional area and the wetted diameter of the collection pipe.

$S$ (–) is the slope of the collection pipe.

Manning's equation may also be used to estimate the size of the collection pipes. However, due to the potential for biological or chemical clogging of the drainage system, including both the granular component and the pipe, provision for cleanout of at least the main headers and transmission pipes should be provided. This shall include a riser at defined spacing to allow insertion of a jetting pipe for cleaning. This alone may define the minimum pipe size required, in order to allow insertion of the cleanout equipment, and hence the minimum pipe size for this purpose may exceed those requirements determined for flow capacity, and typically a minimum of 30 cm.

## 10.8.8   Leachate Collection and Drainage System: Construction and Operation

Efficient drainage of the leachate reduces the risk of instability-related problems and is often a precondition for other measures against instability. Leachate is effectively drained when the drain layer is not fully saturated with leachate, so that the percolation of leachate through the waste mass towards the drainage layer is vertical and the waste mass is stable. Conversely, when the drainage layer becomes saturated, leachate starts to flow within the waste mass, parallel to the slope (see Figure 10.8.8). This leads to instability of the waste body.

The risk of saturation of the drainage layer can be reduced by utilizing a drainage material with high liquid storage capacity. Typically, storage capacity of a granular drainage layer is greater than that of a synthetic drainage layer. Giroud (2005) estimates that, on average, the storage capacity of a granular drainage layer is 20× that of a geonet.

The performance of the leachate drainage and collection systems can be compromised due to clogging of the drainage layer and tubes/pipes for leachate collection (Brune *et al.*, 1991), as shown in Figure 10.8.9 (Levine *et al.*, 2005). This circumstance leads to a progressive decrease in the pore space available for leachate flow and a decrease in hydraulic

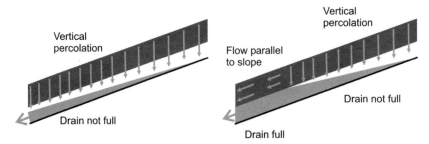

**Figure 10.8.8**   *Schematic of leachate percolation and drainage with drainage layer saturated or unsaturated (Giroud, 2005). Reprinted with permission from The Mercer Lecture 2005–2006, Contribution of geosynthetic to the geotechnical aspects of waste containment by J.P. Giroud © (2005) J.P. Giroud.*

***Figure 10.8.9*** *Effect of clogging in leachate collection pipes (Levine et al., 2005). Reprinted from http://bioreactor.org/florida.html; Florida Center for Solid and Hazardous Waste (FCSHW) The bioreator demonstration project © (2003) FCSHW.*

conductivity. When hydraulic conductivity becomes insufficient to divert leachate to the collection pipes, leachate mounds develop (Reinhart and Manoj, 2000). Clogging is caused by a combination of biological, chemical, and physical factors, including formation of biofilms and incrustations, the latter principally by chemical precipitation, sedimentation and deposition (Cooke *et al.*, 2000). Calcite ($CaCO_3$) has been reported as the main constituent of clogging material (Maliva *et al.*, 2000; Rowe *et al.*, 2000a, b).

The size and shape of the drainage media also have an influence on the potential for clogging. In particular, the smaller the drainage layer grain diameter, the larger the specific area available for incrustations and biofilms to develop, and the faster the clogging rate. Small drainage media are therefore more susceptible to clogging (Rowe *et al.*, 2000a, b). Independent of the size of the drainage media, the type of leachate flow (saturated or unsaturated), the chemistry of leachate, the incidence of redox conditions (aerobic or anaerobic), and the temperature in the collection system also influence the clogging rate (Reinhart and Manoj, 2000). Paksy *et al.* (1998) showed that the combination of saturated flow conditions and an anaerobic environment are likely to enhance the formation of incrustations. The same effect is attributed to the rapid alternation between a saturated and unsaturated condition.

To prevent clogging of the leachate collection and drainage systems or, at least, to reduce the rate of clogging, the following measures should be adopted:

- Periodically clean the leachate collection pipes through cleanout ports (Rowe *et al.,* 1997).
- Utilize pipes with large circular holes and drainage layers with large diameter media (Rowe *et al.*, 1997).
- If possible, minimize the disposal of waste containing elevated concentration of gypsum ($CaSO_4$), such as construction and demolition waste. In fact, gypsum dissolution enhances the formation of calcite incrustation (Rowe *et al.*, 1997).
- The granular material utilized should have a grain size distribution capable of limiting siltation of the drainage layer and drain pipes (Cossu and Lavagnolo, 1999).
- When a geotextile is used as a filter layer, it should be placed directly above the leachate collection system in order to reduce the amount of inorganic fines coming from the landfilled waste that is washed into the drainage layer (Bennet *et al.*, 2000).
- Wash the drainage material prior to utilization (Reinhart and Manoj, 2000).
- Biocides may be added to geotextiles and geocomposites in order to kill microorganisms in the leachate; in this way the contribution to clogging due to biological activity can be reduced (Reinhart and Manoj, 2000). Possible leaching of the biocides to leachate should, however, be considered.
- Monofilament woven geotextiles for filters should be used instead of nonwoven geotextiles because they have much smaller specific surface area available for biofilm formation (Reinhart and Manoj, 2000).

- Manholes and cleanout ports have to be appropriately covered in order to avoid wind-blown material from entering collection pipes (Ghassemi, 1986).

## 10.8.9   Regulatory Requirements

Guidelines have been issued by environmental authorities on how landfill liners and drainage systems should be constructed or how they should function. Below are shown the EU and USA regulations in this respect.

### 10.8.9.1   European Community: EU Landfill Directive 99/31

The latest European directive on landfilling of waste was released on 26 April 1999 (CEC, 1999) and the deadline for implementation of the legislation in the Member States was 16 July 2001.

The objective of the Directive is to prevent or, at least, reduce the negative effects on the environment from the landfilling of waste, by introducing stringent technical requirements for waste and landfills. Surface water, groundwater, soil, air, and human health are recognized as the environmental components to be protected (*Article 1*). The Directive identifies three classes of landfills: (1) landfills for hazardous waste, (2) landfills for nonhazardous waste, and (3) landfills for inert waste (*Article 1*). For all classes of landfills the location of the landfill must take into consideration the protection of the surrounding environment, and in particular the existence of groundwater, coastal water or nature protection zones in the area. Water control and leachate management measures must therefore always be adopted, including collection of contaminated water and leachate (*Annex 1*).

More detailed protection of soil, groundwater and surface water is to be accomplished through the combination of geological barriers and bottom liners during the operational/active lifetime of the landfill and through the combination of a geological barrier and a top cover during the passive/postclosure stage (*Annex 1*). Regarding side and bottom barriers, the Directive states: 'The landfill base and sides shall consist of a mineral layer which satisfies permeability and thickness requirements with a combined effect in terms of protection of soil, groundwater and surface water at least equivalent to the one resulting from the following requirements:

- Landfill for hazardous waste: $K \leq 10^{-9}$ m/s; thickness $\geq 5$ m.
- Landfill for nonhazardous waste: $K \leq 10^{-9}$ m/s; thickness $\geq 1$ m.
- Landfill for inert waste: $K \leq 10^{-7}$ m/s; thickness $\geq 1$ m (*Annex 1*).'

These requirements can be expressed in terms of time needed for leachate to migrate through the liner by advection (gradient 1): 158.6 years (hazardous waste), 31.7 years (nonhazardous waste) and 0.3 year (inert waste), respectively.

If the above conditions cannot be obtained naturally or by conditioning naturally present soil materials, the Directive suggests that a clay liner with similar hydraulic properties and at least 50 cm thickness be installed or that synthetic liners be used.

A drainage layer at least 0.5 m thick and an artificial bottom sealing liner are required in both hazardous and nonhazardous landfills classes. In addition to the drainage layer, if the local authority judges the potential hazard posed by leachate emissions as 'serious', then measures to minimize leachate generation must be adopted, including: topsoil cover at least 1 m thick, impermeable mineral layer, artificial sealing liner (only for hazardous waste landfill) and gas drainage layer (only for nonhazardous waste landfills). Member States have faculty for setting these requirements for inert waste landfills whenever judged necessary (*Annex 1*).

### 10.8.9.2   USA: Subtitle D Landfill

In the 1980s, the EPA and state regulatory agencies adopted the 'dry tomb' landfilling approach. This approach is based on the concept of isolating the waste from water that can generate leachate and lead to groundwater pollution. In theory, if the waste could be isolated from the water that leads to the formation of leachate, then groundwater pollution by landfills could be avoided. The dry tomb landfilling approach, however, leads to a situation where the waste that is isolated from the environment in a compacted soil and plastic sheeting tomb will remain a threat to cause groundwater pollution and to generate landfill gas.

In 1991, the EPA promulgated the Subtitle D regulations based on a single composite liner and equivalent landfill cover, which went into effect in October 1993. Various aspects of this regulation have subsequently been revised (EPA, 1996). Leachate collection and drainage systems shall be constructed in order to keep a maximum 30 cm depth of leachate on the top of the liner. A composite liner must be constructed to achieve this requirement. The upper component of the liner must be a geomembrane, at least 0.8 mm thick. The lower component must be made of a layer of compacted soil, at least 60 cm thick and with a hydraulic conductivity less than $10^{-9}$ m/s. These technical requirements may become more stringent whenever judged necessary. In particular, hydrogeological characteristics of the site, climatic factors, and the quality and quantity of leachate should also be taken into account, and may eventually lead to prescriptions more stringent than those above mentioned.

# References

Abramson, W.L., Lee, T.S., Sharma, S. and Boyce, G.M. (1995): *Slope stability and stabilization methods*. John Wiley & Sons, Ltd, New York, USA.

Allen-King, R.M., Groenevelt, H., Warren, C.J. and Mackay, D.M. (1996): Non-linear chlorinated-solvent sorption in four aquitards. *Journal of Contaminant Hydrology*, 22, 203–221.

Baun, D.L. and Christensen, T.H. (2004): Speciation of heavy metals in landfill leachate: a review. *Waste Management and Research*, 22, 3–23.

Bedient, P.B., Rifai, H.S. and Newell, C.J (1999): *Ground water contamination – Transport and remediation*, 2nd edn. Prentice Hall PTR, Upper Saddle River, USA.

Bennett, P.J., Longstaffe, F.J. and Rowe, R.K. (2000): The stability of dolomite in landfill leachate collection systems. *Canadian Geotechnical Journal*, 37, 371–378.

Benson, C.H. and Daniel, D.E. (1990): Influence of clods in hydraulic conductivity of compacted clay. *Journal of Geotechnical Engineering ASCE*, 116, 1231–1248.

Benson, C.H., Barlaz, M.A., Lane, D.T. and Rawe, J.M. (2004): State-of-practice review of bioreactors landfills. Geo engineering report No. 04-03. Geo Engineering Program, University of Wisconsin–Madison. Madison, USA.

Bonaparte, R., Giroud, J.P. and Gross, B.A. (1989): Rates of leakage through landfill liners. In: IFAI (ed.) *Proceedings of Geosynthetics '89*, vol. 1, pp. 18–29. IFAI, San Diego, USA.

Brune, M., Ramke, H.G., Collins, H.J. and Harnert, H.H. (1991): Incrustation processes in drainage systems of sanitary landfills. In: CISA (ed.) *Sardinia 91, Proceedings of the Third International Landfill Symposium*, pp. 999–1035. CISA, Environmental Sanitary Engineering Centre, Cagliari, Italy.

Buss, S.E., Butler, A.P., Johnston, P.M., Sollars, C.J. and Perry, R. (1995): Mechanisms of lakage through synthetic landfill liner materials. *Journal of the Chartered Institution of Water and Environmental Management*, 9, 353–359.

CEC (1999) Council Directive 1999/31/EC of 26 April 1999 on the landfill of waste. *Official Journal of the European Communities*, L 182, 1–19.

Cheng, H.F. and Reinhard, M. (2006): Sorption of trichloroethylene in hydrophobic microphores of dealuminated Y Zeolites and natural minerals. *Environmental Science and Technology*, 40, 7694–7701.

Chiou, C.T., Lee, J. and Boyd, S.A. (1990): The surface area of soil organic matter. *Environmental Science and Technology*, 24, 1164–1166.

Christensen, T.H., Cossu, R. and Stegmann, R. (1994): *Principles of landfill barrier systems*. In: Christensen, T.H., Cossu, R. and Stegmann, R. (eds) Landfilling of waste: Barriers, pp. 3–10. E&FN Spon, London, UK.

Cooke, A. J., Rowe, R.K. and Rittman B.E. (2000): Modeling clogging of landfill drainage systems. In: SCE (ed.) *Sixth Environmental Engineering Specialty Conference of the CSCE and Second Spring Conference of the Geoenvironmental Division of the Canadian Geotechnical Society*, pp. 74-81. SCE, London, Canada.

Cossu, R. and Lavagnolo, M.C. (1999): Leachate drainage systems: an overview. In: CISA (ed.) *Sardinia 99, Proceedings of the Seventh International Waste Management and Landfill Symposium*, vol. III, pp. 205–214. CISA, Environmental Sanitary Engineering Centre, Cagliari, Italy.

Daniel, D.E. (1987): *Earthen liners for land disposal facilities.* In: Woods, R.D. (ed.) Geotechnical practice for waste disposal. Geotechnical special publications No. 13, pp. 21–39. ASCE, New York, USA.

Daniel, D.E. (1993): *Clay liners.* In: Daniel, D.E. (ed.) Geotechnical practice for waste disposal, pp. 137–163. Chapman and Hall, New York, USA.

EPA (1989): *Requirements for hazardous waste landfill design, construction, and closure.* EPA/625/4-89-022. Center for Environmental Research Information, Cincinnati, USA.

EPA (1993): *Solid waste disposal facility criteria.* EPA530-R-93-017. US Environmental Protection Agency, Washington, D.C., USA.

EPA (1996): *Solid waste disposal facility criteria: Final rule, part II.* Federal Register, 40 CFR, parts 257 and 258. US Environmental Protection Agency, Washington, D.C., USA.

EPA (2001): *Geosynthetic clay liners used in municipal solid waste landfills.* EPA530-F-97-002. US Environmental Protection Agency, Washington, D.C., USA.

Farquhar, G.J. (1994): *Experiences with liners using natural materials.* In: Christensen T.H., Cossu R. and Stegmann R. (eds) Landfilling of waste: Barriers, pp. 37–53. E&FN Spon, London, UK.

Ghassemi, M. (1986): Leachate collection systems. *Journal of Environmental Engineering*, 112, 613–617.

Giroud, J.P. (2000): Lessons learned from failures and successes associated with geosynthetics (Keynote lecture, Eurogeo 2). *Proceedings of the Second European Conference on Geosynthetics*, 1, 77–118.

Giroud, J.P. (2005): *The Mercer Lecture 2005–2006: Contribution of geosynthetic to the geotechnical aspects of waste containment.* Tensar International/International Society for Soil Mechanics and Geotechnical Engineering/International Geosynthetic Society. Located on the internet: 19 October 2009. http://www.ags-hk.org/download/Giroud-MercerLectureHandoutApril20061.pdf.

Gray, D.H. (2002): Optimizing soil compaction and other strategies. Balancing engineering requirements and plant-growth needs in slope protection and erosion control work. *Erosion Control*, 9, 34–41. Located on the internet: 19 October 2009. http://forester.net/ecm_0209_optimizing.html.

Haxo, H.E. and Lahey, T.P. (1988): Transport of dissolved organics from dilute aqueous solutions through flexible membrane liners. *Hazardous Wastes and Hazardous Materials*, 5, 275–294.

Hughes, K.L., Christy, A.D. and Heimlich, J.E. (2003): *Landfill types and liner systems. Extension fact sheet.* Ohio State University, Columbus, USA.

Koerner, R.M. and Soong, T-Y. (2005): Analysis and design of veneer cover soils. *Geosynthetics International*, 12 (1), 28–49.

Korfiatis, G.P., Rabah, N. and Lekmine, D. (1987): *Permeability of compacted clay liners in laboratory scale models.* In: Woods, R.D. (ed.) Geotechnical practice for waste disposal, pp. 611–624. Geotechnical special publications No. 13. ASCE, New York, USA.

Lechner, P. (1994): *Design criteria for leachate drainage and collection systems.* In: Christensen, T.H., Cossu, R. and Stegmann, R. (eds) Landfilling of waste: Barriers, pp. 519–530. E&FN Spon, London, UK.

Levine, A.D., Harwood, V.J., Cardoso, A.J., Rhea, L.R., Nayak, B.S, Dodge, B.M., Decker, M.L., Dzama, G., Jones, L., Haller, E. and Saleh, A.R.M. (2005): *Assessment of biogeochemical deposits in landfill leachate drainage systems.* Report No. 0332006-05. Florida Center for Solid and Hazardous Waste Management, University of Florida, Gainesville, USA.

Maliva, R.G., Missimer, T.M., Leo, K.C., Statom, R.A., Dupraz, C., Lynn, M., and Dickson, J.A.D. (2000): Unusual calcite stromatolites and pisoids from a landfill leachate collection system. *Geology*, 28, 931–934.

McBean, E.A., Rovers, F.A. and Farquhar, G.J. (1995): *Solid waste landfilling engineering and design.* Prentice Hall, New York, USA.

Paksy, A., Powrie, W., Robinson, J.P. and Peeling, L. (1998): A laboratory investigation of anaerobic microbial clogging in granular landfill drainage media. *Geotechnique*, 48, 389–401.

Qian, X., Koerner, R.M and Gray, D.H. (2002): *Geotechnical aspects of landfill design and construction.* Prentice Hall, Upper Saddle River, USA.

Quigley, R.M., Fernandez, F. and Row, R.K. (1988): Clayey barrier assessment for impoundment of domestic waste leachate (Southern Ontario) including clay–leachate compatibility by hydraulic conductivity testing. *Canadian Geotechnical Journal*, 25, 574–581.

Quigley, R.M. and Fernandez, F. (1994): *Effect of organic liquids on the hydraulic conductivity of natural clays.* In: Christensen, T.H., Cossu, R. and Stegmann, R. (eds) Landfilling of waste: Barriers, pp. 203–218. E&FN Spon, London, UK.

Ramke, H.G. (1989): *Leachate collection systems.* In: Christensen, T.H., Cossu, R. and Stegmann, R. (eds) Sanitary landfilling: Process, technology and environmental impact, pp. 343–364. Academic Press, London, UK.

Reinhart, D.R and Chopra, M.B. (2000): MSW landfill leachate collection systems for the new millennium. Report 00-13. Florida Center for Solid and Hazardous Waste Management, University of Florida, Gainesville, USA.

Rowe, R.K. (2005): Long-term performance of contaminant barrier systems. *Geotechnique*, 55, 631–678.

Rowe, R.K., Fleming, I.R., Armstrong, M.D., Cooke, A.J., Cullimore, D.R., Rittmann, B.E., Bennett, P. and Longstaffe, F.J. (1997): Recent advances in understanding the clogging of leachate collection systems. In: CISA (ed.) *Sardinia 97, Proceedings of the Sixth International Landfill Symposium*, pp. 383–390. CISA, Environmental Sanitary Engineering Centre, Cagliari, Italy.

Rowe, R.K., Armstrong, M.D. and Cullimore, D.R. (2000a): Mass loading and the rate of clogging due to municipal solid waste leachate. *Canadian Geotechnical Journal*, 37, 355–370.

Rowe, R.K., Armstrong, M.D. and Cullimore, D.R. (2000b): Particle size and clogging of granular media permeated with leachate. *Journal of Geotechnical and Geoenvironmental Engineering*, 126, 775–786.

Sharma, H.D. and Lewis, S.P. (1994): *Waste containment systems, waste stabilization, and landfills: Design and evaluation.* John Wiley & Sons, Ltd, New York, USA.

Williams, P.T. (1998): *Waste treatment and disposal.* John Wiley & Sons, Ltd, Chichester, UK.

# 10.9

# Landfill Top Covers

**Charlotte Scheutz and Peter Kjeldsen**

*Technical University of Denmark, Denmark*

The purpose of the final cover of a landfill is to contain the waste and to provide for a physical separation between the waste and the environment for protection of public health. Most landfill covers are designed with the primary goal to reduce or prevent infiltration of precipitation into the landfill in order to minimize leachate generation. In addition the cover also has to control the release of gases produced in the landfill so the gas can be ventilated, collected and utilized, or oxidized *in situ*. The landfill cover should also minimize erosion and support vegetation. Finally the cover is landscaped in order to fit into the surrounding area/environment or meet specific plans for the final use of the landfill.

To fulfill the above listed requirements landfill covers are often multicomponent systems which are placed directly on top of the waste. The top cover may be placed immediately after the landfill section has been filled or several years later depending on the settlement patterns. Significant differential settlements may disturb the functioning of the top cover. The specific design of the cover system depends on the type of waste landfilled (municipal, hazardous, or inert waste) and the requirement for protection of the local environment/situation (climatology, morphology, etc.). At modern landfills the cover system is only one of the environmental protection measures which often also include leachate and gas collection and subsequently treatment or recovery. At old abandoned landfills however, top covers may be the only environmental protection measure.

In some landfill regulations (for instance the Subtitle D landfills receiving municipal solid waste in the USA) it is required to minimize infiltration into the waste layers. Therefore top covers containing liner components such as low-permeability clay soils and geomembranes are required. The avoidance of water input to organic waste may impede the microbial stabilization processes including gas generation. Therefore watertight top covers may be in conflict with the purposes of reactor landfills (see Chapter 10.6). At some sites covers sometimes are made to include components for recirculation of landfill leachate (see Section 10.9.2 for more details).

The top cover is an important factor in the water management of landfills. Details about water infiltration through top covers and its influence on the hydrology of the landfill is covered in Chapter 10.3 on landfill hydrology. This chapter provides an overview on the components of traditional top covers used at landfills and the alternatives used for meeting other requirements (such as leachate recirculation or passive gas management by microbial oxidation processes).

| Layer | Function | Materials | Thickness |
|---|---|---|---|
| Surface layer | Support for vegetation Erosion prevention | Topsoil, natural subsoil, compost | 15 - 150 cm |
| Protection layer | Store water for evapotranspiration, protect barrier layer from intrusion, desiccation, freeze/thaw. Maintain stability. Sometimes the surface layer and the protection layer is combined into one layer | Natural subsoil | 60 - 100 cm |
| Drainage layer | Reduce infiltration. Reduce pore water pressures in the cover soil to improve slope stability. | Sand, gravel, geotextile, geonet, geocomposite. Can incorporate pipes for drainage | > 15 cm |
| Barrier layer | Minimize water infiltration and gas escape | Compacted clay liners, geomembrane, geosynthetic clay liners | > 60 cm (compacted clay) |
| Gas collection layer | Collect gas and route it to collection points | Sand, gravel, geotextile, geocomposite or other gas transmitting materials | > 30 cm |
| Foundation layer | Support and found the base of the cover system. Sometimes the gas collection layer and the foundation layer is combined into one layer | Sand, gravel | > 20 cm |
| **Waste** | | | |

*Figure 10.9.1*    *Typical components of a landfill cover system.*

## 10.9.1   Cover System Components

Figure 10.9.1 illustrates the typical components or layers of a final landfill cover system, including a surface layer, a protection layer, a drainage layer, a barrier layer, a gas collection layer, and a foundation layer. In the following the various layers are described in terms of function, material, thickness, and general considerations and concerns. Some of the principles and materials used in landfill top covers are the same as for the bottom lining of landfills. Chapter 10.8 about bottom lining provides details of the different materials which are also used in top covers. Figure 10.9.2 shows two examples of a top cover containing both clay and geomembrane elements to reduce infiltration to the underlying waste layers. Alternative top covers systems containing other functional elements are shown in Figure 10.9.3. The components are:

- Surface layer.
- Protection layer.
- Drainage layer.
- Barrier layer.
- Gas collection layer.
- Foundation layer.

### 10.9.1.1   Surface Layer

The main function of the surface layer is to minimize erosion and function as support for vegetation. Vegetation will promote transpiration of water back to the atmosphere, reduce rainfall impact and decrease wind velocity on the soil

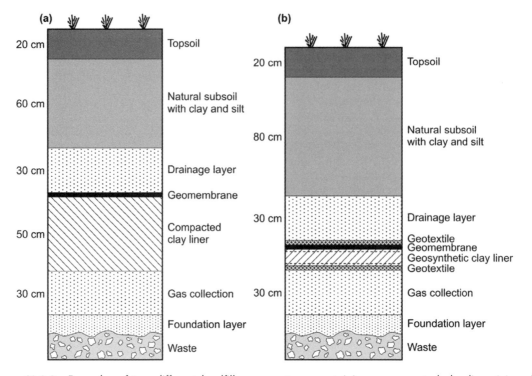

**Figure 10.9.2**  *Examples of two different landfill cover systems containing a compacted clay liner (a) and a geosynthetic clay liner (b) both allowing minimum percolation to the underlying waste layers.*

surface. The surface layer may be landscaped or contoured for ease of surface water runoff, this is particularly important in the presence of steep slopes. The thickness of the surface layer should be a minimum of 15 cm in order to support growth of shallow-rooted plants like grasses (Koerner and Daniel, 1997). However, at sites where trees have to be planted a thicker layer of top soil is needed, up to 150 cm (Cossu, 1994). If the underlying protection layer can accommodate plant roots the surface layer can be reduced. Sometimes the surface layer and the protection layer are combined into one single layer made of the same material. A combined layer should have a thickness of at least 45–80 cm (Cossu, 1994, Koerner and Daniel, 1997, Hauser, 2009). The material used as surface layer has to promote and sustain plant growth and are typically loamy soils. The cost of fertile top soil can be significant in comparison to other materials included in the final cover system. In arid areas where vegetation cannot be sustained other materials like geosynthetics or cobbles can be used as surface layers to minimize erosion from wind and water. The final top slope, after allowance for settling and subsidence should be maintained at 2–5 % to promote runoff of surface water while minimizing erosion (Koerner and Daniel, 1997). Slopes steeper than 5 % promote erosion and require erosion control to be included in the cover design. Slopes steeper than 5–10 % increase instability with a risk of slides.

### 10.9.1.2   Protection Layer

The protection layer underlies the surface layer and serves to protect the underlying drainage layer and barrier layer system from intrusion of plant roots, burrowing animals, and implements. The protection layer also protects the barrier layer from excessive wetting, desiccation, and freeze/thaw. The protection layer will help to store water during wet seasons and increase evapotranspiration minimizing the overall infiltration of water. Often locally available materials are used for construction of the protection layer. Medium textured soils like loams have the best overall characteristics as too finely textured soils tend to accumulate water on the cover surface due to low permeability of the materials, whereas

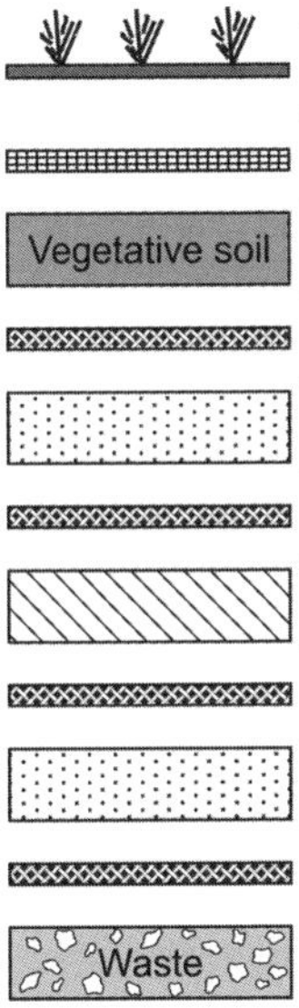

| | 1 | 2 | 3 | 4 | 5 | 6 | 7 | 8 | 9 | 10 |
|---|---|---|---|---|---|---|---|---|---|---|
| Vegetation | | | | | | | | | | |
| Erosion control and vegetation support | | | GG | GG | GG | | | | | |
| Vegetative soil | X | X | X | X | X | X | X | X | X | X |
| Separation | | GT | | GT | | GT | | | | |
| Drainage layer | GV | GV | GD | GN | GD | GV | GV | GD | | |
| Separation | SD | GT | | | | SD | | | | |
| Barrier layer | CL | CL | GM | GM | GB | GB | CL | GM | CL | GB |
| Separation | SD | GT | | | | | | | | |
| Gas collection layer | GV | SD | GD | SD | GD | SD | | | SD | GD |
| Separation | | | SD | | SD | | | | | SD |

**Figure 10.9.3**  *Different examples of potential component sequence for a landfill top cover. Vegetation types: 1: all natural cover; 2: natural cover with geotextile for separation; 3: all synthetic cover; 4: synthetic cover with natural biogas drain; 5: geocomposite cover; 6: natural drainage with geocomposite liner; 7: natural cover without biogas drainage; 8: synthetic cover without biogas drainage; 9: natural cover without water drainage; 10: synthetic cover without water drainage. Abbreviations: CL: clayey soil; SD: sand; GV: gravel; GM: geomembrane; GT: geotextile; GN: geonet; GG: geogrid; GB: geocomposite bentonite; GD: geocomposite drain. Configurations 7 to 10 do not offer the best performance (modified from Cossu 1994). Reprinted with permission from Cossu, R. (1994): Engineering of landfill barrier systems. In: Christensen, T.H., Cossu, R. and Stegmann, R. (eds) Landfilling of waste: Barriers, Chapter 1.2. E&FN Spon. London, UK.*

very sandy soils with a low water retention tend to dry out. If the protection layer is combined with the surface layer it is important that the layer support growth of vegetation. In this case a sandy soil may not contain sufficient moisture and nutrients for plant growth. The required thickness of the protection layer will depend on a number of the site-specific factors. However, if all the above-listed requirements are to be fulfilled a thickness of approximately 1 m is needed.

### 10.9.1.3   Drainage Layer

A drainage layer is often placed below the protection layer and above the barrier layer. The drainage layer primarily serves to drain away infiltrating water and to reduce the head of water on the barrier layer and to dissipate seepage forces. The drainage layer also help to reduce and control pore water pressures in the cover soil and thus improve slope stability. The most commonly used material is sand or sometimes gravel if a high hydraulic transmissivity is needed. A granular drainage layer composed of sand or gravel should have a minimum thickness of 300 mm and a minimum slope of 3 % at the bottom of the layer (Cossu, 1994; Koerner and Daniel, 1997). The hydraulic conductivity of drainage should be no less than $1 \times 10^{-2}$ cm/s (hydraulic transmissivity no less than $3 \times 10^{-5}$ m$^2$/s) at the time of installation (Koerner and Daniel, 1997). Sometimes the water is drained off through porous pipes set in a porous sand or gravel layer or through geotextile or a geonet. If the drainage layer is composed of synthetic materials the same minimum flow capability as a granular drainage layer is required. Alternatively crushed course materials or even shredded tiers can be used. The infiltrating water will drain by lateral run off in the drainage layer. The water from the drainage layer is discharged at the base of the cover system through what is often called a 'toe drain'. Thus it is important that the water can run freely and the drainage layer is not clogged by infiltrating material from the overlying protection layer or biomass from bacterial growth. To prevent clogging of the drainage layer a filter layer of soil (typically well graded sand) or a geotextile between the drainage layer and the overlying protection layer can be necessary.

### 10.9.1.4   Barrier Layer

The barrier layer (also often called the hydraulic barrier) is generally viewed as the most critical component of the landfill cover system. The barrier layer serves a twofold purpose: it prevents percolation of water through the cover system into the waste and prevents the egress of landfill gas from the waste to the atmosphere. Percolation of water is minimized directly by blocking water and indirectly by promoting storage or drainage in the overlying layers, where water is removed by run off, evapotranspiration, or internal drainage. The barrier system forces generated gas to migrate beneath the barrier and must therefore be used in combination with a gas venting or collection system. The barrier layer is a low permeability layer such as a plastic polymer geomembrane, geosynthetic clay liner of bentonite/geotextile fabric or compacted natural clay liners. The barrier layer may be composed of one material or several of the listed materials forming a composite liner. The geomembrane may be used together with either a geosynthetic clay liner or a compacted clay liner or in combination with both. A geomembrane should have a minimum thickness of 1.0 mm. The compacted clay liner should have a minimum thickness of 600 mm and a hydraulic conductivity of $< 1 \times 10^{-7}$ cm/s (Koerner and Daniel, 1997). Most geosynthetic clay liners have hydraulic conductivities in the vicinity of $1 \times 10^{-9}$ to $5 \times 10^{-9}$ cm/s (Koerner and Daniel, 1997). Protection of the barrier layer is very critical for optimal performance. All barrier liners are susceptible towards settlement of the waste mass as this will induce stress on the barrier liner and risk of tear/failure. Clay liners have to be protected towards drying and freezing as this might result in cracking of the liner and increase the permeability of the liner. Extreme wetting could lead to stability problems. To minimize the risk of slope instability it is important that adequate drainage is provided above the barrier layer. The drainage system should conduct flow and discharge of water to the toe of the cover or to other suitable outlets.

### 10.9.1.5   Gas Collection Layer

The gas collection layer is a porous material through which gas can easily migrate to the gas collection and control system. The material used for construction of the gas collection layer should be coarse-grained highly permeable soils like sand or geosynthetics including geotextiles, geonet composites or related drainage geocomposites. When using natural soils the layer should be of minimum 300 mm thickness (Koerner and Daniel, 1997). In a gas collection layer consisting of natural soils such as sand, horizontal perforated pipes for gas collection might be incorporated to ease the gas transport to the gas collection and treatment system.

### 10.9.1.6   Foundation Layer

The foundation layer is the lowest layer forming the sole of the cover system. The foundation layer is placed directly on top of the waste or on top of the daily soil covering layer. In cases where a granular material is used as foundation layer it can also act as a gas collection layer. The disposed waste mass will settle over time due to decomposition of the organic waste. This settlement will cause the subsidence of the upper surface. The total settlement can amount to 10 –20 % of the depth of the landfilled MSW. In addition to total settlement also differential settlement might occur causing stress in the cover system components, e.g. pipes, geomembranes, geotextiles, genets, etc. To prevent failure of the cover system performance due to settlement it is important that the foundation layer is well compacted and graded to accommodate later settlement of the final cover. Also when top covers are constructed the risk of total and differential settlements of the waste layers should be analyzed.

## 10.9.2   Alternative Landfill Top Cover Concepts

Within the last 10–15 years other ideas of how to finally cover landfills have emerged. These alternative cover concepts have been put forward either to use other materials, to optimize the evaporation from the cover resulting in a lower infiltration, to allow water addition for waste stabilization purposes, or for avoiding landfill gas emissions especially at landfills with relatively low gas generation rates. These alternatives are briefly presented in this section.

### 10.9.2.1 Capillary Barrier Concept

One alternative cover can include a capillary barrier to hinder infiltration of water into the waste. The principle of a capillary barrier is based on the difference in grain size between two materials. A capillary barrier consists of a layer of fine material (capillary layer), which is build above a layer of coarse material (capillary block) on a slope. The sealing effect is based on capillary forces and the principles of unsaturated flow. Due to capillary forces, the infiltrating water is held back in the fine pores of the capillary layer while the coarse pores in the capillary block stay nearly dry. As long as the capillary attraction at the interface remains strong, the hydraulic conductivity of the fine sand is higher that that of the underlying gravel, thus limiting infiltration; if the fine sand is sufficiently well drained, the sand maintains enough capillary force to maintain this effect. It is important that the interface is an inclined plane with a slope between 5° and 10° (Barres and Bonin, 1996), so that water can flow laterally along the inclined interface of the layers and be collected. Infiltrated water will accumulate just above the interface without penetrating into the underlying gravel until the force of gravity becomes greater than the forces of capillary pressure. At this point water will pass through the interface and flow rapidly through the coarse material. It is therefore critical to maintain a relatively short distance (5–50 m depending on the infiltration rate and the hydraulic properties of the material used) between drain pipes to avoid breakthrough of water collected by the fine layer into the coarser layer (Ross, 1990). Plant cover increases evapotranspiration and thus reduces water infiltration in the fine sand layer. It is crucial to the performance of the capillary barrier that the fine-textured material of the upper layer is not allowed to migrate down into the coarse-textured layer. To prevent this from happening, a geotextile should be considered as separator between the two layers (Koerner and Daniel, 1997). In periods with high and intense rainfall the coarse-textured layer might become moist and lose its water-impeding capability, at least temporarily. The capillary barrier might replace the use of hydraulic barriers in arid or semiarid areas. One advantage of the capillary barrier might be that local materials can be used. At landfills with low gas production, capillary barriers combined with a top layer for methane oxidation (see later) has been suggested as an alternative to conventional gas collection and treatment (Fornés *et al.*, 2003). In this case the capillary layer prevents water infiltration into the landfill body whereas the capillary block works as a gas distribution layer (Wawra and Holfelder, 2003). The use of capillary barriers at landfills have been studied in various pilot-scale systems (Nyhan *et al.*, 1990; Fayer *et al.*, 1992; Wagner and Schnatmeyer, 2002), however full-scale experiments in the field are still rare (Barres and Bonin, 1996; Warren *et al.*, 1996; Gee and Ward, 1997; Albright *et al.*, 2004).

### 10.9.2.2 Evapotranspiration Landfill Cover Concept

The evapotranspiration landfill cover is an alternative cover design aiming at reducing water infiltration into the landfill. Evapotranspiration covers for landfills and waste sites have recently been described in detail by Hauser (2009). The characteristics of an evapotranspiration cover are that it utilizes a layer of soil covered with native grasses and contains no barrier system. The soil layer provides a water reservoir, which is emptied mainly by evaporation from the soil and plant transpiration. Evapotranspiration is the sum of evaporation of water from the soil surface and plant transpiration (primarily through the stomata of the plants leaves). Infiltrating rainfall will move downward through the cover as a front as it advances deeper into the soil. Evaporation and water extraction by plant roots establishes an upward gradient reversing the direction of movement of the soil water, thus reducing the amount of water moving downward. The plants growing on the cover surface are an important feature of the evapotranspiration cover as they remove water faster than evaporation alone. However, if more water infiltrates through the surface than the soil layer can hold at field capacity, some of it moves through the soil profile and leach into the waste generating leachate. It is therefore critical to the cover function that the soil can hold enough water to minimize water movement below the cover. The soil should therefore have a high water holding capacity and preferably the pore volume should be contained within the midsize pores because they hold much water against the force of gravity, yet plants easily and quickly can remove water from them. The soil should support rapid and prolific root growth in all parts of the cover. The vegetation established on the cover should be native to the site and adapted to the soil. The plants will also reduce erosion of the soil cover and can contribute positively to the final landscaping of the landfill. More details of the hydrological processes in an evaporative landfill cover can be seen in Chapter 10.3 on landfill hydrology.

The advantages of the evapotranspiration cover are that it reduces water infiltration by the use of natural processes, is simple in terms of technological level and can be implemented at many sites. It has been claimed that because evapotranspiration covers are natural these are self-renewing, less prone to failure leading to a long service life (Hauser, 2009). Often it will have low costs for both construction and maintenance. The evapotranspiration cover can due to the plant cover be an esthetic benefit to the site. The evapotranspiration cover can be designed to accommodate gas control by installation of a gas collection layer beneath the soil cover combined with horizontal or vertical collection pipes. Gas collection pipes might be installed during or even after cover placement as the cover does not contain any barrier, the integrity of the cover is not threaten by drilling and installation of gas control pipes and wells. The evapotranspiration cover has successfully been combined with a methane oxidation layer reducing the gas emission from the landfill (Fornés *et al.*, 2003; Berger *et al.*, 2005). The disadvantages are that each site needs site-specific design because of differences in climate, soil properties, and plant cover. The design of an evapotranspiration cover is best obtained by the aid of a model, which based on site-specific input data such as weather (precipitation, solar radiation, temperature), soil data (particle size distribution, bulk density, pore volume, etc.), and plant properties (biomass–energy ratio, optimal and minimum temperature required for growth, maximal potential leaf area index, etc.) can evaluate the numerous interactions among soils, plants, and climate (Hauser, 2009). The construction of an evapotranspiration cover requires that relative large volume of adequate soil is locally available.

### 10.9.2.3   Leachate Recirculation Landfill Cover System

Landfill covers are commonly designed with the main objective to prevent infiltration of precipitation into the waste in order to minimize leachate production and consequently collection and treatment. An alternative landfilling concept is to recirculate the collected leachate in order to accelerate the degradation of the organic waste to reach stabilization in a shorter timeframe. The leachate is withdrawn continuously at the base of the waste through the leachate collection system and reintroduced back at the upper surface of the waste. In this case settlement of the waste will occur faster and the phase with high gas and leachate production and required treatment is shortened. The leachate may be reintroduced into the waste by direct surface application by spraying, through gravity infiltration using open trenches or vertical wells, or by active subsurface injection from manifold systems or vertical wells.

### 10.9.2.4   Methane Oxidation Landfill Cover System

Historically proper landfill cover design has primarily focused on prevention of leachate production, and only secondary on gas control mainly aiming at avoidance of on-site fires and explosions. In recent years it has become clear that methane emitted from landfills to the atmosphere is a significant anthropogenic methane emission source contributing to global climate change. Landfill gas extraction and utilization plants have been made mandatory at new waste disposal sites in many countries. At the same time, research has focused increasingly on development of low-cost technologies that limit landfill gas release from existing landfills where gas collection systems have not been implemented and/or are not economically feasible. Much of that research has focused on biocovers designed to optimize and sustain methane oxidation as a cost-effective technology for controlling emissions from waste disposal sites. Microbial methane oxidation as a technology to reduce landfill gas emissions has been reviewed recently by Scheutz *et al.* (2009).

A biocover is a landfill cover system that has been designed to optimize environmental conditions for biological methane consumption so that the system functions as a vast biofilter. The cover typically consists of a gas distribution layer with high gas permeability to homogenize landfill gas fluxes (e.g. gravel, crushed construction waste, shredder waste, etc.) and an overlying oxidation layer designed to support the methanotrophic populations that consume the methane for carbon and energy. Alternatively to a biocover system, the methane oxidation might take place in a biowindow system. Whereas biocovers are designed to cover all or large sections of a landfill, biowindows are relatively small units at the cover where the biofilter material is integrated into the landfill cover (Figure 10.9.4). Both the biocover and the biowindow system are operated in a passive way, where the landfill gas flows from the waste directly to the filtration material governed by pressure gradients or by diffusion when pressure gradients are low. In biowindows the gas is routed to the biowindow

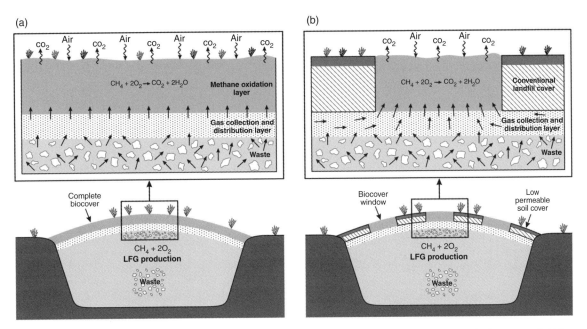

***Figure 10.9.4***   *Designs of landfill covers intended for methane oxidation. (a) The concept of a biocover. (b) The concept of a biowindow cover system.*

naturally as the filter material has higher permeability in comparison to the surrounding landfill cover. A gas collection layer beneath the hydraulic barrier layer leading to the biowindow facilitates gas transportation from the waste to the biowindow (Figure 10.9.4). At landfills with gas collection systems consisting of perforated pipes, the gas might be treated in an actively operated methane oxidation system often referred to as a biofilter. This might be operated in an active or passive way and may be incorporated into the cover system or placed above ground on the site. Various materials have been tested in methanotrophic biofilters, such as composts including composted wastes, wood chips, bark mulch and peat, inorganic materials such as glass beads, bottom ash or porous clay pellets, as well as mixtures of organic and inert materials (Scheutz *et al.*, 2009). In general, the filter material should offer high gas permeability (a large permeable pore space), a large surface area and good environmental conditions suitable for proliferation of methanotrophs. The filter material should provide sufficient water-holding capacity at high gas permeability, support microbial growth, support evapotranspiration, and hinder desiccation and cracks. A high pore space will also minimize the risk of clogging due to accumulation of biofilm or exopolymeric substances. If organic materials like composts are used, high stability should be ensured to prevent microbial degradation and settlement, thereby avoiding potential reductions in gas permeability. Using mature compost is important to minimize oxygen competition between methane oxidizers and other heterotrophs and also avoid anaerobic conditions, which could result in methane production. Favorable environmental conditions are provided by materials with circumneutral pH values and offering an adequate supply of nutrients. A good filter material is also homogenous to prevent preferential flow causing overload to some areas of the filter increasing the risk of hotspot with high methane emission on the cover. Finally it is important that the cover system is designed in a way so infiltrating water can drain away and does not allow to accumulate in the filter material causing clogging and anaerobic conditions. In the case with biowindows, runoff of surface water into the biowindows should be prevented.

Engineered landfill covers systems designed for methane oxidation may serve as a complement to gas collection systems at large landfills or as a sole removal mechanism at smaller or older landfills or landfills containing inert solid waste or mechanically and biologically pretreated waste with low methane production potential. As 'low technology' systems, biocovers may offer many economic advantages, including low operation and installation expenses, and low (or no) maintenance requirements, which make them particularly suitable for developing countries. Finally, these systems foster the use of materials like sewage sludge, municipal solid waste, biowaste, or yard waste that would otherwise be disposed off as wastes. In addition to mitigating methane emissions, it has been demonstrated that properly designed landfill cover materials can degrade a wide range of volatile organic compounds, including halogenated hydrocarbons, some of which are more than a 1000-fold more greenhouse-active than methane (Scheutz and Kjeldsen, 2003, 2005; Scheutz *et al.*, 2003, 2004, 2008).

---

**Box 10.9.1    Methane Oxidation in Landfill Cover Systems – Environmental Factors, Oxidation Rates and Capacities.**

*Methane oxidation.* In landfill covers, methane and oxygen counter gradients may appear due to emission of methane from the waste and influx of oxygen from the ambient air, which provides the necessary conditions for the development of methanotrophic bacteria. Their activity depends on the presence of sufficient concentrations of both methane and oxygen, and thus they tend to be confined to fairly narrow horizontal bands within their habitat, limited in their distribution by the downward diffusion of atmospheric oxygen and the upward diffusion of methane. In general, the methanotrophic active zone is located in the upper 30–40 cm of the soil cover profile with maximal oxidation activity in a zone between 15–20 cm below the surface (Scheutz *et al.*, 2009).

  *Environmental factors.* Methane oxidation is controlled by a number of environmental factors: soil texture, temperature, soil moisture content, methane and oxygen supply, nutrients, etc. Thus, the climatic conditions are of importance for the actual methane oxidation rate. In landfill soil covers temperature and soil moisture are very important parameters controlling methane oxidation. Optimum temperatures are around 25–35 °C for methane oxidation in soil environments although methane oxidation can occur down to 1–2 °C (Scheutz *et al.*, 2007; seed references provided therein). The soil moisture optimum for landfill covers a range of 10–20 % w/w (Scheutz *et al.*, 2009). However, in some cases higher soil moisture optimums are observed (Scheutz *et al.*, 2007; and references provides herein). The oxidation activity is significantly reduced when soil moisture content decreases below 5 % (Scheutz *et al.*, 2007). In landfill covers, the oxygen penetration depth will often be the limiting factor of the methane oxidation process, making soil composition, particle size, and porosity important controlling parameters.

  *Methane oxidation rates and capacities.* Landfill cover containing organic material can develop a high capacity for methane oxidation by selection of methanotrophic bacteria. The table below shows a short summary of methane oxidation rates from column experiments with organic material reported in the literature. In general, steady-state methane oxidation rates for landfill covers are 100–150 g $CH_4/m^2$/day (30–60 % removal) with maximum rates up to 250 g $CH_4/m^2$/day (80–100 % removal; Scheutz *et al.*, 2009). It is important to note that only a few column experiments have been run for more than 250 days and the performance over years as well as the seasonal influence of temperature and precipitation remains unknown.

| | | Methane oxidation rate | | |
| | | Steady state | | Maximum |
| Reference | Filter material | (g $CH_4/m^2$/day) | (%) | (%) |
|---|---|---|---|---|
| Powelson *et al.* (2006) | Mix of compost and polystyrene pellets | 242 | 69 | 72 |
| | Coarse sand | 203 | 63 | 58 |
| Kettunen *et al.* (2006) | Mix of mature sewage sludge compost, deinking waste, and sand (4:2:4) | 31.4 | 97 | |
| | Mix of mature sewage sludge compost, deinking waste, and bark chips (4:2:4) | 25.7 | 74 | |

| | | | | |
|---|---|---|---|---|
| Stein and Hettiaratchi (2001) | Sedge peat moss | 88 (low load), | 55 | 90 |
| | | 93 (high load) | 29 | 50 |
| Humer and Lechner (1999) | Sewage sludge-compost/sand mix (70:30) | 135 | 75 | 100 |
| Du Plessis *et al.* (2003) | Composted pine bark/perlite mixture (1:3) | 38 | 70 | |
| Berger *et al.* (2005) | 30 cm of compost/sand mix on top of 90 cm of sand | 52–54 | 94–98 | 98 |
| Haubrichs and Widmann (2006) | Yard waste compost | 583 | 96 | 100 |
| | Yard waste compost mixed with wood chips (1:1) | 476 | 93 | 100 |
| Scheutz *et al.* (2008) | Compost/wood chips (1:1) | 161 | 58 | 100 |
| | Compost/sand (1:1) | −31 | −10 | 48 |
| | Compost/sand (1:5) | 29 | 12 | 60 |
| | Supermuld® | 110 | 48 | 84 |
| Wilshusen *et al.* (2004) | Compost – leaves | 100 | 19 | 77 |
| | Compost - garden | 0 | 0 | 10 |
| | Compost – wood chips | 100 | 19 | 19 |
| | Compost – MSW | 100 | 19 | 52 |

# References

Albright, W.H., Benson, C.H., Gee, G.W., *et al.* (2004): Field water balance of landfill covers. *Journal of Environmental Quality*, 33, 2317–2332.

Barres, M.and Bonin, H. (1994): *The 'capillary barrier' for surface capping*. In: Christensen, T.H., Cossu, R. and Stegmann, R. (eds) Landfilling of waste: Barriers, Chapter 2.8. E&FN Spon, London, UK.

Berger, J., Fornés, L.V., Ott, C., Jager, J., Wawra, B. and Zanke, U. (2005): Methane oxidation in a landfill cover with capillary barrier. *Waste Management*, 25, 369–373.

Cossu, R. (1994): *Engineering of landfill barrier systems*. In: Christensen, T.H., Cossu, R. and Stegmann, R. (eds) Landfilling of waste: Barriers, Chapter 1.2. E&FN Spon. London, UK.

Du Plessis, C.A., Strauss, J.M., Sebapalo, E.M.T. and Riedel, K.-H.J. (2003): Empirical model for methane oxidation using a composted pine bark biofilter. *Fuel*, 82, 1359–1365.

Fornes, L., Ott, C. and Jager, J. (2003): Development of a landfill cover with capillary barrier for methane oxidation – methane oxidation in a compost layer. In: Christensen, T.H., Cossu, R., Stegmann, R. (eds) Proceedings of the ninth international waste management and landfill symposium. CD-ROM. CISA, Environmental Sanitary Engineering Centre, Cagliari, Italy.

Gee, G.W. and Ward, A.L. (1997): Still in quest of the perfect cap. In: Reynolds, T.D. and Morris, R.C. (eds) Landfill capping in the semi-arid west: Problems, perspectives and solutions, pp. 145–164. Environmental Science and Research Foundation, Idaho Falls, USA.

Haubrichs, R. and Widmann, R. (2006): Evaluation of aerated biofilter systems for microbial methane oxidation of poor landfill gas. *Waste Management*, 26, 408–416.

Hauser, V.L. (2009): Evapotranspiration covers for landfills and waste sites. CRC Press, Miami, USA.

Humer, M. and Lechner, P. (1999): Alternative approach to the elimination of greenhouse gases from old landfills. *Waste Management and Research*, 17, 443–452.

Kettunen, R. H., Einola, J. K. M. and Rintala, J. A. (2006): Landfill methane oxidation in engineered soil columns at low temperature. *Water Air and Soil Pollution*, 177 (1/4), 313–334.

Koerner R.M. and Daniel, D.E. (1997): Final covers for solid waste landfills and abandoned dumps. ASCE Press, Richmond, USA.

Nyhan, J.W. (2005): A seven-year water balance study of an evapotranspiration landfill cover varying in slop for semiarid regions. *Vadose Zone Journal*, 4, 480–499.

Nyhan, J.W., Hakonson, T.E.and Drennon, B.J. (1990): A water balance study of two landfill cover designs for semiarid regions. *Journal of environmental Quality*, 19, 281–288.

Nyhan, J.W., Schofield, T.G. and Starmer, R.H. (1997): A water balance study of four landfill cover designs varying in slope for semiarid regions. *Journal of Environmental Quality*, 26, 1385–1392.

Powelson, D.K., Chanton, J., Abichou, T. and Morales, J. (2006): Methane oxidation in water-spreading and compost biofilters. *Waste Management and Research*, 24, 528–536.

Ross, B. (1990): The diversion capacity of capillary barriers. *Water Resources Research*, 26, 2625–2629.

Scheutz, C. and Kjeldsen, P. (2003): Capacity for biodegradation of CFCs and HCFCs in a methane oxidative counter-gradient laboratory system simulating landfill soil covers. *Environmental Science and Technology*, 37, 5143–5149.

Scheutz, C. and Kjeldsen, P. (2004): Environmental factors influencing attenuation of methane and hydrochlorofluoro-carbons in landfill cover soils. *Journal of Environmental Quality*, 33, 72–79.

Scheutz, C. and Kjeldsen, P. (2005): Biodegradation of trace gasses in simulated landfill soil cover. *Journal of the Air and Waste Management Association*, 55, 878–885.

Scheutz, C., Bogner, J., Chanton, J., Blake, D., Morcet, M. and Kjeldsen, P. (2003): Comparative oxidation and net emissions of methane and selected non-methane organic compounds in landfill cover soils. *Environmental Science and Technology*, 37, 5150–5158.

Scheutz, C., Pedersen, G.B. and Kjeldsen, P. (2008): Biodegradation of methane and halocarbons in simulated landfill biocover systems containing compost materials. *Journal of Environmental Quality*, 38, 1363–1371.

Scheutz, C., Kjeldsen, P., Bogner, J., De Visscher, A., Gebert, J., Hilger, H., Huber-Humer, M. and Spokas, K. (2009): Microbial methane oxidation processes and technologies for mitigation of landfill gas emissions. *Waste Management and Research*, 27, 409–455.

Stein, V.B. and Hettiaratchi, J.P.A. (2001): Methane oxidation in three Alberta soils: influence of soil parameters and methane flux rates. *Environmental Technology*, 22, 101–111.

Wagner., J.-F. and Schnatmeyer, C. (2002): Test field study of different cover sealing systems for industrial dumps and polluted sites. *Applied Clay Science*, 21, 99–116.

Warren, R.W., Hakonson, T.E. and Bostik, K.V. (1996): Choosing the most effective hazardous waste landfill cover. *Remediation*, 1996, 23–41.

Wawra, B. and Holfelder, T. (2003): Development of a landfill cover with capillary barrier for methane oxidation – the capillary barrier as gas distribution layer. In: Christensen, T.H., Cossu, R., Stegmann, R. (eds) *Proceedings of the ninth international waste management and landfill symposium*. CD-ROM. CISA, Environmental Sanitary Engineering Centre, Cagliari, Italy.

Wilshusen, J. H., Hettiaratchi, J. P. A. and Stein, V. B. (2004): Long-term behaviour of passively aerated compost methanotrophic biofilter columns. *Waste Management*, 24, 643–653.

# 10.10

# Landfilling: Gas Production, Extraction and Utilization

**Hans Willumsen**

*LFG Consult, Denmark*

**Morton A. Barlaz**

*North Carolina State University, USA*

Landfill gas (LFG) generation and management have a significant effect on the environmental profile of solid waste landfills. Methane is a greenhouse gas with an infrared activity 25 times that of $CO_2$ for a 100-year time horizon (Solomon *et al.*, 2007). In addition, methane is flammable when present at 5–15 % in air. LFG migration in the subsurface has led to its accumulation in buildings and subsequent explosions that have resulted in death and serious injury. In contrast, when captured, this $CH_4$ can be flared, in which case it is converted to biogenic $CO_2$. If the LFG is converted to energy in the form of heat or electricity the methane is also converted to biogenic $CO_2$ and in addition may offset the utilization of fossil fuels and the resulting emissions.

Previous chapters have presented the microbial ecology of methane generation in anaerobic waste (Chapter 9.4) and how the organic waste is degraded and generates LFG in the reactor landfill (Chapter 10.6). This chapter presents a model for estimating LFG generation over time and discusses various approaches to gas extraction and utilization as an energy source.

## 10.10.1 Modeling of LFG Production

Models to predict the amount of LFG that can be expected are used in at least three applications:

- To determine whether, and at what time, the amount of gas produced will be sufficient to support a beneficial use as described later in this section.
- To size gas collection systems.
- For national inventories of $CH_4$ production from landfills for purposes of national greenhouse gas reporting.

*Solid Waste Technology & Management*   Edited by Thomas Christensen
© 2011 Blackwell Publishing Ltd

In describing LFG modeling, it is critical to differentiate between gas production and gas collection. While only gas collection is typically measured, the fraction of the generated gas that is collected will vary with both the cover permeability and the extent of coverage of the gas collection system. In addition, it is important to clearly state whether calculated flowrates represent $CH_4$, or total LFG ($CH_4$ and $CO_2$).

### 10.10.1.1   Methane Production Model

Methane production from landfills is typically modeled using some form of the following equation:

$$G = W_a L_o k e^{-kt}$$
(10.10.1)

where

$G$ = $CH_4$ production rate from a single year's refuse ($m^3$ $CH_4$/year);
$W_a$ = annual waste acceptance rate (t/year);
$L_o$ = ultimate methane yield ($m^3$ $CH_4$/t);
$k$ = decay rate (per year);
$t$ = years since the landfill started receiving waste.

The LandGem model provided by the United States Environmental Protection Agency (EPA, 2005) is based on this equation. The integrated form of Equation (10.10.1) is:

$$G^* = W L_o (e^{-kc} - e^{-kt})$$
(10.10.2)

where

$G^*$ = $CH_4$ production rate from all years' refuse ($m^3$ $CH_4$/year);
$W$ = the average annual waste acceptance rate (t/year);
$c$ = time since waste burial ceased;
$L_o$, $t$ and $k$ are as defined for Equation (10.10.1).

Equation (10.10.1) is more flexible as it readily accommodates annual variation in the waste disposal rate.

The degradation rate is also some times expressed as the halflife of the organic fraction. The relation between $k$ (decay rate, per year) and the halflife (years) is determined by:

$$k = \ln 2 / halflife \ (years)$$
(10.10.3)

Note that there is no lag time between refuse burial and the onset of $CH_4$ production in the equations. Thus, the peak $CH_4$ production rate is predicted to occur at the time of refuse burial. While this does not appear to be the case based on actual field observation, the error associated with the absence of a lag term does not seem to be significant given the uncertainties in the parameters used in predicting LFG production. However, it would be a simple matter to add a lag time to these equations.

The most common variation on Equation (10.10.1) is to divide the waste into a series of fractions and to specify a unique $k$ and $L_o$ for each fraction. Some divide the waste into rapid, moderate and slowly degradable fractions, while others consider each biodegradable material fraction of the waste individually (e.g. newsprint, grass, office paper, etc.). This is typically referred to as a multi-phase model. However, in the absence of good information on each waste fraction, the multiphase model may not result in an improved prediction relative to Equation (10.10.1) used on the mixed waste.

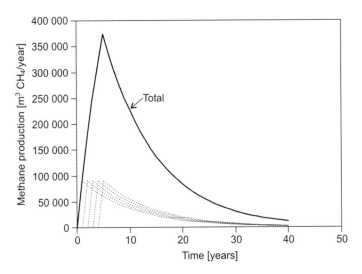

**Figure 10.10.1** *Predicted total methane production from a landfill cell that over 5 years received 10 000 t/year waste (moisture content = 20 % w/w, $L_o$ = 125 m³ CH₄/t of dry waste, k = 0.1 per year). The methane production is for each year predicted by Equation (10.10.1).*

Figure 10.10.1 shows the total methane production from a landfill cell that over 5 years received 10 000 t/year waste (moisture content = 20 % w/w, $L_o$ = 125 m³ CH₄/t dry waste, $k$ = 0.1 per year).

As a first estimate, an average of 5 m³ LFG/year/t landfilled waste can be used (assumed moisture content = 20 % w/w, $L_o$ = 125 m³ CH₄/t dry waste, 66 % CH₄ in LFG, 20 years production, about two-thirds capture of $L_o$). Figure 10.10.2 shows actually measured LFG capturing rates at landfills of different age. Although 5 m³ LFG/t landfilled waste/year can be used as a first estimate, large variations among landfills are observed.

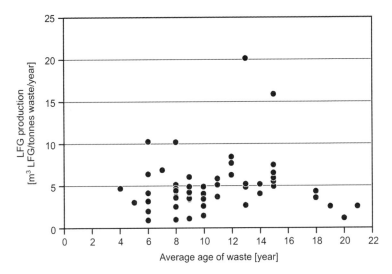

**Figure 10.10.2** *Actually measured LFG capturing rates (m³ LFG/year/t landfilled waste) at landfills of different age (after Willumsen and Bach, 1991). Reproduced with permission from CISA, Italy. © (1991) EuroWaste srl, Italy.*

**Table 10.10.1**  *Summary of LFG production default parameters.*

| Source | $k$ (per year) | $L_0$ (m$^3$ CH$_4$/t dry) | Comments |
|---|---|---|---|
| EPA (1998) | 0.04 (wet), 0.02 (dry) | 125 | Assuming waste at 20% moisture on a wet weight basis |
| EPA (1996) | 0.05 | 340 | Assuming waste at 20% moisture on a wet weight basis |
| Coops *et al.* (1995) | 0.185 (rapidly degradable), 0.1 (moderately degradable), 0.03 (slowly degradable) | 184 | Assuming waste at 20% moisture on a wet weight basis, and waste with an organic carbon content of 136 kg C/wet tonnes |
| SCS Engineers and Augenstein (1997) | 0.06–0.08 (rapidly degradable) 0.04–0.06 (slowly degradable) | 82–97 | Assuming waste at 20% moisture on a wet weight basis; range based on alternate techniques for data analysis |
| IPCC (2006) | 0.05–0.17 | | See Table 10.10.2 |
| Faour *et al.* (2007) | 0.3 | 125 | For a bioreactor landfill, assuming waste at 20% moisture on a wet weight basis |

## 10.10.1.2  Methane Production Default Parameters

There is no single definitive set of defaults for Equation (10.10.1) and selected values are summarized in Table 10.10.1 (Barlaz, 2004). There is significant uncertainty in all these estimates. For example, the EPA (1998) states in a document on LFG modeling that 'Although the recommended $k$ and $L_0$ are based upon the best fit to 21 different landfills, the predicted methane emissions ranged from 38 to 492% of actual, and had a relative standard deviation of 0.85.' Sources of uncertainty include: (1) knowledge of the exact mass of refuse (and its age) under gas collection, (2) the fraction of the produced gas that is collected, (3) variation in waste composition from year to year, since only the organic fraction contribute to methane generation. IPCC (2006) estimated a range of $k$ values for various climate zones and organic waste fractions as shown in Table 10.10.2.

Finally, as described in Chapter 10.6, landfills may be operated to accelerate refuse decomposition, which results in more rapid CH$_4$ production. In theory, this should increase $k$ for the same value of $L_0$. In practice, as relatively short periods of LFG production data are fit to existing models, both $k$ and $L_0$ are adjusted (Faour *et al.*, 2007).

**Table 10.10.2**  *Default k values for various climate zones and organic waste fractions (IPCC, 2006).*

| | Nontopical (MAT < 20°C) | | Tropical (MAT > 20°C) | |
|---|---|---|---|---|
| | MAP/PET < 1 | MAP/PET > 1 | MAP < 1 m | MAP > 1 m |
| Slowly degradable | | | | |
| • Paper and textile | 0.02 | 0.06 | 0.045 | 0.07 |
| • Wood and straw | 0.02 | 0.03 | 0.025 | 0.035 |
| Moderately degradable | | | | |
| • Other nonfood | 0.05 | 0.10 | 0.065 | 0.17 |
| • Garden and park | 0.05 | 0.10 | 0.065 | 0.17 |
| Rapidly degradable | | | | |
| • Food waste | 0.06 | 0.185 | 0.085 | 0.4 |
| Average | 0.05 | 0.09 | 0.065 | 0.17 |

MAT: Mean annual temperature.
MAP: Mean annual precipitation.
PET: Potential evapotranspiration.

## 10.10.2 LFG Management Systems

Having established that gas will be produced, attention must now be given to its collection and management. Proper gas management is critical to avoid explosion risks as well as uncontrolled pollution of both the local area and the atmosphere. LFG management systems can be classified as either passive or active.

### 10.10.2.1 Active Gas Management System

In an active gas management system, gas is removed from a landfill by the application of suction to a series of wells placed in the landfill. The most common system is based on the extraction from vertical perforated pipes, as illustrated in Figure 10.10.3. Vertical wells are normally installed after waste placement by drilling through the refuse, or in some cases installed during filling of the landfill. Alternatively, if gas collection is to be initiated early in the landfill life, then a horizontal system of perforated pipes can be installed while filling the landfill, with additional layers of pipe added as the refuse height increases. Gas can also be extracted from the leachate collection system, including the leachate collection pipes as well as manholes and sumps.

The efficiency of gas collection is affected by the permeability of the landfill cover. A low permeability cover can increase gas collection efficiency relative to a high permeability cover, because with low permeability covers the vacuum induced will extend wider. Equally important is the extent of coverage of the well field. If the well field is only installed once a low permeability cover is in place, then there will be sections of the landfill where gas is not collected for several years. Thus, the overall landfill collection efficiency may be maximized by the early installation of gas wells, even with

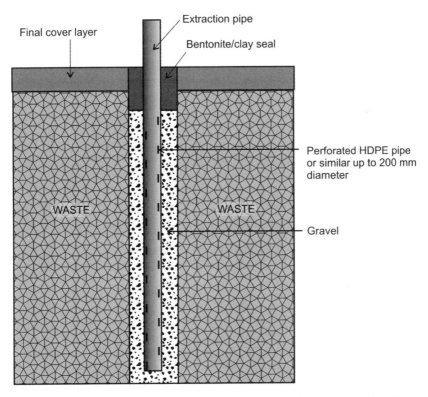

**Figure 10.10.3** *Active gas venting by vertical gas extraction well. Note that the cap in this illustration is conceptual, as the final cap will have multiple layers, as described in Chapter 10.9.*

intermediate (higher permeability) covers. There is also a trade-off between cover permeability and moisture infiltration. To the extent that the cover system restricts moisture infiltration and the refuse is relatively dry, this may reduce the rate of gas production. One way to address this trade-off is to design a landfill with a horizontal gas collection system, and a biologically active intermediate cover that allows moisture infiltration but also promotes $CH_4$ oxidation. Biologically active covers are described in Chapter 10.9.

Gas extraction wells and/or trenches are connected to a pump or compressor that pulls gas from the landfill and conveys it to the utilization system. There are several alternatives for the connection of a single well or trench to a utilization system. The oldest and most common system is to connect individual wells and/or trenches to a main collection pipe. The major limitation to this is the difficulty in regulating both the quality and the quantity of the gas. When there is just one point of control, the vacuum to individual wells and trenches is difficult to control. A second problem is the difficulty in identifying leaks when all the wells are connected in one large system. Thus, a better extraction system is one in which single wells are connected individually to regulation stations so that different sections of the gas collection system can be isolated if problems arise in the individual wells. The regulation of individual wells can even be automated by continuous monitoring of gas quality (Lagerkvist *et al.*, 2000). A major issue is the infiltration of oxygen and nitrogen that results from too much vacuum on the system that can lead to air intrusion through the cover.

The collected gas can be flared, used in energy production or upgraded to a product for distribution, as described later in this chapter.

### 10.10.2.2   Passive Gas Management System

In a passive system, the gas is vented in the absence of a vacuum extraction system. Passive systems are more typical used at smaller landfills with little gas production and at sites where active systems are not required by regulation. Passive systems can be used to better control the location of an emission so to avoid migration to areas adjacent to the landfill.

Figures 10.10.4 illustrates passive systems that can be installed during waste filling inside and outside the landfill boundary, respectively. If the passive venting system is built during waste filling, it is common to install a permeable layer of gravel on slopes above the landfill liner for placement of a perforated pipe that vents directly to the atmosphere (Figure 10.10.4a). The principle is that LFG will flow towards a relatively permeable layer and then into the perforated pipe.

When the gas venting system is installed after the completion of waste disposal and final cover installation, then a trench can be installed outside of the periphery of the landfill. The trench is backfilled with a permeable material, as shown in Figure 10.10.4b. In this system, it is critical to insure that the trench does not fill up with rain or groundwater, but remains unsaturated so that gas can flow into the trench. Trenches are only useful at shallow landfills as it is difficult to excavate deeper than 5–8 m for the trench. At deeper landfills or where more thorough passive venting is needed, gas wells similar to those used for active venting can be utilized. In any case where there is an explosion risk near a landfill, an active venting system is required.

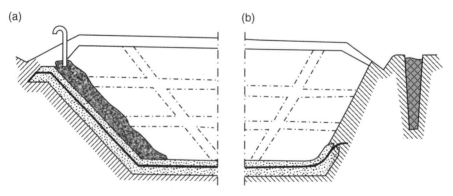

*Figure 10.10.4*   *Passive gas venting system inside alternatively outside the landfill.*

## 10.10.3 Design of Active LFG Collection Systems

Design of an active LFG collection system involves the design of the extraction system and of the pumping and regulation system.

### 10.10.3.1 Extraction Systems

The most common design for an active gas collection system is extraction through vertical perforated pipes. The wells are typically drilled with an auger with a diameter of 50–100 cm. After drilling, a 10–15 cm diameter perforated polyethylene pipe is placed in the middle of the hole and gravel is filled around the pipe. Vertical extraction wells are typically placed 40–80 m apart, depending on the landfill depth. In landfills less than 10 m in depth, even further spacing is acceptable. Pipes are connected to a system of horizontal pipes to convey the gas to a pump station. A common operational problem with gas collection systems is for a well to flood, meaning that water accumulates in the well and it is not functional. Thus, pipes should be sized so that pumps can be lowered into the well for water removal. The moisture content of the warm LFG condenses at the lower temperature encountered in the gas collection system in the top of the landfill. Traps for taking out the condensate must be located in critical points in the collection to avoid stored condensate to prevent effective gas collection.

The application of a vacuum to a well results in a reduction in pressure in the waste around the well. The zone of influence refers to that portion of waste around the well where a vacuum is induced by pumping from the gas well. Theoretically, this can be thought of as iso-pressure curves (Gendebien *et al.*, 1992). In a landfill with a permeable cover, the iso-pressure curves are ellipsoidal while if the landfill has an impermeable cover membrane, the iso-pressure curves have a cylindrical shape (Figure 10.10.5a, b). In actuality, the iso-pressure curves differ from the theoretical curves, as the vacuum around the wells is influenced by a number of synergistic variables such as barometric pressure, gas availability, migration of water and leachate to the extraction wells induced by the applied negative pressure, gas

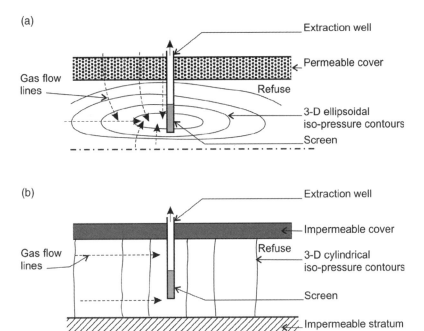

**Figure 10.10.5** *Iso-pressure curves induced by LFG extraction in a landfill with a permeable landfill cover (a) and with impermeable landfill cover (b).*

movement in preferential pathways as a result of waste heterogeneity, waste compaction and irregular leachate water tables. Nonetheless, mathematical models of LFG flow have been published (Findikakis and Leckie, 1979; Lofy, 1996; Townsend *et al.*, 2005).

An alternative to a model of the zone of influence for gas extraction wells is to perform a pumping test. In such a test, a test well is installed and a vacuum is applied to the test well. The vacuum is then measured in a series of well pressure probes placed at increasing distances from the test well. The resulting data can be used to estimate the zone of influence as a function of the applied vacuum. Unfortunately, there is uncertainty in the extent to which one measurement can be applied to the entire landfill due to waste heterogeneity.

A number of landfill sites have horizontal gas extraction systems that allow for gas collection earlier in the life of the landfill. In a horizontal system, a perforated pipe is placed in the middle of a gravel filled trench. The trench must be sloped to insure that water and leachate can be drained either to the leachate system or by separate leachate pumps installed in the gas system. It is common for horizontal gas trenches to be tied together with vertical wells when they are present. Reliable mathematical models of the zone of influence for horizontal systems suffer from the same limitations as described for vertical systems. Experience from the field suggests typical trench spacing of 30–60 m in the horizontal direction and 10–25 m vertically. The upper layer of the pipes must be at least 3–4 m under the surface of the landfill to avoid atmospheric air infiltration. A horizontal gas extraction system is illustrated in Figure 10.10.6.

Assuming the use of a low permeability final cover, gas accumulates just under the cover and a gas collection system should just under the low permeable cover include a high permeability gas collection layer tied into vertical wells or a collection pipe header.

### 10.10.3.2   Pump and Regulation System

In an active gas collection system, the gas is conveyed either to an energy recovery facility or to a flare, as discussed in the following section. The gas extraction system should have the capability to regulate the level of vacuum and subsequent

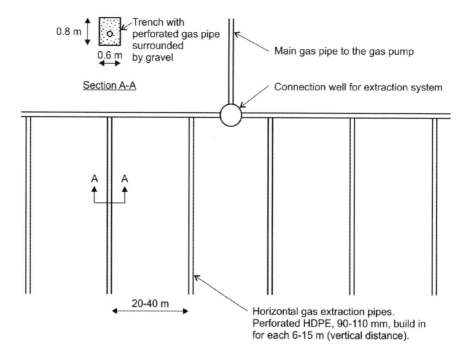

*Figure 10.10.6*   *Horizontal perforated pipes used as extraction system.*

gas flow from individual wells. This is particularly important when the gas is to be utilized for energy, in which case the energy content is directly proportional to the $CH_4$ content. Regulation of gas flow can take place by gas analysis and regulation at the individual wellhead or in a centralized regulation module. If the $CH_4$ concentration decreases below 40–45 %, depending on the energy recovery system requirements, the vacuum on the well can be decreased by using a regulation valve. Similarly, if the oxygen content is over 3 % and/or the nitrogen content is over 25 %, this would indicate that the vacuum is too high, resulting in air intrusion. In this case, the vacuum should be decreased.

LFG is extracted by a gas pump or compressor, which provides sufficient vacuum to pull gas from the landfill. A normal vacuum is 20–100 mbar at the wellhead. Pump or compressor selection depends on site-specific requirements, with the pressure required for gas transport and the inlet pressure for the gas combustion device being the most important factors. The most common type of gas pump is a radial blower, which is often used when only low pressure (up to 400 mbar) is required. A screw compressor or a roots compressor is typically used if higher pressure (up to 4 bar) is required.

The entire gas collection system, including the pump or compressor and the pipe system has to be gas tight to avoid air infiltration on the vacuum side and a gas leak on the pressure side. The presence of oxygen and/or nitrogen in the gas suggests that gas is being drawn in from the atmosphere and leaks in the pipe system must be repaired or the well field must be adjusted to eliminate this situation. A building, which houses a pump or compressor, should have a monitoring system to sound an alarm if the $CH_4$ concentration exceeds 1 % in the building. In addition, flame arrestors are required in the piping system both upstream and downstream of the pump/compressor to avoid backflash of a flame into the pipe system. Finally, the gas collection system must include traps and drains for condensate removal.

## 10.10.4 LFG Flaring Systems

When the LFG cannot be used for energy purposes the gas has to be flared off in a torch. Flaring reduces the methane emission and contribution to the greenhouse effect. Odors and the risk of fire or explosion is reduced, as well. Even when the LFG is used for energy purposes it is normal to have a flare for use if the energy utilization system has a breakdown or is out of work for maintenance.

Flare types can be divided into two main types, open flares and enclosed flares. The principle for a flaring is for both types, however, to mix LFG with atmospheric air and then ignite this mix of oxygen and methane in the gas, having the following combustion process:

$$CH_4 + 2O_2 \rightarrow CO_2 + 2H_2O + \text{heat}$$

### 10.10.4.1 Open Flares

An open flare is a burner where the LFG is mixed with air on top of the burner. The flame is protected by an open windshield, but a poor mixing and different temperatures inside and in the edge of the flame result in incomplete combustion reactions (see Figure 10.10.7). Emissions from the open flare are hard to measure and measurements are hard to average over time. Open flares can usually not meet emission requirements, as introduced in some countries. Open flares have the advantages of being inexpensive and relatively simple, which are important factors when there are no emission standards. But open flares should only be used for test periods, start up and running in, or for temporary use, if the energy utilization system is out of order for a short period.

### 10.10.4.2 Enclosed Flares

An enclosed flare usually consists of a single or array of burners within a cylindrical enclosure lined with refractory material. This construction prevents quenching and as a result the burning is much more uniform and the emissions are low. The flare may be designed to provide a certain retention time and be insulated to maintain a particular combustion temperature in the burning chamber. Requirements for retention time and temperature vary, but the most common is a minimum of 0.3 s at 1000 °C. The destruction and removal efficiency is normally 98.0–99.5 % for an enclosed flare. An enclosed flare is shown in Figure 10.10.8.

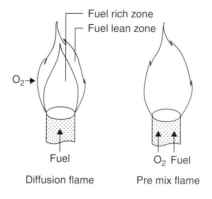

**Figure 10.10.7**    *Open flare for LFG.*

**Figure 10.10.8**    *Picture of an enclosed flare for LFG (Photo: Willumsen).*

**Table 10.10.3** *Emission limits (mg/Nm³, dry gas 3 % O₂) for flares in different countries. See also Gillett (2002).*
*Reproduced with permission from Environmental Protection, UK. © (2002) Environmental Protection UK.*

| Component | Germany TA Luft, 1996 | UK SEPA | Switzerland LRV, 1992 |
|---|---|---|---|
| Particles | 10 | — | 20 |
| CO | 50 | 50 | 60 |
| NO$_x$ | 200 | 150 | 80 |
| SO$_2$ | 50 | — | 50 |
| HCl | 10 | — | 20 |
| HF | 1 | — | 2 |
| Hydrocarbon | — | 10 | — |
| C$_{org}$ | 10 | — | 20 |
| Cd | 0.05 | — | 0.1 |
| Hg | 0.05 | — | 0.1 |
| Total metals | 0.5 | — | 1 |
| PCDD/PCDF | 0.18 | — | — |

### 10.10.4.3 Emission Standards

The emission standards vary from country to country as shown in Table 10.10.3. A typical limit value for CO is 50 mg/Nm³, for NO$_x$ 150 mg/Nm³ and for unburned hydrocarbons 10 mg/Nm³ (IEA Bioenergy, 2000; Rettenberger and Schreier, 1996). The actual emission depends of course on the gas input as well as the efficiency of the flare. Measurements and analyses for different flares have been published by Wiemer and Widder (1996), Rettenberger and Schreier (1996) and Baldwin and Scott (1996).

## 10.10.5 LFG Utilization Systems

Where sufficient gas is produced it may be beneficial to recover the energy content in the LFG. The minimum amount of LFG of interest for utilization depends on local factors, such as ease of extraction and revenues to be obtained from sale of energy. Also the alternative costs to install and operate other means of LFG control must be considered, if extraction and utilization are not introduced. Willumsen (2005) reports that landfills as small as 130 000 t in total capacity may have LFG utilization systems, although the average landfill with LFG utilization is in the order of 1 000 000 t in total capacity.

Figure 10.10.9 shows an estimate of how LFG is utilized at landfills across the world. LFG is most often used as fuel in a gas engine, which drives a power generator and produces electricity. The LFG can also be used in a gas boiler for production of hot water for heating or process heat. These uses are predominant because under normal conditions it is not necessary to clean the gas, except for the removal of particles, if the gas is used in a gas engine or a gas boiler. Table 10.10.4 lists the possible emissions from a gas engine per tonne of MSW landfilled.

In a few cases, where hydrogen sulfide and/or siloxanes are above acceptable concentrations for engines, it may be necessary to treat the LFG for these trace components. Corrosion issues of LFG are presented in Box 10.10.1. The different utilization methods are described below.

### 10.10.5.1 Electricity and Heat Production

The most common use of LFG is in a gas engine that runs an electric generator producing power. Typical gas engine plants produce between 350 and 1200 kW electricity per engine for which 210 m³ LFG/h and 720 m³ LFG/h, respectively, are needed (Figure 10.10.10). In a number of European countries, it is also typical to use the waste heat from the cooling water, exhaust and oil system of the engine. This system is called a combined heat and power plant (CHP). A CHP is feasible when there is a demand for hot water which is the case in areas that provide heat to a city or factory. The CHP system has a total energy efficiency of approximately 87 % compared with approximately 37 % when only electricity is produced. Figure 10.10.11 illustrates the principles of a system for CHP utilization.

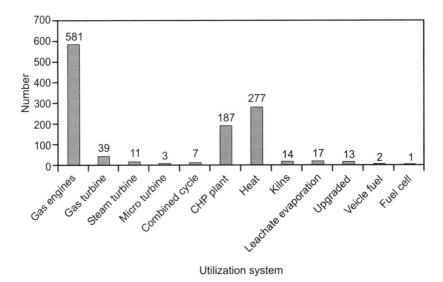

**Figure 10.10.9**   *Estimated distribution of LFG utilization systems worldwide (Willumsen, 2004). Reproduced with permission from Waste Management World. © (2004) Pennwell Coorporation.*

As the gas flow rate approaches 2400 m³ LFG/h, which results in the production of 4 MW power, gas turbines are sometimes used instead of generators. A turbine is similar to a jet engine, which takes huge amount of air in, compresses and heats it by burning the LFG. Then the air expands in a power turbine and electricity is produced in a generator. The largest steam plant in operation in 2003 delivered 45 MW power from a steam turbine. On the low end, small gas turbines, referred to as microturbines, have also been used for beneficial energy recovery at gas flow rates as low as 18 m³ LFG/h and production of only 30 kW of electricity. The development of microturbines is significant as it may reduce the size of a landfill that is needed for profitable gas utilization.

**Table 10.10.4**   *Emissions (g/t waste) from gas engines (after NSCA, 2002). Reproduced with permission from* Environmental Protection, UK. © *(2002) Environmental Protection UK.*

| Substance | Emissions (g/t)[a] | Uncertainty range (± g/t) |
|---|---|---|
| $CO_2$ | 400 000 | 200 000 |
| $CH_4$ | 2000 | 1000 |
| $NO_x$ | 1000 | 1000 |
| $SO_x$ (as $SO_2$) | 100 | 100 |
| HCl | 9 | 10 |
| HF | 10 | 20 |
| NMVOC | 80 | 100 |
| Particles | 20 | 40 |
| As | 0.002 | 0.002 |
| Cd | 0.3 | 0.4 |
| Hg | 0.002 | 0.002 |
| Ni | 0.01 | 0.008 |
| Tetrachloroethene | 0.2 | 0.1 |
| PCDD/PCDF | 400[b] | 1000[b] |

[a] Assessment based on emission concentrations at 273 K, 101.3 kPa, 5 % $O_2$, dry.
[b] The unit is ng TEQ/t.

**Box 10.10.1    Corrosion Problems in LFG Utilization.**

Some trace gas constituents may cause corrosion problems, especially upon combustion. This is therefore a main operational issue in landfill gas flaring. Halogenated hydrocarbons, halocarbons and other organic acids may in particular cause corrosion. The combustion process in fact generates very reactive byproducts, such as HCl, HF, $H_2S$ and water vapour, which may lead to corrosion of the metal parts they get in contact with. $H_2S$ may cause corrosion even at concentrations as low as 100 ppm (Reinhart, 1994), since its combustion in the presence of water forms the highly corrosive sulfuric acid ($H_2SO_4$). As a result of corrosion, all the metal components are progressively damaged (Dernbach, 1985).

In order to prevent, or at least reduce the problem of corrosion, one possibility is to remove the above-mentioned trace compounds from the gas by gas treatment. Halogenated hydrocarbons can be removed from the gas through absorption/adsorption processes, catalytic processes, high-temperature combustion and biological oxidation. $H_2S$ can also be removed from the gas, for instance through reaction with ferrous material or biofiltration (Rettenberger, 1996). However, purification of landfill gas may substantially augment the operational costs and/or not be as effective as needed to prevent corrosion. Therefore, corrosion-resistant materials should always be utilized for all those metal parts that get in contact with the gas during combustion (Dernbach, 1985). Lichti *et al.* (1999) carried out experiments in which different qualities of steel were exposed to landfill gas engine exhaust gas at 400 °C for 25 days. It was observed that HCl leads to blistering and spallation on low-alloy steels and carbon steels, leading to a progressive corrosion of the protective oxide films. The best performance was achieved by a 12 % Cr stainless steel, the corrosion rate of which was < 1 μm/year.

## 10.10.5.2    Boiler System

The second most common use of LFG is a gas furnace, in which the gas is used for heating of water in a boiler. This is the simplest system because the gas does not require any pretreatment prior to utilization. The disadvantage is that the price for heat energy is normally much lower than the price for electrical energy. Thus, where there is sufficient gas and the potential to make the capital investment required for an electric generator, this is more common. Another reason why electrical energy production is more common is the wide distribution of electrical lines relative to locations where there

**Figure 10.10.10**   *Gas engine and generator plant for utilization of LFG (Photo: Willumsen).*

is a boiler. Boilers are common in heat stations for block and district heating, manufacturing plants, and are also used at some greenhouses for heat. A boiler fired with LFG is shown in Figure 10.10.12.

### 10.10.5.3   Direct Use

LFG can be used for direct burning in a process burner. In several cases in the UK, LFG is used as a substitute for natural gas in a kiln firing bricks. Kilns are used in other industrial processes as well, e.g., cement production, and any kiln is a candidate for burning LFG directly.

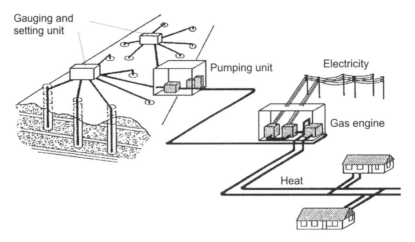

**Figure 10.10.11**   *Utilization of LFG in a gas engine for CHP production (after Willumsen, 1988). Reprinted with permission from Final report to the Directorate General Environment, Costs for municipal waste management in the EU by D. Hogg © (2002) European Commission.*

***Figure 10.10.12***   *Boiler system for utilization of LFG (Photo: Willumsen).*

Another application of the direct use of LFG is leachate evaporation. In this application, the gas is burned on-site and the heat is used to evaporate leachate. The residue that remains after evaporation is high in salts and recalcitrant organic matter (humic and fulvic acids). This residue is a nonhazardous waste and can be buried in a landfill. However, this may again dissolve and result in leachate if not properly managed.

### 10.10.5.4   Injection into the Natural Gas Distribution Network

LFG can be upgraded to natural gas quality after which it can be injected into the natural gas distribution network. While this use eliminates the need for an electric generator, an investment is required for a gas purification plant. Prior to sale as natural gas, LFG must be treated to remove particles, liquid, $CO_2$ and some trace components like $H_2S$, as natural gas in most cases is nearly 100 % $CH_4$ similar high calorific gasses. The major step in the treatment process is the removal of $CO_2$, which must be removed to meet the high energy quality of natural gas, for which the various natural gas using burners and stoves are adjusted for. The most common treatment processes include chemical absorption, pressure swing adsorption and membrane separation. The chemical absorption process is based on physical or chemical absorption of $CO_2$ and the process gives a $CH_4$ quality of approximately 95 %. Pressure swing adsorption is the most common method and results in a product that is approximately 98 % $CH_4$. Membrane separation is the cheapest and simplest to operate but the gas quality is only 75–85 % $CH_4$.

### 10.10.5.5   Use of Gas in Vehicles

There are a few landfills where the gas is compressed and then utilized in vehicles such as compactors, refuse collection trucks, buses or even ordinary cars. The requirements for gas quality and the methods for gas treatment are the same as those described for upgrading to natural gas quality. The profitability depends on the chosen system, but it will be relatively expensive when investing in a system using only a few vehicles. However, if all the gas from a typical landfill has to be used in vehicles, it will require a large number of busses or cars running on the gas (Willumsen, 2004).

### 10.10.5.6 Fuel Cells

LFG can also be used in fuel cells. Fuel cells may be compared to large electric batteries that provide a means to convert chemical energy to electricity. The difference between a battery and a fuel cell is that in a battery, all reactants are present within the battery and are slowly being depleted during battery utilization. In a fuel cell, the reactant (LFG) is continuously supplied to the cell. Oxygen (from air) passes from the cathode through an electrolyte, which allows the passage of oxygen, but not electrons, and combines with hydrogen and carbon at the anode (derived from a hydrogen-rich fuel such as $CH_4$) to form water and $CO_2$. There are several advantages to a fuel cell including electricity conversion efficiencies of 40–50 %, low air emissions, low labor and maintenance requirements and low noise. Fuel cells based on LFG have been tested in the United States with production of 25–250 kW. The limiting factor in the implementation of fuel cells is the high initial investment that limits profitability.

## References

Baldwin, G. and Scott, P.E. (1996): *Predicted effectiveness of active gas extraction*. In: Christensen, T.H., Cossu, R. and Stegmann, R. (eds) Landfilling of waste: Biogas, pp. 521–534. ISBN 0-419-19400-2.

Barlaz, M.A. (2004): *Critical review of forest products decomposition in municipal solid waste landfills*. Technical Bulletin 872. National Council for Air and Stream Improvement, RTP, USA. Located on the internet: 19 October 2009. http://www.ncasi.org/Publications/Detail.aspx?id=97.

Coops, O., Luning, L., Oonk, H. and Weenk, A. (1995): Validation of LFG formation models. *Proceedings of the International Landfill Symposium*, 5, 635–646.

Dernbach H. (1985): Landfill gas utilization in braunschweig – quality of gas and damages due to corrosion. *Waste Management and Research*, 3, 159–159.

EPA (1996): Standards of performance for new stationary sources and guidelines for control of existing sources: municipal solid waste landfills. Code of Federal Regulations, Title 40, Sections 9, 51, 52 and 60. *Federal Register*, 61 (49).

EPA (1998): AP-42 Emission factors for municipal solid waste landfills – supplement E, November 1998. United States Environmental Protection Agency, Washington, D.C., USA. Located on the internet: 19 October 2009. http://www.epa.gov/ttn/chief/ap42/ch02/final/c02s04.pdf.

EPA (2005): Landfill Gas Emissions Model (LANDGEM), version 3.02. EPA/600/R-05/047. United States Environmental Protection Agency, Washington, D.C., USA.

Faour, A.A., Reinhart, D.R. and Huaxin, Y. (2007): First-order kinetic gas generation model parameters for wet landfills. *Waste Management*, 27, 946–953.

Findikakis, A. and Leckie, J. (1979): Numerical simulation of gas flow in sanitary landfills. *AMCE Journal of the Environmental Engineering Division*, 105, 927–945.

Gendebien, A., Pauwels, M., Constant, M., Ledrut-Damanet, M.J., Willumsen, H.C., Butson, J., Fabry, R., Ferrero, G.L. and Nyns, E.J. (1992): LFG: From Environment to Energy. Office for Official Publications of the European Communities, Brussels, Belgium. ISBN 92-826-3672-0.

Gillett, A.G., Gregory, R.G., Blowes, J.H. and Hartnell, G. (2002): Landfill gas engine exhaust and flare emissions. The University of Nottingham, UK.

IPCC (2006). Intergovernmental panel on climate change: 2006 IPCC guidelines for national greenhouse gas inventories: Reference manual (volume 3). IPCC National Greenhouse Gas Inventory Program, London, UK.

Lagerkvist, D.A., Meadows, M., Meijer, J.E. and Willumsen, H.C. (2000): *International perspective on energy recovery from LFG*, pp. 30–33. Report from IEA Bioenergy Task XIV (LFG activity) and IEA CADDET renewable energy programme. Located on the internet: 19 October 2009. http://www.caddet-re.org. ISBN 1 9 00683 05 9.

Lichti K.A., Levi T.P., Swann S.J., Ferguson S.L. and McIlhone P.G.H. (1999): Corrosion behaviour of four high temperature engineering materials exposed to landfill gas engine flue gas. *Material at High Temperatures*, 16, 109–116.

Lofy, R.J. (1996): *Predicted effectiveness of active gas extraction*. In: Christensen, T.H., Cossu, R. and Stegmann, R. (eds) Landfilling of waste: biogas, pp. 395–431. ISBN 0-419-19400-2.

NSCA (2002): Comparison of emissions from waste management options. National Society for Clean Air and Environmental Protection, Brighton, UK.

Reinhart D. (1994): *Beneficial use of landfill gas*. Report No. 94-7. State University of Central Florida – Florida Center for Solid and Hazardous Waste Management, Miami, USA.

Rettenberger G. (1996): *Landfill gas upgrading: removal of hydrogen sulphide*. In: Christensen, T.H., Cossu, R. and Stegmann, R. (eds) Landfilling of waste: Barriers. Elsevier, London, UK. ISBN 0-419-19400-2.

Rettenberger, G. and Schreier, W. (1996): *Predicted effectiveness of active gas extraction*. In: Christensen, T.H., Cossu, R. and Stegmann, R. (eds) Landfilling of waste: Barriers, pp. 505–520. Elsevier, London, UK. ISBN 0-419-19400-2.

SCS Engineers and Augenstein, D. (1997): *Comparison of models for predicting landfill methane recovery*. Final Report to SWANA. File No. 0295028. SCS Engineers, Silver Spring, USA.

Solomon, S., Qin, D., Manning, M., *et al.* (2007) Technical summary. In: Solomon, S., Qin, D., Manning, M., et al. (eds): Climate change 2007. The physical science basis. Contribution of working group 1 to the fourth assessment report of the intergovernmental panel on climate change, pp. 19–91. Cambridge University Press, Cambridge, UK.

Townsend, T.G, Wise, W.R. and Jain, P. (2005): One-dimensional gas flow model for horizontal gas collection systems at municipal solid waste landfills. *Journal of Environmental Engineering*, 131, 1716–1723.

Wiemer, K. and Widder, W. (1996): *Predicted effectiveness of active gas extraction*. In: Christensen, T.H., Cossu, R. and Stegmann, R. (eds) Landfilling of waste: Biogas, pp. 521–534. ISBN 0-419-19400-2.

Willumsen, H.C. (1988): *Recovery of landfill gas*. Final report for demonstration project BM/741/83. European Economic Community, Directorate General XVII, Brussels, Belgium.

Willumsen, H.C. (2004): LFG recovery plants – Looking at types and numbers worldwide. *Waste Management World*, 2004, 125–133.

Willumsen, H.C. (2005): *Optimering af gasindvinding på deponeringsanlæg I Danmark* (Optimization of gas recovering at landfills; in Danish). Arbejdsrapport Nr. 13 fra Miljøstyrelsen. ISBN 87-7614-764-9.

Willumsen, H.C. and Bach, L. (1991): Landfill gas utilization overview. In: *Proceedings of the Third International Waste Management and Landfill Symposium*. CD-ROM. CISA, Environmental Sanitary Engineering Centre, Cagliari, Italy.

# 10.11

# Landfilling: Leachate Treatment

**Hans-Jürgen Ehrig**

*University of Wuppertal, Wuppertal, Germany*

**Howard Robinson**

*Enviros Consulting, Shrewsbury, UK*

Leachate is a specific wastewater with relatively variable composition. Leachate composition depends in principle on the composition of the landfilled waste, but in practice the influence of landfill technology and more complex biological, chemical and physical processes in the landfill body is very strong. As a result, it may sometimes be difficult to predict leachate composition precisely. The rate of flow of leachate is a result of climatic conditions (e.g. precipitation, evaporation) and landfill conditions (e.g. infiltration, storage). During the filling period leachate production is highly variable and depends on the geometry of the landfill and the filling strategy. Therefore, accurate prediction of the leachate volume is only possible for completed sections of the landfill over longer periods of time, e.g. one year (Chapter 10.3).

The philosophy of leachate treatment has changed during recent decades. Until 1990 the most important feature of leachate treatment was the reduction of biologically degradable components, such as biodegradable COD and nitrogen, using the simplest possible technologies. Gradually nonbiodegradable organic substances such as the remaining COD and organic halogens became increasingly important. As a consequence, today many leachate treatment plants are combinations of different treatment steps for biodegradable and nonbiodegradable substances. With the use of chemical and/or physical processes, treatment systems have changed in many cases to more sophisticated and technical systems. The continuing development of leachate treatment has not taken place in a logical manner, thus no single treatment process can be described universally as the 'best available technology', and instead, preferred processes in individual countries reflect the different approach that regulators have taken.

This chapter describes the specific issues about leachate treatment and provides an introduction to the major treatment processes available for leachate treatment.

## 10.11.1 Basic Data for Leachate Treatment

Landfill leachate is a wastewater with complex characteristics. However, in most situations a few specific parameters are of most importance for the treatment processes. Other parameters are often only of limited importance for the environment,

*Solid Waste Technology & Management*   Edited by Thomas Christensen
© 2011 Blackwell Publishing Ltd

or their concentrations are lower than required effluent discharge standards. Municipal solid waste landfill leachates are mainly polluted by organic substances and nitrogen.

In order to describe organic substances, in most cases the summarised parameter chemical oxygen demand (COD) is used. Many investigations to characterise organic substances in detail have been unsuccessful, because they comprise a very complex mixture. For many wastewaters the summarised parameter biochemical oxygen demand (BOD) is used to provide an indication of the biologically degradable portion of the COD. For leachate this is only useful if the relationship of BOD to COD is greater than 0.4. Then the degradable organic substances are in the region of 1.7–1.9× the $BOD_5$ value. At lower values of this relationship the biodegradable part of COD is greater than twice the $BOD_5$ value. In most instances the residual COD after biological treatment is relatively high, and can only be further reduced by chemical or physical treatment steps. In countries such as Germany, organic halogens comprise only a small specific proportion of overall organic substances, but may be an important parameter for leachate treatment. These components are often measured using the surrogate parameter of 'adsorbable organic halogens' (AOX), although there is very little, if any, evidence to demonstrate any relationship between AOX values (measured in mg/l) and concentrations of hazardous organic compounds (measured in μg/l).

Although COD values can be very high, especially in leachates from recently landfilled wastes, in many cases nitrogen may be the most important risk for the environment. Most of the nitrogen comprises ammoniacal-N [the sum of ammonia ($NH_3$) and ammonium ($NH_4^+$) nitrogen] and in many cases only this parameter is measured. Nevertheless, in some instances, significant concentrations of organic nitrogen can be measured, which have potential to be converted to ammonium in a biologically active environment.

Heavy metal concentrations in most municipal solid waste landfill leachates are much lower than values found in household sewage, and rarely exceed specified discharge values, primarily because at these sites a high proportion of the metals are immobilised by the organic matter in the landfill. However, at landfills which receive increased proportions of inorganic wastes, such as incinerator ashes, heavy metals can become more important parameters.

Similarly, organic micropollutants, often defined as priority pollutants, are rarely of great importance for leachate treatment. However, with the reduction of organic waste in future landfills, they may become more important, because of less degradation and attenuation capacity when less organic waste is present in the landfill. Micropollutants in terms of dioxins are in Japan considered a major problem in leachates from incineration slag and ash landfills.

Leachates from incineration slag and ash landfills may contain high concentrations of chloride and sulfate. In most cases, other substances in such leachates are present at relatively low concentrations, and do not require specific treatment systems. However, at higher COD or nitrogen concentrations, their treatment often becomes a more serious problem. One solution may involve membrane separation, with additional evaporation or direct evaporation. Co-disposal of the concentrate or 'brine' from the process with municipal solid waste, can result in serious problems as inorganic salts in leachate increase in concentration, where they can hinder treatment processes. Both parameters could become more important in the future, with the reduction of organic fractions in municipal solid wastes being landfilled. Sulfur compounds are readily reduced to sulfide, and although this is readily removed from leachate in the presence of ferrous iron, where iron is lacking, free sulfide can be emitted to produce noxious odours and extremely toxic hydrogen sulfide concentrations. Sulfide concentrations in leachate can reach concentrations that are toxic to biological treatment processes.

Leachate alkalinity is an important parameter for biological treatment processes, particularly for the buffering of acidity produced during the biological nitrification of ammoniacal-N. Changes in pH values and/or calcium–carbonic acid equilibrium can affect many treatment processes.

The leachate flow depends on climatic conditions, the situation of the landfill, the surrounding strata, the type of waste, the water storage capacity of waste, landfilling techniques, the landfill operational conditions etc. For new landfills it is extremely difficult to estimate future leachate flow rates accurately. Only at existing landfills with flow measurements over a number of years can future leachate flow rates be estimated accurately, but often then only as an average value over a year or a season. For leachate treatment the daily flow rate is an important parameter, but can often only be estimated with a large element of uncertainty. As a consequence, the provision of raw leachate storage capacity, and the consideration of appropriate treatment process capacity are very important points.

The degree of leachate treatment required depends on the specific regulations in different countries. In Austria and Germany the discharge limits are applied on a national basis (Table 10.11.1), although values may be reduced as result of the specific circumstances. In other countries, such as the United Kingdom, every situation is considered on a site specific basis, in order to determine locally appropriate discharge limits.

***Table 10.11.1***    *Effluent limits for treated leachate in Austria and Germany.*

| Parameters | Units | Austria (to surface water) | Germany (to surface water) | Germany (sewer) |
|---|---|---|---|---|
| pH | — | 6.5–8.5 | — | — |
| COD | mg/l | 50 | 200 | — |
| $BOD_5$ | mg/l | 10 | 20 | — |
| Ammonium ($NH_4^+$) | mg N/l | 10 | | — |
| Ammonia ($NH_3$) | mg N/l | 0.5 | 70–100 | |
| Nitrate ($NO_3^-$) | mg N/l | 35 | | — |
| Nitrite ($NO_2^-$) | mg N/l | — | 2 | — |
| Total phosphorus | mg/l | — | 3 | — |
| Sulfide | mg/l | 0.5 | — | 1.0 |
| Cyanide | mg/l | — | — | 0.2 |
| AOX | mg/l | 0.5 | — | 0.5 |
| Volatile hydrocarbons | mg/l | 5 | — | — |
| Hydrocarbons | mg/l | — | 10 | — |
| BTX | mg/l | 0.1 | — | — |
| Arsenic | mg/l | — | — | 0.1 |
| Lead | mg/l | 0.5 | — | 0.5 |
| Cadmium | mg/l | 0.1 | — | 0.1 |
| Chrome | mg/l | 0.5 | — | 0.5 |
| Chrome (VI) | mg/ | — | — | 0.1 |
| Copper | mg/l | 0.5 | — | 0.5 |
| Nickel | mg/l | 0.5 | — | 1.0 |
| Mercury | mg/l | 0.01 | — | 0.05 |
| Zinc | mg/l | 0.5 | — | 2 |
| Toxicity against fish | dilution factor | 3 | 2 | — |

## 10.11.2    Specific Problems in Leachate Treatment

Leachate treatment faces three specific problems, usually not met in sewage treatment: large variations in flow, changes in composition over time, and potential changes with new landfill section because of changes in waste management.

### 10.11.2.1    Flow Variations

As the result of local climatic conditions, uneven distribution of rainfall, and other influences, in most cases daily leachate flow rates show wide variations. To reduce the influence of such variations, and to balance flow rates requiring to be treated, storage capacity for untreated leachate is generally necessary. Most treatment processes, especially those involving biological treatment, generally operate better and more reliably at relatively constant treatment rates.

As a result of the difficulties of flow rate estimation, the long-term risk of treatment plants being overloaded or underloaded is great. If landfills receive an engineered capping layer after closure, then leachate flow generally decreases, depending on the type and quality of the liner system, and possible short- or long-term liner failures. Measurements at different closed landfills have shown that clay covers of up to 60 cm thickness, or normal soil cover up to more than 1 m, may not necessarily reduce leachate flow rates (Krümpelbeck *et al.*, 1999). But such cover systems can often produce more uniform leachate flow rates, although it is difficult to estimate the impact they do have.

### 10.11.2.2    Changes in Composition Over Time

Unlike most other wastewaters, leachate quality changes over time at each landfill site. In the long term, over decades and centuries a slow but significant reduction of most components occurs. The reduction rate varies between parameters, and it may require 200–400 years before a direct discharge of leachate being produced can be accepted without a requirement for treatment. Critical parameters for most landfills are COD and ammoniacal-N.

Additionally, the organic components of leachate change after some years of landfill operation, with a change in biological conditions within the waste body, from the more acidic phase to a more methanogenic phase. During this transition the concentration of COD decreases, but simultaneously the proportion of biodegradable organic substances (BOD) decreases substantially. The resulting effect is a decrease in concentrations of organic substances, but an increase in the proportion of these that are not readily biodegradable. The rate and extent of this change depends on the landfill operational regime and inputs, and on the intensity of the initial acidic phase of decomposition.

### 10.11.2.3   Changes in Composition for New Landfill Sections

A number of changes in waste composition, and in landfill operation, can have significant effects on leachate quality. The composition of wastes being deposited not only has initial influences on leachate quality but also, together with the extent of compaction of wastes, affect the rates of flow of liquids through the landfill. Often the biological, chemical and physical processes are also affected, which will all influence leachate quality. The recycling of specific waste fractions such as paper, plastic, glass, metals, compostable organic substances, demolition wastes etc. will increasingly result in new types of landfill. But each of these new landfills will be the result of specific regional waste management conditions, and the influence of waste types remaining to be landfilled will become more and more important. In the past, the total mixture of all types of waste in a landfill produced more uniform landfill types, with less variation in leachate quality between different sites. In future, landfills will change to become more specific, with very different leachate qualities and flow rates. With increasing separation, removal and pretreatment of organic waste fractions from waste streams, in the future such effects will inevitably increase. As a result, inorganic leachate parameters will become more important, as problems for the existing treatment processes or in meeting required discharge limits. For example, heavy metals may not be immobilised by organic waste components, allowing concentrations to increase in the leachate. Another parameter likely to increase might be sulfide in leachates from such sites.

## 10.11.3   Leachate Treatment Processes: On Site and/or Combined with Sewage Treatment

Until now leachate treatment development has responded to the major pollutants in leachate from municipal landfills, and to increases in environmental protection during recent decades. With more restrictive effluent standards, and a focus on additional parameters, leachate treatment is becoming more complex, and today in many cases leachate treatment plants are a combination of different treatment steps. Figure 10.11.1 presents a flow sheet of frequently used leachate treatment options.

### 10.11.3.1   On Site Leachate Treatment

To select an optimum on site leachate treatment system, not only treatment results and costs must be considered. Other important factors are production of sludge and other wastes, and the type of treatment process. Different leachate treatment processes consume different amounts of energy. Some processes require additional materials with associated variations in consumption of resources. All issues must be considered when choosing a leachate treatment process.

For the removal of organic leachate contaminants, an important factor is the means by which these are removed. For oxidation processes such as biological or chemical oxidation, the organic compounds are oxidised to $CO_2$. Other processes such as membrane separation or flocculation only separate the organic substances, and do not actually provide any breakdown of them. A third group of processes is represented by activated carbon, which is also primarily only a separation process, but has the benefit that during thermal reactivation of the carbon, the separated organic substances are also oxidised. These differences can be an important point to consider for the philosophy of leachate treatment.

The first stage of any leachate treatment plant should be an adequate flow balancing system for raw leachate, which in some circumstances might involve temporary storage of leachate within the landfill itself. Experience has shown that

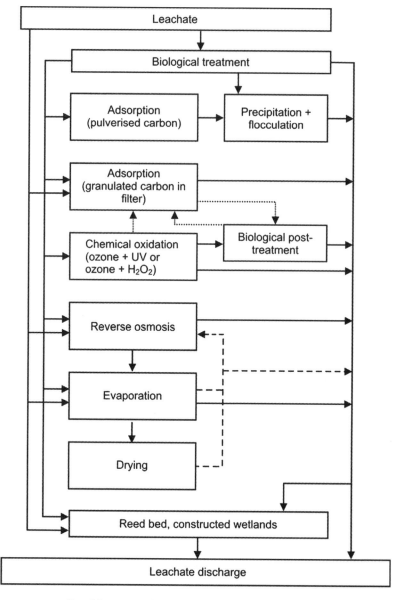

**Figure 10.11.1**   *Flow scheme of possible leachate treatment systems.*

very large volumes of external storage may be necessary, in order to adequately balance normal leachate flow variations over the course of a year, if the landfill cannot be used. Under the climatic conditions of central Europe, long periods of external storage, for example in open lagoons, during winter months, can rapidly cool leachate resulting in problems for subsequent biological treatment.

As a consequence of the need to achieve specific discharge limits, the composition of effluents at most treatment plants is often determined in detail. But for the solution of leachate treatment problems it is also important to analyse raw leachates in detail. This is also necessary for parameters which are not affected by the treatment process, or which are not regulated by discharge consents. Such parameters include alkalinity, calcium, phosphorus, chloride, sulfate, etc.

### 10.11.3.2  Treatment Combined with Sewage Treatment

In the past, a very widespread option for leachate treatment has been co-treatment with domestic sewage in a biological treatment plant. With co-treatment, biologically degradable organic substances and nitrogen can generally be removed to the same extent as in an on site biological leachate treatment plant. Normally, the quantity of leachate is not a problem for the sewage treatment plant but the overall load of contaminants, for example, kg/day ammoniacal-N, can be. Because of the very high concentrations of ammoniacal-N found in many leachates, the addition of just a little leachate into a domestic sewer can result in a doubling or greater of the overall loading to the treatment plant. Flow balancing of leachate flows may be very important, and where possible, pumping of leachate during the night, when flows of sewage reduce, may be beneficial. After co-treatment there also remains the problem of nonbiodegradable components. Remaining COD and AOX values may be reduced and low, but this is primarily as a result of dilution by the high sewage flow rate. Secondary treatment of the combined effluent to reduce COD and AOX is less effective and very expensive, as a consequence of the large dilution. However, co-treatment can provide a solution able to handle high leachate flow rates which could not be treated efficiently by an on site plant, in circumstances where the sewage treatment works has adequate spare capacity.

Potentially also the combination of some on site treatment of leachate followed by discharge to a municipal sewage treatment facility could be attractive, but this would in most cases suggest that a significant fraction of nitrogen is removed on site prior to discharge to the sewage treatment plant.

## 10.11.4  Stripping Processes

One option for reducing ammoniacal-N concentrations is stripping of ammonia gas. At raised pH values or temperatures an increased proportion of the total ammoniacal-N (ammonium + ammonia) is present as gaseous ammonia (Box 10.11.1). During intensive contact with gases (e.g. with air) concentrations of dissolved ammonia gas adjust to an equilibrium between liquid and gaseous phases. Using this effect ammonia can be stripped from the liquid within the gas stream (usually air). The efficiency of this process increases with increasing pH value or air temperature. With increasing efficiency the quantity of air required will decrease, and the concentration of ammonia gas in the air increases. In most cases ammonia gas will be removed from air subsequently by scrubbing with sulfuric acid, although it can also be converted to nitrogen by combustion within a landfill gas flare (see Knox, 2001). The remaining end product of sulfuric acid sorption (ammonium sulfate) can be used as a fertiliser, or for chemical processes.

Ammonia stripping is possible in aeration tanks or in stripping towers. Aeration tanks are mostly operated as cascades, with fresh air in every cascade, or air flow in opposite direction to the water flow to reduce air flow. The equilibrium between concentrations of ammonia gas in water and air is reached within a short distance (10–20 cm). As a consequence, the required depth of stripping tanks can be relatively small. Aeration tanks can be accurately sized, by a combination of theoretical design and practical operation (Ehrig, 1987). Stripping towers are filled with material which has a large free space and surface area. In contrast to aeration tanks, basic data for theoretical calculations and design must be measured in pilot-scale experiments, using a comparable tower, the same packing material and identical water and air loads. The complex chemistry of leachate can result in serious clogging with both inorganic and organic precipitation within such stripping towers.

**Box 10.11.1    Principle of Ammonia Stripping Process.**

Ammonium ($NH_4^+$) and ammonia ($NH_3$) exchange according to pH. The acid dissociation constant is $pK_a = 9.3$ at 25 °C, but decreases with increasing temperature:

$$NH_4^+ \rightarrow NH_3 + H^+$$

Below is a plot of the percentage of ammonia out of the sum of ammonium + ammonia in water depending on temperature and pH value.

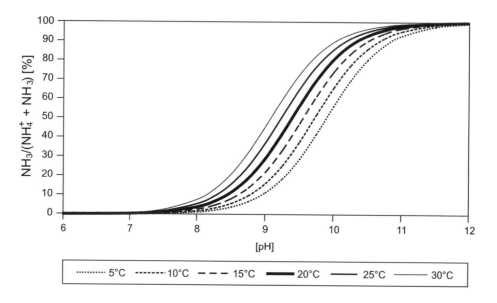

The gaseous ammonia can diffuse to gaseous phases as aeration air until equilibrium between liquid and gaseous phases is reached. But with every molecule of removed ammonia a part of ammonium is changed to ammonia until the old proportion is reached. As example at pH = 10 and $T = 20$ °C approximately 80 % of (ammonium+ammonia) exists in the form of ammonia and only 20 % as ammonium. But with every removed molecule of ammonia the remaining sum of ammonium + ammonia changes again to the relationship 80 : 20. This dynamic changing is the reason why the elimination can be much higher than 80% in this case.

Under similar operation conditions the same amount of aeration removes the same percentage of ammonium independent on influent concentration. Increasing influent concentrations makes stripping more and more economic. But if the removal rate is to be increased from 90 to 99 % the amount of required air must be doubled.

Unlike many other treatment processes, the required air volume removes a constant percentage of the stripped substance, regardless of the influent concentration in leachate. Removal efficiencies of 90 % (temperature 20–40 °C, pH ≥ 11) need several thousand volumes of air per volume of water. Damhaug (1981) presented data for a pilot-scale ammonia stripping plant, with five cascades, and with air flow in a countercurrent direction to water flow. With operating temperatures of 10 to 19 °C and pH values of between 9.5 and 12.0, the removal rate varied between 78 and 95 %. The required air to water ratio varied between 2000 and 3700 $m^3$ air/$m^3$ water. But at very high concentrations of ammoniacal-N, stripping processes are increasingly cost effective. Knox (2001) reported data for a pilot-scale leachate treatment plant

**Table 10.11.2** *Technical and operation data of a pilot stripping tower (Knox, 2001). Reproduced with permission from CISA, Italy. © (2001) EuroWaste srl, Italy.*

| Parameter | Data |
|---|---|
| Leachate flow rate | 4.3 m$^3$/day |
| Temperature | 70 °C |
| Air flow rate | 210 m$^3$/h |
| Air:liquid ratio | 1167 |
| Influent ammonia | 4000–4500 mgN/l |
| Effluent ammonia | 50–62 mgN/l |

with an air stripping tower in Hong Kong, for a very strong methanogenic leachate (Table 10.11.2). This stripping tower pretreatment was combined with a subsequent biological treatment process in order to achieve removal of organic substances.

A cost-effective alternative to air stripping can be steam stripping of the ammonia gas. The advantage of steam stripping is the naturally high temperature of steam. The condensed steam, usually within an acid solution, concentrates the removed ammonia.

Stripping systems are sometimes used in combination with reverse osmosis plants. In such cases the reverse osmosis plant does not need to be optimised for ammonium retention.

Significant removal of a number of trace organic components often present in landfill leachates, can be achieved during air stripping treatment processes. Robinson and Knox (2001, 2003) provided removal efficiencies for some volatile compounds (see Table 10.11.3). Stripping may also provide significant removal for many other organic compounds in landfill leachate. It is unlikely that an air stripping treatment system would be employed specifically to reduce concentrations of trace organics in landfill leachates because of the costs.

## 10.11.5 Evaporation

The aim of evaporation is the separation of solidified leachate contaminants from the remaining water. A number of systems have been used:

- Direct evaporation of raw leachate.
- Evaporation of reverse osmosis concentrate.
- Two-step evaporation of reverse osmosis concentrate (first step: slurry production up to 10–30 % dry solids; second step: production of dry material with more than 90 % dry solids).

**Table 10.11.3** *Stripping of volatile components from leachate (Robinson and Knox, 2001, 2003). LOD = Limit of detection achievable routinely in leachate samples. Presence (%) represents the amount of sample in which the compound was above the limit of detection.*

| Parameter | LOD (µg/l) | Presence (%) | Median value (µg/l) | Removal (%) |
|---|---|---|---|---|
| Ethylbenzene | 10 | 15 | 10 | 40 |
| Mecoprop (MCPP) | 0.1 | 98 | 11 | 50 |
| Naphthalene | 0.1 | 70 | 0.46 | 40 |
| Toluene | 10 | 54 | 21 | 25 |
| Xylenes | 10 | 35 | 35 | 40 |

Very different technologies are required for evaporation. Initial plant designs encountered serious problems with corrosion. High costs as a consequence of material problems, and high energy consumption, lead to the development of several new evaporation processes. Only very few plants of each type have been constructed and operated. As a consequence, it is difficult to compare treated wastewater quality, residuals and costs of the different plants.

Important features of evaporation processes are energy consumption, quality of solids produced and the quality of treated leachate. Treated leachate quality is mainly affected by volatile components in leachate. A critical parameter is ammoniacal-N, which can be vaporised as ammonia, and subsequently recondensed.

Evaporation can be separated into direct and indirect processes. In direct processes, the exhaust gas or high-temperature gas has direct contact with the leachate. Using indirect evaporation processes energy is transferred into the leachate using a heat exchanger. To reduce corrosion problems, indirect processes are often operated with low evaporation temperatures and partial vacuums.

Residuals from evaporation comprise solids of variable water content. If their disposal as solid waste is not possible, an additional solidification step may be needed. The raw solids from evaporation processes often have a high solubility, with a consequently high environmental risk. As a result, in Germany underground storage (abandoned mine shafts) is the only disposal option for these solids. Solidification can also be used, in order to reduce the solubility of these solid components to allow their disposal in landfills. Without additional components, the solid residuals from evaporated leachate are typically in the range 5–20 kg/m$^3$ , depending on the leachate strength. Evaporation after biological treatment can produce solids with ammonium and nitrate content. In such a mixture exothermic reactions are possible.

The gases that have been evaporated are condensed as treated leachate. This stream may be polluted by ammonium and volatile hydrocarbons. During the evaporation process, detergents and anti-foaming agents are often required, and residuals of these products may remain in the treated leachate, or in the solid residuals.

Ettala (1997) presented data for an experimental evaporation plant in Finland. The plant comprised a sand filter, pH adjustment to 4, a degasifier to remove carbon dioxide, and an evaporator with 1500 m$^2$ of plastic heat transfer surface. The plant operated with an absolute pressure between 15.8 and 17.6 kPa, at temperatures of between 55.0 and 57.3 °C. The energy consumption was 13 kWh/m$^3$ of leachate treated. Nearly 18 % of leachate was recovered as concentrate and disposed back into the landfill. The COD was reduced from 227 mg/l to < 30 mg/l, and ammoniacal nitrogen from 120 mg/l to 0.1 mg/l.

## 10.11.6   Membrane Processes

Membrane processes comprise a physical system, which separates a treated permeate stream, from a highly polluted concentrate stream. The permeate is the treated effluent from the membrane process. The concentrate (often termed 'brine') is a residual liquid stream which requires additional treatment or disposal. A common practice is the conversion of concentrate into solid material, generally by subsequent evaporation and/or drying. In some cases, high-pressure separation can produce an extremely high-strength concentrate, and later solidification can be used to produce a solid material. At several landfills concentrates are simply recirculated into the landfill. Several studies have suggested that no increase in leachate pollution has occurred over several years of operation. However, since contaminants are merely being concentrated, and not treated, degraded, or removed from the site, ultimately over a period of decades they must reappear in leachate, and the environmental risks at the landfill will remain the same.

Typical membrane processes include reverse osmosis, nanofiltration and ultrafiltration (Box 10.11.2). Ultrafiltration is mostly used for separation of solids, for example during biological treatment stages. However, a small proportion of large organic molecules may also be retained by ultrafiltration. These molecules are removed from the system together with the excess sludge. To date, the nanofiltration has rarely been used during leachate treatment. Smaller molecules such as chlorides can pass through the nanofiltration membranes. Therefore, the remaining residuals are lower, however many problematic molecules can also pass through the membranes. At a few plants, nanofiltration has been used after biological treatment. The concentrate may be treated by adsorption or chemical oxidation, and then recirculated to the start of the biological treatment stage.

**Box 10.11.2   Principle of Membrane Processes and Cross-Flow Filtration.**

Membrane processes are a kind of filtration or better separation of different components in a solution. Membrane processes are characterised according to the dimension of the component which must be separated:

- Microfiltration: 0.05–2.0 µm (macromolecular range to microparticle range).
- Ultrafiltration: 0.005–0.2 µm (molecular to macromolecular range).
- Nanofiltration: 0.001–0.01 µm (molecular range).
- Reverse osmosis: < 0.002 µm (ionic range).

The membranes are of synthetic or specific ceramic materials. The membranes of reverse osmosis systems are comparable to cell membranes. If such semipermeable membranes separate two liquid phases with different concentrations water flows from the low concentrated side to the high concentrated side as long as the concentration on both is different (see figure below: Osmosis). If the concentrations on both sides are identical the water level on one side is increased and the difference of water level is called osmotic pressure ($\Delta\pi$). But the idea of reverse osmosis is to concentrate the high concentrated solution (or polluted solution) on one side of the membrane and push only the water through the membrane. Therefore it is necessary to apply pressure on the concentrated side. This pressure must be stronger than the osmotic pressure ($\Delta\pi + \Delta p$). With every increase of concentration (see figure below: Reverse osmosis) the osmosis pressure increases and consequently the necessary applied pressure. With decreasing amount of concentrate the necessary pressure increases.

Many membrane processes are operated as cross-flow systems (see figure below: Cross flow principle). As consequence of membrane composition back-washing is impossible. From this it follows that it is necessary to prevent the formation of a layer on the membrane. The most used solution is a high flow rate across the flow through the membrane by recirculation of the concentrated solution widely.

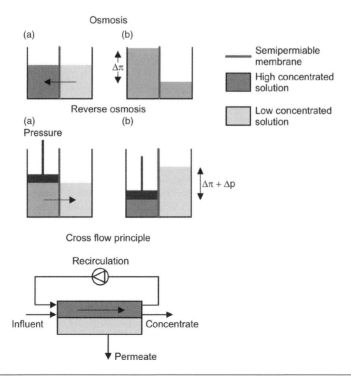

The most commonly used membrane process is reverse osmosis. An important advantage of reverse osmosis can be production of a high-quality treated permeate. However, all separated leachate pollution components have not been treated at all, but remain in the concentrate. If chemicals are required for effective operation of a reverse osmosis process (such as acids), then the amount of residual solid materials is increased. Most of the solid material is soluble, and a highly engineered containment facility is required for safe disposal. In Germany, the residuals are generally stored in barrels and in old mines.

Separation efficiency depends on the type and molecular size of leachate contaminants, the membranes used, and the operational pressure. The efficiency of separation is also influenced by the type of membrane modules used, with very different flow conditions and different risk of coating, and the possibility of membrane cleaning. Coating of membranes by fouling and scaling depends on the mixture of leachate components and on various physical changes which take place during membrane separation. To reduce coating and clogging, cross-flow processes are used in most cases (Box 10.11.2). Physical changes include change in solubility, temperature, calcium–carbonic acid equilibrium etc. Sulfuric acid may be added to reduce coating and to improve ammonium elimination. Disinfectants are used to reduce biological coating, and detergents can remove coatings.

Typical modules for reverse osmosis processes are tubular modules, spiral wound modules, hollow fibre module and disc tube modules. Although every reverse osmosis influent should be solid-free, spirally wound modules and especially the hollow fibre modules are extremely sensitive to presence of solids in raw leachate. An advantage of disk tube modules is the higher area per volume, when compared with tubular modules. The operational pressures of reverse osmosis systems up to 7000 kPa and for high-pressure reverse osmosis systems up to 15 000 kPa.

In practice, leachates with similar composition can show very different behaviour during reverse osmosis processes. This often demands pilot-scale trials for membrane separation. During these tests it is necessary to determine flow rates, specific cleaning requirements for membranes, and very importantly the operational life of individual membranes. An important factor is the permeate recovery rate. This is the proportion of incoming leachate which can be recovered as permeate. Values of 75–95 % may be possible. However, the higher values require high-pressure reverse osmosis. The remaining proportion of liquid represents the volume of concentrate that will be produced. Peters (1999) stated that flux rates depend on many parameters and reported average flux rates of 13–15 l/h permeate/$m^2$ membrane.

Hanashima *et al.* (1999) reported tests with reverse osmosis membranes (disc tube modules) to treat leachate from landfills which contained incineration residuals. Although these contained very high concentrations of chlorides (6500 mg/l) a recovery rate of 95 % could be observed. In addition to chloride removal, concentrations of dioxins in permeate could be reduced by up to 99.8 %.

## 10.11.7   Precipitation

In Germany, some leachate treatment plants use precipitation steps to reduce concentrations of nonbiodegradable organic substances. Many of these organic substances have a relative high molecular weight. Such molecules, for example humic like substances, can be removed from leachates by precipitation and flocculation (Box 10.11.3). Iron and aluminium sulfates or chlorides can be used as precipitation agents. The main precipitation requirements are low pH values, in the area of pH 4.5–5.0 for iron and pH 5.0–5.5 for aluminium. These conditions can be achieved by using the acid precipitation agents alone. However, as increasing quantity of chemicals are used, the quantity of sludge produced by the process increases too. To prevent excess sludge production, over approximately 10 mol/$m^3$ of iron or aluminium, the agents can be replaced by acid. With precipitation after biological leachate treatment, COD removal rates of 60–70 % are possible (Ehrig, 1987, 1998a). The addition of iron or aluminium solutions and acid increases the salt content of treated leachate (chloride and/or sulfate)

Heavy metal removal from other wastewaters by precipitation is common practice. However, concentrations of heavy metals in municipal solid waste landfill leachates are relative low (Robinson, 1996) and therefore removal is not normally necessary. However, as waste compositions change, this situation may also change. The chemicals most commonly used for precipitation include calcium, sodium carbonate and sodium hydroxide. The benefits of this process in leachate treatment are limited as a result of the complex matrix of leachate contaminants. In many cases, complex bonding between heavy metals and organic substances hinders the effectiveness of precipitation as a treatment process. Some improvement

---

**Box 10.11.3   Principle of Precipitation.**

The change of water-soluble substances into insoluble substances is called precipitation. As example many soluble heavy metals in solution react preferably with sulfide to metal sulfide (e.g.: $Fe^{2+} + S^{2-} \rightarrow FeS$). The solubility of most metal sulfides is very low. But in most cases the produced solids are very small. To remove them from the liquid phase by settling, coagulation to bigger flocs is necessary.

Precipitation is not constricted to heavy metals. Several other soluble substances in water can be removed with this process. The treatment of potable water uses precipitation in combination with coagulation to remove large organic molecules. Most slightly or not biodegradable organics in leachate have also a high molecular weight and can be removed with this process.

In some cases precipitation can only occur during pH changes. But for technical use in most cases precipitation agents are used. Such precipitation agents can be sulfide, lime, sodium hydroxide, acid, iron or aluminium salts. Some of these precipitation agents can in addition react as flocculation agents to produce settable flocs.

---

may be possible if sulfides are used for precipitation. Nevertheless, all precipitation processes are very strongly influenced by the pollution matrix of leachates. As a consequence, laboratory- and pilot-scale tests are absolutely essential.

Ammonium can also be removed by precipitation. Precipitation agents for this process are magnesium and phosphate (Ehrig, 1998a). The process is only effective where high ammonium concentrations are present in leachate. Until recently, the precipitation of magnesium/ammonium phosphate has rarely been used, but with increased pressures to recycle more substances, usage may increase because the precipitation end product can be directly used as a fertiliser. Contamination of this end product by other leachate components is generally low.

## 10.11.8   Adsorption

Adsorption is the transfer of organic substances from a liquid phase onto a solid phase. The efficiency of adsorption processes depends on the chemical and physical properties of the soluble substances and of the solid surface. Several adsorption materials can be used for the process. However, for organic leachate substances only activated carbon has shown an acceptable relationship between adsorption capacity and process cost. Activated carbon consists of carbon materials from various sources (coal, wood, peat etc.) which have a very large surface area ($800–1200\,m^2/g$). In the field of domestic wastewater treatment, or treatment of potable waters, activated carbon is frequently used to remove organic micropollutants and many nonbiodegradable organic substances. Activated carbon can be thermally reactivated, with less than 10 % loss in mass during each reactivation. During this process, the adsorbed organic substances are oxidised by gasification and subsequently incinerated.

Activated carbon can be used in the form of either pulverised or granulated carbon. Pulverised carbon is cheaper, but cannot be reactivated. Pulverised carbon must be dosed into a mixing tank, and then subsequently separated by flocculation and settling. Granulated carbon is used within filtration tanks (Figure 10.11.2). Unlike a sand filter, the carbon material must be changed once its maximum adsorptive capacity has been reached. In order to allow the contents of carbon filter tanks to be replaced easily, a specific pipe system with gate valves is necessary to operate the plant in a progressive manner, as shown in Figure 10.11.2.

The effectiveness of the adsorption process can be described by the adsorption isotherm, which presents the mass of adsorbed material per mass of activated carbon (e.g. mg COD/g activated carbon) as a function of the equilibrium concentration in the leachate (for example: mg COD/l). An isotherm is shown in Figure 10.11.3 and the mathematical representation is provided in Box 10.11.4 (Ehrig *et al.*, 1998b). Because of the unknown composition of many leachate organic substances, the isotherm must be determined experimentally; in many cases five to ten measurements are enough to estimate the isotherm. For a given equilibrium value, the adsorption capacity and thereafter the carbon consumption can be calculated using the isotherm (Box 10.11.5).

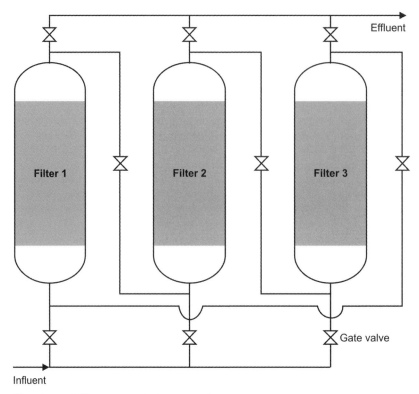

**Flow through filtersystem during operation**

Flow direction after start of operation:          filter 1 - filter 2 - filter 3
after change of carbon in the first filter = filter 1:   filter 2 - filter 3 - filter 1
after change of carbon in the first filter = filter 2:   filter 3 - filter 1 - filter 2
and so on

***Figure 10.11.2***   *Schematic view of an activated carbon filter system with three filters.*

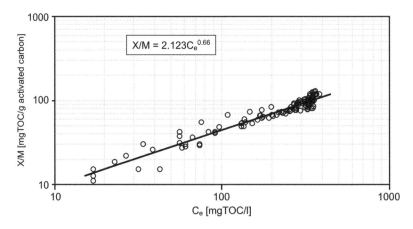

***Figure 10.11.3***   *Measured adsorption data and linearised Freundlich isotherm for TOC on activated carbon determined in pilot-scale experiments and calculated isotherm.*

---

**Box 10.11.4    Freundlich Isotherm Used in Organic Pollutant Sorption to Activated Carbon.**

The Freundlich isotherm is usually used to model the sorption of dissolved organic pollutants onto activated carbon:

$$X/M = K \cdot C_e^n$$

where $X/M$ = load of organics per weight of activated carbon (e.g. mg COD or TOC/g activated carbon) and $C_e$ = equilibrium concentration in the solution = remaining concentration after adsorption (e.g. mg COD or TOC/l). $K$ and $n$ are the characteristic parameters of the isotherm.

The Freundlich isotherm may for experimental purposes conveniently be transformed into a linear equation:

$$\log X/M = \log K + n \cdot \log C_e$$

In a regression analysis, $\log K$ is the interception and $n$ the slope.

---

The different operational techniques for mixing tanks with pulverised carbon, and filters with granulated carbon, are very important for the effectiveness of the adsorption process. The effluent concentration from a mixing tank is simply the equilibrium concentration of the isotherm. However, with optimal operation of filter systems, the equilibrium concentration at the start of the first filter approximates the influent concentration. The adsorption capacity with this high influent concentration is much higher than at the lower effluent concentrations. In most cases the total cost of mixing systems with cheaper pulverised carbon is higher, as a consequence of the low equilibrium concentration. Box 10.11.6 explains how carbon filters are operated to obtain the maximum adsorption capacity.

## 10.11.9   Chemical Oxidation

Chemical oxidation modifies the structure of pollutants in wastewater to similar, but mostly less harmful, compounds through the addition of an oxidising agent. During chemical oxidation, one or more electrons transfer to the oxidant from the targeted pollutant, causing its destruction. The optimal end products of organic oxidation are

---

**Box 10.11.5    Example of the Calculation of Amount of Activated Carbon Needed for a Mixing Tank System and a Filter System Meeting the Same Effluent Values.**

The amount of activated carbon needed per liter of leachate to removed TOC from leachate is calculated for a system using a mixing tank and one using a plug flow filter. The basic TOC adsorption data following the Freundlich isotherm are found in Figure 10.11.3. the influent concentrations is 310–410 mg TOC/l (average 350 mg/l) and the effluent criteria 70 mg/l. To be on the save side, the system is calculated for an effluent concentration of 50 mg/l.

**Mixing tank:**
Equilibrium concentration = effluent concentration = 50 mg TOC/l.
$X/M = 2.123 * 50^{0.66} = 28$ mg TOC/g activated carbon.
Carbon requirement = (influent TOC – effluent TOC)/($X/M$) = (350–50)/28 = 10.7 g carbon/l leachate.

**Filter system:**
Assumption: filter system is optimally operated.
Activated carbon will be changed if the effluent of first filter is a little les than the influent, here assumed = 340 mg TOC/l.
$X/M = 2.123 * 340^{0.66} = 99.5$ mg TOC/g activated carbon.
Carbon requirement = (influent TOC – effluent TOC)/($X/M$) = (350–50)/99.5 = 3 g carbon/l leachate.

**Box 10.11.6   Operation of Carbon Filters.**

In filter systems, the effluent values after carbon changes are very low, and increase gradually until effluent limits are reached. This effect is presented in the figure below, with two filters in sequence. If the carbon from filter 1 is removed to be replaced by new carbon, the effluent concentrations are reduced. At the same point, the circles of filter 2 change to triangles, which indicates that filter 2 is changed to filter 1 in the row. However, the advantage of filter systems is only achieved if the total quantity of carbon in each filter is adequate. Based on results from pilot-scale trials at individual landfills, accurate calculation of the necessary number of filters is possible. The slope of equilibrium concentrations along the filter length was measured in the range of 300–600 mg COD/l or 100–200 mgTOC/l per meter of carbon height (Ehrig and Hagedorn, 1998). With these slope the minimum of remaining filter height during carbon change can be calculated. Near influent concentrations most isotherms underestimate the adsorption capacity. To use the total adsorption capacity additional 1.5 to 2.0 m adsorber high should be added. At most leachate treatment plants in Germany, two filters are used, with a total carbon height of between 4 and 8 m. For the concentration gradients shown above, a carbon height of 4–8 m may in many cases not be sufficient for an optimum usage of activated carbon. In the figure it can be seen that two filters are not enough, because the effluent concentrations of filter 1 at the point of carbon change are significant lower than influent concentrations.

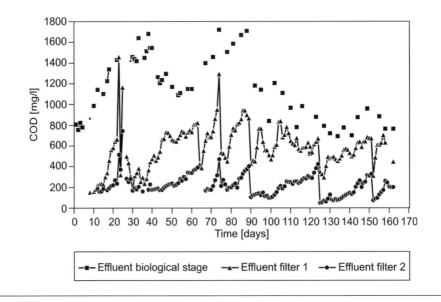

water and carbon dioxide. An old oxidation process is alkaline chlorination. But with the complex organic matrix of leachate the production of hazardous substances can be very high. For chemical oxidation at normal temperature and pressure, ozone and hydrogen peroxide are generally used. Combinations of both agents and UV light are possible alternatives. Many combinations of oxidation processes have at some stage been investigated for use in leachate treatment. However, few types and combinations have provided realistic treatment options which can be used at full scale.

Hydrogen peroxide has been widely used as an efficient treatment to remove low concentrations of sulfide from leachates, by converting the sulfide to elemental sulfur. This is a very specific reaction, and so required dosing rates can be predicted accurately based on chemical stoichiometry. A contact period of about 30 min has widely been found to be adequate, for complete removal to be achieved. However, the fact that the sulfide removal process is so specific, indicates that at normal temperatures and pressures, hydrogen peroxide is very slow to oxidise organic components in leachates. Adequate oxidation rates can only be obtained by addition of UV light, and at pH values of approximately 4. The production of UV light consumes a lot of energy, and colour and turbidity in leachates increase requirements because the penetration of UV light is reduced. Additionally, low pH values consume large amounts of acid, and increase the salt content of the treated leachate. During the contact of hydrogen peroxide with organic polluted water the peroxide breaks down very rapidly to so called radicals which are strong oxidation agents. These radicals react with the organic substances in water. To prevent a breakdown of hydrogen peroxide before usage during storage and transport (pumping) several restrictions for the material must be considered. Chemical oxidation with hydrogen peroxide is mostly carried out in mixing tanks. The detention time in the mixing tank for leachate oxidation should be investigated with laboratory-scale experiments. In most cases several hours are needed.

Most full-scale chemical oxidation plants use ozone for oxidation of leachate components. Ozone is a gaseous component with a very rapid decay rate and must be produced directly before usage. Ozone can be produced from the oxygen in the air by electric discharge but as result of decreasing costs of pure oxygen and easier handling in most cases pure oxygen is used. Ozone can react directly with organic pollutants similar to hydrogen peroxide during breakdown with radicals. At several plants ozone is supported by UV light. Experiments have shown that the benefits of UV light are limited, and addition of hydrogen peroxide is easier and cheaper, and in most cases has the same effect. Overall the enhancement effect has been found to be limited in many cases, and benefits should be compared with the additional costs involved. Ozone is a toxic gas and should only be used in closed tanks with an intensive mixing of water and gaseous ozone. In most cases a detention time of several hours is necessary for leachate treatment. A destruction of ozone before discharge to the environment is necessary.

Other oxidation systems can use ozone and a catalyst. The catalyst is a carbon-like material, with the advantage that some substances may be adsorbed and then oxidised subsequently. Overall, ozone consumption is lower than without catalyst, but the operational life of these catalysts is poorly understood at present.

Chemical oxidation is a nonspecific oxidation process. This means that all available organic substances, biodegradable and nonbiodegradable, are oxidised. In some circumstances, it may also be possible to achieve oxidation of ammoniacal-N, although this will be a costly process, when compared with alternatives. To prevent excess ozone consumption, a pretreatment stage is necessary to remove biodegradable organic substances and to remove or oxidise ammoniacal-N. The intensity of oxidation depends on the contact time between leachate and ozone. During a first phase with low oxidation efficiency, organic halogens can be produced if chloride is available in high concentrations, which is the case in most leachates. During a second phase, the organic substances can be oxidised, primarily to organic fragments. Many of these fragments are biodegradable. Only by use of a third phase of oxidation, can a significant proportion of organic substances be oxidised to carbon dioxide.

If leachate is recycled between chemical oxidation and biological treatment stages, significant reductions in ozone consumption are possible. An increasing effect, up to between three and five cycles was noted by Steensen (1998). In such systems the intermediate biologically degradable products are removed from leachate biologically, before the next chemical oxidation phase. With such a procedure the consumption of expensive ozone can be reduced.

Ozone cannot be stored and must be produced on site as it is required by the plant. Today's common practice is to use oxygen for the production of ozone, however only a part of oxygen is converted to ozone by electricity. To save costs, some oxidation systems work with oxygen recirculation, to increase the ozone yield. However, conversion of oxygen to ozone requires a very clean gas, and therefore gas cleaning of recirculation gas is necessary. Any exhaust gases containing ozone must be cleaned, because ozone is very toxic. If the gas cannot be recycled, use of the exhaust which contains high oxygen levels should be encouraged, for example in biological processes. Theoretical ozone consumption is 3 kg ozone/kg COD. In practice, actual consumption may be lower – in the area of 2.3–3.0 kg ozone/kg COD. Steensen (1998) estimated that, for several leachates, ozone requirements depended on COD reduction (Box 10.11.7) and demonstrated that the activation of ozone by UV light or hydrogen peroxide had only a limited effect. The recirculation of leachate through oxidation and biological treatment can reduce ozone consumption to between 1.5 and 2.0 kg/kg COD. As mentioned above, only in the last stage of chemical oxidation is conversion of significant amounts of organic substances to carbon dioxide achieved. As a consequence, the reduction of COD is faster than the reduction of carbon (measured as TOC) during chemical oxidation. The average relation between COD and TOC of leachate in Germany is 3. This was the reason why a limit for effluent TOC was fixed at 70 mg/l, calculated from an equivalent COD limit of 200 mg/l. However, in oxidised leachate the COD-TOC relationship is generally < 3. Future investigations are necessary to estimate the environmental impact of such intermediate carbon substances.

---

**Box 10.11.7   Ozone Consumption and Biological Posttreatment (Steensen, 1998). Reprinted with permission from Chemische Oxidation und biologische Nachreiningung zur weitergehenden Sickerwasserbehandlung by Martin Steensen, Vol 63, University of Braunschweig © (1998) Martin Steensen.**

Ozone requirement at processes with the only usage of ozone:

$$f_{COD} = [1.35 + 0.005 * \eta(COD)]/\{1 - e^{-0.1*[100-\eta(COD)]}\}$$

where $f_{COD}$ = ozone consumption (g ozone/g COD):

$$\eta(COD) = COD \text{ reduction} = \text{percentage reduction}/100$$

Biological posttreatment after oxidation:

$$\eta(COD, \ bio) = 0.6 * \eta(COD, \ oxidation).$$

---

The use of ozonation as a specific effluent polishing stage has been extremely successfully applied as a means of removing pesticides (such as isoproturon) which are not readily biodegradable. Robinson *et al.* (2003) reported nearly ten years experience in great detail, including operational costs, at a plant where dosing of ozone at rates of up to 150 mg/l, following biological treatment in SBRs and reed beds, was successful in removing up to 0.5 mg/l of isoproturon – but could only remove about 10 or 15 % of a residual hard COD of about 400 mg/l (following degradation of ozone breakdown products in a final polishing reed bed). Presence of brominated compounds in intermediate stages of the process was noted, and at this site ozone was generated using air, not oxygen.

In principle chemical oxidation is a residual free process. But at several oxidation plants the production of calcium oxalate during oxidation has been observed with variable degrees of incrustation problems – which can significantly reduce the effectiveness of UV light treatment.

The effluent from chemical oxidation, with COD values below 200 mg/l, can contain $BOD_5$ values of up to 50–70 mg/l. Such high values may require a small biological polishing stage. Convenient options may be a rotating biological contactor, as has been used in Germany. Nevertheless, although biomass is fixed onto the material of an RBC, problems with underloading can be observed. More natural systems such as reed beds, as above, may provide a better option.

Very detailed calculations of chemical oxidation processes and additional biological posttreatment processes are given by Steensen (1998).

## 10.11.10   Wetlands

Constructed wetlands and reed beds are designed and manmade systems which attempt to simulate treatment that has been observed to take place when polluted water passes through natural wetlands. In the UK these systems tend to be called 'reed beds' or 'reed bed treatment systems' (RBTS), but internationally they are usually called 'constructed wetlands' (CW).

Systems are able to treat wastewaters by degrading organic matter (BOD, COD) and oxidising ammoniacal-N, removing suspended solids, and to a lesser extent reducing concentrations of nitrate and phosphorous. Treatment mechanisms are complex and involve bacterial oxidation, filtration, nitrification and chemical precipitation, but they are popular because they are seen to have the following advantages:

- Relatively low capital and operating costs.
- Simplicity of operation (low requirement for operator supervision).
- Suitable treatment for low flows, previously untreated.
- Seen as a natural and therefore 'green' process.
- They are attractive and provide wildlife habitat.
- As a polishing stage they can improve and enhance performance of initial treatment processes.

A constructed wetland has the potential advantage of long-term, sustainable treatment, with very low costs of operation and maintenance, whether used alone for treatment of weak leachates from older landfills, or for polishing of stronger leachates that have been pretreated by other biological processes. This is especially important for leachate control, where (unlike sewage treatment works) large landfill sites may be remote and unmanned, and require indefinite treatment timescales. Passive constructed wetlands can offer very long lifetimes, with little or no equipment replacement.

The treatment technology relies on processes similar to those used extensively in gravel 'filter beds', enhanced by the extensive rhizomatons root system of the reed plants (*Phragmites australis*) which can transfer limited quantities of oxygen into the surrounding media, stimulating bacterial communities. Other constituents of the effluent can be immobilised, or absorbed by the plants themselves (on some occasions alternative plants such as bulrush have been used successfully). Although aerial growth of the reed plants (which can reach 2–3 m high) dies down during the winter months, treatment has been demonstrated to continue effectively. In situations where effluent from a previous treatment process, such as an SBR, may be warm (20–25 °C), the dead reed stems typically mat down on the gravel surface of the bed, providing heat insulation and maintaining temperatures within the root zone of the reeds.

The European Water Pollution Control Association (EWPCA, 1990) concluded that nitrification of ammonia to nitrate has not generally occurred in temperate reed bed treatment systems, because of oxygen limitations, but noted that it has been reported in some 'polishing' treatment schemes. In a summary of UK reed bed performance, Findlander *et al.* (1990) concluded that neither soil beds nor beds containing coarse granular media, removed significant amounts of nitrogen. This suggests that permanent nitrogen removal is not likely when leachate is treated in wetlands.

A variety of basic constructed wetland designs have been used, and performance varies significantly from one design type to another: Horizontal flow systems and vertical flow systems. For both design types, most successful applications involve subsurface flow within the (generally gravel and sand) media in which the reeds are planted – avoiding surface flow which would bypass the main treatment surfaces.

The majority of all of the constructed wetlands that have been built in the UK have been subsurface, horizontal flow systems. Such systems typically comprise lined beds, containing depths of gravel of 0.6–0.9 m, with a horizontal surface, and are longer than they are wide. Leachate is introduced along one of the shorter edges of the bed, and discharges from the bed at the far end, generally via buried pipework. Liquid level in the bed is adjusted by a simple overflow mechanism outside the bed – usually this will be maintained at or close to the gravel surface, but the horizontal surface of the bed allows intermittent flooding to be effected. Figure 10.11.4 shows a typical arrangement for a horizontal-flow reed bed

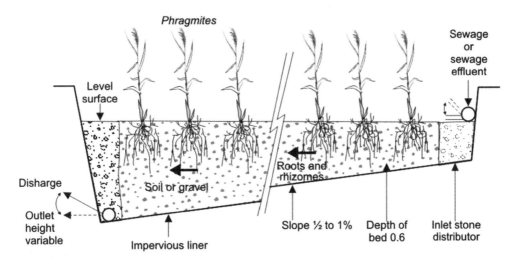

**Figure 10.11.4**  *Typical arrangement for a horizontal-flow reed bed treatment system.*

treatment system. For use in leachate treatment, single-sized gravel of up to 10 mm in size has most commonly been used, and is commonly emplaced at a depth of 600 mm. Provision of a basal slope can assist in draining of the bed, but gravel depth should not exceed 1 m.

Horizontal-flow reed beds have been used with success for the direct treatment of relatively weak leachates – for example, from very old closed landfills, where low concentrations of ammonia and remoteness of sites can make this an ideal solution. Their ability to reduce concentrations of residual BOD, suspended solids, iron and also trace organic compounds such as mecoprop is significant. Increasingly, however, horizontal flow reed beds are being used as a second stage of polishing, following a biological first treatment stage, before discharge of effluent to a watercourse. More than 20 such combined systems have been constructed in the UK since 1990 (e.g. Robinson *et al.*, 2003a, b). Box 10.11.8 presents a plant in the UK.

Vertical flow reed bed systems have not been described in terms of successful application for treatment of leachates in the UK. Data regarding the loadings of ammoniacal-N that can be treated reliably and consistently remain the greatest uncertainty. Nevertheless, such schemes have potential to achieve improved rates of nitrification per square metre of reed bed area, compared to traditional horizontal flow systems.

---

**Box 10.11.8   Case: Use of Reed Beds to Polish Effluent from Biological Leachate Treatment.**

The Efford Landfill leachate treatment scheme (Hampshire, UK) includes a typical horizontal flow reed bed for polishing of biologically pretreated leachate. A SBR system treats up to 150 m³/day of strong leachate, effluent being polished by passage through a reed bed, before discharge into a small, rural stream. The reed bed is able to provide excellent removal of residual SS, ammoniacal-N and BOD in SBR effluent, as well as up to 35 % removal of residual COD (Robinson and Olufsen, 2004).

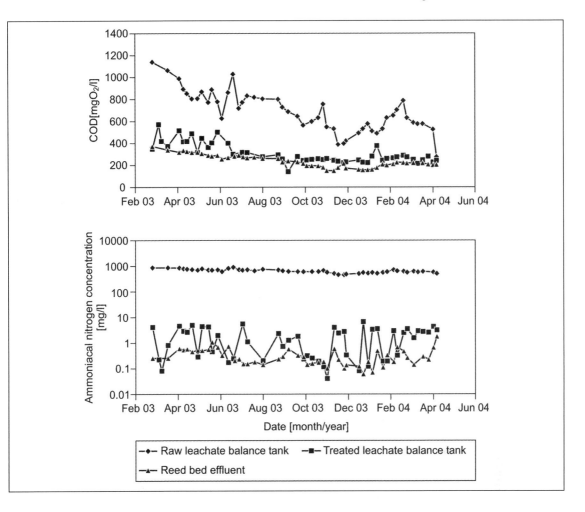

## 10.11.11   Biological Processes: General Features

Biological processes are the most commonly used municipal solid waste landfill leachate treatment options. These processes are able to reduce biodegradable organic substances and nitrogen. Typical removal efficiencies are shown in Table 10.11.4. In many cases, effluents from biological treatment may not be adequate on their own to allow direct discharge of treated effluents, but biological pretreatment is a very useful and sometimes necessary step for several chemical and physical treatment processes.

Two different techniques for biological degradation of organic substances are used in wastewater treatment:

- Aerobic treatment: degradation of organic substances by oxygen.
- Anaerobic treatment: degradation of organic substances without usage of oxygen.

Anaerobic degradation of many organic substances can be observed in most landfills containing household or other organic wastes. During intensive methane production stages in a landfill, these anaerobic degradation processes are fully developed. In most cases an additional anaerobic treatment step for treatment of leachate has only a small effect.

**Table 10.11.4**   *Typical performance of aerobic biological leachate treatment schemes in removal of COD and BOD (in mg/l, concentrations of ammoniacal-N provided for comparison).*

| Site | Parameter | Influent (mg/l) | Effluent (mg/l) | Removal (%) |
|---|---|---|---|---|
| Arpley, UK[a] | COD | 5990 | 1470 | 75 |
| | $BOD_{20}$ | 1720 | 67 | 96 |
| | $BOD_5$ | 688 | 20 | 97 |
| | $NH_4/NH_3$-N | 1460 | 3.7 | >99 |
| Pitsea, UK[b] | TOC | 365 | 337 | 8 |
| | $BOD_5$ | 28 | 19 | 32 |
| | $NH_4/NH_3$-N | 281 | 0.4 | >99 |
| Bryn Posteg, UK[c] | COD | 5518 | 153 | 97 |
| | $BOD_5$ | 3670 | 18 | >99 |
| | $NH_4/NH_3$-N | 130 | 9 | 93 |
| Llanddulas, UK[d] | COD | 3400 | 1310 | 61 |
| | $BOD_{20}$ | 682 | 15 | 98 |
| | $BOD_5$ | 88 | 5 | 94 |
| | $NH_4/NH_3$-N | 1330 | < 0.3 | >99 |

[a]Robinson *et al.* (2003a)
[b]Knox (1991)
[c]Robinson and Grantham (1988)
[d]Robinson *et al.* (2005)

As discussed earlier, ammoniacal and organic nitrogen are the most important polluting parameters in leachate, next to organic substances. Organic nitrogen can be converted to ammoniacal-N under either aerobic or anaerobic conditions, whereas ammoniacal-N can only be oxidised to nitrate (nitrification) under strictly aerobic conditions. Nitrate can be reduced to gaseous nitrogen (denitrification) under so-called 'anoxic' conditions. The term anoxic refers to a situation where absence of oxygen means that nitrate is used as the electron acceptor for the oxidation processes. To degrade the remaining organic substances in landfill leachate as far as possible, and to oxidise ammoniacal-N to nitrate, aerobic biological treatment is generally the preferred option for biological treatment.

Biological leachate treatment also removes trace organic and heavy metals. Robinson and Knox (2001, 2003) provided indicative median removal rates for those trace compounds based on an extensive survey of UK and Irish leachate treatment plants, as shown in Table 10.11.5.

In contrast to chemical oxidation or incineration, biological oxidation converts only a proportion of organic carbon to carbon dioxide. Other carbon is used for the production of new biomass. In most cases, sludge production is lower than in sewage treatment plants, because organic loading rates are also lower. Nevertheless, 20–30 % of leachate organic carbon can be converted into biomass. Many analyses of excess sludge from different biological leachate treatment plants have shown that this sludge is not polluted by organic micropollutants or heavy metals, but this should be checked at specific treatment plants.

Although aerobic biological treatment processes have been widely applied to the treatment of domestic wastewaters, and of industrial effluents, there are several specific characteristics of leachates which must be recognised in the design of appropriate facilities:

- The 5-day biochemical oxygen demand ($BOD_5$) test is rarely an adequate measure of the extent to which organic compounds can be degraded within a well-designed treatment plant, containing an acclimatised bacterial population. The $BOD_5$ test uses a standard (sewage-based) bacterial seed, and is limited to a 5-day incubation period. Although the 20-day BOD test ($BOD_{20}$) may be more helpful, generally the COD value is used for design purposes. Most, if not all, successful aerobic biological leachate treatment plants, achieve far higher removal of organic compounds than is predicted by BOD results. Figure 10.11.5 shows typical removal of COD by biological treatment as a function of the BOD/COD ratio of the influent leachate.

**Table 10.11.5**  *Median concentrations of a range of trace organic compounds and heavy metals detected in more than 5 % of UK landfill leachates, and median % removal during aerobic biological treatment (after Robinson and Knox, 2001, 2003).*

| Parameter | % detected | LOD (µg/l) | Median concentration (µg/l) | % removal |
|---|---|---|---|---|
| Chromium | 33 | 20 | 50 | 30 |
| Nickel | 86 | 10 | 60 | 20 |
| Copper | 60 | 5 | 11 | 50 |
| Zinc | 100 | 5 | 135 | 70 |
| Lead | 8 | 20 | < 50 | No data |
| Arsenic | 94 | 1 | 8 | 70 |
| Aniline | 17 | 1 | < 1 | 80 |
| AOX | 97 | < 8 | 177 | No data |
| Biphenyl | 51 | 0.1 | 0.1 | 60 |
| DEHP | 25 | 1 | < 1 | 90 |
| Ethylbenzene | 15 | 10 | < 10 | 80 |
| MCPA | 12 | 0.1 | < 0.1 | 95 |
| MCPP (mecoprop) | 98 | 0.1 | 11 | 99 |
| Methylene chloride | 12 | 1 | < 1 | No data |
| MTBE | 8 | 1 | < 1 | 90 |
| Naphthalene | 70 | 0.1 | 0.46 | 95 |
| Nonyl-phenol | 83 | 0.2 | 1.0 | 95 |
| Organotin | 50 | 0.02 | 0.20 | No data |
| Pentachlorophenol | 8 | 0.1 | < 0.1 | 50 |
| Phenols (mono) | 54 | 20 | 0.03 | 99 |
| PAH (Borneff 6) | 29 | 5 | < 5.25 | 50 |
| Toluene | 54 | 10 | 21 | 80 |
| Xylenes | 35 | 10 | 35 | 60 |

- High concentrations of ammoniacal-N, generally > 500 mg/l and regularly in excess of 1000 mg/l at modern landfills, are many times stronger than levels of 25–30 mg/l typically encountered in domestic wastewaters. Although direct toxicity of ammoniacal-N to nitrification processes is not a significant issue in sewage treatment, wide experience in leachate treatment systems has demonstrated that (at typical pH values in the range 7–8), concentrations of 80 mg/l of ammoniacal-N or above significantly inhibit the nitrification process. Actual levels of toxicity are related to presence

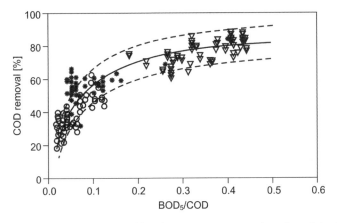

**Figure 10.11.5**  *COD removal in biological leachate treatment as a function of the BOD/COD ratio.*

of free ammonia, which in turn is a function of concentration of ammoniacal-N and pH value (at higher pH values, a higher percentage of free ammonia is present).

- The acidity produced during nitrification will often require very large additions of alkalinity to buffer pH values – requirements to add between 10–15 l of 32 % w/v NaOH to every cubic metre of leachate treated are not uncommon.
- High concentrations of other contaminants can lead to problems not generally encountered in treatment of domestic or other weaker wastewaters. High salinity may require that bacteria being used to effect treatment are gradually acclimatised to the leachate being treated, and this may take several months.
- Degradation of high (> 5000 mg/l) BOD and COD values can generate high volumes of organic sludge, which may clog attached growth systems and require routine removal from suspended growth systems. High concentrations of iron, calcium and other metals may lead to similar problems with inorganic sludges or deposits.
- Foaming can be a serious problem at biological leachate treatment plants. Foaming can be reduced by antifoaming agents, and most antifoaming agents are biodegradable oily substances. These substances have an extremely high COD, and where high rates of dosing are necessary, they can comprise a significant proportion of the total load of the treatment plant.

### 10.11.11.1  Nitrification

The conversion of organic nitrogen to ammoniacal-N occurs under several conditions, and partly within the landfill. Oxidation of ammoniacal-N to nitrate (called 'nitrification') is a very sensitive process, but for the removal of nitrogen it is an essential stage.

Nitrification is the biological oxidation of ammoniacal nitrogen to nitrate nitrogen by autotrophic bacteria, which derive energy from the oxidation reaction and utilise inorganic carbon. The nitrification reaction is a two-stage oxidation, each stage being performed by a distinct group of bacteria. The first stage, oxidation of ammoniacal nitrogen to nitrite nitrogen, is performed by bacteria of the genus *Nitrosomonas*. The second stage, where this nitrite nitrogen is further oxidised to nitrate nitrogen, is performed by species of *Nitrobacter*. The reactions are shown empirically below:

- First stage: $55 \, NH_4^+ + 76 \, O_2 + 5 \, CO_2 \rightarrow 54 \, NO_2^- + 109 \, H^+ + 52 \, H_2O + C_5H_7O_2N$.
- Second stage: $115 \, NO_2^- + 52.5 \, O_2 + 5 \, CO_2 + NH_4^+ + H_2O \rightarrow C_5H_7O_2N + H^+ + 115 \, NO_3^-$.

From the above empirical reactions it may be calculated that for 1 kg of ammoniacal nitrogen that is nitrified:

- 4.27 kg of dissolved molecular oxygen are used.
- 7.14 kg of alkalinity, as CaCO3, are destroyed.
- 0.22 kg of new cells are synthesised.

Both groups of bacteria are relatively sensitive (compared with those groups which oxidise organic substrates) to environmental conditions, and either one or both stages can be easily inhibited by:

- Low pH values (below about 6.5).
- Insufficient dissolved oxygen (below about 2 mg/l).
- Low temperatures ($< 5 \,^\circ$C), or high temperatures ($> 35 \,^\circ$C).
- Toxic inhibition.

A wide range of chemicals are known to inhibit nitrification by toxic effects, but such inhibition is rare, and treatment can be readily achieved, particularly in methanogenic leachates. However, the range of toxic chemicals includes both ammoniacal nitrogen itself, and the intermediate oxidation product, nitrite nitrogen, both of which can potentially inhibit the second stage of the reaction, leading to a build up of nitrite nitrogen in effluent.

The optimum pH value for biological nitrification is typically between 7.5 and 8.0 (often quoted as pH 8.4), and nitrification rates decrease very sharply at pH < 5.5. Nitrifying bacteria can, however, sometimes adapt to acidic environments, but rarely operate efficiently within them. Therefore, unless the wastewater being treated contains sufficient alkalinity, the nitrification reaction will ultimately prove to be self-inhibitory as it releases hydrogen ions and depresses pH values. This process, together with maintenance of insufficient concentrations of dissolved oxygen, is the most common cause of failure in full-scale treatment plants trying to accomplish nitrification.

In strong methanogenic leachates it has routinely been found necessary to add alkalinity, at rates up to 50 % of that originally present in raw leachate (e.g. Robinson and Luo, 1991). Lower concentrations of ammoniacal nitrogen may also require addition of alkali, albeit in lesser amounts. Nutrient requirements in leachate treatment are well understood, for example the need to add phosphorus (e.g. Robinson, 1990; Robinson and Grantham, 1988).

### 10.11.11.2 Denitrification

Biological denitrification is the reduction of nitrate nitrogen to nitrogen gas by facultative heterotrophic organisms that use organic carbon for energy and as a carbon source. These organisms can use molecular oxygen as the electron acceptor under aerobic conditions, but use nitrate nitrogen as the electron acceptor when no oxygen is present.

Denitrification of nitrate nitrogen to nitrogen gas occurs in many wastewater treatment plants and has become the most widely used nitrogen removal process in municipal wastewater treatment. It has also been reported to be a significant process at full-scale landfill leachate treatment plants (Robinson, 1990). A large number of bacterial species which occur naturally in the activated sludge process, or in extended aeration treatment systems, are capable of denitrification.

Denitrification can sometimes be incorporated into a treatment process for reasons other than a need to reduce nitrate levels – one example can be as a means of reducing requirements to add external alkalinity.

In a treatment plant where nitrification is the main process being achieved, it is generally necessary to add an organic substrate to allow denitrification to proceed. A variety of organic substrates are suitable. Methanol ($CH_3OH$) is widely used. The overall reaction, including cell synthesis, may be expressed by the following equation:

$$NO_3^- + 1.08\,CH_3OH + 0.24\,H_2CO_3 \rightarrow 0.06\,C_5H_7O_2N + 0.47\,N_2 + 1.68\,H_2O + HCO_3^-$$

The $N_2$ gas is released to air. From this empirical reaction it may be calculated that for 1 kg of nitrate nitrogen that is denitrified:

- At least 2.47 kg of methanol are used.
- 0.45 kg of new cells are synthesised.
- 3.57 kg of alkalinity are formed.

The presence of dissolved oxygen inhibits the denitrification process, concentrations of more than 0.2–0.5 mg/l have been shown to significantly reduce denitrification activity. Denitrification activity is reduced at low temperatures and enhanced by an increase in temperature up to an optimum of 40 °C. Temperatures above this level prevent denitrification.

As alkalinity is formed during denitrification, the addition of an acid such as sulfuric acid to maintain the pH value within a narrow optimal range of pH 7.0–7.5 may possibly be needed. It has been shown that for pH < 6 and pH > 8 there is a rapid decrease of denitrification activity.

Figure 10.11.6 shows a typical scheme for predenitrification and postdenitrification in biological treatment plants. A predenitrification configuration should be preferred if during at least some years, organic compounds present in incoming leachate can be used as a carbon source for denitrification. Presence of inhibitory levels of ammoniacal-N may make this less attractive, and with increasing age of landfills, as the nonbiodegradable proportion of leachate organic carbon increases, it is generally less useful for denitrification. Denitrification by dosing of an external carbon source is necessary in most cases.

## 10.11.12 Biological Processes: Aerated Lagoons

Experiences of leachate treatment during the 1970s and 1980s began with aerated lagoons, as a very simple and easy technology. Aerated lagoons can be designed to look like natural lagoons, or like a hollow with a plastic liner. As an

## Pre-denitrification

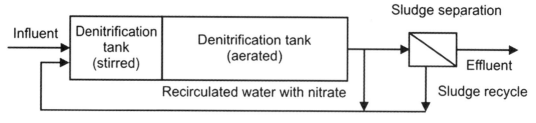

## Post-denitrification

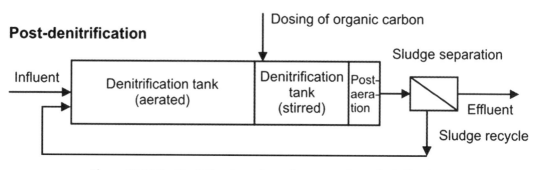

***Figure 10.11.6***   *Denitrification scheme for pre- and postdenitrification.*

alternative to plastic liners, some aerated lagoons in Germany consist of concrete tanks. For direct discharges into surface waters, an additional lagoon may be necessary to settle residual sludges. Reed beds have been widely used for this purpose, as a final polishing step. The operation of aerated lagoons is relatively simple, because it is a self-controlling system. The main technical installation is the aeration system of the lagoon.

Aerated lagoons are strongly influenced by climatic conditions. In many central and northern Europe conditions, as in Germany or the United Kingdom, water temperatures during winter routinely fall below 10 or 5 °C. Below 5 °C, with volumetric loadings of 1–2 g $BOD_5$/m$^3$/day, it is difficult to achieve $BOD_5$ values of below 25 mg/l. Below 10 °C nitrification rates decrease significantly. At water temperatures below 10 °C, over a longer period of time, nitrification stops totally and may require several weeks to fully recover. Only at average temperatures $\geq$ 15 °C, with volumetric loads < 20 g $BOD_5$/m$^3$/day, can organic substances be reduced to values below 25 mg/l $BOD_5$, and full nitrification be expected (Albers, 1985). At such volumetric loading rates, an oxygen demand of approximately 100 g $O_2$/m$^3$/day is necessary. Denitrification with aerated lagoons is very difficult to achieve or control. One reason is that lagoons are not easily mixed without aeration. Similar results have been published by Maehlum and Haarstad (1997) with relative good treatment rates achieved at temperatures over 10 °C, but poor nitrification at low temperatures. Variation in effluent data during these observations can only partly be explained by variations in influent loading.

At landfills in warmer climates, where stronger leachates are generated, consistent and efficient nitrogen removal can be maintained throughout the year. Chen et.al. (1997) reported results from a very large lagoon leachate treatment plant in Hong Kong, operated as an SBR, and designed following extensive and detailed treatability trials (Robinson *et.al.*, 1995). This plant achieved not only a high reduction in COD values and ammoniacal-N (from levels of up to 150 000 mg/l and 8400 mg/l, respectively), but also a significant reduction of nitrate (Table 10.11.6). One reason for this effect is the high production of biomass, during degradation of the very high COD values. The sludge flocs produced have an anaerobic/anoxic core for denitrification, and an aerobic surface. These simultaneous processes were able to achieve an overall nitrogen reduction of approximately 55 %.

**Table 10.11.6**    *Operation data of a lagoon leachate treatment plant (Chen et al., 1997). Reproduced with permission from CISA, Italy. © (1997) EuroWaste srl, Italy.*

| Parameter | Unit | Influent (average) | Influent (peak values) | Effluent |
|---|---|---|---|---|
| pH |  | 7 | 4.8–7.8 | 7.7–8.5 |
| COD | mg/l | 40 000 | 153 300 | 600–1000 |
| BOD$_5$ | mg/l | 25 000 |  |  |
| Ammonia | mg N/l | 3500 | 8400 | <2 |
| Total N | mgN/l |  |  | <1500[a] |

[a] Estimate to correspond to an average 55 % removal of nitrogen

## 10.11.13    Biological Processes: Activated Sludge

The most common aerobic processes used for treatment of domestic wastewater are activated sludge systems. Their main difference from aerated lagoons is a separate settling tank after the aeration reactor, to settle biomass and return this as return sludge back into the aeration tank. The result of this process is an increase in concentrations of biomass within the aeration tank, resulting in a higher treatment capacity. Settlement of biomass for separation, during effluent clarification is the usual technique adopted in classic activated sludge plants. A more recent technology is the separation of returned biomass from effluent by use of membranes. This can be a microfiltration or an ultrafiltration process. Nearly all activated sludge plants in Germany now use ultrafiltration membranes, although the process has not yet been widely adopted in other countries Biomass separation by ultrafiltration has several advantages and few disadvantages, compared with settling tanks:

- Ultrafiltration can separate higher concentrations of biomass (up to 50 kg suspended solids/m$^3$).
- Ultrafiltration retains the total amount of biomass. In comparison with traditional settling tanks, there should be no loss of any biomass. As a consequence, the sludge age in the plant can be longer, with the same concentration of biomass in the aeration tank, allowing treatment rate to be increased, and the stability of the treatment process to be improved.
- Ultrafiltration destroys the sludge flocs of biomass, which are necessary for the settling process. Very small and poorly settling flocs have a larger surface area, allowing increased contact between leachate substrates and oxygen. Both effects have potential to increase treatment efficiency. These effects are very important for the nitrification process.
- Ultrafiltration can have problems treating leachate, which contains much higher concentrations of inorganic components. The flow of effluent through an ultrafiltration membrane must be high to allow high sludge recycling rates, and precipitation of inorganic substances can damage membranes.

The possibility of operation of ultrafiltration membrane activated sludge plants with high suspended solids concentrations can reduce the size of these plants. As a consequence, most of these plants use indoor steel tanks as aeration tanks, which can reduce cooling problems. In addition, the high recirculation rate through the ultrafiltration membranes, the high activity of biomass, and the temperature of the aeration air increases the temperature. Normally this increases the treatment efficiency. But during summer periods temperatures can increase above 40 °C. At such temperatures, treatment efficiency falls very rapidly, and cooling of the aeration reactor can be necessary. At some activated sludge plants, reactors are operated at increased pressures of typically 500 kPa. An advantage of such plants is the higher solubility of oxygen in water, although the required higher pressures of air for aeration require more energy.

A very important part of biological leachate treatment is the denitrification of nitrate nitrogen which is produced during nitrification. In contrast to domestic wastewater treatment, the carbon content of raw leachate is not sufficient for denitrification in the long term. If denitrification is required, dosing of additional readily-biodegradable organic substances is necessary. At least 4.5 kg of biodegradable COD is required for every 1.0 kg of nitrate nitrogen to be removed. In some circumstances it may be possible to make use of industrial wastewaters and liquid wastes. Strong biodegradable organic substances are needed, in order to minimise transport costs to the leachate treatment plant. The effect of the total pollution mixture of dosing agents on biological treatment must be considered. Readily degradable substrate is not only important

**Table 10.11.7**  *Typical performance data from MBR leachate treatment schemes (mg/l). Datacourtesy of Wehrle Environmental, Germany.*

| Location | Freiburg, Germany | | Lorrach, Germany | | Lüneburg, Germany | |
|---|---|---|---|---|---|---|
| Dates | 1999– | | 1997– | | 1996– | |
| Flow | 65–120 m$^3$/day | | 264 m$^3$/day | | 70 m$^3$/day | |
| MBR volume | 98 m$^3$ | | 2 × 120 m$^3$ | | 114 m$^3$ | |
| Membrane area | 84 m$^2$ | | 226 m$^2$ | | 160 m$^2$ | |
| **Parameter** | **Influent** | **Effluent** | **Influent** | **Effluent** | **Influent** | **Effluent** |
| COD | 1500 | 400 | 2000 | < 400 | 3000 | < 200 |
| BOD$_5$ | — | — | 150 | < 5 | — | — |
| Ammoniacal-N | 900 | 100 | 600 | < 5 | 1250 | < 150 |

to maximise process rates, but also to minimise any increases in effluent COD. If the use of waste products is not possible, then biodegradable organic chemicals must be used. In most countries methanol and acetic acid are often used. Methanol is less expensive than acetic acid, but the storage and dosing of methanol requires a storage and dosing system which takes account of its flammable and explosive nature. Microorganisms require an initial period of time (some weeks) for adaptation to methanol. If there are periods of time without methanol addition, or long maintenance phases at the plant, with for each start of dosing, a new adaptation phase is required. The use of specific chemical substances as a carbon source for denitrification often produces very light sludge, with poor settling characteristic, which requires the improved separation provided by ultrafiltration. Wens *et al.* (2001) described pilot-scale experiments into sludge separation by ultrafiltration. During the tests, two different types of ultrafiltration systems were used. One scheme was a normal ultrafiltration system, with sludge pumped through the modules (cross flow). The system operated at a pressure of 200–400 kPa, with energy consumption of 1–2 kWh/m$^3$ treated, and a resulting flux rate through the membrane of 60–90 l/m$^2$/h. The other system was a submersed membrane installed within the activated sludge reactor. Permeate was sucked through the membrane, with a negative pressure of 50 kPa. The energy consumption was 0.5 kWh/m$^3$ treated, and the resulting flux was 8–12 l/m$^2$/h. Operation results of different biological treatment plants with membrane sludge separation are presented in Table 10.11.7.

Only during the initial years of a landfill operation is COD in leachate likely to be sufficient for denitrification. During this phase, predenitrification is the optimal operational procedure in order to make use of this COD. However, predenitrification requires a high recirculation rate from the nitrification tank into the denitrification tank. The maximum possible nitrate elimination rate depends on the recirculation rate, because only nitrate which is recycled to influent could be eliminated. The maximum nitrate elimination rate is: Elimination (%) = 100 − 100/(1 + RR), with RR = [(sludge recycle volume) + (recirculation water volume)]/(influent volume). Recirculation rates (RR) > 10 can transport large amounts of oxygen back into the denitrification tank, with the consequence of reduced denitrification rates, and increased consumption of added carbon substrate. If the denitrification tank does not have a good hydraulic design, then such high recirculation rates can produce short-circuited flows, which further reduce denitrification rates. With recirculation rates > 10, the calculated denitrification volume must be enlarged. If the leachate COD is not sufficient to allow required denitrification rates, then chemicals or residuals must also be dosed as above. In most cases a combination of leachate COD and organic agents is not possible, since the more readily biodegradable substance will be used preferentially by the microorganisms, generally that being provided separately. This effect can be very important for the operation of biological treatment processes. Where COD values are slightly lower than required, the entire required COD must be dosed as an additional chemical. That means the COD load may be twice as high as that present initially in the raw leachate.

With postdenitrification, the leachate COD cannot be used for denitrification. However, overall, the required COD is slightly lower than for predenitrification. In addition, no recirculation of biomass is necessary, and thereafter the elimination of nitrate is not limited. To prevent the discharge of untreated organic compounds added as a carbon source, however, an additional final aeration tank is required. With a combination of pre- and postdenitrification, it may be possible to use leachate COD for the predenitrification, and an added external carbon source for the subsequent postdenitrification.

Chang (1998) published a dimension procedure for leachate-activated sludge plants with denitrification and nitrification. The parameters used in this dimension procedure are the result of laboratory-scale experiments. To date, the results could not be verified at larger pilot scale plants. Particularly at very low reactor temperatures ($< 5\,°C$) and for high sludge concentrations ($> 12\,kg/m^3$), the results should be used with caution. The effluent values are also input parameter of this calculation.

## 10.11.14 Biological Processes: Sequencing Batch Reactors

A modification of the traditional activated sludge process is the sequencing batch reactor (SBR). The SBR treatment process has been developed as a readily automated, extended aeration system that is well-suited to the higher organic strength and concentrations of ammoniacal-N in landfill leachates. The larger volume of the SBR main tank makes for efficient aeration, high rates of dilution of incoming leachates and high resistance to shock loading. The great majority of well-engineered aerobic biological leachate treatment systems successfully installed in the UK, make use of SBR technologies.

An SBR is a cyclically operated, suspended growth, activated sludge process. The only conceptual difference between the SBR and a conventional activated sludge system is that each SBR tank carries out functions such as aerobic biological treatment, equalisation, settlement of solids, effluent clarification and decanting, over a time sequence rather than in spatially separate tanks. The ability to vary the time sequence, compared to the inflexibility of specific volumes of separate tanks) enables a very robust and flexible treatment system to be provided.

The operating cycle of a typical SBR comprises four main phases, nominally: fill, react, settle and decant.

Although in treatment of domestic and other relatively dilute wastewaters, the fill stage – when wastewater is pumped into the SBR – may be a relatively rapid stage, for leachate treatment feeding of leachate generally takes place throughout the 'react' stage, in order to balance oxygen demand and oxygen supply, to avoid shock loadings of microorganisms and to avoid toxic inhibition from contaminants such as ammoniacal-N.

In leachate treatment, the process is readily automated, and generally operated within a 24-h cycle, in a tank which typically provides a mean hydraulic retention time (HRT) of ten days or longer when treating strong leachates. In general terms, for such leachates, selection of a shorter HRT does not reduce operational costs at all, and may only result in marginal reduction in capital costs. A significant benefit of a 24-h cycle is a standard time of day (or night) when a discharge of effluent is made.

A typical cycle of operation for SBR treatment of landfill leachate is therefore:

1. **Fill and React:** Over a period of 18–20 h, leachate is gradually fed into the SBR, during which period the reactor is aerated and the pH value is controlled.
2. **Settle:** Aeration is stopped for 1–2 h, during which period sludge flocculates and settles, and supernatant liquor is clarified.
3. **Decant:** Effluent is decanted from the surface of the SBR, by means of one of a number of options (bellmouth overflow, floating decant – either gravity or pumped, etc.), typically during a period of one or two hours depending on volume involved. The treatment cycle then recommences.

In many instances, where the relatively high flow rate from the SBR cannot be fed directly to the disposal route (e.g. to sewer, surface water, or for further treatment such as in a reed bed), then an effluent balance tank is generally used to balance flows and allow discharges to be made evenly, at lower rates over a 24-h period.

Successful operation of any dispersed biomass/activated sludge biological treatment system requires development and acclimatisation of a biomass that will flocculate well, settle rapidly and compact well in the base of the SBR during the settlement/clarification stage. These features are generally assisted in a leachate SBR by the fact that sludge may contain up to 40 or 50 % of inorganic solids – primarily iron and calcium – and which in solution are also beneficial to flocculation.

Effluents from tank-based SBR systems may typically contain between 50–200 mg/l of suspended solids in effluent, but simple polishing processes such as dissolved air flotation (DAF; see Box 10.11.9) or reed beds (each discussed elsewhere) can readily provide effluents suitable for discharge into surface water courses. About half of all tank-based SBR systems presently treating leachate in the UK, discharge effluents into such watercourses.

**Box 10.11.9  Dissolved Air Flotation After SBR Treatment: the Arpley Plant, UK (Robinson *et al.*, 2003a). Reproduced with permission from CISA, Italy. ⓒ (2003) EuroWaste srl, Italy.**

Dissolved air flotation is a process for the removal of fine suspended material from an aqueous suspension, in which solid particles are attached to small air bubbles, causing them to float to the surface. Attraction between the air bubbles and the particles results from adsorption forces, or physical entrapment of bubbles within the particle, colloid or floc. Chemical conditioning is generally used to increase the effectiveness of the DAF process, and optimisation of coagulation processes prior to DAF is key to improving effluent quality and minimising unit costs.

The most commonly adopted method of producing microbubbles of the optimum size (20–70 µm) is by recycling a proportion of the treated water through a pressurised (typically 2–5 bar) air saturation system, where it is saturated with air at the high pressure. Water then passes through a pressure relief nozzle in the base of the DAF tank within which air precipitates as tiny bubbles, with an enormous surface area. A key benefit of this process of producing bubbles is that it produces a very positive attachment between air bubbles and the particles it is required to remove. Particles, colloids or flocs act as nucleation sites for the bubbles to precipitate on – which is a much more effective means than relying on contact between particles and larger bubbles introduced by some other means.

The rising particles float to the surface of the water, forming a scum/sludge layer which is removed, usually by means of mechanical scrapers or scoops. Treated water flows out from a lower level.

The first UK application of DAF to a leachate treatment system was at Arpley Landfill in Warrington, during 2001/2002. Here, effluent from biological treatment of very strong leachate (ammoniacal-N 2500 mg/l, COD to 10 000 mg/l, conductivity 20 000 µS/cm) within an SBR, is treated using DAF, before receiving final polishing in a reed bed and then being discharged into the River Mersey. A relatively small DAF unit is able to treat effluent at the required rate of up to 20 $m^3$/h (450 $m^3$/day) and incorporates initial polyelectrolyte dosing. The table below provides typical operating data for the DAF unit. The Arpley unit has demonstrated not only the effective reduction in concentrations of suspended solids, typically from 250 mg/l to < 40 mg/l, but also the associated reductions in values of organic materials, many of these being present within colloidal materials which are effectively removed by the DAF process.

| Parameter | Leachate | SBR effluent | DAF effluent |
|---|---|---|---|
| COD | 5990 | 1470 | 1060 |
| $BOD_{20}$ | 1720 | 67 | 6 |
| $BOD_5$ | 688 | 20 | <1 |
| TOC | 1240 | 356 | 281 |
| Ammoniacal-N | 1460 | 3.7 | 3.2 |
| Nitrate-N | 1.9 | 1490 | 1238 |
| Iron | 13.0 | 5.51 | 0.72 |
| Sodium | 2560 | 3490[a] | 3770 |
| Chloride | 2710 | 2300 | 2650 |

[a]Related to dosing with NaOH for alkalinity.

Smaller, tank-based SBRs can readily be constructed more simply, for application at small rural landfills, where glass-coated steel tanks may be more appropriate.

The effluent of SBRs and conventional activated sludge plants can be improved by membrane separation (see above) and also by dissolved air flotation (DAF).

### 10.11.14.1  Lagoon-Based SBR

In the UK, initial SBR leachate treatment systems began to be designed and constructed during the early 1980s that were based on engineered HDPE-lined lagoons (e.g. Robinson, 1987). The term 'SBR' was not generally used, although in hindsight this is exactly the process that was being applied.

**Table 10.11.8** *Typical performance data from lagoon-based SBR leachate treatment schemes in the United Kingdom (mg/l).*

| Parameter | Chapel Farm 1990 (Robinson, 1992) | | Harewood Whin 1990 (Last *et al.*, 1993) | |
|---|---|---|---|---|
| | Influent | Effluent | Influent | Effluent |
| COD | 10 600 | 491 | 33 800 | 292 |
| BOD$_5$ | 4100 | 3.0 | 21 800 | 4.0 |
| Ammoniacal-N | 412 | 0.3 | 603 | 4.3 |
| Nitrate-N | < 0.2 | 27.2 | 0.2 | < 0.2 |
| Nitrite-N | 0.3 | < 0.1 | < 0.1 | < 0.1 |

Treatment plants were characterised by oval lagoons, with shallow sloping sides, a concrete base slab and an operating depth of typically 3 m. Two or more low-speed floating surface aerators provided efficient oxygen transfer and resuspension and turbulent mixing of solids at appropriate times in the operating cycle. The lagoon configuration encouraged optimum mixing patterns and, when needed, additions of phosphoric acid as a nutrient were made manually. Addition of alkalinity was only occasionally required, often because these plants were either installed at new landfills where significant volumes of acetogenic leachate were available, or at sites where leachates were relatively diluted (ammoniacal-N of 300–600 mg/l).

Often, the need for treatment was to enable discharges of treated effluent to be made into small sewage treatment works, which were unable to accept untreated leachates, and treatment objectives involved provision of an effluent similar in quality to domestic wastewater (eg BOD < 100 mg/l, ammoniacal-N 30 mg/l). Up to 20 such plants were constructed during the 1980s and most, if not all, continue to operate successfully and reliably, using simple adjustable timers to control settings of the cycle, and they were typically operated by weighbridge or similar staff, on an 'hour per day' basis. Very reliable treatment has nonetheless been achieved and maintained, over extended periods, by such treatment systems. Table 10.11.8 presents typical operating data.

Lagoon-based SBRs, while providing reliable treatment of a wide range of leachates, suffer from two drawbacks. A minor limitation is that where a single SBR lagoon is being used, provision must be made for short-term storage of leachate being generated using the period in which the cycle is in 'settle' and 'decant' phases. This represents at most only 4–6 h of production each day. Simple blending of leachates from different tipping cells, of different ages, is preferable to give a more consistent feed quality.

The most important constraint in UK applications is that during winter months, the water temperatures within the lagoon fall, often down to 3–4 °C or less, and these low temperatures limit the rates of treatment, especially the rates at which ammoniacal-N can be nitrified. The use of floating surface aerators, although extremely energy efficient for provision of oxygen and mixing, may encourage cooling during very cold weather. Nevertheless, significant rates of nitrification have been achieved and observed at temperatures down below 10 °C, and many operators have been able to adjust treatment rates on a seasonal basis – storing leachate during colder winter months, often within landfilled wastes, and maximising treatment rates during warmer summer months, to 'catch up' any backlogs. An alternative option, tried at a few sites with varying degrees of success, has been addition of heat into the lagoon, where this is available cheaply either directly from landfill gas, or as waste heat from landfill gas power generation. Care must be taken not to expose bacteria responsible for treatment, to excessive temperatures (> 40 °C) during the heat transfer process.

### 10.11.14.2  Tank-Based SBR

As leachate treatment experience increased in the UK, during the late 1980s and early 1990s, and as venturi-type high-efficiency aerators became more widely available and more reliable, increasingly SBR leachate treatment plants were designed with treatment taking place within a circular, or other tank. The main benefits of such systems, compared with lagoon SBRs, included temperature control, reduced load requirements, increased structural strength and energy efficiency.

Treatment is generally carried out in circular tanks, 5–6 m deep, with a simple roof structure, to contain the SBR, and with high-efficiency venturi aerators, meaning that all energy consumed by the aerators and all heat energy generated during biological treatment can be retained and used to ensure that treatment takes place within a temperature range of

**Table 10.11.9** *Typical performance data from tank-based SBR leachate treatment schemes in the United Kingdom (mg/l).*

| Parameter | Efford 2003 (Robinson and Last, 2003) | | Llanddulas 2002 (Robinson and Last, 2003) | | Arpley 2001 (Robinson *et al.*, 2003a) | |
| --- | --- | --- | --- | --- | --- | --- |
| | Influent | Effluent | Influent | Effluent | Influent | Effluent |
| COD | 942 | 462 | 3410 | 762 | 5990 | 1060 |
| BOD$_5$ | 72 | 22 | 1520 | 9 | 688 | < 1 |
| Ammoniacal-N | 820 | 1.59 | 965 | 1.2 | 1460 | 3.2 |
| Nitrate-N | 0.21 | 423 | < 0.3 | 668 | 1.9 | 238 |
| Nitrite-N | 0.04 | 0.56 | < 0.1 | < 0.1 | < 0.1 | 0.1 |

20–30 °C, at all times. Burial of SBR tanks (only applicable where concrete structures are used) can improve insulation further and also reduce visual impact. Experience has demonstrated that, even in extremely exposed locations (e.g. Robinson, 1999), very stable temperatures and treatment rates can be maintained at all times.

Reinforced concrete is generally the preferred material for tank-based SBR systems, having a long life expectancy and a structural integrity that generally removes any need for secondary containment of the SBR to be provided (although simple leak detection systems may be appropriate where tanks are buried). Alternative materials such as epoxy-coated steel have proved inadequate to prevent corrosion in some instances, and glass coated steel tanks require annual emptying, steam cleaning, inspection and repair if a warranty is to be provided – which is not practicable if an increasingly acclimatised biological population is to be maintained. Very high treatment efficiencies can be achieved by such SBR systems – typical examples are presented in Table 10.11.9.

The warmer and more stable operating temperatures within a tank-based SBR system generally result in higher rates of treatment, although as in a lagoon SBR, short-term storage and blending of influent leachates is beneficial – one or two days flow typically being stored within a raw leachate balance tank.

Venturi aerators have proved to be efficient, reliable, and are readily installed to provide flexibility in optimising mixing patterns for optimum treatment. Only when installed at excessive power (greater than about 80 watts/m$^3$ aeration volume) have any adverse effects on floc formation been observed in leachate treatment applications.

## 10.11.15  Biological Processes: Fixed Film Reactors

Fixed film reactors provide an alternative biological treatment system. In this type of reactor biomass growth is attached to solid material. Nitrification bacteria can become strongly fixed on such materials, in the absence of high concentrations of degradable COD. Therefore, nitrification can be much more stable than in an activated sludge plant, if all other conditions are optimal. However, because fixed film reactors are not completely mixed, unlike aeration tanks, and so if the concentrations of ammoniacal-N in leachate are too high for the microorganisms (ammonia is toxic for nitrifying bacteria), then the nitrification rate falls rapidly, and this breakdown moves through the entire reactor. As a consequence, the design of a fixed film reactor must guarantee a high nitrification rate in the initial part of the reactor. As a first estimate, the concentration of ammoniacal-N in the first part or first stage of a fixed film reactor, should be below 100 mg N/l. In completely mixed reactors, operated at typical pH values of 7.5–8.0, toxic effects have been observed at concentrations as low as 50–80 mg/l of ammoniacal-N, and these values are also likely to apply to bacteria operating in fixed film reactors. If such levels cannot be achieved in leachate being dosed onto the bacteria, then a proportion of treated effluent could be recycled to dilute the influent leachate.

In leachates from older, methanogenic landfills, relatively low levels of suspended solids and biodegradable organic substances make it possible to use solid material with high surface area and smaller space. If these factors are considered, very effective leachate treatment plants for nitrification can be designed for strongly methanogenic leachates. High rates of sludge production during denitrification make it difficult to use fixed film reactors for this process, because clogging of media will be a serious problem. If both types of fixed film are combined, it is essential to prevent sludge flow from

**Table 10.11.10**   *Operation data of rotating biological contactors.*

| Parameter | Unit | Influent | Effluent |
|---|---|---|---|
| Wittek *et al.* (1985) | | | |
| COD | mg/l | 481 | 278 |
| BOD$_5$ | mg/l | 71 | 10 |
| Ammonium | mgN/l | 201 | 0.28 |
| Nitrite | mgN/l | 0.05 | 0.02 |
| Nitrate | mgN/l | 0.3 | 211 |
| Knox (1992) | | | |
| Average effluent values | | | |
| BOD$_5$ | mg/l | | 14 |
| Ammonium | mg N/l | | 2.96 |
| Data collected during loading rate of 3.1 g N/m$^2$/day | | | |
| BOD$_5$ | mg/l | 27 | 23 |
| TOC | mg/l | 303 | 264 |
| Ammonium | mg N/l | 153 | 0.1 |
| NO$_X$ | mg N/l | 34 | 105 |
| Nitrite | mg N/l | 11 | 0.1 |

the denitrification reactor, from entering the nitrification reactor. Such a system would rapidly reduce nitrification rates to low values.

To date, typically constructed fixed film reactors for leachate treatment are rotating biological contactors (RBCs). Ehrig (1983) reported data from laboratory experiments with a 4-stage RBC for leachate nitrification. In contrast to activated sludge plants, high nitrification rates could be measured over a wide loading range. But at ammoniacal-N loading rates > 1–2 g N/m$^2$/day, the nitrite production increases to values > 1 mgN/l. Wittek *et al.* (1985) presented data from a full-scale RBC (plastic material with a specific area of 220 m$^2$/m$^3$, total surface area 28 500 m$^2$) at loadings of 1–2 gN/m$^2$/day (Tables 10.11.10, 10.11.11). Knox (1992) published data from an RBC plant at Pitsea landfill (plastic material with a specific area of 180 m$^2$/m$^3$, total surface area 30 000 m$^2$). Operational results are shown in Table 10.11.4. Using these limited results, loading rates of only a few grammes of ammoniacal-N/m$^2$/day may be possible. At such loading rates not more than three stages are likely to be needed for RBCs.

At some leachate treatment plants in Germany, RBCs are used after chemical oxidation has been used to reduce biodegradable organic substances. Biomass production from such pretreatment processes is too small and variable for activated sludge plants. But on an RBC, the attached biomass has a slower decay rate. As a consequence, under such circumstances RBCs (and other fixed film reactors) can provide more stable operation than activated sludge plants. Results from two plants are presented in Table 10.11.12 and Table 10.11.13. For other posttreatment plants in Germany, final COD values are reduced to lie in the range of 50–70 % of those following chemical oxidation.

Submersed fixed film reactors filled with coarse materials have been introduced in Germany. The materials have a surface area of 240 m$^2$/m$^3$. At loading rates of 3.6–4.8 g N/m$^2$/day, full nitrification was achieved (Spiess, 1995). At another plant in Germany, a submerged fixed film reactor was used as a second biological treatment step (as security step) to guarantee ammoniacal-N effluent values of < 3 mg/l (Ehrig, 1998c). Under normal operation, this second stage was not loaded with ammoniacal-N because the first stage (activated sludge plant) reduced values to less than 1 mg/l. During this phase, the fixed biomass film was underloaded and decayed at slow rates. During occasional phases where the first stage process became overloaded, there was not enough biomass on the solid material in the second stage to reduce the overflow from the first stage. Such a system can only be operated if the security stage receives a continuous, small but consistent load, maybe as bypass direct from the influent.

Fixed film reactors of most types can be dimensioned for nitrification, using the above values for the load per unit surface area of media.

**Table 10.11.11**   *Influent and effluent values of leachate treatment plants (A–F) with biological treatment (activated sludge) and activated carbon adsorption (Ehrig, 2001). n.d. = No data available. Plant A corresponds to the plant shown in Figure 10.11.7.*

| Plant | Parameter | Unit | Influent | Effluent biological treatment | Effluent activated carbon |
|---|---|---|---|---|---|
| A | BOD$_5$ | [mg/l] | 1500–6000 | < 5 | < 5 |
|   | COD | [mg/l] | 1000–12 000 | 500–800 | < 200 |
|   | AOX | [mg/l] | 1.0–2.5 | 0.5–2.0 | 0.1–0.7 |
|   | NH$_4$ | [mg N/l] | 400–800 | < 10 | < 10 |
| B | BOD$_5$ | [mg/l] | n.d. | n.d. | n.d. |
|   | COD | [mg/l] | 3000–6000 | 500–900 | < 110 |
|   | AOX | [mg/l] | 0.8–2.0 | 0.6–1.4 | 0.1–0.48 |
|   | NH$_4$ | [mg N/l] | 400–800 | < 10 | < 10 |
| C | BOD$_5$ | [mg/l] | 280 | < 20 | < 20 |
|   | COD | [mg/l] | 1700 | 900–1200 | < 400 |
|   | AOX | [mg/l] | 1.1 | 0.5–0.9 | 0.1 – < 0.5 |
|   | NH$_4$ | [mg N/l] | 900 | < 10 | < 10 |
| D | BOD$_5$ | [mg/l] | 90–210 | < 20 | < 20 |
|   | COD | [mg/l] | 1100–1600 | 300–900 | < 200 |
|   | AOX | [mg/l] | 0.8–1.7 | 0.9–1.2 | < 0.5 |
|   | NH$_4$ | [mg N/l] | 500–900 | < 10 | < 10 |
| E | BOD$_5$ | [mg/l] | 50–150 | < 10 | < 10 |
|   | COD | [mg/l] | 450–800 | 250–320 | < 200 |
|   | AOX | [mg/l] | n.d. | n.d. | n.d. |
|   | NH$_4$ | [mg N/l] | 250–400 | < 50 | < 50 |
| F | BOD$_5$ | [mg/l] | 500–800 | < 10 | < 10 |
|   | COD | [mg/l] | 2000–5600 | 750–1200 | < 200 |
|   | AOX | [mg/l] | 0.6–2.2 | 0.5–1.5 | 0.1–0.87 |
|   | NH$_4$ | [mg N/l] | 600–1800 | 3–52 | < 20 |

**Table 10.11.12**   *Influent and effluent values of leachate treatment plants with biological treatment (activated sludge), chemical oxidation (Plants A–B) and at one plant a biological posttreatment stage (Plant B; Ehrig, 2001). n.d. = No data available. Plant A corresponds to the plant shown in Figure 10.11.8.*

| Plant | Parameter | Unit | Influent | Effluent biological treatment | Effluent oxidation with ozone | Effluent biological posttreatment |
|---|---|---|---|---|---|---|
| A | COD | [mg/l] | 320–5796 | 236–854 | 66–176 | — |
|   | NH$_4$ | [mg N/l] | 125–1350 | < 1.2 | < 0.9 | — |
| B | COD | [mg/l] | 1200–4000 | 300–1000 | 50–200 | 18–150 |
|   | NH$_4$ | [mg N/l] | 600–1900 | 0.1–9.0 | n.d. | n.d. |
|   | AOX | [mg/l] | 1.0–3.8 | 0.5–2.0 | 0.08–0.25 | 0.04–0.18 |
|   | BOD$_5$ | [mg/l] | n.d. | < 20 | 20–40 | < 10 |

**Table 10.11.13**   *Influent and effluent values of a leachate treatment plant (Plant A) with biological treatment (activated sludge), chemical oxidation, a biological posttreatment and an activated carbon adsorption (Ehrig, 2001). n.d. = No data available.*

| Plant | Parameter | Unit | Influent | Effluent biological treatment | Effluent oxidation with ozone | Effluent biological posttreatment | Effluent activated carbon adsorption |
|---|---|---|---|---|---|---|---|
| A | COD | [mg/l] | 758–1332 | 461–669 | 4–138 | 2–28 | 1–85 |
|   | NH$_4$ | [mg N/l] | 375–885 | n.d. | n.d. | n.d. | 0.1–0.6 |
|   | AOX | [mg/l] | 0.85–2.1 | n.d. | n.d. |  | 0.17–0.43 |

Hippen *et al.* (2001) observed a reduction of nitrogen (ammoniacal-N + nitrate-N + nitrite-N) at some RBC plants for leachate treatment without, or with very low, consumption of organic carbon, as would be necessary for denitrification. Further investigations showed that this could be an aerobic/anoxic deammonification within the biofilm on the RBC surface. This process is defined as a combination of preceding nitritation and subsequent oxidation of the remaining ammonium with nitrite as an electron receptor. It is a very sensible process, because the outer part of the biofilm on a fixed bed surface is likely to be aerobic and the inner part must be anoxic. The nitritation on the outer part of the biofilm is controlled by ammonia inhibition and the oxygen concentration. The thickness and the constant stability of the biofilm is a very important part of this process. Overall nitrogen reduction rates of 60–80 % are possible. As mentioned above, there is no addition of external organic carbon necessary, but carbon is important for the control of biofilm thickness. With very low organic substances in the influent stream, the produced biofilm is not thick enough to produce aerobic and anoxic upper and lower parts.

## 10.11.16   Treatment Combinations

As shown in Figure 10.11.1, in most cases leachate treatment is a combination of different treatment steps. The reduction of biodegradable and nonbiodegradable substances is one reason for using several treatment processes, but economic and environmental influences can also encourage the use of combinations of different types of treatment process. Unlike experiences with sewage treatment, experience and publication of results from leachate treatment are much smaller. To date, for several processes, it is not possible to precisely specify full-scale treatment systems. It can be even more difficult to dimension a combination of different treatment processes. Influent and effluent values from different treatment combinations in Tables 10.11.11 to 10.11.16 may provide an initial tool to assist in this. Some of the plants included in the tables are shown in Figures 10.11.7 to 10.11.10.

In many cases an important consideration in the selection of appropriate leachate treatment processes, could be the production of residues, and the consumption of energy by specific treatment steps or combinations, which are presented in Table 10.11.17.

**Table 10.11.14**   *Influent and effluent values of leachate treatment plants with biological treatment (Plant A only) and a reverse osmosis system (Ehrig, 2001; Poitel et al., 1999). Plant A corresponds to the one shown in Figure 10.11.9. n.d. = No data available. Kjeldahl nitrogen = ammonium + organic nitrogen.*

| Plant | Parameter | Unit | Influent | Effluent biological treatment | Permeate reverse osmosis |
|---|---|---|---|---|---|
| A | COD | [mg/l] | 446–872 | 353–697 | 5.3–27.0 |
| | $BOD_5$ | [mg/l] | 10–220 | 5–110 | 1–19 |
| | $NH_4$ | [mg N/l] | 80–396 | 0.02–25.2 | 0.03–10.1 |
| | AOX | [mg/l] | 0.4–1.4 | n.d. | < 0.01–0.05 |
| B (1995) | COD | [mg/l] | 1255 | | 9 |
| | $BOD_5$ | [mg/l] | 120 | | < 1 |
| | Kjeldahl nitrogen | [mg N/l] | 330 | | 2 |
| B (1996) | COD | [mg/l] | 2011 | | 12.5 |
| | $BOD_5$ | [mg/l] | 160 | | < 1 |
| | Kjeldahl nitrogen | [mg N/l] | 598 | | 3 |

**Table 10.11.15**   *Influent and effluent values of leachate treatment plants with biological treatment and a two-stage reverse osmosis system (Ehrig, 2001). n.d. = No data available. For details of plant A see Figure 10.11.10.*

| Plant | Parameter | Unit | Influent | Effluent biological treatment | Permeate reverse osmosis I | Permeate reverse osmosis II | |
|---|---|---|---|---|---|---|---|
| | | | | | | Average | Maximum |
| A (without biological pretreatment) | COD | [mg/l] | 1366–3010 | — | n.d. | < 15 | < 15 |
| | NH$_4$ | [mg N/l] | 130–854 | — | n.d. | 6.3 | 17 |
| | AOX | [mg/l] | 1.09–2.24 | — | n.d. | 0.045 | 0.09 |
| | BOD$_5$ | [mg/l] | 112–740 | — | n.d. | < 2 | < 2 |
| A (with biological pretreatment) | COD | [mg/l] | 1543–2305 | 736–1145 | 36–55 | 4.3–22 | |
| | NH$_4$ | [mg N/l] | 230–659 | 0.1–48.0 | 0.1–9.0 | 0.1–4.2 | |
| | BOD$_5$ | [mg/l] | 145–253 | 21–40 | 1.7–6.7 | 0.8–4.8 | |

| | | | Average/ maximum | Effluent biological stage | | Permeate reverse osmosis | |
|---|---|---|---|---|---|---|---|
| | | | | Average | Maximum | Average | Maximum |
| B (Hippen et al., 1997) | COD | [mg/l] | 3561/4590 | 1301 | 1755 | 15.8 | 45 |
| | NH$_4$ | [mg N/l] | 833/1651 | 0.1 | 0.2 | 0.1 | 0.12 |
| | Organic N | [mg N/l] | 251/1103 | 82 | 125 | 5.3 | 9.4 |
| | NO$_3$ | [mg N/l] | 1.3/2.9 | 84 | 140 | 9.8 | 21 |
| | NO$_2$ | [mg N/l] | 0.15/5.25 | 1.2 | 30.9 | 0.12 | 0.79 |
| | BOD$_5$ | [mg/l] | 1512/2860 | 23.7 | 64 | 2 | 5.3 |

**Table 10.11.16**   *Influent and effluent values of leachate treatment plants (Plants A–C) with a two-stage reverse osmosis system (Ehrig, 2001). n.d. = No data available.*

| Plant | Parameter | Unit | Influent | | Permeate reverse osmosis I | | Permeate reverse osmosis II | |
|---|---|---|---|---|---|---|---|---|
| | | | Range | Average | Range | Average | Range | Average |
| A | COD | [mg/l] | 1590–2980 | 2285 | 18–130 | 74 | 4–25 | 14.5 |
| | NH$_4$ | [mg N/l] | 900–1800 | 1350 | 50–220 | 135 | 4.4–8.8 | 6.6 |
| | AOX | [mg/l] | 1500–1900 | 1700 | n.d. | n.d. | 0.002–0.021 | 0.0115 |
| | BOD$_5$ | [mg/l] | 230–1320 | 775 | n.d. | n.d. | < 2–3 | 2.5 |
| B | COD | [mg/l] | 2619–5498 | 4124 | 9–25 | 20 | n.d. | n.d. |
| | NH$_4$ | [mg N/l] | 398–794 | 577 | 1–18 | 8 | n.d. | n.d. |
| | BOD$_5$ | [mg/l] | 70–185 | 122 | < 1–3 | 1 | n.d. | n.d. |
| C | COD | [mg/l] | 1301–1707 | 1504 | n.d. | n.d. | 5.2–16.8 | 11 |
| | NH$_4$ | [mg N/l] | 0.5–1.9 | 1.2 | n.d. | n.d. | < 0.05 – 0.48 | 0.26 |
| | AOX | [mg/l] | 0.78–2.43 | 1.6 | n.d. | n.d. | < 0.01 – 0.02 | 0.015 |

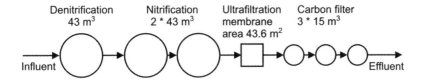

- Average flow rate: 80 m³/day
- Denitrification and nitrification tanks are closed tanks under a pressure of 300 kPa
- Carbon source for denitrification: wastewater from biowaste composting
- Carbon consumption: 3 kg/m³ leachate

**Figure 10.11.7**   *Leachate treatment plant A in Table 10.11.11.*

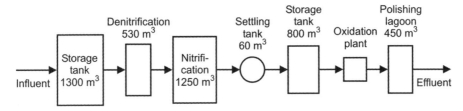

- Average flow rate: 55 m³/day
- Acetic acid for denitrification: 3 l/m³ leachate
- Energy consumption of total plant: 200 kW
- Ozone consumption 2.2-2.8 kg ozone/kg COD
- Energy consumption of oxidation stage: 170 kW with 40 kW for UV-light

**Figure 10.11.8**   *Leachate treatment plant A in Table 10.11.12.*

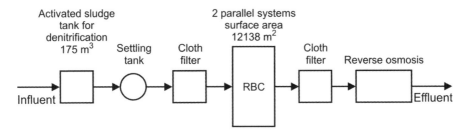

- Average leachate flow: 120 m³/day
- Reverse osmosis: tubular modules with 280 m² membrane area
- Concentrate: 25-30% of leachate (infiltrated back into landfill)
- Energy consumption: 13.5 kWh/m³ leachate
- Sulphuric acid: 1 kg/m³ leachate (pH-adjustment)
- Citric acid: 0.07 kg/m³ leachate (membrane cleaning)
- Cleansing agent: 0.03 kg/m³ leachate (membrane cleaning)

**Figure 10.11.9**   *Leachate treatment plant A in Table 10.11.14.*

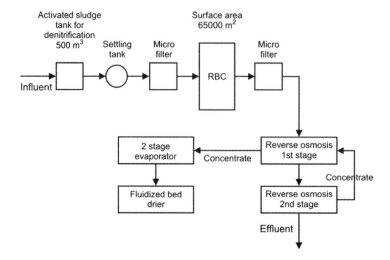

- Average flow rate: 65 m$^3$/day
- Reverse osmosis: 1$^{st}$ stage: tubular modules with a membrane area of 425 m$^2$
- Reverse osmosis: 2$^{nd}$ stage: spiral wound modules with a membrane area of 128 m$^2$
- Evaporator: 2 stage forced circulation evaporator
- Residuals: 0.7% of leachate influent

**Figure 10.11.10**   *Leachate treatment plant A in Table 10.11.15.*

**Table 10.11.17**   *Secondary waste and energy consumption of different leachate treatment steps and combinations (Ehrig, 2001).*

| Process | Quantity | Description | Energy consumption (kWh/m$^3$ leachate) |
|---|---|---|---|
| Biological treatment | 0–0.6 kg sludge/kg COD eliminated | 35 % dry matter | 15–25 |
| Biological treatment (denitrification with leachate COD) | ≈ 3 kg sludge/m$^3$ leachate | 35 % dry matter | |
| Biological treatment (denitrification with organic chemicals) | ≈ 7.3 kg sludge/m$^3$ leachate | 35 % dry matter | |
| Activated carbon adsorption | ≈ 0.6 kg/m$^3$ leachate | In primary filter (sand filter, ultrafiltration, microfiltration) | 0.5–3.0 |
| Chemical oxidation (ozone) | ≈ 0.8 kg/m$^3$ leachate ≈ 2.7 kg sludge/m$^3$ leachate | Production of calcium oxalate. Includes residues from primary filter and postbiological treatment. | 50–100 |
| Reverse osmosis | 150–300 l/m$^3$ leachate | Liquid concentrate | 8–10 |
| High-pressure reverse osmosis | | Liquid concentrate | 13–15 |
| Evaporation (of liquid concentrate) | 20–40 l/m$^3$ leachate | Liquid concentrate | 8–10 |
| Drying | 5–10 kg/m$^3$ leachate | > 95 % dry matter | 20–25 |
| Reverse osmosis + evaporation + drying (other sources) | 5–20 kg/m$^3$ leachate (average 12 kg/m$^3$) | > 90 % dry matter | |
| Ammonia recovery after stripping | 4–6 kg/m$^3$ leachate | Nitrogen fertiliser | 0.1–0.3 |

# References

Albers, H. (1985): Untersuchungen zur Behandlung von Sickerwässern in Belebungsanlagen und belüfteten Teichen (Investigation into leachate treatment with activated sludge plants and aerated lagoons; in German). *Veröffentlichungen des Instituts fur Stadtbauwesen TU Braunschweig*, 39, S228–S254.

Carville, M.S., Last, S.D., Olufsen, J.S. and Robinson, H.D. (2003): Characterisation of contaminant removal achieved by biological leachate treatment systems. In: CISA (ed.) *Sardinia 2003, Proceedings of the Ninth International Waste Management and Landfill Symposium*. CD-ROM. CISA, Environmental Sanitary Engineering Centre, Cagliari, Italy.

Chang, L. (1993): Bemessung von einstufigen Belebungsanlagen zur Behandlung von Sickerwasser aus Hausmülldeponien (Dimension of one stage activated sludge plants for sanitary landfill leachate treatment; in German). *Korrespondenz Abwasser*, 40, 319–327.

Chang, L. (1998): Auslegung von einstufigen Belebungsanlagen zur Stickstoffelimination bei Sickerwässern aus Siedlungsabfalldeponien (Design of one stage activated sludge plants with nitrogen elimination for leachate from municipal solid waste landfills; in German). *Veröffentlichungen des Instituts für Siedlungswasserwirtschaft*, Vol. 62. University of Braunschweig, Germany.

Chang, L. (1999): Dimensioning of single-stage activated sludge treatment plants for nitrogen removal from landfill leachate. In: Christensen, T.H., Cossu, R. and Stegmann, R. (eds) *Sardinia 99, Proceedings of the Seventh International Landfill Symposium*, Vol. II, pp. 207–214. CISA, Environmental Sanitary Engineering Centre, Cagliari, Italy.

Chen, T., Esnault, D. and Koenig, A. (1997): First year operation of the NENT landfill leachate treatment works in Hong Kong. In: Christensen, T.H., Cossu, R. and Stegmann, R. (eds) *Sardinia 97, Proceedings of the Sixth International Landfill Symposium*, Vol. II, pp. 219–228. CISA, Environmental Sanitary Engineering Centre, Cagliari, Italy.

Damhaug, T. (1981): Ammonia stripping from leachate by countercurrent aeration in shallow tanks. In: EAS (ed.) *Documentation Fifth European Sewage and Refuse Symposium EAS*. EAS, Munich, Germany.

Ehrig, H.J. (1983): Biological oxidation of ammonium containing leachate from solid waste deposits. In: EWPCA (ed.) *Documentation of the International EWPCA–IAWPRC Seminar Rotating Biological Discs*, 1983, 291–298.

Ehrig, H.J. (1987): Weitergehende Reinigung von Sickerwäsern aus Abfalldeponien (Advanced treatment of sanitary landfill leachate; in German). *Veröffentlichungen des Instituts für Stadtbauwesen*, Vol. 41. University of Braunschweig, Germany.

Ehrig, H.J. (2001): Sickerwasser aus Abfallablagerungen. In: ATV-DVWK (eds) *ATV-Handbuch Industrieabwasser, Dienstleistungs- und Veredelungsindustrie*, , pp. 347–380. Verlag Ernst and Sohn; ISBN 3-433-01468-X.

Ehrig, H.J.and Hagedorn S. (1998): Activated carbon adsorption. In: *Symposium: Management and treatment of MSW landfill leachate*. CD-ROM. Venice, Italy.

Ettala, M. (1997): Full-scale leachate treatment using new evaporation technology. In: Christensen, T.H., Cossu, R. and Stegmann, R. (eds) *Sardinia 97, Proceedings of the Sixth International Landfill Symposium*, Vol. II, pp. 423–426. CISA, Environmental Sanitary Engineering Centre, Cagliari, Italy.

EWPCA (1990): *European design and operations guidelines for reed bed treatment systems*. Report prepared by the EC/European Water Pollution Control Association emergent hydrophyte treatment systems expert contact group. EWPCA, London, UK.

Findlater, B.C., Hobson, J.A. and Cooper, P.F. (1990): Reed bed treatment systems: performance evaluation. In: *Proceedings of the International Conference on the Use of Constructed Wetlands in Water Pollution Control*, pp. 193–204. Pergamon Press, Cambridge, UK; ISBN 0-08-040784-6.

Hanashima, M., Shimaoka, T., Kobayashi, T., Ushikoshi, K., Suzuki, H., Katsura, K., Toji, A. and Kojima, D. (1999): Treatment of high salinity leachate by reverse osmosis and concentrate treatment technology. In: Christensen, T.H., Cossu, R. and Stegmann, R. (eds) *Sardinia 99, Proceedings of the Seventh International Landfill Symposium*, Vol. II, pp. 367–374. CISA, Environmental Sanitary Engineering Centre, Cagliari, Italy.

Hippen, A., Rosenwinkel, K.-H. and Seyfried, C.F. (2001): Leachate treatment in the process of deammonification. In: Christensen, T.H., Cossu, R. and Stegmann, R. (eds) *Sardinia 2001, Proceedings of the Eighth International Landfill Symposium*, Vol. II, pp. 197–206. CISA, Environmental Sanitary Engineering Centre, Cagliari, Italy.

IWA Publishing (2000): *Constructed wetlands for pollution control: Processes, performance design and operation*. Scientific and Technical Report No. 8. International Water Association specialist group on use of macrophytes in water pollution control, 156 pp. IWA, London, UK; ISBN 1-900222-05-1.

Knox, K. (1991): Full-scale nitrification of a methanogenic leachate using rotating biological contactor and vegetated ditch. *Annual Madison Waste Conference*, 14, 327–356.

Knox, K. (1992): Rotating biological contactors. In: Christensen, T.H., Cossu, R. and Stegmann, R. (eds) *Landfilling of waste: Leachate*, p. 211. Elsevier Science Publishers, London, UK; ISBN 1851667334.

Knox, K. (2001): Development of a novel process for treatment of leachate with very high ammoniacal nitrogen concentrations. In: Christensen, T.H., Cossu, R. and Stegmann, R. (eds) *Sardinia 2001, Proceedings of the Eighth International Landfill Symposium*, Vol. II, pp. 207–213. CISA, Environmental Sanitary Engineering Centre, Cagliari, Italy.

Krümpelbeck, I. and Ehrig, H.J. (1999): Long-term behaviour of MSW landfills in Germany. In: Christensen, T.H., Cossu, R. and Stegmann, R. (eds) *Sardinia 99, Proceedings of the Seventh International Landfill Symposium*, Vol. I, pp. 27–36. CISA, Environmental Sanitary Engineering Centre, Cagliari, Italy.

Last, S.D., Barr, M.J. and Robinson, H.D. (1993): Design and operation of Harewood Whin Landfill: an aerobic biological leachate treatment works. *Waste Management Proceedings*, 1993, 9–16.

Maehlum, T. and Haarstad, K. (1997): Leachate treatment in ponds and wetlands in cold climate. In: Christensen, T.H., Cossu, R. and Stegmann, R. (eds) *Sardinia 97, Proceedings of the Sixth International Landfill Symposium*, Vol. II, pp. 337–344. CISA, Environmental Sanitary Engineering Centre, Cagliari, Italy.

Noma, Y., Matsufuji, Y., Takata, M. and Tomoda, K. (1999): Study on the mass flow of dioxins in a landfill site disposed of municipal solid waste incinerator ash. In: Christensen, T.H., Cossu, R. and Stegmann, R. (eds) *Sardinia 99, Proceedings of the Seventh International Landfill Symposium*, Vol. II, pp. 375–381. CISA, Environmental Sanitary Engineering Centre, Cagliari, Italy.

Nuttall, P.M., Boon, A.G. and Rowell, M.R. (1997): *Review of the design and management of constructed wetlands.* Construction Industry Research and Information Association Report 180, 267 pp. CIRIA, London, UK.

Peters, T. (1999): Past and future of membrane filtration for the purification of landfill leachate. In: Christensen, T.H., Cossu, R. and Stegmann, R. (eds) *Sardinia 99, Proceedings of the Seventh International Landfill Symposium*, Vol. II, pp. 335–344. CISA, Environmental Sanitary Engineering Centre, Cagliari, Italy.

Poitel, D., Courant, P., Primi, C. and Mandin, JM. (1999): Various leachate treatment plants in France. In: Christensen, T.H., Cossu, R. and Stegmann, R. (eds) *Sardinia 99, Proceedings of the Seventh International Landfill Symposium*, Vol. II, pp. 135–142. CISA, Environmental Sanitary Engineering Centre, Cagliari, Italy.

Robinson, H.D. (1987): Design and operation of leachate control measures at Compton Bassett Landfill Site, Wiltshire, UK. *Waste Management and Research*, 5, 107–122.

Robinson, H.D. (1990): On-site treatment of leachates from landfilled wastes. *Journal of the Institution of Water and Environmental Management*, 4 (1), 78–89.

Robinson, H.D., Chen, C.K., Formby, R.W. and Carville, M.S. (1995): Treatment of leachates from Hong Kong landfills with full nitrification and denitrification. In: Christensen, T.H., Cossu, R. and Stegmann, R. (eds) *Sardinia 95, Proceedings of the Fifth International Landfill Symposium*, Vol. I, pp. 511–534. CISA, Environmental Sanitary Engineering Centre, Cagliari, Italy.

Robinson, H.D. (1996): A review of the composition of leachates from domestic wastes in landfill sites. Report number CWM 072/95 in the series: The Technical Aspects of Controlled Waste Management. Wastes Technical Division, UK Environment Agency, London , UK, 550 pp.

Robinson, H.D. and Harris, G. R. (2001): Use of reed bed polishing systems to treat landfill leachates to very high standards. *Alternative Uses of Constructed Wetland Systems Conference*, 2001, 55–68. Aqua Enviro Technology Transfer, Doncaster, UK.

Robinson, H.D., Carville, M.S. and Harris, G.R. (2003): Advanced leachate treatment at Buckden landfill, Huntingdon, UK. *Journal of Environmental Engineering and Science*, 2, 255–264.

Robinson, H.D., Farrow, S., Last, S. and Jones, D. (2003a): Remediation of leachate problems at Arpley Landfill Site, Warrington, Cheshire. In: CISA (ed.) *Sardinia 93, Proceedings of the Ninth International Waste Management and Landfill Symposium*, CD-ROM, 11 pp. CISA, Environmental Sanitary Engineering Centre, Cagliari, Italy.

Robinson, H.D. and Grantham, G. (1988): The treatment of landfill leachates in on-site aerated lagoon plants: experience in England and Ireland. *Water Research*, 22, 733–747.

Robinson, H.D. and Knox, K. (2001): *Pollution inventory discharges to sewer or surface waters from landfill leachates.* Final Report REGCON 70, by Enviros and Knox Associates, 19 pp + Appendices. Environment Agency, London, UK.

Robinson, H.D. and Knox, K. (2003): *Updating the landfill leachate pollution inventory tool*. R&D Technical Report No. PI-496/TR(2),by Enviros and Knox Associates, 56 pp. Environment Agency, London, UK.

Robinson, H.D. and Last, S.D. (2003): On-site leachate treatment expands rapidly. *Local Authority Waste and Environment*, 2003, 12 pp.

Robinson, H.D. and Luo, M.M.H. (1991): Characterisation and treatment of leachates from Hong Kong landfill sites. *Journal of the Institution of Water and Environmental Management*, 5 (3), 326–335.

Robinson, H.D., Olufsen, J.S. and Last, S.D. (2005): Design and operation of cost-effective leachate treatment schemes at UK landfills: Recent cast studies. *CIWM Scientific and Technical Review*, 6 (1), 14–24.

Steensen, M. (1998): Chemische Oxidation und biologische Nachreinigung zur weitergehenden Sickerwasserbehandlung (Chemical oxidation and biological post-treatment for advanced leachate treatment; in German). *Veröffentlichungen des Instituts für Siedlungswasserwirtschaft*, Vol. 63. University of Braunschweig, Germany.

Spiess, A. (1995): Biologische Sickerwasserbehandlung in Festbettanlagen (Biological leachate treatment with fixed film reactors; in German). In: ATV (ed.) *Symposium Behandlung von Deponiesickerwasser*. ATV, Magdeburg, Germany.

Van Dijk, L. and Roncken, G.C.G. (1997): Membrane bioreactors for wastewater treatment: The state of the art and new developments. *Water Science and Technology*, 35, 35–41.

Wens, P., de Langhe, P. and Staelens, B. (2001): Biological treatment of leachate by means of a biomembrane reactor. In: Christensen, T.H., Cossu, R. and Stegmann, R. (eds) *Sardinia 2001, Proceedings of the Eighth International Landfill Symposium*, Vol. IV, pp. 169–178. CISA, Environmental Sanitary Engineering Centre, Cagliari, Italy.

Wittek, J. and Fleischer, B. (1985): Erfahrungsbericht über die Sickerwasserbehandlung in Tauchtropfkörpern (Operation report of a rotating biological contactor treating landfill leachate; in German). In: *Sickerwasser aus Mülldeponien – Einflüsse und Behandlung. Veröffentlichungen des Instituts für Stadtbauwesen*, 39, 401. University of Braunscheig, Germany.

# 10.12

# Landfilling: Planning, Siting and Design

**Roberto Raga and Raffaello Cossu**

*University of Padua, Padua, Italy*

**Anders Lagerkvist**

*Luleå University of Technology, Luleå, Sweden*

Landfill planning, siting and design are complex processes transferring understanding of the environmental issues of landfills and their technical features into a full-scale facility meeting the actual regional demands for landfill volume for a range of years. Landfills are usually planned and designed for a long period, often 10–30 years, in order to distribute the costs on a large volume of waste and to reduce the troublesome and politically sensitive process of siting a new landfill. A landfill usually is seen as an undesirable construction by the neighbouring community and often difficult to get approved. Proper planning, siting and design, including adequate consideration of the use of the area after closing of the landfill, are mandatory in obtaining public acceptance of a landfill. Planning of a landfill involves:

- Capacity needs: Define types and estimate quantity of waste to be landfilled for the considered planning period.
- Political framework: National and regional legislation defining classes of landfills and their minimum requirements.

Siting of a landfill involves:

- Geographical and geological considerations: Identifying areas not suitable for landfills because of physical instability, flooding risks and the presence of protected ecological habitat.
- Physical planning: Paying attention to conflicting physical planning and zoning and to accessibility.
- Hydrogeological considerations: Assessing hydrological barriers and vulnerability to local surface water bodies and groundwater.
- Landscaping and capacity estimates: A preliminary landscaping of the landfill in order to make a first estimate of actual capacity of the site for landfilling.
- Selection of preferred site: Establish selection criteria and identify best option.
- Public involvement.
- Acquisition of land.

*Solid Waste Technology & Management*   Edited by Thomas Christensen
© 2011 Blackwell Publishing Ltd

Design of a landfill involves:

- Defining landfill classes and waste acceptance criteria.
- Landfill technology.
- Landscaping and final use.
- Layout of landfill and facilities.
- Capacity sectionizing (which types of waste to which section) and time-phasing.
- Earth works and soil balancing.
- Permits and authorization.

## 10.12.1   Landfill Planning

Landfill planning means to determine the need for landfill volume and to identify the main criteria to be used in siting and designing of landfills. This planning may take place on a national level focusing on the more strategic and political aspects of landfilling or on a regional level leading to the actual construction of a landfill. Modern landfills are engineered constructions with substantial investment cost, operational costs and long-term environmental and maintenance costs. Thus, it is desirable to distribute these cost on a large volume of waste and also preferably over a long period of years. Most modern landfills have a total capacity of 1–50 million $m^3$, with a few exceptions larger than 100 million $m^3$. Technical facilities may have lifetimes of 10–30 years suggesting that landfills from a technical–economical point of view should be designed for such long periods as well. This indicates that landfills preferably should receive annually in the range of 50 000–5 000 000 $m^3$ of waste, which again suggest how big a region the landfill can serve. In contrast, in rural areas with simple landfill facilities and only few neighbours to the landfill, the annual amount of landfilled waste may be low, governed by practicality rather than by engineering considerations.

The need of landfill capacity is determined by the waste management system and how large a population that it serves. The amount of waste generated and the treatment and recycling schemes introduced determine the amount of waste that has to go to landfill. This includes wastes that go directly to landfill, wastes treated prior to landfilling and residues from other treatments. The latter could include ashes from incineration plants or sewage sludge from wastewater treatment plants. Since landfills are planned for long periods, the estimation of needed landfill capacity also must pay attention to current national and local waste management plans as well as possible future developments. This suggests that time phasing may provide vital flexibility.

### 10.12.1.1   Capacity Needs

The estimation of needed landfill capacity is based on:

- Number of waste generation units, e.g. population, industrial activity and the expected development over the planning period.
- Unit generation rates of all waste types and their development over time.
- Mass flows in the waste management system identifying the types and amounts of waste going to landfill.
- Conversion of mass of waste to be landfilled to landfill volume needed.

Estimating the amount of waste going to landfill must be related to waste type at least in the early stages of the planning. In a modern landfill all waste types may not be landfilled in the same landfill section. For example hazardous wastes, homogeneous industrial wastes and incineration ashes may be landfilled separately or eventually routed to another landfill. When the decision has been made as to which types of landfills will be available at the actual site, the appropriate waste types provide the basis for estimating needed landfill capacity.

Converting mass of waste to volume is done via the bulk density of the waste. This value depends on the type of waste, the landfilling technology (in particular compaction) and the depth of the landfill. However, in the planning phase general density data can be used:

- Mixed waste, including organic waste:
  - 550 kg/m$^3$, compacted by tracked bulldozer;
  - 650 kg/m$^3$, compacted by wheeled loader;
  - 950 kg/m$^3$, compacted by steel wheeled compactor.
- Mechanically biologically pretreated waste: 1400 kg/m$^3$.
- Incineration ashes: 1500–2200 kg/m$^3$.
- Sewage sludge: 700 kg /m$^3$, dried.
- Construction and demolition waste: 800 kg/m$^3$.

The actual landfill volume is larger than the waste capacity wanted since space for berms for sectioning, intermediate soil covers, and final soil cover also must be accounted for.

### 10.12.1.2   Political Framework

The facts that landfills historically have a bad reputation and that the cost of remediation of old uncontrolled dumps have been very high, have fostered much political focus on landfill permitting, which also affects the planning process. In several countries landfills have been categorized into classes defined by the waste they can receive and the technical and environmental requirements that they have to meet. There may be more that one class of a landfill on a specific site. In general the more complex or hazardous the waste is, the more restrictions are introduced in the siting and design of the landfill. Usually this also means that it is more difficult to obtain public acceptance. This suggests that the need for landfill capacity is determined paying attention to the classification introduced by legislation.

While landfills traditionally have accepted a wide range of waste types, the landfill classification and the associated quality criteria introduced in many countries are changing the focus in the direction that the waste has to satisfy certain criteria in order to be landfilled. The landfill no longer accepts anything that cannot be handled otherwise. The landfill sets certain criteria that the waste must meet to be landfilled. This again may introduce additional treatment of the waste prior to landfilling, which in addition also may make other disposal schemes attractive. Through this process the landfill becomes an integrated part of the waste management system.

### 10.12.2   Landfill Siting

Landfill siting is a very complex process, which may involve a large number of factors and parties. Furthermore, conflicts can occur among criteria and among involved parties. Therefore, landfill siting needs articulated criteria, use of decision support tools and a multidisciplinary approach in order to consider the interests of all stakeholders. Main issues of this articulated procedure are:

- Regulatory aspects on site selection (restrictions).
- Identification and assessment of siting criteria.
- Establishment of site selection procedure and methodologies.
- Public participation strategies.

The procedure is usually a stepwise one often involving five steps:

- Identification of feasible locations of a landfill.
- Selection of the most promising candidate locations according to established criteria.
- Supplementary data collection on the candidate locations.
- Final selection of the preferred location.
- Acquisition and legal approval of site.

The process may be iterative if no ideal candidates can be identified and criteria used early in the screening process may be modified to develop more alternative locations. In densely populated areas, however, the process may be very complicated because no site is close to ideal and many locations are politically infeasible. In such cases the siting procedure quickly becomes a political issue leaving little room for the procedures described below. However, the various factors described still are of importance. The landfill siting is further discussed with respect to:

- Geographical and geological considerations.
- Physical planning.
- Hydrogeological considerations.
- Landscaping and capacity estimates.
- Selection procedures.
- Public involvement.

### 10.12.2.1   Geographical and Geological Considerations

The locations of a landfill must allow the landfill to maintain its physical integrity until the waste is so stable that it can be considered of little harm to the environment. The location should have little vulnerability to any accidental migration of pollutants, and the construction and operation of the landfill should have no irreversible effects on land features outside the landfill boundaries. Therefore landfills should not be located in landscapes subjects to:

- Landslides.
- Avalanches.
- Seismic activity.
- Flooding.
- Geotechnical instability.
- Habitats for protected or endangered species.

These restrictions are often specified in the general or regional regulation.

### 10.12.2.2   Physical Planning

As population densities grow, the pressure on the land resource increases and siting of a landfill must pay attention to this, including speculations on future development since the landfill will constitute a feature in the landscape for a century or maybe even longer. Considerations that often restrict the siting:

- Distance to urban areas, in particular residential areas.
- Distance to recreational areas, nature reserves.
- Distance to historical sites.
- Distance to airports.
- Sensitive surface water bodies and wetlands.
- Sensitive drinking water infiltration areas.

The siting of landfill, however, would benefit from:

- Proximity to where the waste is generated.
- Proximity to transport facilities, usually roads but potentially also railroads and waterways.
- Proximity to supply lines for water and power as well as sewers.

### 10.12.2.3   Hydrogeological Considerations

The hydrogeological condition of a site is a main feature of the protection of the environment. The hydrogeology is also referred to as an important barrier between the landfill and the environment. The hydrogeological evaluation of a site aims at assessing water flows to the site and the potential migration of any leachate that may escape the leachate collection system. These evaluations can be based on existing geological data, borehole information on sediment type, water level and geophysical logs, geophysical surface-based mapping methods and local reconnaissance. Key questions are: Where will leaking leachate migrate, how fast are major groundwater or surface water resources at risk and, if leachate leakage should appear, is it possible to remediate the situation? The presence of a substantial unsaturated zone (for example > 30 m between landfill bottom and the groundwater table) is considered important in providing attenuation capacity in case of leachate leakage.

The hydrogeological conditions should be assessed in the first screening phase in order to rule out the most vulnerable locations. When the selection has been narrowed down to a few alternatives a more thorough evaluation may be advisable, likely including fieldwork on the site. Hydrogeological evaluations are always associated with significant uncertainty unless the data behind the evaluation comprise a dense grid of observations surrounding the actual site. The hydrogeological assessment of the top candidate sites should be performed by experts experienced with the local conditions.

One effective way of evaluating the data and information collected in the hydrogeological study and of providing a context for selecting the optimum site is to define the ideal and the worse locations for a landfill as presented in Box 10.12.1. In reality, of course, compromises have to be made in the selection of a site since all of the ideal conditions are seldom found at a given location (Attar, 1989). However, the identification and assessment of all relevant factors enable landfill developers and designers to take the factors into account and to include appropriate measures when the actual conditions fall short of the ideal.

### 10.12.2.4   Landscaping and Volume Estimation

The construction of a landfill will change the topography of the site for centuries. Shaping the final contours of the landfill will determine the future landscape. In densely populated areas this will be an important feature of the landfill and an attractive landscape, which even may add value to the community, may improve the public acceptance of the landfill. Active landscaping, involving recreational areas, golf courses and the like have been reported. Landscaping of the completed landfill should also reflect the future use of the land and fit with the features of the surrounding landscape.

---

**Box 10.12.1   Ideal and Nonideal Hydrogeological Conditions for a Landfill Site.**

The best-case scenario for a landfill with regard to hydrogeology is:

- Geologically stable area (*i.e.*, no active fault movement, no settling subsoils).
- Impermeable strata at the base of the landfill.
- Large unsaturated layer below landfill base (> 30 m).
- Far from surface water bodies (> 1000 m).
- Low hydraulic conductivity in the first underlying aquifer to minimize the migration of potential contamination.
- Nearest aquifer far below the landfill base.
- Nearest aquifer not used for drinking purposes.
- No surafce water enters site during rain storms or snow melts.

The worst-case scenario for a landfill with regard to hydrogeology is:

- Many permeable layers (gravel pits, fractured zone, fault zone).
- Bottom of the landfill is close to the water table, or depth from ground surface to groundwater < 10 m.
- Local groundwater used for drinking.
- Close to surface water body or flood plain (< 100 m).
- Adjacent areas allow surface water to enter site during rain storms or snow melts.

---

Having established the preliminary contours of the completed landfill and assessed the volume obtainable by excavating or levelling of the bottom of the landfill, the overall volume of the site can be estimated. However, a coarse estimate of the volume of the site should be done at the screening stage, while a more detailed estimate must be done before final selection. This should include consideration of actual area to be filled, slope contours and volume of landscaping soil.

### 10.12.2.5   Selection Procedure

The selection procedure can be based on graphical representation of the region or on numerical schemes ranking potential sites. The two approaches address the same criteria and combinations of them can be applied.

***Graphical procedures.***    Graphical procedures are based on illustration of the main features and criteria on a map of the region. This may be geographical and geological features as well as human-activity related features such as urban areas, recreational areas and roads. Probably, the first and most famous of these approaches is the 'land suitability analysis' developed by Ian L. McHarg in 1969 (McHarg, 1969; Steiner, 1991). He utilized a technique named 'overlay maps', where each feature (criterion) is plotted on a map of the region by means of colors; different color intensities across the map denote variation in fulfilling the criterion. By superimposing all the thematic maps, it is possible to identify the most suitable areas for landfilling and those that should be avoided.

The geographic information system (GIS), originating directly from the 'overlay maps', is widely used in computer-aided decision support tool for site selection. These tools combine large amounts of georeferenced data with other statistical information to assist in evaluating siting locations. Basically, complex sources of information, such as national survey maps, aerial or satellite images are dismantled into files of homogeneous elements that could be easily classified and stored in a database. Classification can relate to land cover, land use, topography, population density, waste generation, etc. Querying the database can generate thematic maps. These maps are then combined by means of Boolean functions that 'add' or 'subtract' these thematic features, or search for particular patterns. Again, the main output of this procedure is an image of region (digitized map, usually) showing areas that are suitable for landfill siting and those unsuited. Box 10.12.2 presents an example of such an application.

---

**Box 10.12.2   An Example of a Multicriteria Evaluation of Landfill Sites Using GIS (McFadden and Whitehead, 2003). Reproduced with kind permission from Professor J. Smithers, University of Guelph, Canada. © (2003) University of Guelph, Canada.**

A GIS program (ArcView) was used to site a landfill in Macoupin County, Illinois, USA; see figure, part (a).
The major features of the County are:

- 26 towns.
- 48 924 inhabitants with a population density of 22.1 persons/km$^2$.
- Total land area of 2237.8 km$^2$.
- The predominant land cover is agricultural crops with large sections of forest and grasslands.
- The criteria used in this landfill siting were social, spatial, economic, political and ecological ones.
- The main social factor was distance from populated areas.
- Economic feasibility was represented by the proximity of the potential sites to roadways and to waste source.
- Wetlands, wells, towns, streams, roads, railroads and land cover were considered constraining criteria.

Seven constraint maps were generated to indicate which areas are suitable and not suitable for landfilling. The nonsuitable areas were represented as 'buffer zones' around major features (wells, wetlands, etc.) and special land covers. Then three factor maps were generated to indicate which areas are the most and least suitable for the siting of a landfill. These maps took into account the distance of each potential site from streams, wells and towns. Finally, a multicriteria evaluation was applied to classify the criteria, using a binary scale for constraints (denoting suitable or nonsuitable areas) and a scale of one to nine, to highlight the suitability degree in relation to each factor. It allowed to mathematically aggregate all the maps and to produce a final constraint map representing restricted and available areas for landfilling (b), that of the factors representing different degrees of suitability for landfill siting (c) and, by superimposition, a final image with the most suitable location (d).

The best location was identified in the northern part of the County, where the dominant land cover is row crops, close to two roads and quite far from the major towns, railways, streams, lakes or wetlands.

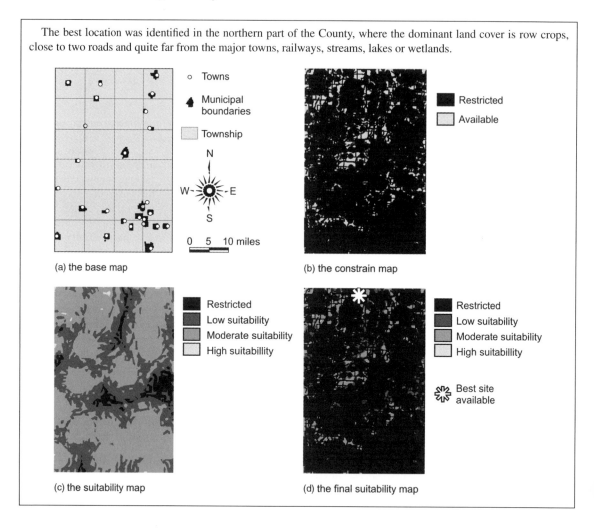

(a) the base map

(b) the constrain map

(c) the suitability map

(d) the final suitability map

***Ranking procedures.***   Ranking procedures basically use similar criteria as the graphical procedures, but are most often used on a limited number of alternative siting locations subject to many criteria. Multicriteria ranking procedures are based on the simple idea that complex problems can be divided into more manageable parts to be analysed and then reassembled into the final result.

The most important advantages of these models are:

- The decision maker does not need specific theoretical knowledge about the procedure.
- The procedure is direct and transparent and does not require sophisticated computer facilities.
- The results are particularly clear and simple making them suitable for communication with the public.

The procedure involves a two-dimensional matrix, named the 'performance matrix': the rows represent the alternatives, and the columns are the factors (called 'criteria') by which the alternatives are judged. The elements of the matrix represent a standardized measurement of the decision maker's satisfaction with each alternative, in relation to the criteria. The relative priority of each criterion is also determined by means of 'weights' (e.g. numbers used to rank the criteria) that are associated to every factor. The various methodologies differ from each other in the way they compute the 'satisfaction values' and the 'weights,' and the method for combining them into the final score. The most common approaches are based on direct assignation of weights; these weights multiply the satisfaction values of a specific alternative with respect to every criterion. Data are then aggregated by using simple calculations (such as addition, subtraction, etc.). As most of these models are based upon the 'utility theory,' if the criteria have positive value, the more the score, the most suitable is the site (Nijkamp *et al.*, 1990; Voogd, 1983).

Thus, site selection is performed by:

- Assessing the relative importance of each regulatory constraint and that of each siting criterion.
- Assessing the performance of each area with respect to every regulatory constraint and site criterion.
- Combining all these values in final 'suitability' scores.
- Ranking the sites with respect to these scores.

Box 10.12.3 illustrates an example of a ranking procedure.

---

**Box 10.12.3   An Example of a Multicriteria Evaluation of Landfill Sites Using a Ranking Procedure (Manoliadis and Sachpazis, 2001). Reprinted with permission from Electronic Journal of Geotechnical Aspects of a Landfill Site Selection Study in North Evia – Greece, volume 6, by O.G.K. Manoliadis and I. Sachpazis © (2001) Mete Oner.**

In the early 1990s, the district of Northern Evia (Central Greece) decided establish a new landfill. The district comprises 13 municipalities and has a population of 10 500 habitants. The waste production is 3500 t/year.

The site selection procedure was very complex as the territory is partially mountainous and most of the nonmountainous part is not suitable for landfill construction because of landscape protection zones, water sources protection zones, potential flood-plain areas and densely populated areas. Therefore, a two-stage, multicriteria evaluation was applied. In the fist stage, a rough list of promising sites was identified merely based on geotechnical criteria. In the second stage a ranking of four sites was selected by using environmental, socioeconomic and technical criteria (see below). In detail, the relative importance of every factor (or better, the trade-off among criteria) was investigated and weights were associated to each criterion ($a_i$) and criteria cluster ($A_j$); weights were standardized to sum up to one. Then, the four site options (Elitsa, Aidipsos, Ellinika, Mantoudi) were assessed with respect to every cluster and criterion; all the data were standardized by assigning a simple 1:10 scale to each criterion magnitude.

Finally, the overall scores were computed by combining the factor estimations and the relative criterion and cluster weights, for each site, by using a multiobjective approach. It is simply based on the idea that an optimal 'solution' (choice option) can be ideally defined and, therefore, using this solution as reference, the other options can be rated by computing the difference between each of them and the 'ideal point' of satisfaction (represented by the maximum score). The preferred site is the option nearest to the ideal point (i.e. the higher the score the closest the option to the ideal one). Results are displayed below.

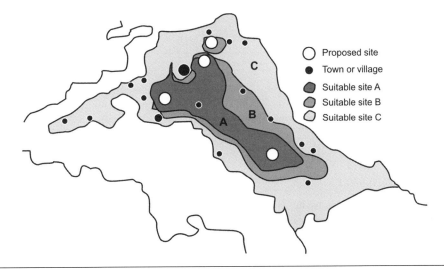

| Clusters | Weights | Criteria | Weights |
|---|---|---|---|
| Geotechnical | 0.2 | Aquifer thickness | 0.33 |
| | | Permeability | 0.33 |
| | | Infiltration–percolation | 0.33 |
| Environmental | 0.4 | Land cover and ecological character | 0.25 |
| | | Absence of optical intrusion | 0.25 |
| | | Odors nuisance | 0.25 |
| | | Public health, safety and nuisance | 0.25 |
| Economical | 0.3 | Net present value | 0.20 |
| | | Depreciation cost | 0.20 |
| | | Impacts on housing/tourist development | 0.60 |
| Technical–operational | 0.10 | Landfill capacity | 0.20 |
| | | Cover material availability | 0.30 |
| | | Access roads | 0.10 |
| | | Distances from the waste consumption | 0.40 |

| Criteria | Elitsa | Aidipsos | Ellinika | Mantoudi |
|---|---|---|---|---|
| Aquifer thickness | 8 | 7 | 8 | 8 |
| Permeability | 8 | 7 | 7 | 8 |
| Infiltration–Percolation | 7 | 8 | 8 | 7 |
| Land cover and ecological character | 10 | 5 | 7 | 7 |
| Absence of optical intrusion | 10 | 6 | 8 | 7 |
| Odors nuisance | 10 | 5 | 7 | 7 |
| Public health, safety and nuisance | 9 | 7 | 7 | 7 |
| Net present value | 10 | 7 | 9 | 8 |
| Depreciation cost | 9 | 7 | 7 | 8 |
| Impacts on housing/tourist development | 10 | 5 | 7 | 7 |
| Landfill capacity | 7 | 8 | 8 | 7 |
| Cover material availability | 7 | 5 | 7 | 7 |
| Access roads | 6 | 7 | 6 | 7 |
| Distance from the waste consumption | 10 | 7 | 5 | 4 |

Results (the higher the score, the closer the option is to the ideal one)

| Criteria clusters | Elitsa | Aidipsos | Ellinika | Mantoudi |
|---|---|---|---|---|
| Geotechnical | 7.59 | 7.26 | 7.59 | 7.59 |
| Environmental | 7.25 | 5.75 | 7.25 | 7.75 |
| Economical | 9.20 | 7.00 | 7.40 | 7.80 |
| Technical–Operational | 6.70 | 5.00 | 4.70 | 4.40 |
| Overall score | 7.848 | 6.352 | 7.198 | 7.398 |

### 10.12.2.6   Public Involvement

The point at which the developer of a landfill should seek public input or present a proposal to the public is a matter for debate. If the consultation of the public is sought too late in the process, the public may doubt that its comments and advice will be taken seriously. When several realistic alternatives emerge from the selection process that has been described previously, the time is appropriate for notification of the public and for public review and comment.

The concerns of the general public with regard to the siting and operation of a landfill include the following:

- Proximity: The traffic and landfill operations will increase noise, dust, odour and accident levels in the neighbourhood.
- Real estate values: Those living near the proposed landfill are concerned that property values will decline.
- Leisure use: The view that the presence of the proposed landfill will negatively impact leisure activities (e.g., fishing, hiking).
- Aesthetical: The perception that the presence of a landfill diminishes the quality of life of those living or working near the site.
- Health: The presence of the waste will influence air quality and potentially affect people living in the neighbourhood.
- Environmental: Worry about disturbances in ecosystems and freshwater resources.

The objective of public involvement is to ascertain that the public understands the proposals, the purpose and need for developing and operating the landfill and to provide opportunities for public involvement in the decision-making process.
  Some of the advantages of public involvement are:

- An increase in the likelihood of gaining the support of the public for the landfill plan.
- The possibility of providing useful information which otherwise may have been missed.
- Familiarization of the public with the final design and intended use of the landfill which may directly benefit the community (e.g. land use as a park).
- Assurance to the public that all views have been considered, increasing the accountability of decisions.

Involving the public in landfill siting may include the folllowing steps:

- Provide the public with all details of the plan (newspaper announcement, direct mail, public displays at libraries etc.).
- Establish the need for the new site by explaining the situation, including the status of the existing facilities and the reason that a new disposal site is needed.
- Explain the alternatives that have been considered and the reasons that they have been rejected or selected.
- Explain the operations, the process for managing the site, the control of leachate and gas, and the closure and postclosure procedures.
- Describe the impacts of the development and operation of the landfill on the local environment and on the people who may be affected.
- Make every effort to clearly understand the concerns of people who live near the site.
- Keep options open. If new information becomes available as a result of the consultation with the public, then the selection process should have the flexibility to incorporate the public's comments into the final selection.
- Review previous assessments of environmental impacts of the potential sites as new or more information is collected as a consequence of the discussions of the situation with people and businesses affected by the proposal.

## 10.12.3   Landfill Design

Design of a landfill involves:

- Defining landfill classes and waste acceptance criteria.
- Landfill technology.
- Landscaping and final use.
- Layout of landfill and facilities.
- Capacity sectionizing (which types of waste to which section) and time-phasing.
- Earth works and soil balancing.
- Permits and authorization.

### 10.12.3.1   Defining Landfill Classes and Waste Acceptance Criteria

The first step in designing a landfill is to categorize the various waste types as to their landfilling compatibility, i.e. which waste goes to which class of landfill. It may not be practical to have many different landfill classes on a single site, because each class of landfill should have reasonable capacity to be practical to construct and to operate. Each class should also have separate leachate control systems and potentially also be different as to the landfill technology introduced. The classes of landfills on a site could include:

- A landfill classified for inert waste and constructed and operated with a minimum of engineering. This section could receive waste that might be defined as inert according to the current legislation (e.g. demolition waste).
- A landfill classified for inorganic waste and constructed and operated with engineering measures focused on control of leachate. This section could contain waste as for example inorganic industrial waste and incineration bottom ashes, but could also contain stabilized hazardous waste.
- A landfill classified for waste containing degradable organic matter and constructed and operated with engineering measures focused on gas as well as leachate control.
- A landfill classified for certain hazardous waste and with highly engineered control measures.
- A landfill classified for special wastes that according to regulations cannot be handled otherwise than by landfilling. This could, for example, be asbestos, impregnated wood or PVC waste.

Many countries have their own classification of landfills, while others only distinguish between municipal landfills receiving a mix of nonhazardous waste and hazardous waste landfills.

Acceptance criteria must be developed for each of the landfill classes in order to ensure that only waste compatible with the landfill class are accepted. The criteria may, in addition to the distinguishing between hazardous and nonhazardous waste, include total contents of critical pollutants, content of organic carbon, methane potential and/ or respiration potential, and leachability of selected substances. The relevant criteria will depend on national regulation. Acceptance of waste may also be based on positive lists of waste from specific sources. For example, it may not be necessary define exact criteria for residential waste collected by public services. Box 10.12.4 presents the leaching criteria used in the European Union (EU). The EU applies three levels of test as part of the waste acceptance: (1) A characterization test which is thorough but rare and describes the origin and composition of the waste, its total substance content and determines its behaviour (degradation and leaching), (2) a compliance test (typically a batch leaching test) which is simple and done regularly to check that the waste is within the acceptable range, (3) an inspection test assessing each load entering the landfill. This may be a visual inspection only (CEC, 2003).

---

**Box 10.12.4   Acceptance Criteria for Landfill Waste Regarding Leaching as Introduced in The European Union.**

Waste acceptance criteria for landfills are based on a procedure comprising tests for basic characterization, for compliance and for on-site verification.

A specific type of waste can be accepted in a specific landfill class according to the results of the basic characterization test, that have to be carried out with focus on the following main aspects:

- Contaminant release as a function of time (long-term leaching characteristics).
- Contaminant release as a function of the main leaching controlling parameters such as pH, redox potential and complexation among others (Van der Sloot *et al.*, 1999).
- Compliance testing is carried out periodically on waste deemed acceptable for a specific landfill class and the tests to be considered are one or more of those used for the basic characterization.
- On-site verification procedures should be set up by Member States and can comprise rapid test methods and/or simple visual inspection.
- Leaching tests that can be considered for the characterization and compliance testing have been developed and standardized by the European Committee for Standardization (CEN).

---

Leaching tests for characterization.

| Test name | CEN/TS14405 | CEN/TS14429 | CEN/TS14997 |
|---|---|---|---|
| Type of test | Percolation test (Column leaching test) | Batch tests at specified conditions (pH dependence leaching test) | Batch tests at specified conditions (pH dependence leaching test) |
| Reference | CEN TC 296 WG6 | CEN TC 292 WG6 | CEN TC 292 WG6 |
| Particle size | < 4 mm | < 2 mm | < 2 mm |
| Leachant | Demineralized water | Demineralized water with pH control using $HNO_3$ or NaOH (pH 3–12) | Demineralized water with pH control using $HNO_3$ or NaOH (pH 3–12) |
| Amount of solid | 0.5–0.7 l | 100 g | 100 g |
| Liquid:solid ratio (L/S; l/kg) per step | Cumulative L/S: 0.1, 0.2, 0.5, 1, 2, 5, 10 | 10 | 10 |
| Number of steps | 7 | 8 | 8 |
| Contact time per step | Total: maximum 3 weeks | 48 h | 48 h |
| Filter size | 0.45 µm | 0.45 µm | 0.45 µm |

Leaching test in four steps for compliance tests.

| Test name | EN 12457-1 | EN 12457-2 | EN 12457-3 | EN 12457-4 |
|---|---|---|---|---|
| Type of test | Compliance test for leaching of granular waste materials | Compliance test for leaching of granular waste materials | Compliance test for leaching of granular waste materials | Compliance test for leaching of granular waste materials |
| Reference | EU CEN TC292 | EU CEN TC292 | EU CEN TC292 | EU CEN TC292 |
| Particle size | < 4 mm | < 4 mm | < 4 mm | < 10 mm |
| Leachant | Demineralized water | Demineralized water | Demineralized water | Demineralized water |
| Amount of solid | 100 ± 5 g | 100 ± 5 g | 100 ± 5 g | 100 ± 5 g |
| L/S (l/kg) per step | 2 | 10 | 2 in first step; 8 in second step | 10 |
| Number of steps | 1 | 1 | 2 | 1 |
| Contact time per step | 24 h | 24 h | 6 h for first step; 18 h for second step | 24 h |
| For serial batch tests: Renewal of leachant or solid | | | New leachant | |
| Method of agitation | End over end or roller-table rotation (10 rpm) | End over end or roller-table rotation (10 rpm) | End over end or roller-table rotation (10 rpm) | End over end or roller-table rotation (10 rpm) |
| Filtration/filter size | 0.45 µm | 0.45 µm | 0.45 µm | 0.45 µm |
| Comments | Validated in EU ECN-C-01-117 | Validated in EU ECN-C-01-117 | Validated in EU ECN-C-01-117 | Validated in EU ECN-C-01-117 |

### 10.12.3.2 Landfill Technology

The landfill technology must be defined for each of the classes of landfills at the site. National regulation often determines the minimum level of landfill technology in terms of liners, leachate collection, leachate treatment, gas collection and utilization and final cover.

For most of the landfill classes the range of technical options is limited although the actual solutions may vary. Two issues must in any case be assessed:

- The deeper the landfill, the slower the leaching: for a given water infiltration, a deep landfill will within a certain time period, e.g. 30 years, reach a lower liquid to solid ration than a shallow landfill, which means that the deep landfill has leached less of the leachable fraction of the waste at the end of the period. This may be an important aspects in determining the amount water to enter the landfill (landfill top covers that limit water infiltration or addition of water to enhance leaching) and in defining the life expectancy of the bottom liner and leachate collection system.
- Landfills receiving organic waste may employ various technologies for enhancing methane generation, carbon degradation and removal of ammonia. These technologies include the bioreactor landfill, the flushing bioreactor and the semiaerobic landfill, as described in Chapter 10.6.

The technological elements of the landfill are described in separate chapters: bottom liners and leachate collection (Chapter 10.8), top covers (Chapter 10.9), gas collection and utilization (Chapter 10.10) and leachate treatment (Chapter 10.11).

### 10.12.3.3   Landscaping and Final Use

An early step in landfill design is to define what the landfill most likely will be used for after its closure. Although the landfill may be active for 20–30 years the final use should receive some considerations, because the final use may define acceptable slopes, acceptable settlements and topcover quality for vegetation. Chapter 10.14 describes issues about closing of landfills and their final use.

Important issues are:

- Final slopes: The slopes most likely vary over the area, but should be planned so that surface runoff is easy and any local ponding of storm water is reduced to a minimum. Very steep slopes may limit the use of the site after closure, may be vulnerable to soil erosion and may cause instability for the vegetation.
- Deep landfills which are filled fast will have the largest absolute settlements, which may be detrimental to constructions etc.
- The final landscaping of the landfill should be designed to blend in with the countryside.

The landscaped landfill surface constitutes the upper boundary of the landfill body and hence is an important parameter in calculating the landfill volume available at the site

### 10.12.3.4   Layout of Landfill and Facilities

The physical layout of the landfill must be defined considering the area available and the final landscape. An example of a landfill layout is given in Figure 10.12.1. Only part of the area is used as actual fill area, the rest being occupied by facilities, such as buildings for offices, maintenance garage, leachate treatment facility, gas management facilities etc. The landfill may include also a public access area where private cars may deliver waste to the landfill.

The infrastructure as well as the fill areas must be planned. Box 10.12.5 briefly describes the main facilities and infrastructures on a landfill.

In order to optimize investment costs the actual fill area is divided in sectors and the preparation of each sector is carried out at the moment the same sector is needed for waste deposition.

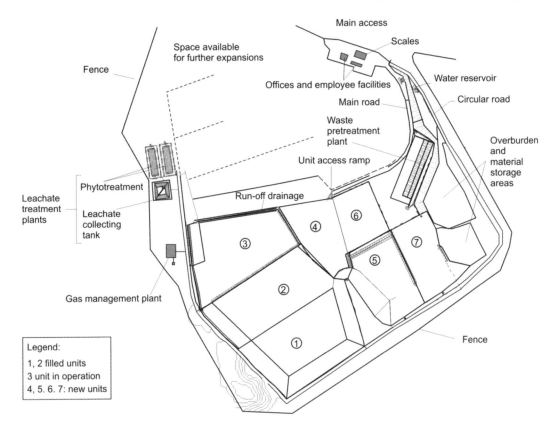

**Figure 10.12.1** *Example of landfill layout.*

---

**Box 10.12.5  Landfill Infrastructures and Facilities.**

Physical planning of a landfill site should include a number of infrastructures and facilities necessary for the landfill operation. The most important are:

***Roads:*** An all-weather access road should consist of two lanes of sufficient width and strength to carry the delivery vehicles. Waiting space should be provided near the scales and parking areas should be provided as well. Separate dirt roads should be prepared for the compactors and heavy machinery that otherwise would ruin the main roads.

***Drainage:*** Appropriate permanent and temporary drainage facilities must be provided to control surface runoff at the site. Surface waters must be diverted away from the areas to be filled with solid wastes.

***Overburden and material storage:*** A stockpile area should be provided for storage of material excavated during construction of a landfill, of the various materials that may need to be imported to the site for operation, and of any other imported materials.

***Fencing and signs:*** Fencing limits access to the landfill site. A gate is required at the site entrance and should be locked when the site is unattended. A sign prominently located should identify the landfill site, the hours of operation, fees and any restrictions on users or materials acceptable for delivery.

***Buffer zone:*** A buffer zone around the site should be required to buffer adjacent property, to limit access by unauthorized persons and reduce nuisance to surrounding residents. The site should be visually separated from adjacent property by the use of plantings and soil berms, which also reduce the problems of litter, noise and dust.

***Buildings:*** A building should be provided for office space, sanitary facilities and employee facilities; however, very small sites may not require all these facilities.

***Garage facilities:*** Garage facilities should be available for the maintenance of heavy machinery and compactors and for protecting machinery not in use.

***Utilities:*** Power is required for maintenance of on-site operating equipment and for lighting. An electric generator may also be installed. Water is needed in the event of a fire, for machine maintenance and for dust control.

***Scale:*** A scale is required for recording the weights of solid waste. The scale type and size will depend on the size of the landfill operation and the vehicles using the facility.

***Truck wash facilities:*** In order to prevent mud and debris from dropping from the vehicles when they return to the public roads, consideration should be given to truck wash facilities. These facilities can also be used to clean the landfill equipment.

***Public access and transfer station:*** In order to avoid public access to the working area of the landfill, which could create unsafe conditions, consideration should be given to providing a drop-box container or transfer station near the entrance to allow for wastes delivered by individual persons to be deposited safely.

### 10.12.3.5   Capacity Sectionizing and Time-phasing

For landfills with life times beyond 5 years it is often recommendable to divide each landfill class into sections providing volume for 2–5 years of landfilling and to develop the fill area and install bottom liners and leachate collection system as the sections are filled. This has potentially several benefits:

- The investment cost are distributed in time, putting less pressure on the investment needs and on the cash flow.
- Avoid maintaining fill areas not in use and avoid collecting runoff from lined but not yet filled areas.
- Provides flexibility as to future needs for landfill volume.
- Provides flexibility if landfill technology develops or the official technical prescriptions change.

The area of each section should not be so small that instalment costs become too high or the landfilling operation become impractical.

Depending on the design of the landfill it may not be possible to finish one section completely before starting to fill another section. This could be the case if access to one section passes over another section or if slopes and geotechnical stability require that adjacent sections rise sequentially. Therefore landfilling of a section may be subject to time-phasing; e.g. the lower lifts of waste are put in over a period of maybe some months where after a temporary cover is established and vegetated until landfilling is continued on the same section, maybe two years later. Also the final cover and the final vegetation may be established in stages in order to improve the appearance of the landfill and stabilize slopes (vegetated slopes are less vulnerable to soil erosion than bare slopes).

### 10.12.3.6   Earth Works and Soil Balancing

The amount of soil necessary for daily and final cover must be accounted for in the design of the landfill. The refuse:soil ratio usually used in landfill for daily cover ranges from 2:1 to 5:1 on a volumetric basis. In general, a ratio of 3:1 (refuse to soil) can be used to plan for the operation of most sites. This ratio can be modified upward or downward, depending on any special cover requirements, phasing requirements, or final cover requirements. The volume of the final cover can be estimated considering a minimum depth of 0.6 m and a maximum depth of 1.5 m. It is here important to remember that the soil takes up some of the landfill capacity. In addition the need for soil for embankment and for noise barriers must be estimated.

Once the basic data has been obtained, the required landfill volume as well as soil volume estimated, then the soil balance should be calculated. The key question is here, if soil is in demand, to estimate the needed excavations and levelling of the sites to provide a good base for the liner as well as sufficient soil for the design. A series of trials and adjustments will be necessary to optimize the design. Because import of soil and gravel may be very costly, it is generally better to have a slight excess of material, rather than a shortage. It is not uncommon that a 10 m deep landfill will need 3–4 m$^3$ soil/m$^2$ fill area.

### 10.12.3.7   Permits and Authorization

The siting and operation of landfills are subject to legal restrictions and permitting procedures. The regulations are chiefly based on the concern for possible adverse impact of the landfills on human health and environment. The form these regulations takes is different in different regions of the world and also varies with the wastes to be landfilled. For example there may be different rules for industrial waste and municipal waste. The generic permitting procedures include environmental impact assessments and the subsequent formulation of accepted impacts. The permit may also prescribe measures to be taken in order to limit emissions and impact on human health, including aftercare and monitoring. Permits or licences may also include contingency plans and provisions for future damages in the form of financial guarantees.

## References

Attar, A.M. (1989): Siting a landfill: The first step. *World Wastes*, 32 (13), 14–20.

Bagchi, A. (1994): *Design, construction and monitoring of landfills*, 2nd edn. John Wiley & Sons, Ltd, New York. ISBN 0-471-30681-9.

CEC (2003): Council Decision 2003/33/EC of 19 December 2002 establishing criteria and procedures for the acceptance of waste at landfills pursuant to Article 16 of and Annex II to Directive 1999/31/EC. *Official Journal of the European Communities*, L11/27–49.

Manoliadis, O.G.K. and Sachpazis I. (2001): Geotechnical aspects of a landfill site selection study in north Evia – Greece. *Electronic Journal of Geotechnical Engineering*, 6.

McFadden, C. and Whitehead, S. (2003): *Locating a landfill site in Macoupin County, Illinois, using GIS*. University of Guelph, USA. Located on the internet: 19 October 2009. http://www.uoguelph.ca/geography/research/geog4480_w2003/index.htm.

McHarg, I.L. (1969): *Design with nature*. Doubleday and Company, New York, USA.

Nijkamp, P., Rietveld, P. and Voogt H. (1990): *Multicriteria evaluation in physical planning*. North-Holland, Amsterdam, The Netherlands.

Steiner, F. (1991): *The living landscape*: An ecological approach to landscape planning, McGraw-Hill, New York, USA.

Van der Sloot, H.A., Hjelmar O., Mehu J. and Blakey N. (1999): Waste characterization by means of leaching tests to assess treatment, reuse and disposal options. In: *Proceedings of the Seventh Waste Management and Landfill Symposium*. CD-ROM. CISA, Environmental Sanitary Engineering Centre, Cagliari, Italy.

Voogd, H. (1983): *Multicriteria evaluation for urban and regional planning*. Pion Limited, London, UK.

# 10.13

# Landfilling: Operation and Monitoring

**Nicole D. Berge, Eyad S. Batarseh and Debra R. Reinhart**

*University of Central Florida, Orlando, USA*

A successful landfill requires, in addition to good siting and good design, also good operation and monitoring.

The technical operation of a landfill requires the integration of several different and necessary components, such as equipment, waste filling sequences and placement methods, waste compaction, and daily cover placement. Additionally, aspects such as temporary road placement, safety and security, waste input control and storm water management must be addressed and/or implemented. The technical operation of efficiently placing the waste in the landfill is accompanied by environmental procedures controlling noise, odors, fire, litter, dust and vectors at the site.

Modern landfills are usually requested, as part of their permit, to monitor their performance with respect to leachate, air, groundwater, liner and settlement. The monitoring, in addition to supplying the authorities with mandatory information, provides an account of how well the design, engineering and operation of the landfill meet the expectations defined in the planning and permitting documents of the landfill. Monitoring is thus the only way of documenting that the landfill is a reliable element in the waste management plan. Monitoring is usually intense during the actual operation of the landfill, but many aspects of the monitoring may have to continue for decades after closure of the landfill as long as the landfill potentially can impact the environment.

This chapter presents the basic aspects of the technical landfill operation, the environmental procedures and the monitoring.

## 10.13.1 Technical Operation

Technical operation of a landfill typically involves moving equipment, waste filling sequences and placement methods, waste compaction, daily cover placement, temporary roads, safety and security, waste input control and storm water management.

*Solid Waste Technology & Management*   Edited by Thomas Christensen
© 2011 Blackwell Publishing Ltd

#### 10.13.1.1    Equipment

A variety of equipment is required to operate a landfill. Machinery to move, place, spread, and compact the waste and daily and final cover soils is necessary. Service or support equipment, such as excavators, service and water trucks and grinders, are also necessary to maintain a landfill.

Crawler tractors, scrapers, compactors, draglines (or excavators) and motorgraders are types of equipment that have been used at landfills. Figure 10.13.1 presents pictures of some landfill equipment, including a crawler tractor, compactor, excavator, motorgrader and an off-road truck. Of these, the crawler tractor is the most versatile. It is excellent at spreading both waste and cover soil. Additionally, the crawler tractor is able to compact both materials well, achieving waste densities between 475 and 725 kg/m$^3$ (Tchobanoglous *et al.*, 1993). Another advantageous attribute of the crawler tractor is its ability to excavate, which may be necessary in instances in which a cover soil source is a borrow pit on site or must be excavated at a nearby off-site location. Crawler tractors are also able to prepare the landfill site and build temporary roads.

Compactors designed specially for landfill operations are almost universally used and preferred (Vesilind *et al.*, 2002). They are heavy and equipped with knobbed steel wheels, which aid in effectively spreading and compacting large amounts of waste. Densities between 725 and 950 kg/m$^3$ are often achieved (Vesilind *et al.*, 2002). Compactors, however, are poor at excavating and hauling materials. Thus, at a site in which cover soil must be excavated from on-site borrow pits or hauled long distances, additional equipment is necessary.

Pans and scrapers have been used to excavate and place daily cover soils at landfills. These machines are specially designed to scrape up cohesive dirt that can be removed in layers, haul it to the landfill, and then run down the compacted slope, effectively placing dirt evenly over the waste (Vesilind *et al.*, 2002). Pans and scrapers are inefficient in instances where haul distances are long, the amount of soil being moved is large, and the material is difficult for a scraper to load (i.e. saturated soil, sand, gravel). In these instances, hydraulic excavators and trucks have been shown to move dirt faster and at lower cost than scrapers (Bolten, 2000a). The articulated (or artic) dump truck has become the preferred hauling vehicle at many landfills (Bader, 2003). This truck is able to move swiftly over the rough terrain usually encountered at a landfill, and is able to load and unload excavated materials quickly. Additionally, the artic truck is much more maneuverable than traditional pans and scrapers (Bader, 2003).

The exact type, number and size of equipment needed is highly dependent on individual landfill requirements. The type and amount of waste, hauling requirements and distances, as well as the size of the landfill and method of operation, determine the specific equipment needs. Generally, more equipment is needed at larger landfills when compared to smaller sites. The mode of operation of the landfill may also dictate special equipment needs, such as transfer equipment. Climatic conditions, budget constraints, and expected growth also are factors that may dictate the equipment needs of an individual landfill (Vesilind *et al.*, 2002). More details about equipment needs may be found elsewhere (Tchobanoglous *et al.*, 1993). Just to give a rough estimate, a medium-sized compactor may over an 8-h day have an annual capacity of 20 000 t of waste.

#### 10.13.1.2    Filling Sequences and Waste Placement

After the site has been prepared (i.e. excavation and placement of liner), the landfill is ready to receive waste. There are many different waste placement methods. The three most common methods are the excavated cell method, the area method and the canyon/depression method, as illustrated by Figure 10.13.2:

- When filling via the excavated cell method, cells are excavated, lined with synthetic and/or low permeability clay liners to prohibit subsurface movement of leachate and gases, and then filled with waste. This method of landfilling is best suited in areas in which the water table is not near the surface and where the excavated soil may function as cover material.
- The area method differs from the excavated method in that there is no excavation; the waste is placed above grade. This method is preferred when the terrain is unsuitable for the excavation method or when the groundwater table is near the surface. A liner and leachate collection system is constructed above grade. Cover material can either be hauled to the landfill from an off-site source or excavated on site (Tchobanoglous *et al.*, 1993). The area

(a) Crawler tractor

(b) Compactor

(c) Excavator

(d) Motorgrader

(e) Off-road truck

**Figure 10.13.1**   *Landfill equipment: (a) crawler tractor, (b) compactor, (c) excavator, (d) motorgrader, (e) off-road truck. Reproduced with permission from Cenage Learning, Inc., Chicago. © (2002) Cengage Learning.*

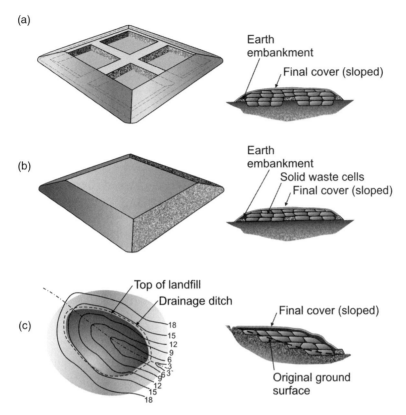

**Figure 10.13.2** *Landfill waste placement methods: (a) excavated cell, (b) area and (c) canyon/depression (from Vesilind et al., 2002). Reprinted with permission from Solid Waste Engineering 1E by Vesilind, Worrell and Reinhart, 9780534378141 © (2002) Cengage Learning.*

method is mandatory in Austria in order to ensure natural drainage of leachate to ditches also after closure of the landfill.

- The canyon/depression method involves the placement of waste into an existing canyon, ravine, borrow pit, or quarry. This technique allows the geometry of the depression to dictate filling and compaction procedures (Tchobanoglous *et al.*, 1993).

The placement method (how) and the filling sequence (where) are determined during the design of the landfill and is a function of site topography, local hydrology and geology and future development plans. Generally, filling of the site begins in one corner and moves outward. Waste is generally placed in lifts, or thin layers of waste. The area in which the waste is currently being unloaded, placed, and compacted is called the working face. The working face must be large enough to allow for unloading of several vehicles simultaneously. Typically, 4–6 m is required per vehicle (Vesilind *et al.*, 2002).

The first layer placed consists of preselected waste, with sharp and heavy objects removed, and is placed directly above the leachate collection system. It must be placed and compacted in a manner in which all equipment is kept from damaging the leachate collection system. This layer functions to protect the leachate collection system and is often called the operational layer (Vesilind *et al.*, 2002). In some instances, the operational layer may consist of a layer of soil or other alternative material. Subsequent filling continues by spreading the waste in 50- to 70-cm lifts and then compacting it (Tchobanoglous *et al.*, 1993). After placement of one or more lifts, horizontal gas collection system construction can begin as part of an active gas extraction plan during landfilling. Trenches are excavated and filled with gravel and perforated pipes, allowing gas to be extracted as it is produced. Additionally, horizontal liquid injection trenches can be constructed during filling, allowing for liquid injection to occur while filling and for a more uniform liquid distribution system after final landfill capacity is attained.

Subsequent lifts and gas extraction systems are continually added until final design grade has been met, after which a final cover is constructed above the waste. Vertical gas wells may be installed at that time for long-term gas extraction, as can any vertical liquid injection lines if operating as a bioreactor.

### 10.13.1.3   Compaction

As waste is placed in the landfill, it is compacted to allow for placement of a maximum quantity of waste, thus increasing the life of a landfill cell, allowing more waste to be placed in the same volume. In addition to compaction, other factors such as cell geometry and depth impact efficiency of waste placement. Box 10.13.1 presents a more detailed discussion about the relationship between cell geometry and compaction. Compaction increases waste density and can be thought of in terms of a compaction ratio. The compaction ratio (CR) is defined as (Tchobanoglous *et al.*, 1993):

$$CR = V_i / V_f \tag{10.13.1}$$

where:

$V_i$ = volume of waste prior to compaction;
$V_f$ = volume of waste after compaction.

The CR is a function of several factors, including slope, moisture content, height of waste layer, waste composition, number of passes a compactor makes over the waste and compaction vehicle type (Vesilind *et al.*, 2002). Slope of the landfill affects the compaction ratio; flatter slopes generally result in higher compaction ratios than steeper slopes because they are easier to compact by the landfill equipment. Higher moisture contents result in higher compaction ratios than drier waste, as wet waste compacts more efficiently. Additionally, as lift thickness increases, the degree of compaction attainable decreases. Diversity of the waste in each lift also affects the CR. If all materials are similar, especially in size, a higher compaction ratio may result. As one would expect the more passes a compacting device makes over the waste and the heavier the compactor, the higher the CR and thus waste density.

---

**Box 10.13.1   Cell Geometry (After Bolten, 1998). Reproduced with permission from MSW Management. © (1998) and (200) Forester.**

Maximizing the use of landfill space by ensuring the landfill cell contains as much waste as possible in as small a volume as possible is a goal of many landfill operators. Achieving high compaction ratios is critical in order to meet this goal. However, compaction ratio alone will not yield the optimal solution. Landfill space is optimized when a minimum amount of surface area is used, allowing for a minimum amount of necessary cover soil. Thus, cell geometry is also a critical factor to consider. Besides a sphere (the geometric shape with the maximum amount of volume and minimum amount of surface area), a slanted cube shape depicted below can result in the most efficient volume to surface area ratios. In most cells, daily cover soil is placed on three sides of the cell; the top, face and side.

Both compaction ratios and depths should be considered when determining the dimensions of the cell to minimize the volume of air space filled and cover soil required. Determining the optimum dimensions of the cell is an iterative process in which the cell dimensions are varied until the minimum surface area is achieved. Generally, the cell width is chosen first, as it is rarely varied because it is based on the amount of space required to allow the maximum expected number of trucks to unload at once. Next, an arbitrary depth is chosen, allowing the height to be calculated based on the slope of the cell, and the horizontal distance taken up by the waste can also be calculated. Finally, the surface area of the cell is calculated, using the following set of equations (see figure):

$$A_T = V_{waste}/D$$
$$A_F = W*L$$
$$A_S = [V_{waste}/(D*W)]*L$$
$$SA = A_T + A_F + A_S$$

Where:

$W$ = width of cell (m)
$V_{waste}$ = volume of waste added to the cell each day (m$^3$)
$D$ = depth of waste in the cell (m)
$L$ = length of slope, $\sqrt{(D^2 + H^2)}$ (m)
$H$ = the base of the slope $L$; often $H = 3*D$ (m)
$A_T$ = area of top of cell (m$^2$)
$A_S$ = area of side of cell (m$^2$)
$A_F$ = area of face of cell (m$^2$)
$SA$ = total surface area of cell requiring cover (m$^2$)

Once the optimal cell dimensions yielding the minimum surface area are determined, the volume of cover soil required can be calculated using the following equation:

$$V_{coversoil} = D_{coversoil}*SA$$

Where:

$V_{coversoil}$ = volume of cover soil required
$D_{coversoil}$ = depth of cover soil that is being applied

The ratio of cover soil to waste ($R_{C:W}$) can also be calculated:

$$R_{C:W} = V_{coversoil}/V_{waste}$$

If using an alternative daily cover on the face of the cell, such as wood chips or compost, the most efficient cell design is not one in which the minimum surface area is achieved, rather one in which the top and side surface areas are minimized, as that is where the soil will be placed. In those instances, a much deeper cell results as the optimal design (Bolten, 2000b).

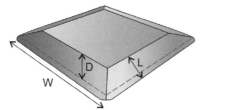

W = width of cell

D = depth of waste

L = length of slope

For a 500 m$^3$ daily cell (maybe corresponding to 400t/day received) and assuming $W = 30$ m, $D = 2.5$ m, and $H = 3*2.5$ m $= 7.5$ m, the total surface of the cell is 489.9 m$^2$ and with a soil depth of 0.15 m ($D_{coversoil}$) the soil volume is 73 m$^3$, corresponding to a volumetric ratio of soil cover to waste of 0.146 ($R_{C:W}$).

### 10.13.1.4    Waste Specific Weight

Specific weight or bulk density of the waste placed in the landfill is often determined using landfill tipping records and surveys; the tipping records are an account of the weight of waste received by the landfill and the landfill surveys indicate the volume in which the waste was placed. Landfills generally contain 15–20 % by volume of cover soil. The density of the cover soil is approximately 2600 kg/m$^3$ .Thus, when calculating the in-place specific weight of an entire landfill, both the soil and waste specific weights should be accounted for. Typical specific weights of landfilled waste (no cover soil), rang from 600 kg/m$^3$ for loosely compacted waste to more than 1000 kg/m$^3$ for compacted mineral waste as waste incineration bottom ash. In bioreactor landfills, the weight of added moisture also serves to increase the inplace specific weight. Compaction should be minimized when operating the landfill as a bioreactor to allow for adequate moisture movement throughout the waste mass.

### 10.13.1.5    Daily Cover

After the last load of waste is placed in the landfill for the day, it is recommended (and in many places required) that a layer of soil (10–25 cm) or alternative cover material be placed on the waste. This cover functions: (1) to control the amount of water infiltrating into the waste and thus reduces the early generation of leachate, (2) to minimize the amount of litter, dust and air pollution, and (3) to control the burrowing and emerging of small animals and insects, such as rodents and flies. Box 10.13.1 contains a detailed example of calculating the amount of daily cover soil required and the ratio of cover soil to waste in the landfill.

Soil is generally used as daily cover material, as it is effective at intercepting and retaining rainfall, minimizing excessive leachate production. When soil is unavailable or not economical for use as daily cover material, alternative cover materials may be used (Tchobanoglous *et al.*, 1993). Compost and mulch have been used quite successfully as daily cover material. Additionally, geosynthetic clay liners, old carpets, construction and demolition wastes and agricultural residues have been used as alternative cover materials. Table 10.13.1 contains generalized ratings of the suitability of different daily cover materials (Tchobanoglous *et al.*, 1993). The daily cover material must be sloped to promote surface water runoff. To operate a bioreactor landfill, selection of permeable or removable daily cover is essential and slopes should be provided to encourage retention of moisture. Box 10.13.1 contains information about how to optimize the design of a cell when using an alternative cover.

### 10.13.1.6    Temporary Roads

Quite often, temporary access roads are necessary to allow collection vehicles and other landfill equipment to continue filling of the landfill as the working face changes location. Road construction is usually completed by landfill employees

*Table 10.13.1*    *Suitability of various materials for daily cover material (Tchobanoglous* et al.*, 1993). Reprinted from Integrated Solid Waste Management – Engineering principles and management issues by G. Tchobanoglous, H. Thiesen and S. Vigil © (1993) McGraw-Hill.*

| Function | Yard waste – mulch | Yard waste – compost | MSW compost | Geosynthetic clay liner | Typical native soil | Clayey silty sand | Clay |
|---|---|---|---|---|---|---|---|
| Provides pleasing appearance and control blowing of paper | G–E | G–E | G | E | E | E | E[a] |
| Prevents rodents from burrowing | P | P | P | G-E | P | F–G | P |
| Keeps flies from emerging | F | F–G | F | E | G | P | E[a] |
| Minimizes water infiltration | P | G–E | F–G | E | F–G[b] | P | E[a] |
| Retains rainfall and snowmelt | P | G–E | F–G | G | F–G[b] | P | G[a] |
| Minimizes landfill gas from venting | P | P | P | F–G | P | P | P–F[a] |

E = excellent; G = good; F = fair; P = poor.
[a]Except when allowed to dry and cracks form.
[b]When a thick layer of soil is used, rating in G–E.

using equipment on-site. As the working face is moved, road location is changed accordingly. Often materials received at the landfill can be used to construct the road, such as concrete rubble.

### 10.13.1.7   Safety and Security

The health and safety of both the landfill employees and the general public is a critical factor to consider when operating a landfill. Potential hazards include contact with hazardous materials (pathogens, toxic air pollutants, asbestos), high accident risk, allergies due to environmental hazards (gas, dust, litter), and excessive noise and side effects caused by excessive vibration due to landfill equipment operation (Heasman, 1999; Wilhelm, 1989). Necessary protective equipment should be supplied to employees. Additionally, access to landfills is generally restricted (often surrounded by a fence) to control trespassing and unnecessary hazards to the general public. A health and safety plan must be developed and prominently displayed.

### 10.13.1.8   Waste Input Control

Waste input control is important when operating a landfill. Detailed records of the quantity of waste entering the landfill are necessary. Additionally, inspections of the waste must ensure no hazardous waste is being landfilled.

An entrance scale and gatehouse are necessary to determine the quantity of waste being disposed. Landfill employees operate the gatehouse, recording the weights of the incoming and outgoing trucks. The weights are tabulated to determine the amount of daily (and monthly, yearly) waste acceptance. The weights may also be used to calculate in-place densities to monitor the performance of the compaction process. Additionally, the weights may be used to charge individual agencies for their waste contribution to the landfill.

Load inspection is conducted by spreading out the contents of a collection vehicle to visually determine if any hazardous waste is present. This is usually completed near the working face. If hazardous wastes are found, they must be removed and it is generally up to the collection company to dispose of them. Additionally, a fine may be imposed on the collection company (Tchobanoglous *et al.*, 1993).

### 10.13.1.9   Stormwater Management

Because infiltration of all rainfall, stormwater runoff, intermittent streams and artesian springs add to the total amount of leachate generated and thus the amount that must be managed, an effective stormwater management plan is necessary for landfills. In addition, improper stormwater control can result in erosion of cover material. The most effective control measure is the use of a properly designed cover (or daily cover) with an appropriate slope (generally 3–5 %; Taylor, 2000). Runoff can then be effectively diverted off of the top of the landfill via concrete lined trapezoidal ditches. If operating the landfill as a bioreactor, it can be advantageous to design slopes to capture and retain stormwater, allowing for it to infiltrate into the waste and act as a moisture source.

## 10.13.2   Environmental Procedures

Because of the nature of the municipal waste managed at a landfill, nuisances result that must be controlled to protect public health, worker safety and environmental quality. These include noise, odor, litter, dust, fire, small animals and insects.

### 10.13.2.1   Noise Control

Noise may be excessive at landfill sites because of routine movement of collection trucks and the loud engines of the landfill equipment, causing a nuisance to both landfill employees and any residences located near the site. Thus, it is imperative that a noise control program be implemented (McBean *et al.*, 1995). Landfill employees should be equipped with hearing protection devices. To prevent noise from reaching adjacent residences, several control methods may be used including tree planting to create a buffer zone to mitigate noise, maintaining equipment properly, regulation of hours and ensuring the working face (the loudest area of the landfill) is located at a maximum distance from nearby residences (McBean *et al.*, 1995).

## 10.13.2.2   Odor Control

Odor control is another issue that must be addressed when operating a landfill. Odors exist due to putrescible wastes, disposed sludge, landfill gases and leachate seeps. The placement of daily cover, and even the immediate covering of materials that have an offensive odor, aid in controlling odor (McBean *et al.*, 1995). Chemicals can be employed to attenuate odor including ozone and mixtures containing plant oils and surfactants (O'Malley, 2003). Additionally, research has been conducted using antibacterial oils to inhibit microbial growth and thus odor (Otieno and Magagula, 1999). Odor neutralizing chemicals are often dispensed in a perimeter misting system. Also, air from enclosed areas may be vacuumed into chambers, where neutralizing chemicals may be applied prior to discharge of air (O'Malley, 2003). Employing an effective gas control system will also aid in mitigating odor. Collected gas can be deodorized via thermal oxidation, wet gas scrubbing, activated carbon filtration and biofiltration (Feinbaum, 2000; Frechen, 1989).

## 10.13.2.3   Fire Control

*In situ* landfill conditions are generally anaerobic. Because anaerobic processes are exothermic, as biological activity increases, internal temperatures also rise. Waste has high insulating properties, thus heat generally accumulates. *In situ* temperatures as high as 65 °C have been observed. At high temperatures, low flashpoint components can easily combust, starting a fire. Additionally, anaerobic processes generate methane. When methane concentrations of 5–15 % are present in air, it is explosive and thus can initiate internal fires (Tchobanoglous *et al.*, 1993). Generally, this is not an issue in younger landfills, as methane concentrations are greater and little air is present. However, in active landfills in which the gas extraction system is not properly maintained, oxygen intrusion may occur, resulting in an explosion hazard. Methane explosions can be a more prevalent issue in older landfills, after methane production has decreased and air begins to infiltrate into the landfill.

Subsurface fires are difficult to detect and often last for long periods of time. In areas void of oxygen, pryolysis occurs, resulting in what is referred to as glowing fires. In areas containing oxygen, fires are easily started and generally last until the oxygen is consumed. Conditions often causing fires include poorly engineered and maintained caps, inadequate compaction, and poor maintenance of gas systems (allowing oxygen to enter). Fires can be prevented by adequately maintaining and managing the gas collection system and cap and monitoring appropriate *in situ* temperatures (Lewicki, 1999).

## 10.13.2.4   Litter Control

Litter must also be controlled and is the main complaint of nearby residents. Litter results from delivery of uncovered loads, wind and operational practices. The most common control method is the placement of portable screens near the working face. Other mitigation methods may include more frequent covering of waste on windy days, requiring all delivered loads to be covered, collecting litter as necessary, unloading waste on a minimal surface area and operating the working face so as to minimize wind disturbance (creating it in the opposite direction of the wind). Whenever fences are used, frequent cleaning of the fences is needed in order to minimize their vulnerability to being damaged from strong winds.

## 10.13.2.5   Dust Control

Dust problems are caused by earthmoving and compaction activities, wind and traffic on roadways. Excessive dust can cause health problems for landfill employees (allergic reactions), increases in equipment maintenance costs and frequency and nuisances for nearby residents. The most common control measure is to spray access roads with water. Other methods to control dust include using dust-free roads (i.e. asphalt), enforcing speed limits via speed bumps, using vegetation to reduce wind speed at ground level and selecting an appropriate time to move soils to minimize the amount of dust (McBean *et al.*, 1995).

### 10.13.2.6   Small Animal and Insect Control

Small animals and insects such as birds, rodents, and flies can cause problems at landfills and thus must be controlled. Seagulls frequent landfills and may cause problems by picking up components of the waste and dropping them in nearby fields or surface waters (McBean *et al.*, 1995). Birds are particularity difficult to control. Methods that have been used include loud noises or recordings of birds in distress, netting systems, dogs, firing devices and chemical poisons. A landfill site in the UK has been experimenting with a netting system to totally enclose the filled landfill to protect it from birds (Terry and Willett, 1999). This technique has been shown to be quite effective; however, general upkeep of the netting systems can become costly.

Rodents are also problematic. They are sometimes brought to the landfill in waste loads and will remain if a food source and shelter are present. As with birds, rodents are difficult to control. The only method suggested is total elimination of the food source, which is difficult to achieve (McBean *et al.*, 1995). Larger animals such as pigs, donkeys, and deer can also be dangerous and disrupt the cover. Site-specific control measures should be implemented if disruption by larger animals is expected.

Insects, such as flies and mosquitoes, are also a problem. They spread disease and are usually brought in with the waste load. Covering the waste will help eliminate the food source. Extermination has also been effective, though measures should be taken to ensure the health of the employees is not compromised (McBean *et al.*, 1995).

## 10.13.3   Landfill Monitoring

Monitoring of landfills is necessary to ensure that landfill integrity is maintained and no release of harmful contaminants occurs, protecting both the environment and public health and safety. Monitoring during both pre- and postclosure periods is necessary and generally is designed to observe leachate and gas quality and migration. Regulations generally require monitoring of leachate (quantity, quality, leakage), local air quality, groundwater quality both upstream and downstream of the landfill and landfill performance parameters, such as head on the liner and cap integrity. The EU Landfill Directive in addition requires that meteorological data (precipitation, air temperature and wind) are monitored (or obtained from a nearby weather station) in order to provide data for water balance calculations.

In addition, as landfill design transitions to bioreactor technology, monitoring of *in situ* parameters such as temperature and moisture content is necessary for adequate system operation and process control. Bioreactor landfills are operated to accelerate stabilization of the waste by controlling the chemical and biological processes occurring within the landfill. Process control requires the observation of critical conditions and the ability to respond to changes in these conditions through adjustment of operating parameters. Box 10.13.2 presents a case study of *in situ* monitoring of head on the liner, temperature and moisture at a bioreactor landfill located in Florida (USA).

---

**Box 10.13.2    A Case Study: Landfill Monitoring at the New River Regional Landfill, Florida, USA (FCSHW, 2003). Reprinted from http://bioreactor.org/florida.html; Florida Center for Solid and Hazardous Waste (FCSHW) The bioreator demonstration project © (2003) FCSHW.**

The New River Regional Landfill (NRRL) in Florida, USA, hosted a bioreactor demonstration project. *In situ* monitoring was applied in this landfill for measuring many important parameters such as temperature, head on the liner, moisture content and gas composition. Leachate and gas flow rates and settlement were also monitored.

To measure the moisture content, temperature and gas, a moisture/temperature/gas (MTG) sensor was built to operate in conjunction with a data logger. The sensor's body is a 20-cm section of 5-cm diameter PVC well screen, described in more detail elsewhere (Gawande *et al.*, 2003). These sensors were installed at different depths within the 4-ha test area of the landfill. The sensors were designed to measure the electrical resistance of the moisture between two electrodes embedded in a granular insoluble media inside the MTG sensor. The resistance has been found to be inversely proportional to moisture content. A Type-T thermocouple was for temperature measurements. An example of a horizontal temperature profile (°C) obtained from these sensors is shown below for the upper 4 m of the landfill. The size of the area was 200 * 200 m. Gas was sampled through a rigid tube brought up to the surface and a portable gas meter was used to measure gas quality.

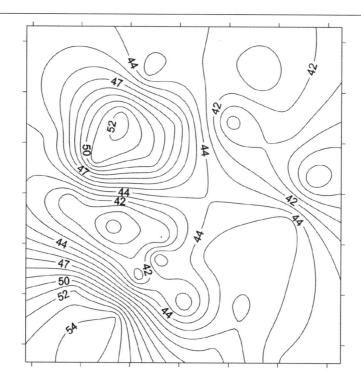

To measure head on the liner, pressure transducers were placed in the leachate collection system of two of the four cells operated at the time of the study. A total of 216 pressure transducers were installed in cells where waste had not yet been placed. Pressure measurements were taken using a standard differential voltage signal. The cables from each transducer ran outside of the landfill area and connected to a data logger and relay multiplexer. The data logger interpreted electrical signals emitted by the transducers and stored them. The multiplexer increased the number of transducers that can be scanned by the data logger from 16 to 64. A cellular phone was attached to the data logger to enable downloading data from long distances. A 12-volt DC battery that was continuously recharged by a solar panel powered the station. An example of a head on the liner profile (cm water) is shown below. The size of the area is approximately 30 * 30 m.

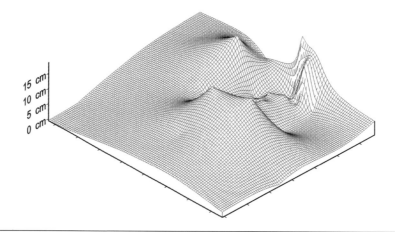

### 10.13.3.1 Leachate Monitoring

In order to minimize adverse environmental impacts from leachate, it is important to closely monitor leachate quality, quantity, leakage through the liner and outbreaks on side slopes. Landfill operators are generally responsible for ensuring leachate is managed effectively and the leachate collection and removal system is maintained throughout the design period of the landfill.

Leachate quality analyses are necessary to ensure leachate water quality standards are not violated. All analyses must be conducted on leachate sampled from a location at which the leachate is representative of the *in situ* landfill conditions, such as in the leachate collection and removal system, and not changed or altered as a result of sampling techniques. Common quality parameters measured include ammonia-N, chloride, COD, BOD, heavy metals, solids (total, suspended, and dissolved), pH, dissolved oxygen and conductivity (McBean *et al.*, 1995). In bioreactor landfills, leachate quality parameters are often used to aid in determining the extent of degradation occurring *in situ* and may define when operational changes should be made. Leachate quantity measurements should also be conducted. All leachate removed from the landfill must be tabulated on a daily basis.

Leachate leakage is also critical to monitor. In landfills with a double liner system the leachate collection system on the bottom liner is considered a leak detection system. The leak detection system generally consists of a highly permeable material, such as coarse sand, gravel, or geosynthetic, which allows for efficient drainage of all leachate collected. Leachate collected from this system is indicative of the integrity of the primary liner system. If significant amounts of leachate are collected, it can be concluded that the primary liner system has failed. It is important to note that double liner systems have been shown to be effective at prohibiting leachate leakage into the groundwater, thus use of such a liner system may in some instances replace the need for groundwater monitoring wells (Koerner, 2003).

Additionally, leachate leakage monitoring can be conducted in the zone between the landfill liner and the groundwater table. Direct monitoring may be conducted by placing lysimeters under the liner, allowing for collection of leachate samples. Indirect monitoring of leachate leakage includes instruments placed near or under the liner system that detect changes in moisture content or chemical concentration (salinity in most cases; Bagchi, 1994). It is recommended that lysimeter placement be under positions with maximum leakage potential and, at the same time, near the edge of the landfill to minimize the length of transfer pipes (Bagchi, 1994).

Monitoring of leachate seeps is also important. Leachate outbreaks may occur on side slopes because of local waste heterogeneities containing impermeable objects, impermeable cover soil placement, or leachate injection. Conducting routine visual inspections of side slopes is the most common detection method. If seeps are found, measures should be taken to install drains, slope cover material towards the inside of the landfill, reduce leachate injection volumes, or prevent infiltration of stormwater.

### 10.13.3.2 Air Monitoring

Monitoring of both air/gas quality and migration is important at and around landfill sites to protect the environment and to ensure public health and safety is not compromised. Escape and migration of gaseous contaminants from landfills can cause several negative environmental impacts such as explosions (if methane concentrations are 5–15 % in air), acute and chronic toxicity (especially if hydrogen sulfide is present), asphyxiation, smog formation, greenhouse effects, odors, vegetation damage and groundwater contamination. Air monitoring programs at landfills generally include monitoring of ambient air quality, extracted gases from the landfill, gases present in the vadose zone and any off gases from a treatment or energy recovery facility.

Ambient air quality is monitored to detect concentrations and migration of harmful gases, such as methane and hydrogen sulfide. Locations around the landfill can be equipped with explosive gas meters, hydrogen sulfide meters and sample collection equipment for samples to be taken at different locations around the site. Additionally, the ambient air quality in buildings located on-site must be measured routinely. Global positioning systems (GPS) can be used to aid in developing contour maps depicting plume concentrations of gases measured around the landfill site, allowing determination of gas migration patterns to be more accurate (Curro, 2000). Gas samples can be collected in either an evacuated canister, syringe, or air collection bag and stored for subsequent analysis by gas chromatography (Tchobanoglous *et al.*, 1993) or can be analyzed using a portable infrared gas meter.

Gas concentrations must also be monitored in the vadose zone to detect any lateral migration of gases and is often accomplished by using soil probes. The soil probes should be installed around the perimeter of the landfill and anywhere in which vegetation appears distressed. Landfill gas has a negative impact on vegetation growth, thus an indirect measure of gas migration may include a visual inspection of vegetation (Tchobanoglous *et al.*, 1993). Gas composition may be measured using portable equipment or by collecting a sample and storing for subsequent analysis via gas chromatography.

Extracted gas is monitored to quantify gas composition and to ensure that hazardous constituents are not exiting the landfill. Additionally, determining gas composition can aid in observing the extent of *in situ* waste degradation, particularly in a bioreactor landfill. Off gases are monitored at treatment and energy recovery facilities as well to ensure compliance with all air pollution control laws.

### 10.13.3.3   Groundwater Monitoring

Monitoring groundwater around landfill sites is necessary to aid in detecting leachate leakage of harmful contaminants (see leachate monitoring section). Typically, groundwater monitoring wells are positioned both upgradient and downgradient of the landfill. The exact number, location, and depth of wells is controlled by site hydrology and geology. The wells function to measure groundwater levels (to determine flow direction), test the hydraulic properties of the geological formation and measure groundwater quality (to determine if leakage has occurred; McBean *et al.*, 1995). Generally, after the site hydrology has been characterized, background measurements are taken from the upgradient wells and compared with measurements taken at the downgradient wells. Differences may be indicative of contamination by the landfill. Another approach is to monitor downgradient wells over time and use control charts to identify any change in quality over time caused by leachate contamination (Christensen *et al.*, 1992). Additionally, if a potential source of pollution is nearby, the region between that source and the landfill should be monitored.

Samples collected should be representative of the water in the formation, thus all wells should be purged prior to sample collection. In addition, all equipment should be cleaned and maintained to ensure samples are not contaminated during sampling. Sampling frequencies depend on factors such as goundwater velocity, the resource value of the groundwater, signs of contamination and local regulations.

Box 10.13.3 provides an example of groundwater monitoring.

---

**Box 10.13.3   Groundwater Monitoring. Reprinted from Christensen, T.H., Kjeldsen, P. and Jansen, J.L.C.: Groundwater control monitoring at sanitary landfills. Chapter 4.4. In: Christensen, T.H., Cossu, R. and Stegmann, R. (eds) Landfilling of waste: leachate, pp. 497–514. © (1992) CRC Press.**

Groundwater control monitoring is performed to ensure that the landfill as designed and operated does not affect the groundwater quality around the landfill. The groundwater control monitoring program attempts to detect an unexpected event (the landfill releases leachate into the groundwater) as early as possible (so that time is available to remediate the situation before too much groundwater is affected) at a reasonable cost (the monitoring may continue for decades after the landfill has stopped receiving waste).This suggests that a statistical approach focusing on indicator parameters should be used to signal an alarm (Christensen *et al.*, 1992). If an 'alarm' is observed, follow-up monitoring must be made to determine if the alarm is real (actual leachate has entered the groundwater) or a statistical error. If real, mapping of the leakage must be done quickly and remediation established.

Sampling of groundwater is the most common approach to groundwater control monitoring at landfills because of its simplicity and well established technology. Two main approaches exist: Down-gradient wells may be compared with up-gradient wells or time series in down-gradient wells may be monitored. Christensen *et al.* (1992) describe how a program based on time series can be established. Early in the life of the landfill, only limited data are available for calculation of the statistical parameters and hence for setting control limits in a control chart for each well and parameter. This suggests that the control program is adjusted as more data become available. Equally important is the selection of analytical parameters that have a large contrast between leachate and groundwater and that migrate without attenuation in order to obtain as early a warning as possible. Christensen *et al.* (1992) also point out the need for a contingency plan allowing a quick and targeted action when an alarm is signaled from the monitoring program.

The figure below, based on Christensen *et al.* (1992), shows a typical control chart (here specific conductivity in a selected well) with three control limits. Five consecutive observations above the lower alarm criteria [average (*x*) + one standard deviation (*s*)] identify a small and slow increase in a parameter, while the upper limit [average (*x*) + three standard deviations (3*s*)] quickly identify a large sudden event. It is suggested that 3–5 parameters are monitored in order to minimize the statistical error.

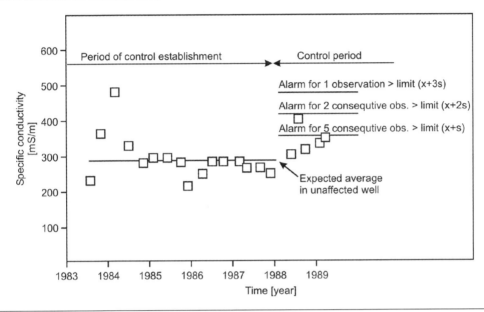

### 10.13.3.4   Head on the Liner Monitoring

Monitoring head on the liner is necessary to evaluate the performance of the leachate collection system. Although head on the liner measurements provide the most direct method of monitoring the performance of the leachate collection system, it is perhaps the most difficult parameter to monitor. The leachate collection system is the ultimate barrier between the environment and leachate produced within the landfill. The ability of the leachate collection system to protect groundwater from contamination is dependent primarily upon the structural integrity of the liner and the rate at which leachate can be removed from the landfill.

The primary techniques used to monitor leachate head on the liner include measurement of leachate levels in external leachate sumps, piezometers which reach the leachate collection system, pressure transducers installed in the leachate collection system and time domain reflectometry (TDR) sensors. Box 10.13.2 presents a case study in which pressure transducers were used to monitor head on the liner in a bioreactor landfill. Additional information describing head on the liner measurement can be found elsewhere (Reinhart and McCreanor, 1998).

### 10.13.3.5   Settlement Monitoring

Settlement is an important parameter to monitor because of its impact on landfill cap integrity. The cap functions to ensure long-term integrity and to support postclosure uses of the landfill site, such as maintaining a vegetative layer. The cap must be maintained to prevent loss of soil via an effective stormwater management plan.

Settlement is a result of waste decomposition or overburden pressures exerted on the landfill due to added moisture, waste, or cover soil. Settlement rarely occurs uniformly in the landfill, thus cracks and breaks in the cap may occur if there is excessive differential settlement. In bioreactor landfills, settlement measurement is more critical than in conventional landfills; as leachate is recirculated or moisture is added, significant amounts of settlement may occur, adversely impacting injection and gas collection systems. Additionally, settlement may be used as an indirect measure of waste degradation and thus bioreactor landfill performance. The extent of settlement is dependent on compaction, characteristics of the waste, amount of decomposition, amount of water in the landfill and height of the landfill. Generally, the majority of settlement occurs during the first five years of landfill operation (Tchobanoglous *et al.*, 1993). A case study illustrating typical settlements measured in bioreactor landfills is presented in Box 10.13.4.

Settlement can be measured using a variety of techniques. The most common include settlement plates, GPS surveying and flyover aerial photography. Routine surveys of the landfill surface using the GPS technology is becoming more common and can be used to accurately determine the amount of settlement. Depending on the number of survey points used, both differential and average settlements can be measured with great accuracy.

---

**Box 10.13.4    A Case Study of Settlement at a Pilot-Scale Landfill in California, USA. Reprinted with permission from Monitoring the performance of anaerobic landfill cells with fluids recirculation Final report to the Western Regional Biomass Energy Program, Yolo County, Planning and Public Works Department, Woodland, CA, USA by R. Yazdani, J. Kieffer, H. Akau and D. Augenstein © (2003) Ramin Yazdani.**

Settlement is an important parameter to monitor as it indirectly measures the integrity of the final cap on top of the landfill. Additionally, as the waste settles an air space may be recovered and subsequently used to add additional waste to the landfill. A case study of a landfill, measuring settlement over time was presented.

Yolo County, California, is operating a pilot-scale landfill project to evaluate the bioreactor landfill concept (Yadzini *et al.*, 2003). Bioreactor landfills were operated to accelerate stabilization of the waste by controlling the chemical and biological processes occurring within the landfill by adding liquids and/or amendments to the landfill. The pilot-scale project consisted of two, 930-m$^2$ test cells that began operation in 1996. The combined capacity of the cells was approximately 8000 t of waste. In one cell liquids were added (well water) and leachate recirculated. This cell is referred to as the enhanced cell. The second cell was operated similarly to the enhanced cell, except no liquids or leachate was recirculated. Cell two is referred to as the control cell.

Settlement was observed in both cells, as depicted below (Yadzini *et al.*, 2003). As shown, settlement began almost immediately in the enhanced cell, which was expected because of the liquids being added. Based on data to date, the enhanced cell settled four times more than the control cell. The enhanced cell settled approximately 17.7 % of the initial height (almost 4 m) in over 6 years. The control cell only settled 4.0 % (almost 1 m) in 6 years. Settlement is still being observed in the enhanced cell, though the settlement rate is decreasing.

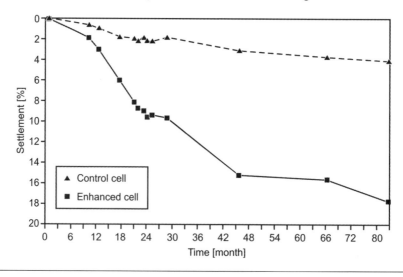

### 10.13.3.6 Temperature Monitoring

Temperature monitoring of *in situ* landfill conditions in conventional landfills is conducted primarily to aid in preventing or detecting subsurface fires. Knowledge of *in situ* temperature profiles can provide landfill operators an indication of what areas of the landfill may be reaching combustible temperatures so prevention measures can be implemented. In addition, temperature monitoring can detect locations of subsurface fires. In bioreactor landfills, temperature monitoring is much more critical, as it is used to monitor and control the biological processes occurring *in situ*. Because waste degradation involves biochemical reactions producing significant amounts of heat, temperature can be used as an indirect measure of waste degradation; higher temperatures indicating an area containing more rapidly degrading waste. The temperature within a landfill is determined by a balance between heat production during the biological degradation of organic waste fractions and the loss of heat to the surrounding soil and atmosphere (Rees, 1980). The microbial processes are capable of significant heat generation, particularly at higher moisture conditions.

Temperature is a relatively simple and inexpensive parameter to monitor. Common temperature measuring devices used in landfills include thermocouples and thermistors. Thermocouples are pairs of dissimilar metal wires joined at one end. A temperature difference between the paired wires generates a net thermoelectric voltage. This voltage is proportional to the temperature difference as well as characteristics of the wire pair, as where thermistors measure temperature as a function of the change in electrical resistance of materials as a result of temperature changes. They are composed of inexpensive semiconductor materials. In order to function, a small electric current is applied which can generate sufficient heat to affect the temperature of the surrounding material. They are accurate over a smaller range than thermocouples, however they are quite stable and accurate over the temperature ranges expected in landfills. Thermocouples tend to be more rugged and less expensive than thermistors, however they are slightly less accurate.

### 10.13.3.7 Moisture Monitoring

Moisture content control is the most critical parameter for successful bioreactor landfill operation, thus its measurement is necessary. Spatial distribution of *in situ* moisture can be determined by removal of solid waste samples and using traditional gravimetric techniques or by placing sensors inside the landfill to detect the spatial distribution of water. Quite often, using a drill rig and auger to remove a solid waste sample is costly and difficult to accomplish. Waste heterogeneities make drilling in some locations difficult. Additionally, as the waste sample is removed, water loss may occur and result in inaccurate moisture content measurement.

Placing water content sensors *in situ* is advantageous, as destructive sampling is not required and measurement can be continuous. *In situ* moisture content devices available for use in landfills were developed primarily for use in the agricultural industry for irrigation purposes. Landfills, by their nature, pose severe challenges to the application of these devices. Although each technology has its unique application difficulties when applied within landfills a universal problem is that of the heterogeneous nature of landfills as compared with the farm soils for which the sensors were originally designed. The most commonly used devices in landfills include, time domain reflectometry (TDR; Davis and Annan, 1977; Topp *et al.*, 1980), neutron probes (Evett, 1998), electrical resistance (McCann *et al.*, 1992), capacitance probes (Dean et *al.*, 1987) and time domain transmissometry (TDT). A more complete description and discussion can be found elsewhere (Thomas, 2001).

## References

Bader, C.D. (2003): The articulated dump truck: Workhorse of the landfill. *MSW Management*, 13 (2), 56–61.
Bagchi, A. (1994): *Design, construction and monitoring of landfills*, John Wiley & Sons, Ltd, New York, USA.
Bolten, N. (1998): Taking it beyond compaction. *MSW Management*, 8 (2), 12–14.
Bolten, N. (2000a): When digging beats scraping. *MSW Management*, 10 (1), 7.
Bolten, N. (2000b): Landfill airspace and waste density: The big picture. *MSW Management*, 10, 4.

Christensen, T.H., Kjeldsen, P. and Jansen, J.L.C. (1992): Groundwater control monitoring at sanitary landfills. Chapter 4.4. In: Christensen, T.H., Cossu, R. and Stegmann, R. (eds) *Landfilling of waste: leachate*, pp. 497–514. Elsevier Applied Science, London, UK.

Curro, J. (2000): Using the global positioning system in landfill gas system monitoring, *MSW Management*, 10, 7.

Davis, J.L. and Annan, A.P. (1977): Electromagnetic detection of soil moisture progress report I. *Canadian Journal of Remote Sensing*, 3, 76–86.

Dean, T.J., Bell, J.P. and Baty, A.J.B. (1987): Soil moisture measurement by an improved capacitance technique, Part 1. Sensor design and performance. *Journal of Hydrology*, 93, 67–78.

Evett, S.R. (1998): Some aspects of time domain reflectometry, neutron scattering, and capacitance methods for soil water content measurement. In: IAEA (ed.) *Comparison of soil water measurement using the neutron scattering, time domain reflectometry and capacitance methods*, pp. 5–49. IAEA Tecdoc 2000. IAEA, Vienna, Austria.

FCSHW (2003): *The bioreactor demonstration project*. Florida Center for Solid and Hazardous Waste. Located on the internet: 19 October 2009. http://bioreactor.org/florida.html.

Feinbaum, R. (2000): Odor control, going the extra mile. *MSW Management*, 10, 2.

Frechen, F.B. (1989): Odour emissions and controls. In: Christensen, T.H., Cossu, R. and Stegmann, R. (eds) *Sanitary landfilling: process, technology and environmental impact*, pp. 509–519. Academic Press, London, UK.

Gawande, N.A., Reinhart, D.R., McCreanor, P., Townsend, T. and Thomas, P. (2003): Municipal solid waste *in situ* moisture content measurement using an electrical resistance sensor design and operation of bioreactor landfills. *Waste Management*, 23, 667–674.

Heasman, L. (1999): The health effects of controlled landfill sites – an overview. In: CISA (ed.) *Proceedings of the Seventh International Landfill Symposium*, Vol. II, pp. 675–678. CISA, Environmental Sanitary Engineering Centre, Cagliari, Italy.

Koerner, R. (2003): Linear system design and materials. *USEPA Workshop on Bioreactor Landfills*. CD-ROM. USEPA, Arlington, USA.

Lewicki, R. (1999): Early detection and prevention of landfill fires. In: CISA (ed.) *Proceedings of the Seventh International Landfill Symposium*, Vol. II, pp. 613–620. CISA, Environmental Sanitary Engineering Centre, Cagliari, Italy.

McBean, E.A., Rovers, F.A. and Farquhar, G.J. (1995): *Solid waste landfill engineering and design*. Prentice Hall PTR, Englewood Cliffs, USA.

McCann, I.R., Kincaid, D.C. and Wang, D. (1992): Operational characteristics of the watermark model 200 soil water potential sensor for irrigation management. *Applied Engineering in Agriculture*, 5, 603–609.

O'Malley, P.G. (2003): Controlling odor at transfer stations and MRFs: there's more than one way to reduce complaints. *MSW Management*, 13 (2), 44–52.

Otieno, F. and Magagula, C.S. (1999): Management strategies of odour problems at landfill sites. In: CISA (ed.) *Proceedings of the Seventh International Landfill Symposium*, Vol. II, pp. 589–596. CISA, Environmental Sanitary Engineering Centre, Cagliari, Italy.

Rees, J.F. (1980): Optimization of methane production and refuse decomposition in landfills by temperature control. *Journal of Chemical Technology and Biotechnology*, 30, 458–465.

Reinhart, D. R. and McCreanor, P.T. (1998): *Prediction and measurement of leachate head on landfill liners*. Report 98-3. Florida Center for Solid and Hazardous Waste Management, Gainesville, USA.

Taylor, M.A. (2000): Landfill final-cover storm drainage. *MSW Management*, 10, 1.

Tchobanoglous, G., Theisen, H. and Vigil, S. (1993): *Integrated solid waste management: engineering principles and management issues*. McGraw Hill, Inc., New York, USA.

Terry, R.J. and Willett, D.B. (1999): Bird control at landfill sites: total enclosure netting systems. In: CISA (ed.) *Proceedings of the Seventh International Landfill Symposium*, Vol. II, pp. 581–587. CISA, Environmental Sanitary Engineering Centre, Cagliari, Italy.

Thomas, P.A. (2001): *The testing and evaluation of a prototype sensor for the measurement of moisture content in bioreactor landfills*. MS thesis, University of Central Florida, Orlando, USA.

Topp, G.C., Davis, J.L. and Annan, A.P. (1980): Electromagnetic determination of soil-water content: measurement in coaxial transmission lines. *Water Resources Research*, 16, 574–582.

Vesilind, P.A., Worrell, W. and Reinhart, D.R. (2002): *Solid waste engineering*. Brooks/Cole, Pacific Grove, USA.

Wilhelm, V. (1989): Occupational safety at landfills. In: Christensen, T.H., Cossu, R. and Stegmann, R. (eds) *Sanitary landfilling: Process, technology and environmental impact*, pp. 509–519. Academic Press, London, UK.

Yadzini, R., Kieffer, J., Akau, H. and Augenstein, D. (2003): *Monitoring the performance of anaerobic landfill cells with fluids recirculation*. Final report to the Western Regional Biomass Energy Program. Planning and Public Works Department, Yolo County, Woodland, USA.

# 10.14

# Landfill Closure, Aftercare and Final Use

**Heijo Scharff**

*NV Afvalzorg Holding, Assendelft, The Netherlands*

Landfill management does not stop at termination of waste acceptance and placement. Before a landfill can be abandoned safely or returned to society a top cover needs to be constructed and financial provisions for aftercare need to be safeguarded. Closure is not the moment that the gate is closed and waste processing is stopped. Closure as defined in regulations is the moment that the competent authority has concluded that the operator has fulfilled all the permit requirements concerning environmental protection measures and provisions for aftercare. It takes several years before all the protection measures at the top of the landfill are installed. Landfills can be closed entirely or in distinct phases. The engineering of the top cover, the nature of the aftercare and the financial provisions depend on the type of final use that is selected for the landfill. Often landfill sites also have old sections that do not meet current standards. Local regulations do not always require remediation to current standards. These old landfill sections are not dealt with here.

## 10.14.1   Regulatory Framework

Closure and aftercare of landfills are often specified by national regulation. The current European regulatory framework concerning closure and aftercare of landfills as described in the European Landfill Directive (CEC, 1999) are briefly presented here, offering insight into the many issues involved in landfill closure and aftercare.

Annex I of the European Landfill Directive provides the technical requirements of the top cover, which is the main technical feature of landfill closure. It states (Article 2) that 'measures shall be taken' … 'to control water from precipitations entering into the waste body'. It is important to note the difference between control and prevent. In a climate with an annual precipitation of 800 mm control can also be realised with a suitable soil cover with adequate vegetation that reduces the infiltration to for example 100 mm/year. Annex I (Article 3.1) furthermore states that 'protection of soil, groundwater and surface water is to be achieved by' … 'a combination of a geological barrier and a top liner during the passive phase/postclosure'. The European Landfill Directive does not clearly define what a top liner is. Annex I (Article 3.3) does however mention that 'if the competent authority after consideration of the potential hazards to the environment finds that the prevention of leachate is necessary, a surface sealing may be prescribed'. Thus the European Landfill Directive offers some flexibility to design isolation measures according to environmental risk at each specific site. If the potential impact due to the nature of the waste or level of stabilisation of the waste is low, then infiltration of a certain amount of precipitation will not harm the environment. Annex I (Article 3.3) provides recommendations for surface sealing construction should the competent authority deem such a surface sealing necessary. Unfortunately most of the European member states seem to have interpreted these recommendations as minimum requirements. A majority of the national regulations in Europe has made surface sealing mandatory on both hazardous and nonhazardous waste landfills. These regulations require that such a surface sealing consists of an 'artificial sealing liner' and an 'impermeable mineral liner'.

Concerning capping and aftercare the Landfill Directive [Article 8(a)(iv)] requires 'that adequate provisions, by way of financial security or any other equivalent' … will be made … 'to ensure that the obligations (including aftercare provisions) arising under the permit' are met, 'and that the closure procedures required by Article 13 are followed'. Article 10 requires that the financial provisions shall cover the cost of closure and aftercare for a period of at least 30 years. Article 13(b) states that 'a landfill or part of it may only be considered as definitely closed after the competent authority has carried out a final on-site inspection, has assessed all the reports submitted by the operator and has communicated to the operator its approval for closure'. Article 13(c) requires that 'after a landfill has been definitely closed, the operator shall be responsible for its maintenance, monitoring and control in the aftercare phase for as long as may be required by the competent authority, …'. Finally Article 13(d) requires that the operator shall be held responsible as long as the competent authority considers the landfill likely to cause a hazard to the environment.

The moment the operator is no longer held responsible or can be released from aftercare is often called completion. Article 13(d) implies that the competent authority needs to be convinced that environmental risks no longer exist before the operator can be released from its responsibilities. In reality the aftercare period can be and most likely will be much longer than 30 years. For operators it is very uncertain for how long the financial provisions need to be made.

The requirements for monitoring procedures are given in Annex III of the European Landfill Directive. The obligations to collect and report data include settlement of the landfill, meteorological data, potential gas emission, leachate volume and composition, surface water volume and composition and groundwater level and composition.

## 10.14.2   Closure

Closure of a landfill usually involves placement of a top cover, establishing of vegetation on the site, securing permanent installations and decommissioning of redundant structures.

### 10.14.2.1   Top Cover Placement

After termination of waste acceptance and placement the landfill is capped with a temporary cover. This is usually a locally available soil in a layer of 0.3–1.0 m thickness. The final cover, in most cases a combination top cover consisting of a mineral liner and a high density polyethylene (HDPE) membrane, can not be installed immediately. Final top covers are described in Chapter 10.9. Several years (maybe 7–10 years) after the landfill (cell) has reached its final volume significant settlement may still occur. When the settlement is irregular and not evenly distributed over the surface, damage to the top cover construction may occur; see Figure 10.14.1. Membranes can be ripped. Cracks in mineral liner can occur. Mineral liners are supposed to be 'self-healing'; however, if the crack is too big, mineral liners are not able to heal. In practice it has been observed that cracks fill up with drainage sand overlying the liner or that plant roots invade the cavities that arise (Rahmke, 2006). When this happens the result is a permanent leak in the mineral liner. After termination of waste placement it is therefore good practice to first be patient and follow the development of settlements.

Settlement refers to the overall volume reduction in the landfill body. Settlement should not be underestimated. For municipal solid waste landfills containing a lot of biodegradable material total settlement can range between 25 and 50 % of the initial fill height (Wall and Zeiss, 1995). Settlement can be due to compression of the soil on which the landfill is situated and to degradation and compression of the waste itself. Weak clay and peat are soil materials that can be compressed considerably by the weight of overlying material. Specific measures prior to the construction of the bottom liner can be carried out to reduce this type of settlement. The volume reduction of the landfill body is caused by the combined effect of compaction during placement and the mass of overlying waste. It strongly depends on the nature of the waste. Secondary settlement in the landfill body is caused by a combination of mechanical creep, physicochemical

**Figure 10.14.1**   *Cracks in the soil cover and damage of the surface sealing of Nuenen Landfill, The Netherlands.*

corrosion and biodegradation (McDougall and Pyrah, 2001). The effect of degradation is highest for waste that contains a high percentage of biodegradable material.

According to the European Landfill Directive (Annex III, Article 5) height measurements should be carried out annually. In practice very few landfill operators actually perform settlement measurements prior to installing a permanent surface sealing. Settlement of the top of the waste can be measured by installing permanent measurement points in the soil cover. This can be a simple concrete tile on top of the soil or a 1 × 1 m steel plate and pipe that is installed at the boundary of soil and waste. The measurement itself can be carried out with the well known surveyor's levelling instrument or the theodolite. The settlement of the soil beneath the liner can be measured in a variety of ways. A simple approach consists of installing a 2 × 2 m reinforced concrete slab on the drainage layer of the bottom liner. Steel pipes with a known length are attached to the concrete slab and periodically extended with the increasing height of the waste. The measurement of the height of the pipes and comparison with a known level outside the landfill indicates the settlement of the subsoil. A disadvantage is that the pipes have a limited lifetime due to corrosion of the steel. Plastic pipes can not be used due to their flexibility especially under increased temperatures that occur in landfills with active biodegradation. Another approach is to insert a pressure sensor into the leachate drains. The pressure difference with a sensor at the bottom of the drainage system collection well indicates the difference in height. Cable length measurement or GPS data enable comparison of the measurement data with the designed height of the drainage system. The data collected gradually shows less and less settlement. There are no guidelines for acceptable settlement, but it can be considered safe to install a surface sealing when the settlement has reached values <5 cm/year. Figure 10.14.2 shows annual settlements (5–30 cm/year) monitored over 7 years at a Dutch landfill.

If settlement occurs gradually in the same rate over the entire surface, it will not damage the surface sealing. The real threat for the liner system is differential settlement. Data collection and evaluation aiming at verifying differential settlement over small distances is quiet laborious. Therefore in practice it is easier to follow the general settlement in the landfill. If the general settlement itself has decreased to a very low level, then differential settlement is small as well.

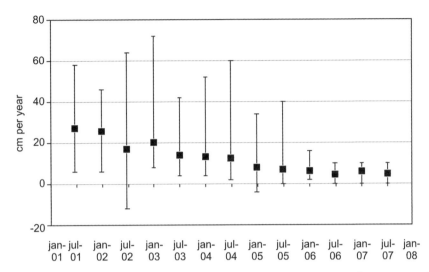

**Figure 10.14.2** *Settlement (average and range of 17 measurements points) on a 22-ha commercial waste landfill (Wieringermeer, The Netherlands) closed in 2000. As it is very unlikely that the level increases, the observed negative settlements are probably due to measurement errors.*

### 10.14.2.2 Vegetation

Revegetation of completed landfills is essential in order to adapt the site to the surrounding environment, to improve public acceptance, to minimise erosion on slopes and to minimise leachate production by increasing evapotranspiration (Neumann and Christensen, 1996). The main aspects to observe for revegetation are gas migration control, soil cover, plant species selection and planting strategies for woody species.

***Vegetation Damage.*** Vegetation on temporary covers or final covers without gas- and watertight lining can become exposed to landfill gas. Vegetation damage due to landfill gas occurs frequently, especially during and shortly after the operational period. Closed landfills in general have properly developed vegetation covers and extensive damage is considered an exception (Neumann and Christensen, 1996). Landfill gas tends to migrate through soil covers through preferential pathways. Very often these are cracks in the soil that arise during a dry period. Close to such a preferential pathway exposure of plants to landfill gas can exceed the plant's tolerance. In the areas between such pathways the migration of landfill gas can often be compensated by microbial oxidation of landfill gas in the soil cover and diffusion of atmospheric air into the soil.

Gendebien *et al.* (1992) found 31 cases of vegetation damage described in literature. Vegetation damage can be observed as dwarf growth, superficial root development, dying leaves, dying branches or plant death. The damage to plants is caused by migration of landfill gas into the root zone and displacement of soil air. This usually results in depletion of oxygen and consequently anoxic conditions in the soil air. The plant may be affected by asphyxiation ('suffocation' due to the lack of oxygen), by the presence of toxic gasses or by changes in pH and composition of the soil pore water. The effect of these aspects may be increased by external stress to the plant such as drought and strong wind.

Trace gasses in the landfill gas may potentially be toxic. This effect has however not been convincingly demonstrated. Asphyxiation is considered a much more dominant aspect (Gendebien *et al.*, 1992). Methane is not considered toxic to plants. The microbial oxidation of methane leads to oxygen depletion in the soil. This adds to the effect of soil air displacement by landfill gas migration. Most plants normally grow at 5–10 % oxygen in the soil air (Gendebien *et al.*, 1992). Several woody species are more demanding and require 12–14 % oxygen. In a landfill cover with a significant landfill gas flux, oxygen sufficient for plant growth may be found only within shallow depths and the roots will not penetrate deep in to the soil. In such a situation there will be limited access to water and nutrients. Plants have a very variable tolerance to carbon dioxide (Gendebien *et al.*, 1992). Most plants show no growth inhibition up to 5 % carbon dioxide in the soil air. The indirect effect of carbon dioxide may be decreased pH and changes in composition of soil pore water.

***Prevention of Vegetation Damage.***   Gas migration control in terms of extraction of gas is necessary to prevent substantial amounts of landfill gas from migrating into the root zone. Depending on the type and age of the landfill active or passive measures could be appropriate. When the gas production has decreased to 100–200 m³ LFG/h or the methane content drops below 50 %, utilisation of active measures are in general no longer considered feasible. Flaring is often considered feasible above 50 m³ LFG/h at 30 % methane. Recently flares have been developed for much lower methane (10–15 %) concentrations (e.g. Jacobs and Scharff, 2003; Steinbrecht and Spiegelberg, 2007). The lower limit for commercially available flares for low calorific landfill gas is 10 m³ methane/h. For such amounts of landfill gas passive measures including microbial oxidation in biofilters, biowindows or oxidising covers are also an option (e.g. Gebert, 2004; Humer, 2001; Kjeldsen, 2007).

Another important aspect for prevention of vegetation damage is the depth, structure and composition of the cover soil. The depth at which microbial oxidation occurs in cover soils varies with varying conditions and is in general between 0.4 and 1.0 m below the surface. Deep rooting plants can therefore not be applied on methane oxidising cover soils. The roots could also create preferential pathways and result in local landfill gas emissions. Grass covers require a minimum of 0.5 m soil and trees require a minimum of 1.5 m soil for proper root development (Neumann and Christensen, 1996). Methane oxidising covers can therefore only be combined with grass vegetation. In addition the recultivation soil should provide suitable structure, sufficient water storage capacity and sufficient nutrients. Plants cannot grow without nutrients. Sufficient water storage is necessary for plants in order to survive dry periods. Structure is necessary for soil aeration. Too much clay can hamper soil aeration and result in cracks and consequently preferential pathways for landfill gas emission during dry periods. Too much sand on the other hand could result in insufficient nutrients and water storage capacity.

***Selection of Vegetation.***   For temporary covers grasses and herbs provide suitable vegetation. A temporary cover needs to be removed in order to install the permanent mineral liners and or membranes. A vegetation of grasses and/or herbs does not require removal. The cover soil can be stockpiled and re-applied on the permanent liner including the remains of grasses and herbs.

The desired vegetation strongly depends on the nature of the final use selected for the landfill. A park with intensive use will have a different vegetation than a landfill that is not intended to be used, but just fitted into the natural environment. It is sometimes considered (but not often executed) to give the completed landfill an economical destination by growing energy crops or fibre crops. Species selection should always include consideration of the local conditions such as climate, soil types, depth of the soil layer and wind-exposed areas (Neumann and Christensen, 1996). Alternatively the soil that is present could be replaced or improved in order to be able to support the vegetation of choice. A well known example is the very precise soil mixture that is required to construct greens on golf courses. In the absence of a gas- and watertight liner the plant species should also be tolerant to low oxygen concentrations as such conditions might develop due to landfill gas migration and oxidation.

Stability to strong wind can be improved if the landfill vegetation is staggered. That is bushes are planted at the wind front, small trees like *Carpinus betulus* behind, and further behind larger trees like *Betula pendula*, *Alnus incana*, *Robinia pseudacacia* or *Acer pseudoplatanus* dependent on local climate. Thus the wind forces on the individual trees are minimised. It may be worth to consider not planting trees on the top of hilly landfills. On the top it may be difficult to ensure stability towards the wind. Increasing the thickness of the recultivation soil locally may improve stability. Areas with bushes and dry meadows, that may develop on top of completed landfills, are today rare elements in the cultivated landscape of Western Europe. From an ecological point of view they can provide an important habitat for wild animals and flora (Neumann and Christensen, 1996).

### 10.14.2.3   Permanent Installations

On the completed landfill several installations have to remain intact and operational. These installations may include leachate collection pits, leachate treatment plants, piezometers, gas wells, gas manifolds, condensate traps, flares and biofilters. When a gas- and watertight liner is installed, transits through the liner for these installations have to be accounted for. In general existing installations are not designed to allow for the liner to be installed on and around it. The installations usually have to be redesigned and refitted. It is considered good practice to try to minimise the transits through the liner.

In redesigning it is also recommended to consider the final use of the landfill. This requires close cooperation between the engineers and the landscape architect. Thus installations can be located where they are least conspicuous. Piezometers and gas wells can be equipped with lids at 'grass root level'. Gas manifolds with control valves can be located in shrubberies. A flare can be located behind a group of trees. At the same time the installations should be accessible for the people that operate, maintain and monitor them. On those sites that are freely accessible to the general public the installations or the access to the installations also needs to be 'vandal proof' in order to guarantee continued functionality.

### 10.14.2.4   Decommissioning

Various structures on a landfill loose their functionality when waste disposal ends. Some of these structures may still be useful during closure. A weighbridge may be used for establishing the correct amount of construction and capping materials. The site office, canteen and sanitary facilities can be used by the contractor responsible for capping and landscaping. It should not be excluded that a building or the foundation of a building can have a function for the new destination of the landfill. Material and equipment sheds for instance can possibly be used for maintenance of the landscape of the new destination. In case the operational period of the landfill is relatively short, it could be considered to install weighbridges and build offices that can be moved to a new location. This is not possible for foundations. Structures that become obsolete should be torn down and taken away. Fencing can be taken down in case the new destination offers unlimited access to the general public. This is however not always the case. For instance in Germany, landfills often are not opened to the public. In this case the fencing needs to be maintained.

## 10.14.3   Final Use

### 10.14.3.1   Final Use Planning

Final use of a landfill in most countries in Europe is essentially determined by the spatial planning scheme of the municipality in which the landfill is located. The destination as defined in the planning scheme is often waste management. Only rarely a change in use has already been accounted for in the planning scheme. In addition to the municipality, other planning authorities could be of influence. National authorities often have bills on spatial planning. Regional authorities often have issued regional planning schemes. In addition other authorities such as water boards can have approved planning schemes concerning water quantity and quality control that could limit future activities on landfills. In most cases the existing municipal planning scheme needs adaptation after closure of the landfill. This adaptation has to be in compliance with the planning intentions and restrictions imposed by other authorities. In general the owner of the landfill must take the initiative to propose a final use for the landfill and start the process to adapt the local planning scheme. This requires consultation of and cooperation with the competent authorities. This is a time-consuming process and should be started well in advance of the closure procedure.

In addition to the authorities it can also be considered to involve the local population in the selection of a final use for a landfill. This can be of particular importance if a village is located nearby or if due to construction of new neighbourhoods a city landfill over the years has become surrounded by housing areas.

The most common types of final use for landfills are no use or use as a park or recreational area. Use as a park in general is limited to planting of vegetation and construction of foot and cycle paths. There is a growing interest in recreational facilities close to housing areas. This extends beyond an afternoon stroll and walking the dog. Closed landfills provide excellent opportunities to create facilities for running, mountain biking, horse riding, (cross country) skiing and many more activities. In many countries there also is a large demand for golf courses. Landfills can be an excellent basis for construction of golf courses. The golf courses and other activities can provide a basis to operate cafés and restaurants.

Many landfills are, at the point of closure, located close to urban areas where it often is difficult to find suitable land for recreational activities and for development. This will result in higher land prices and reduce the public resistance against building offices and houses on closed landfills. The challenge of construction on landfills is in realising an effective protection against potential landfill emissions and to find construction methods that do not require foundation with piles. Piles would perforate the liners and are therefore unacceptable. Liner construction technology is sufficiently evolved to provide adequate security. Recent projects, for example, involving indoor ski halls or office buildings have shown that it is technically possible and environmentally acceptable to construct buildings on landfills.

Selection of a final use that is in high demand could benefit many. The general public can have attractive facilities close by. Authorities can be offered solutions for the high pressure on scarce land. The developer/owner of the final use will usually have a lower landlease than outside the landfill. The landfill operator receives a lease and can thus (partly) finance the cost of site maintenance. Finally the landfill supports an interest of society, which benefits the image of the landfill operator.

### 10.14.3.2   Precautions Regarding Final Use

The most important precaution to consider is effective prevention of contact between the user of the site and the waste. A 1-m thickness soil layer can in general be considered sufficient to make it very unlikely that contact with the waste occurs. Fifteen years experience on the Velsen landfill (see Box 10.14.1) supports this position. Parts of the landfill cover prone

---

**Box 10.14.1   Final Use of The Velsen Landfill, The Netherlands.**

The Velsen Landfill was in operation between 1977 and 1992. Mainly commercial waste and construction and demolition waste were landfilled. The landfill was located in an area with clay soil in a polder below sea level. The cover consists of at least 1 m of top soil. The landfill was designed as an attractive undulating hilly landscape. A landmark was constructed using large items from construction and demolition waste. In the middle of this material two huge concrete circular plates were erected with a staircase in between. Also a concrete cast of a mountain wall from the Belgian Ardennes was build. This wall is used to practice climbing. Apart from walkways, bridle paths and cycle paths, the site has a mountain bike track, which is frequently used for the national championship. The initial design included an outdoor ski slope. In recent years that has been replaced by an indoor ski slope with artificial snow. Every year in August the natural amphitheatres of the hills are used to erect stages for a three-day dance event. This event alone attracts around 40 000 visitors, with a total of 400 000 visitors annually.

***Figure 10.14.3*** *Ventilation underneath an office building on a landfill (Nauerna, The Netherlands).*

to erosion have shown to be downhill sections of the mountain bike track (braking) and storm water drains. Unacceptable erosion can be prevented with regular inspection and maintenance. On landfills with a surface sealing system the chances of contact with waste are negligible. Even with a shovel it is quiet an effort to brake through a combination surface sealing.

Contact with leachate or landfill gas should also be prevented. Leachate and landfill gas need to be controlled as part of the aftercare requirements. The measures to protect the environment simultaneously act as prevention against human exposure. As leachate has a tendency to migrate downwards and the user is on top of the landfill, it is very unlikely that the user will be exposed to leachate. Exposure to landfill gas on landfills capped with at least 1 m of soil and equipped with an active landfill gas extraction is in general not posing any threat to human health or the environment. The amount of landfill gas that could escape through a crack is very small and dilution under normal weather conditions is in the order of 40 000 to 600 000 times within 1–2 m of the crack (Young and Heasman, 1985). But even with active extraction measures landfill gas can migrate into confined spaces. Build up of methane in confined spaces poses a threat that should receive a lot of attention. There are many recorded incidents of fire or explosions due to slow build up of the concentration of methane in wells, closets, basements, etc. followed by exposure to a source of ignition (e.g. electrical switch, cigarette). The most reliable way to prevent migration of methane into a building on a landfill is to provide free ventilation of atmospheric air under the building. This could be done with a layer of aggregate of sufficient thickness and porosity under the foundation or with a building on piles where the area underneath the building could be used as car park (see Figure 10.14.3).

Attention should also be paid to damage caused by the landfill to the building. This is mainly due to settlement. Settlement can be accelerated prior to construction by means of temporarily storing of a couple of metres of soil on top of the waste. Various buildings that have been constructed on landfills have been equipped with jacks in order to correct differential settlements (see Box 10.14.2).

**Box 10.14.2   Final Use of The Nauerna Landfill, The Netherlands.**

The Nauerna landfill was opened in 1985 and is still in operation. The landfill occupies an entire polder of 80 ha. In 2005 the oldest part of the landfill was capped with an HDPE membrane and a mineral liner. Prior to the construction of the liner a few metres of waste were removed from the location where the office building was to be constructed. The weight of the waste removed corresponds to the weight of the building. The finished building will thus not cause any additional settlement compared to the surrounding waste. In case differential settlement does occur, a correction by means of hydraulic jacks up to 30 cm is possible. The building is composed of light materials (FSC approved wood) with a low environmental impact. The building is heated and cooled with landfill gas.

The final use on the landfill should in no way compromise the environmental protection measures or their maintenance. This does not only apply to avoiding deep rooting plants. A foundation for buildings or any constructions in the soil cover could damage the surface sealing. Excavation in soil covers should be carried out with great care. For example before pools or bunkers for golf course are constructed a sufficient layer of soil, sand or clay should be applied to prevent damage of any surface sealing. The same applies to installing fences, gates, sign posts, playground equipment, park benches, tables, etc. Equipment that requires frequent inspection or maintenance such as pump, flow meters, piezometers or automatic sampling stations should be easily accessible.

Finally, service mains like sewers, water pipes, gas pipes, communication cables and electrical cables should preferably be installed in the soil cover. In case maintenance should be required no excavation in waste is necessary. This also prevents diffusion of contaminants through drinking water pipes potentially contaminating the water.

### 10.14.3.3   Landscape Plan

Developing a landscape plan for a landfill is an iterative process involving several stakeholders: the landfill owner, the physical planning authorities and the public. The planning should start years ahead of the closure of the landfill. The

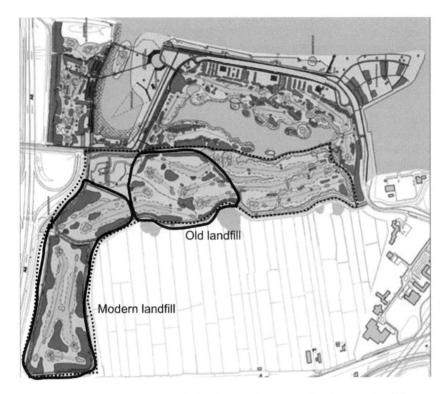

***Figure 10.14.4***   *Example of a landscape plan for Hollandse Brug landfill.*

landscape plan should address the full extent of the landfill after closure of the fill, but implementation may be in stages as sections of the landfill are closed.

The first sketchy landscape plan can be used to create and enhance enthusiasm among stakeholders for a more ambitious final use. Such a plan is often an attractive compilation of plans, cross-sections and aerial views from different perspectives. When sufficient support for various ideas is created a more detailed landscape plan of a limited number of alternatives could serve as starting point for a feasibility study and the selection procedure. Once a certain alternative is selected the final landscape plan is drawn up. This plan can be used for the procedure to amend the municipal planning scheme. It also serves as point of reference for the consultants that make the specifications and plans for the construction works.

An example of a landscape plan is given in Figure 10.14.4. This particular landscape plan encompasses an old landfill of 12 ha, a landfill according to modern standards of 16 ha and a derelict recreational area of 50 ha. The plan aims at upgrading the entire area. It includes a 9-hole and an 18-hole golf course with a clubhouse and a restaurant, a beach with kiosk, bridle paths, foot paths, cycle paths, park vegetation, ecological connections, bridges, areas for shoreline recreation such as fishing and luxury apartments. The golf course, restaurant and apartments provide extra financial resources for financing of the total project.

## 10.14.4   Aftercare Monitoring

The aftercare monitoring of the landfills often includes gas, leachate, groundwater and settlement. The purpose is to document that the landfill behaves as expected and to provide data for determining when the landfill is fully stabilised and all technical measures can be terminated.

### 10.14.4.1  Gas Monitoring and Vegetation Inspection

The European Landfill Directive as well as many national landfill directives require monitoring of potential emissions of landfill gas and/or the efficiency of landfill gas extraction systems. During aftercare the frequency can be reduced from monthly to twice a year (CEC, 1999). The European Landfill Directive mentions that gas monitoring concerns atmospheric pressure and the compounds methane, carbon dioxide, oxygen, hydrogen sulfide and hydrogen. But it does not clearly specify how landfill gas should be monitored.

Some European member states have regulations that require walkover surveys. These surveys pay attention to vegetation damage as an indicator for methane emission and/or require measurements. In most case these measurements consist of so-called spike tests or surface scans. Surface scans are usually carried out with a portable flame ionisation detector (FID). Essentially this determines combustible hydrocarbons. On landfills it is assumed that these hydrocarbons are predominantly methane. The measurement is carried out in ambient air close to the landfill surface. Both with a methane oxidising cover soil and with an impermeable surface sealing very low methane concentrations ($< 5$ ppm) can be detected with the FID. With spike tests a narrow lance is inserted 10–30 cm in the cover soil. With a portable landfill gas monitor a gas sample from the soil is extracted and analysed. In general methane, carbon dioxide and oxygen are analysed. In case of a gas- and watertight surface sealing very little methane ($< 5\,\%$) should be present in the soil gas. In case of a methane oxidising cover soil a gradual decrease of the methane concentration is expected reaching the top of the cover soil. It is important to note that both measurements reveal concentrations. Since the flux is not established, these measurements can not be considered emission measurements.

The European regulation on Pollutants Release and Transfer Registers (E-PRTR; CEC, 2006a) requires estimation and reporting of 91 pollutants including methane. The requirement to report does not stop until the landfill has been released from aftercare. The 2006 IPCC guidelines (Pipatti, 2006) proposes that the emission is calculated with the equation: Emission = (Production – Recovery) (1 – Oxidation Factor). Direct measurement of the gas fluxes from a landfill is very complicated and still in its infancy. It can therefore not be considered a real alternative to the estimation procedure yet (Scharff and Jacobs, 2006).

Since it concerns annual estimates, changes in the storage component of landfill gas in the waste due to variations in atmospheric pressure are considered negligible. The method currently used to determine oxidation for reporting purposes is to apply the IPCC default value for the oxidation factor. This factor is 0 for uncovered landfills and 0.1 for landfills covered with a methane-oxidising material, such as soil or compost (Pipatti, 2006). An accurate annual recovery estimate can be determined with a flow meter and instruments to determine pressure, temperature and methane concentration at the end of the pipe that collects and transports the landfill gas (as long as such a pipe is present). There is no scientifically accepted method to measure methane production inside a landfill. The current method to estimate landfill methane production is based on modelling. The 2006 IPCC guidelines (Pipatti, 2006) require application of first order decay (either single- or multiphase) models. The dominant parameters in these models are the annual amounts and the carbon content of the waste. In Western Europe most landfills are equipped with weighbridges, so the annual amounts should not be a problem. It is however impossible to determine the carbon content of every truckload of waste. Consequently the carbon content will be an approximation based on the judgement of the operator or its consultant. The guidance document to the E-PRTR regulation (CEC, 2006b) mentions six different models that can be applied to estimate the methane production. The various models use different degradation rates, dissimilation factors, lag times and conversion factors. These models consequently give different results for the same data set (Scharff and Jacobs, 2006).

### 10.14.4.2  Leachate Monitoring

Leachate is the agent that can cause groundwater and surface water contamination. Monitoring contamination of groundwater and surface water includes more than just monitoring leachate. Essentially the complete water balance including precipitation and evapotranspiration as well as groundwater conditions should be investigated. If the member state determines that a water balance is an effective instrument to determine potential groundwater contamination, Annex III of the European Landfill Directive (CEC, 1999) proposes monitoring of meteorological data, leachate volume and composition, surface water volume and composition and groundwater level and composition. During the aftercare phase the meteorological data include volume of precipitation, evaporation, temperature and humidity. The data should be monthly averages based on daily values. Essentially this means that during the aftercare phase a meteorological station should be maintained at the landfill. Unless there is an alternative meteorological station nearby of which the data can be considered representative for the landfill.

Evaporation needs to be determined with a lysimeter or through other suitable methods. In determining evaporation it is not considered that plants can increase the effect by means of transpiration. The combined effect is called evapotranspiration. This is hard to determine. But national meteorological services very often have correction factors for evaporation available for different types of vegetation in different climatic conditions.

Precipitation can be measured relatively easy by means of automated collection bottles. Runoff can be collected and measured, but this is rarely done. Often it is only accounted for as a percentage of precipitation. When an impermeable cap is installed within a couple of years the landfill dries out and no leachate is generated anymore. As long as the leachate is collected it can be sampled and measured. According to Annex III of the European Landfill Directive leachate volume and composition should be determined twice a year during the aftercare phase. The parameters to be measured and the substances to be analysed vary according to the composition of the waste deposited. They must be laid down in the permit document and reflect the leaching characteristics of the waste. When the surface sealing has a few leaks, a small amount of leachate could still be generated. Except in the case of an electronic leak detection system it is impossible to determine where such leaks are. Consequently it is also impossible to know where leachate should be sampled. Continuous sampling of a remaining amounts of leachate at the bottom of leachate wells may not be representative for the potential threat to the groundwater should there also be leaks in the bottom liner. All this means that determining a water balance is a rather inaccurate method to determine potential groundwater pollution.

The European Landfill Directive requires that surface water volume and composition be determined at least at two points, one upstream and one downstream of the landfill, twice a year during the aftercare phase. The motivation is not clear, but it can be assumed that it is intended to monitor decrease in surface water quality. Based on local conditions (e.g. no surface water present) the competent authority may exempt the operator from monitoring surface water.

Groundwater in general travels at speeds ranging from decimetres to hundreds of metres per year, depending on the groundwater table gradient and the hydraulic conductivity of the soil. A contamination of the groundwater therefore in general takes several months to years to reach the monitoring points (piezometers) usually placed at the outskirts of the landfill. Groundwater measurements must be such as to provide information on the groundwater likely to be affected by the landfill. Annex III of the European Landfill Directive requires at least one measuring point in the groundwater inflow region and two in groundwater outflow region. This number can be increased on the basis of a specific hydrogeological survey and the need for an early identification of accidental leachate release into the groundwater. At a landfill of approximately 20 ha often 20 to 30 measuring points are installed instead of three. In many cases filters at different levels in the aquifer are required to differentiate impact over depth. In order to have a suitable reference for impact to the groundwater its quality needs to be determined before landfill operation starts. This needs to be done at three or more measuring points. The groundwater level should be determined twice a year during the aftercare phase. The frequency for sampling and analysing groundwater is site-specific. The frequency must be based on the evaluation of local groundwater flow and the possibility for remedial actions between two consecutive samplings if a trigger value is reached. Significant adverse environmental effects should be considered to have occurred when a significant change in groundwater quality is determined. Determination of a trigger level should be done taking into account the local hydrogeological formations and groundwater quality. The trigger values should be laid down in the permit whenever possible. Three times the natural background concentration is often considered a suitable trigger value. When a trigger value is reached the sampling and analysis must be repeated. If this confirms reaching the trigger value then a contingency plan as specified in the permit should be followed. In selecting the parameters to be analysed local groundwater composition, waste and leachate composition and the mobility of the parameter in the groundwater should be considered. Parameters with high concentrations in the local groundwater are not suitable as indicators of leachate contamination. Parameters that are bound in soil particles or easily degraded in soil and groundwater are also not suitable. Suitable parameters are mobile conservative (not reacting with other compounds in soil and groundwater) compounds. Compounds such as ammonia, arsenic, chloride and benzene are potentially suitable. Christensen *et al.* (1992) provides more details about groundwater monitoring at landfills.

### 10.14.4.3  Settlement

Concerning the 'settling behaviour of the level of the landfill body' Annex III of the Landfill Directive only mentions that a 'yearly reading' should be made during the aftercare phase.

## 10.14.5   End of Aftercare

Inside landfills that have been capped with a (semi)permeable cap, such as a capillary barrier or a soil cover, processes in the waste continue. That implies that emissions potentially continue and consequently need to be controlled as long as they are likely to harm human health and the environment. As the processes continue the level of stability increases and the emissions decrease. At a certain point in time the emission reaches a level that no longer harms human health or the environment. At this point the landfill can be released from aftercare. Unfortunately a well established framework to assess landfill stability, current and future environmental and human health risks does not exist.

For example the United Kingdom has regulations concerning the end of aftercare (Environment Agency, 2005). The procedure involves a site-specific risk assessment. The advantage of this approach is that it considers local conditions in the assessment of the risks. The disadvantage is that the risk assessment is submitted at the moment the operator has carried out all the required capping and landscaping works and wants to surrender its permit. The operator will only get certainty about the suitability and acceptability of the implemented measures after approval by the competent authority of the risk assessment and the request to surrender the permit. That implies that during operation and closure the operator can not make any substantiated decisions to enhance stabilisation of the waste and reduce its emission potential.

Germany is considering introducing criteria for the release from aftercare in a new integrated landfill ordinance (Stegmann, 2006). The approach is based on a list of limit values primarily addressing loads instead of concentrations. To a certain extent local groundwater conditions are considered. These criteria make it easier to assess emissions and compare them with an acceptable impact. With this list it is also completely clear to the landfill operator which level of emission should be reached before the landfill is eligible for release from aftercare. It is therefore much easier for the operator to make a founded decision on measures to accelerate stabilisation of the waste. A disadvantage could be that to a smaller extent than the UK regulations site specific conditions can be included.

In case of an impermeable surface sealing the processes inside the waste stop due to a lack of water and the absence of a distribution mechanism. The prevention of long term emission is based on the continued functionality of the sealing. If the sealing fails, the processes, and consequently the emissions, start up again. Sealing methods with an unlimited functionality are not available. The current expectation is that surface sealing systems start losing their functionality after approximately 50–100 years. Some European member states assume that the loss of functionality will appear gradually. Consequently the emissions will also reappear gradually. It is assumed that these emissions will not harm human health or the environment. These assumptions are however not substantiated. A risk avoiding and to a large extent logical approach would be that this type of protection can only work if the surface sealing is periodically replaced. This has led the Dutch government (Ministerie van Volksgezondheid en Milieuhygiëne, 1979) to amend (1996, in force since 1998) existing environmental legislation in such a way that landfill aftercare is everlasting and that complete replacement of the surface sealing every 50 years is mandatory. Also after closure the responsibility and the financial provisions for aftercare are transferred from the landfill operator to the regional authority. A landfill operator could go bankrupt, whereas a regional authority will never disappear. Box 10.14.3 presents further issues on financial aspects of aftercare.

---

**Box 10.14.3   Financial Provisions for Landfill Aftercare.**

The purpose of financial provisions is to ensure that the costs of closure and aftercare are covered even if the landfill operator goes bankrupt during operation. In general a distinction is made between financial provisions for capping and for aftercare. Some European member states (e.g. France) also require a financial provision for accidental pollution. Financial provision may be provided as cash, bond or bank guarantee depending on the member states' regulations. In most European member states there is no standard calculation method and it is uncertain for how long the financial provisions for aftercare need to be made. There is one exception. The Netherlands has required never-ending aftercare by law since 1998. Dutch landfill operators have the certainty to provide an amount of money that will enable aftercare for ever. That means that the accumulated amount of money should keep its value and only part of the interest can be used for carrying out aftercare. In order to create uniformity in aftercare the Dutch competent authorities have drafted a 'checklist for landfill aftercare plans' (Zeegers, 2002). During operation of the landfill the operator annually pays a certain amount of money per tonne of waste landfilled to the competent authority. The total amount of money required is calculated with a model (Geusebroek and de Jong, 2005). The model considers the site specific situation of the landfill. All the measures to protect the environment are quantified. Every square metre of surface sealing, every metre of leachate drain, every gas well, every piezometer, etc. is accounted for with specific costs. The competent authority is bound to a protocol that limits their possibilities to invest this money. The purpose of this protocol is to limit the financial risk. Essentially they are investing the landfill operators' money. At the same time they are required to obtain a high return on investment in order to keep the costs as low as possible. With a separate model financial risks can be calculated (Leeuwen, 2005). The competent authority includes the financial risk and the effect of return on investment and/or interest accumulation before the aftercare starts in determining the amount of money per tonne of waste. Depending on the landfill situation this can vary between 1.5 and 3.0 €/t. The Dutch competent authorities also have the legal possibility to charge 2.2 €/t of waste for financial provisions for capping. Until recently this was however rarely put into practice. The majority of the Dutch landfill operators are publicly owned companies. For publicly owned companies an exemption can be made if the operator can convince the competent authority by means of a bank statement that sufficient provisions have been made and labelled within the company.

Even if aftercare is not everlasting, but for instance lasts 60 or 120 years, we can wonder if this is a realistic and sensible solution. Can we be certain that someone will take care of an emission that will occur after 90 years? Can we be certain that our solutions will be considered acceptable in the future? Many landfill operators believe that everlasting or even long-lasting aftercare is a fictional and utopian dream. In order not to transmit the environmental burden from our generation to the next, it is necessary to adopt landfill methods that result in intrinsically safe constructions within one generation. Additional stabilization of the landfill may be needed.

# References

CEC (1999): *Council Directive 1999/31/EC of 26 April 1999 on the landfill of waste.* Official Journal of the European Communities, Brussels, Belgium.

CEC (2003): *Council Decision 2003/33/EG of 19 December 2002 establishing criteria for the acceptance of waste at landfills pursuant to Article 16 and Annex II of Directive 1999/31/EC on the landfill of waste.* Official Journal of the European Communities, Brussels, Belgium.

CEC (2006a) *Regulation of the European Parliament and of the Council of 18 January 2006 concerning the establishment of a European Pollutant Release and Transfer Register and amending Council Directives 91/689/EEC and 96/61/EC.* Official Journal of the European Communities, Brussels, Belgium.

CEC (2006b) *Guidance document for the implementation of the European Pollutant Release and Transfer Registers.* Official Journal of the European Communities, Brussels, Belgium.

Christensen, T.H., Kjeldsen, P. and Jansen, J.L.C. (1992): *Groundwater control monitoring at sanitary landfills.* In: Christensen, T.H., Cossu, R. and Stegmann, R. (eds) *Landfilling of waste: leachate,* pp. 497–514. Elsevier Applied Science, London, UK.

Environment Agency (2005): *Guidance on landfill completion and surrender.* Report LFTGN09. Environment Agency, Bristol, UK.

Gebert, J. (2004): Mikrobielle Methanoxidation im Biofilter zur Behandlung von Rest-Emissionen bei der passiven Deponieentgassung. *Hamburger Bodenkundliche Arbeiten,* 55, ISSN: 0724-6382.

Gendebien, A., Pauwels, M., Constant, M. *et al.* (1992): *Landfill gas. From environment to energy.* Office for Official Publication of the European Communities, Luxembourg.

Geusebroek H.L.J.and de Jong, F. (2005): *Rekenmodel IPO Nazorg Stortplaatsen Versie 2.0.* Located on the internet: 19 October 2009. http://www.nazorgstortplaatsen.nl/rinas/Handleiding%20RINAS%202.0.pdf; http://www.nazorgstortplaatsen.nl/rinas/install.html.

Humer M. and Lechner P. (2001): Microorganisms against the greenhouse effect – suitable cover layers for the elimination of methane emissions from landfills. *Proceedings from the Solid Waste Association of North America 6th Annual Landfill Symposium.* CD-ROM. San Diego, USA.

Jacobs, J. and Scharff, H. (2005): Combustion of LFG containing less than 5 % methane. In: CISA (ed.) *Proceedings of the Tenth International Waste Management and Landfill Symposium.* CD-ROM. CISA, Cagliari, Italy.

Kjeldsen, P., Fredenslund, A.M., Scheutz, C. and Lemming, G. (2007): Engineered biocovers – Passive mitigation systems for landfill gas: Status for the demonstration project BIOCOVER. In: P. Lechner (ed.) *Waste Matters. Integrating Views. Second BOKU Waste Conference.* CD-ROM. Vienna, Austria.

Mathlener, R.A., T. Heimovaara, H. Oonk, L. Luning, H.A. Van Der Sloot, A. van Zomeren (2006): *Opening the black box: Process based design criteria to eliminate aftercare of landfills.* Dutch Sustainable Landfill Foundation, s'-Hertogenbosch, The Netherlands. Located on the internet: 19 October 2009. http://www.sustainablelandfilling.com/webfiles/DuurzaamStortenNL/files/060501_Report_final_lay_out.pdf.

McDougall, J.R. and Pyrah, I.C. (2001): Settlement in landfilled waste: Extending the geotechnical approach. In: CISA (ed.) *Proceedings of the Eighth International Waste Management and Landfill Symposium.* CISA, Cagliari, Italy.

Ministerie van Volksgezondheid en Milieuhygiëne (1979): *Wet van 13 juni 1979, houdende regelen met betrekking tot een aantal algemene onderwerpen op het gebied van de milieuhygiëne.* Located on the internet: 19 October 2009. http://wetten.overheid.nl/cgi-bin/sessioned/browsercheck/continuation=18986-002/session=054397928553285/action=javascript-result/javascript=yes.

Neumann, U. and Christensen, T.H. (1996): *Effects of landfill gas on vegetation.* In: Christensen, T.H., Cossu, R. and Stegmann, R. (eds) Landfilling of waste: Biogas. E&FN Spon, London, UK; ISBN 0 419 19400 2.

Pipatti, R. and Svardal, P. (2006): Solid waste disposal. In: Eggleston, H.S., Buendia, L., Miwa, K., Ngara, T. and Tanabe, K. (eds) IPCC guidelines for national greenhouse gas inventories, volume 5: waste. Prepared by the National Greenhouse Gas Inventories Programme. IGES, Tokyo, Japan. ISBN 4-88788-032-4. Located on the internet: 19 October 2009. http://www.ipcc-nggip.iges.or.jp/public/2006gl/vol5.htm.

Rahmke, H-G., Witt, K.J., Bräcker, W., Tiedt, M., Düllmann, H. and Melchior, S. (2006): Anforderungen an Deponie-Oberflächenabdichtungssysteme. *Unterlagen der Workshop Arbeitskreis 6.1 – Geotechnik de Deponiebauwerke.* CD-ROM. Deutschen Gesellschaft für Geotechnik und Fachhochschule Lippe und Höxter, Höxter, Germany.

Ritzkowski, M., Heyer, K.-U. and Stegmann, R. (2006): Fundamental processes and implications during in situ aeration of old landfills. *Waste Management,* 26, 4, 356–372.

Scharff, H.and Jacobs, J. (2006): Applying guidance for methane emission estimates for landfills, *Waste Management,* 26, 417–429.

Scharff, H., Jacobs, J., van der Sloot, H.A and van Zomeren, A. (2007): Sustainable landfill and final storage quality. In: Lechner, P. (ed.) *Waste Matters. Integrating Views. Second BOKU Waste Conference.* CD-ROM. Vienna, Austria.

Stegmann, R., K-U. Heyer, K. Hupe and A. Willand (2006): *Deponienachsorge – Handlungsoptionen, Dauer, Kosten und quantitative Kriterien für die Entlassung aus der Nachsorge.* Abschlussbericht UFOPLAN-Nr. 204 34 327. Umweltbundesamt, Dessau, Germany.

Steinbrecht, D. and Spiegelberg, V. (2007): New DN-ES+S landfill aftercare conception based on a fluidised bubbling bed combustion plant burning very poor landfill gas. In: Lechner, P. (ed.) *Waste Matters. Integrating Views. Second BOKU Waste Conference.* CD-ROM. Vienna, Austria.

van Leeuwen, O.D.F. (2005): *Handleiding berekening risicobedrag.* Located on the internet: 19 October 2009. http://www.nazorgstortplaatsen.nl/risicomodel/program/Handleiding%20risicomodel%20(versie%201-12-2005).pdf.

van Zomeren, A., van den Berg, P., Bleijerveld, R. and van der Sloot, H.A. (2007): Identification of in-situ processes controlling emissions of a stabilised waste landfill by field measurements and geochemical modelling. In: CISA (ed.) *Proceedings of the Eleventh International Waste Management and Landfill Symposium.* CD-ROM. CISA, Cagliari, Italy.

Wall, D.K. and Zeiss, C. (1995): Municipal landfill biodegradation and settlement. *ASCE Journal of Environmental Engineering,* 121, 3, 214–233.

Woelders, H., Hermkes, H., Oonk, H. and Luning, L. (2007): Dutch sustainable landfill research program: five years experience with the bioreactor test cell. In: CISA (ed.) *Proceedings of the Eleventh International Waste Management and Landfill Symposium*. CD-ROM. CISA, Cagliari, Italy.

Young, T. and Heasman, H. (1985): An assessment of the odor and toxicity of trace components of landfill gas. In: GRCDA (ed.) *Proceedings of the Eighth International Symposium on Landfill Gas*. CD-ROM. GRCDA, San Antonio, USA.

Zeegers, H.R. (2002): *Checklist nazorgplannen stortplaatsen*. Located on the internet: 19 October 2009. http://www.nazorgstortplaatsen.nl/documenten/IPO-checklist%202002%20stortplaatsen.pdf.

# 11

# Special and Hazardous Waste

# 11.1

# Healthcare Risk Waste

**Ole Vennicke Christiansen**

*Danwaste Consult A/S, Copenhagen, Denmark*

**Peder Bisbjerg**

*EP&T Consultants Sdn. Bhd., Kuala Lumpur, Malaysia*

Healthcare risk waste is also known as clinical waste, infectious waste, medical waste or hospital waste. This waste must be handled with caution, as it can cause the spread of diseases and even deadly epidemics. Most countries define healthcare risk waste as a hazardous or special waste, that is to say a waste stream that is unsuited for treatment together with municipal solid waste.

In most countries, healthcare risk waste is collected separately within hospitals; normally there is already at the ward level a sorting between ordinary noninfectious healthcare waste (in the following called 'ordinary waste') and healthcare risk waste. Usually healthcare risk wastes are treated in dedicated treatment facilities.

Other types of risk waste are also generated by activities in the healthcare sector, like for instance chemical waste (such as laboratory waste and photo chemicals) and unused medicine. These waste types are usually collected separately and handled as hazardous waste, but in some cases, these are handled together with healthcare risk waste. Hospitals also generate small quantities of radioactive waste; particular rules apply to the handling and disposal of this waste.

The healthcare sector also generates ordinary waste types that are not classified as healthcare risk waste or other forms of hazardous waste. These waste types make up 80–90 % of the waste from the healthcare sector and include general waste from bed wards and other patient care arrangements, office waste, kitchen waste, bulky waste and garden waste. This ordinary waste is handled within the municipal solid waste handling system in accordance with the regulations of the local and national authorities.

This chapter briefly describes the healthcare risk waste and its handling and treatment.

*Solid Waste Technology & Management*   Edited by Thomas Christensen
© 2011 Blackwell Publishing Ltd

## 11.1.1   Defining Healthcare Risk Waste

Typical healthcare risk wastes are:

- Cultures of microorganisms and biological materials.
- Human blood and blood products.
- Pathological wastes such as tissues, organs, body parts, containers of body fluids.
- All sharp items, i.e. needles, syringes, scalpel blades, etc. (even if not contaminated).
- Contaminated animal carcasses, body parts, bedding and related wastes.
- Materials (soil, water, or other debris) which result from the cleanup of a spill of any healthcare risk waste.
- Waste contaminated by or mixed with healthcare risk waste.

These wastes present a health risk and are therefore collected and treated separately. Diseases transmitted through waste from hospitals and similar sources are numerous. Some of the infections transmitted through blood are HIV/AIDS, hepatitis B and C and haemorrhagic fevers such as the Ebola virus. Faeces and/or vomit can transmit gastroenteric infections and hepatitis A. Finally, diseases such as tuberculosis are transmitted through saliva and skin infections through pus.

In this chapter, the designation healthcare risk waste defines all waste handled as a biohazard or together with waste posing a biohazard (see Box 11.1.1 for a definition of biohazard).

The largest source of healthcare risk waste is hospitals and these generate well over 50 % of all healthcare risk waste. Smaller sources include polyclinics, clinics and public health centres. Finally, veterinarian clinics, doctor's practices and dentists are small sources of healthcare risk waste.

---

**Box 11.1.1   Definition of Biohazard.**

A biohazard (biological hazard) is a risk that is posed to humans by a biological organism, or by a material produced by such an organism. Some definitions also include biological agents and materials that are potentially hazardous to animals and/or plants.

Biohazardous agents may include but are not limited to certain bacteria, fungi, viruses, rickettsiae, chlamydia, parasites, recombinant products, allergens, cultured human or animal cells and the potentially infectious agents these cells may contain, viroids, prions and other infectious agents.

Biohazards are identified by the internationally recognised warning symbol shown above.

---

For hospitals, the waste generation rate for healthcare risk waste varies according to a number of factors: The type of hospital, the stage of development of the country where it is located, etc. On an international basis waste generation rates for hospitals vary between 0.1 kg/bed/day for poor countries (e.g. Bhutan, Bangladesh) to 0.8 kg/bed/day for rich countries such as Austria or the United States. Waste generation rates are given per occupied hospital bed per day, so for a hospital with less than full occupancy, this has to be factored into the calculation.

The waste quantities vary from source to source: In a hospital, the waste per bed per day thus differs from ward to ward, and the amount of healthcare risk waste depends on any particular specialisation of a hospital. Data from Danish hospitals show that the waste amount may vary from 0.2 to 1.0 kg/bed/day, and similarly in the Philippines it may vary from 0.15 to 0.31 kg/bed/day. For overall planning and system sizing, about 0.5 kg/bed/day can be used for a number of countries, including Denmark, Malaysia and Thailand. However, for any detailed planning, the waste generation rates must be established through field verification. When examining the waste generation rates, it is essential to ensure that the waste separation is done correctly at the ward level. In hospitals where ordinary waste is unintentionally mixed with the infectious waste, the healthcare risk waste quantities will be considerably larger; this can lead to an incorrect basis for the detailed planning.

The World Health Organisation website offers excellent advice (WHO, 2009; see also Jantsch and Vest, 1999; WHO, 2000).

## 11.1.2   Internal Handling of Healthcare Risk Waste

Healthcare risk waste is handled independently from other waste types, due to the risk of infection. This separation applies from generation, during collection and transport internally at the source, during the external collection and transport, right through to the final treatment. Therefore, source separation is necessary to ensure a hygienically safe handling of the healthcare risk waste. If parts of the ordinary waste described above are treated together with healthcare risk waste for practical or safety reasons, these wastes are to be handled according to the same guidelines as the healthcare risk waste.

It is important that personnel handling healthcare waste receive thorough training. They must know the hospital or institution's procedures, packaging types, personal protection measures, first aid and how to respond in case of accidents or damaged packaging.

This section covers the separation, packaging, collection and internal transport of healthcare risk waste. Most focus will be on the handling of healthcare risk waste within an individual hospital or other large institution; but also the handling of healthcare risk waste from a small source such as a clinic of a physician or dentist is described below.

### 11.1.2.1   Separation

Healthcare risk waste requires special measures of separation, packaging, marking/coding, temporary storage and collection. The separation is made at the source in accordance with set requirements and regulations. It is essential that all personnel groups who are exposed to the waste fully understand the separation criteria.

It is a prerequisite that there is a good separation between the infectious waste and the ordinary waste. If some infectious waste is mixed with the ordinary waste, this puts collectors of the ordinary waste at risk. Furthermore, in poor countries where scavengers sift through the garbage in search of recyclable materials, the scavengers are exposed to infectious diseases. If ordinary waste is mixed into the healthcare risk waste, this increases the quantities of healthcare risk waste and thereby the handling and treatment costs.

At the ward level, all waste is generally separated by the medical staff into three basic categories, though most institutions will have more categories: For example, radioactive waste is a distinct waste that is handled on its own. In some countries, plastic materials such as intravenous bottles are collected separately and sold to plastic recyclers. The three common categories are:

- Ordinary waste that is placed in black bags. This waste includes kitchen waste, glass, drinks containers, paper towels, tissue, office paper, packaging and other materials that have not been in contact with possible sources of infectious diseases.

- Healthcare risk waste that is collected in yellow or red plastic bags. This includes all waste that could potentially be infectious, such as bandages, gauze, adult diapers, body parts, human tissue, placentas, etc.
- Sharps that are placed in protective boxes. This is waste such as needles and syringes, ampoules, razor and scalpel blades, intravenous needles and broken glass. Developed countries use a variety of rigid plastic boxes (suitably colour-coded) to store these wastes; in less developed countries any available container will be utilised. For example, most sharps can be collected relatively safely, and at minimal cost, in a 5- or 10-l plastic vessels with a screw top.

### 11.1.2.2  Packaging

Upon separation, the waste must be packed correctly. The choice of packaging depends on a number of factors such as waste type, storage conditions, collection procedures, external transportation and the treatment system. Normally, packaging can be divided into three categories (see Figure 11.1.1):

- Source packaging is the packaging used where the waste is generated; this can include plastic bags, containers for sharps and perhaps recyclable bottles/containers/cans. There are special requirements to waste packaging according to the waste type; for instance the packaging for sharp and pointed objects must be resistant to penetration. Some of the packaging used at source can also be used during collection; in this case it should be clearly labelled.
- Collection packaging is used for transportation of the infectious waste within a hospital or institution. This can be plastic bins, strong plastic sacks, multiple-layer paper sacks lined with plastic or wax-treated cardboard boxes. This packaging must be easily identifiable through a warning colour, usually red or yellow. Furthermore, the 'biohazard' symbol should be visible (see Box 11.1.1) and there may be a text such as 'healthcare risk waste.' Finally, the origin of

**Figure 11.1.1**   *Health risk waste collection in rural hospital, Nghe An Province, Vietnam. All wards use trolleys equipped as the one pictured: The blue bin is for ordinary waste (left, back), the yellow bin is for infectious waste (front) and the red bucket is used for recyclables (right, back). The yellow box for sharps can be seen on the right (on the side of the trolley). Photo: Peder Bisbjerg.*

the waste should be clearly marked on the packaging, for example the name of the department or ward. It is essential that the waste is not handled again and the infectious waste will remain within the collection packaging right through final treatment.

- Transport packaging, which is the designation of the packaging that is used for external transport of the collection packaging. This can be 120- to 800-l containers made of plastic, glass fibre or metal; these are often lined with a strong plastic bag. The packaging should be properly closed during transportation and will often be marked in accordance with the ADR requirements (see 11.1.3).

### 11.1.2.3   Temporary Storage

The healthcare risk waste is temporarily stored in wards, laboratories, clinics and other institutions in the collection packaging in a room specifically prepared for this use. The room should have running water and a spillage collection system. Furthermore, close to the temporary storage room it is advisable to have a small stock of unused packaging for the waste. In hot climates, the healthcare risk waste can be stored in cooled storage rooms. The waste can then be collected less frequently, but this increases both capital and operating costs.

### 11.1.2.4   Internal Collection

Specially trained personnel should collect the waste from the temporary storage room once or twice daily. In order to minimise the infection risk, it is important that the packaging remains undamaged through the whole collection and transportation process. The internal transport is made on wagons that are easy to push and better off hospitals will have electrified transportation vehicles. The waste is transported to a central storage area. This will be close to the waste treatment facility for many larger hospitals and otherwise it will be the collection point for external transportation. The area must be isolated from where other activities take place. It should be fenced and kept locked outside daily working hours. If during the internal handling the packaging is damaged, the damaged item is usually placed into larger packaging; here it is essential that all due care is taken whilst handling the infectious waste. Figure 11.1.2 shows a central storage facility at a hospital in Vietnam.

### 11.1.2.5   Healthcare Risk Waste Handling at Small Sources

Small sources, such as clinics will normally sort their healthcare risk waste into a fraction that can go into a coloured bag held in a small plastic bin and a 'sharps' collection system. In poor countries these wastes pose a considerable risk, as there is frequently no collection system for doctor's practices and clinics. Hence, the bag containing healthcare risk waste will often be thrown into the MSW. Often the bag with healthcare risk waste is a standard bag, i.e. it is not colour-coded nor does it have any other form of warning. In countries with intensive scavenging activities, these bags pose a serious risk to both scavengers and garbage collectors.

### 11.1.3   Collection and Transportation of Healthcare Risk Waste

In developed countries the external transport of healthcare risk waste follows the requirements to transportation of infectious substances set out by the European Agreement concerning the International Carriage of Dangerous Goods by Roads (United Nations Economic Commission for Europe, 1995), hereafter called the ADR requirements. These require that the waste shall only be transported when there is double containment, i.e. if the bag holding the infectious waste ruptures, there must still be a second layer around the waste to prevent any form of spill.

For external packaging, the transport packaging is normally used; this can be standard plastic, glass fibre or steel containers. According to the ADR requirements, they shall be marked with the UN number of the contents as well as the biohazards symbol. The ADR requirements require that during the collection no emptying of the waste into the collection vehicle can take place. The transport packaging must be transported whole and undamaged to the treatment facility. Collection is often done using a closed platform truck with loading at the back. The vehicle should be adapted, so that

***Figure 11.1.2*** *Typical central storage facility for healthcare risk waste at a large hospital in Ho Chi Minh City, Vietnam. From this locked enclosure, the waste is collected daily and taken to a specialised healthcare waste incinerator serving the whole city. Photo: Peder Bisbjerg.*

the packaging can be fastened during transport to minimise the risk of spilling the healthcare risk waste. At the treatment facility, the transportation packaging is emptied, for example directly into an incinerator's feed mechanism (see Figure 11.1.3). After emptying, the transportation container is washed and then returned to the hospital to be reused.

In less affluent countries, the ADR requirements are also often followed. For example, the healthcare risk waste (in plastic bags) is placed in characteristically coloured 140-l wheeled plastic bins and then transported in a closed truck to the treatment facility. Again, the bins are washed prior to being reused for waste.

It is relatively simple and inexpensive to arrange healthcare risk waste collection and treatment for large sources. It is far more costly to collect waste from thousands of doctors or clinics that generate very modest quantities of healthcare risk waste. Therefore, in many developing countries only waste from the large sources is safely collected and treated. In order to ensure the collection of this waste, municipalities may need to establish a collection system. Here, it is important to prepare regulations or guidelines that instruct the primary healthcare sector to make use of such arrangements.

## 11.1.4   Treatment of Healthcare Risk Waste

There are a number of possibilities for the safe treatment and disposal of healthcare risk waste. Highly specialised wards or laboratories can generate waste that is so infectious that it is immediately disinfected, for example with an autoclave (see 11.1.4.2), before it is placed in the healthcare risk waste bin. However, most healthcare risk waste is transported to a treatment facility where it is treated. Autoclaves will operate in a batch mode where waste is disinfected whenever the autoclave is full or at least once per day. Small incinerators will frequently only operated during the daytime, whereas large ($\geq$0.5 t/h) systems will often function continuously, 24 h/day, 7 days/week.

### 11.1.4.1   Incineration

The preferred technology for the treatment of infectious healthcare risk waste is incineration. There are several reasons for this: First, after the thermal treatment all biological organisms are destroyed and no longer pose a risk. Second, the

***Figure 11.1.3***   *Waste being unloaded at a hospital's incinerator in the UK. Note that the yellow bags of healthcare risk waste at this plant are mixed with black bags of ordinary waste. Both are burned in the incinerator. If healthcare risk waste is burned in an incineration plant designed for ordinary waste (municipal waste), the healthcare risk waste must be kept separated from the ordinary waste until fired into the furnace. Photo: Peder Bisbjerg.*

residues from the incinerator are inert and can be placed directly in a secure landfill. Third, the ashes are an obvious indicator that the infectious waste has been successfully treated and furthermore body parts are no longer recognizable. Many larger hospitals have their own incinerator, which will incinerate the hospital's healthcare risk waste and some receive waste from neighbouring institutions. Figure 11.1.3 shows the incineration of a mix of healthcare risk waste (yellow bags) and ordinary waste (black bags) in the United Kingdom. The second common scenario is an independent incineration facility, operated either by the authorities or by a private company. Such a facility receives healthcare risk waste from hospitals, institutions, clinics and so forth within its region. Such facilities usually demand payment for the treatment of any external waste. Figure 11.1.4 shows a commercial healthcare risk waste incinerator in Malaysia.

### 11.1.4.2   Sterilisation

Sterilization can be done by autoclaving, microwave treatment and radiation. Unlike incineration; autoclaving, microwaving and radiation do not reduce the organic content or otherwise change the appearance and structure of the waste during treatment, excepting the methods that require shredding of the waste into smaller parts. Hence it cannot be visually established whether the waste has undergone sufficient treatment and, consequently, the methods require a high degree of care and control. After treatment, the inert waste is handled in the normal waste system; hence in some countries it is incinerated and in most countries it is placed on a MSW landfill.

- Autoclaving is sterilisation utilising steam under pressure to kill bacteria and vira at temperatures above $100\,°C$. Normally the temperature is around $120\,°C$ and the waste is exposed to this temperature for over 1 h. Autoclaving has been practised for many years and is an established, well tested method. The advantage is that it is less costly than incineration and can be used by a small hospital; the main disadvantage is that the waste is not transformed in the autoclave, and hence there is no visual difference between treated and untreated waste.

***Figure 11.1.4***   *Commercial healthcare risk waste incineration facility in Malaysia. Photo: Peder Bisbjerg.*

- Microwave Treatment. Over the past 15 years, several suppliers for the deactivation of healthcare risk waste have introduced microwave treatment to the market. It is necessary to shred the waste prior to the microwaving and the treated waste has an appearance similar to confetti. The method is less costly than incineration but currently not widespread.
- Radiation. For sterilisation, ionised radiation is used as treatment method to deactivate pathogenic microorganisms. The sterilisation reduces the ability of the microorganisms to reproduce but requires division of the waste into smaller parts to a suitable homogeneous size in order that the sterilisation can be effective. Sterilisation by radiation is a well known process from the food industry but for healthcare risk waste it is still being developed. Another method of sterilisation is chemical treatment where typically disinfecting preparations like hypochlorite or formaldehyde are used.

### 11.1.4.3   Landfill

When there is no other form of treatment available, one possibility is to place the healthcare risk waste on a landfill. In this case the waste should be clearly marked and placed in a ringfenced and locked enclosure so that scavengers have no access. In some cases the healthcare risk waste is covered by lime, which increases the pH to above 10, providing some additional barrier to the migration of the infectious materials. Here it is also essential that there is always a guard present on the landfill. This is not a recommended solution, but it does ensure that scavengers are not exposed to the healthcare risk waste.

### 11.1.5   Economy

The expense of participating in a healthcare risk waste management system depends on the waste quantities produced. Thus, the waste is often weighed in connection with the collection. The cost may include:

- A basic rate covering the administration of participating in the arrangement. Typically, the basic rate is € 20–30 per year.

- An emptying price normally based on a full container. The emptying price is typically € 10–15 per emptying, including transport to the treatment plant.
- The cost of treatment. Most collection arrangements include a fee for the treatment of the collected waste. The treatment price is typically in the range € 150–300 per tonne.

Hence for large waste generators, such as hospitals, the treatment and disposal costs may be as low as € 150 per tonne of waste. For small waste generators, collection costs of around € 10–20 per month can be considered reasonable.

The packaging for the healthcare risk waste can also be costly: For a hospital using state of the art packaging for the waste, the source and collection packaging can cost over € 300 per tonne of collected healthcare risk waste. If readily available used containers are used for sharps and coloured plastic bags are used for other infectious waste, the costs can be substantially reduced.

# References

Jantsch, F. and Vest, H. (1999): *Management of solid and liquid waste at small healthcare facilities in developing countries*. GTZ, Germany.

WHO (2000): *Starting Health Care Waste Management in Medical Institutions*, World Health Organization, Copenhagen, Denmark.

WHO (2009): The World Health Organisation website. Located on the internet: 19 October 2009. http://www.healthcarewaste.org/en/115_overview.html.

# 11.2

# Waste Electrical and Electronic Equipment

**Marianne Bigum and Thomas H. Christensen**

*Technical University of Denmark, Denmark*

Waste electrical and electronic equipment (WEEE) is one of the fastest growing special waste types with an estimated growth of 3–5 % per year (Cui and Forssberg, 2003). WEEE is a very heterogeneous waste type that contains many compounds that are considered to be harmful to both humans and the environment, as well as many metals that have the potential of being recycled and reused. This makes the waste fraction (WEEE) very interesting as it is a problematic waste as well as an important secondary resource.

## 11.2.1  WEEE Characterization

WEEE is defined as disposed electrical and electronic equipment (EEE), where EEE is defined as any equipment that is dependent on electric currents or electromagnetic fields in order to work properly, including equipment for the generation, transfer, and measurement of current (Schafer *et al.*, 2003). The European Union (EU) has categorised the different types of WEEE into 10 categories, as seen in Table 11.2.1 (2002/96/EF). The categories and products however are continuously updated because of new technologies and new types of EEE.

### 11.2.1.1  Quantities

The annually produced amounts of EEE can be determined from the sales records for the various countries and in Europe the producers of EEE are obliged by the WEEE directive to report sales records to the national authorities. However the different EEE products have different product life spans and people also tend to store smaller worn out products in their homes (the attic effect). The resulting production of WEEE is therefore not easily determined. Cooling and heating equipments typically have long life spans where as smaller household appliances and consumer products such as mobile phones have shorter life spans. The accumulation of smaller worn out products can be illustrated by old cell phones which have been replaced by new models but which are often not discarded but are kept in the home. Another significant issue that makes it difficult to map the flows and production of WEEE is the export of second hand EEE, and the illegal export of WEEE. Significant amounts of WEEE often end up in developing countries where it is minimally treated without due consideration for either the environment or human health, followed by disposal in unsafe open dump

**Table 11.2.1**  *The different types of WEEE divided into the 10 WEEE categories defined by the EU WEEE directive (2002/96/EF). Reprinted with permission from Final report to the Directorate General Environment, Costs for municipal waste management in the EU by D. Hogg © (2002) European Commission.*

| No. | WEEE category |
|-----|---------------|
| 1  | Large household appliances |
| 2  | Small household appliances |
| 3  | IT and telecommunications equipment |
| 4  | Consumer equipment |
| 5  | Lighting equipment |
| 6  | Electrical and electronic tools (with the exception of large-scale stationary industrial tools) |
| 7  | Toys, leisure and sports equipment |
| 8  | Medical devices (with the exception of all implanted and infected products) |
| 9  | Monitoring and control instruments |
| 10 | Automatic dispensers |

landfills. Overall it is very difficult to determine the flows of WEEE. The following bullets sum up some of the main aspects that make an overall quantification of WEEE difficult:

- Different life spans within WEEE categories.
- Legal export of EEE for second hand use.
- Illegal export of WEEE to second world countries.
- Accumulation of obsolete EEE in households.

The production of WEEE per inhabitant varies significantly and also depends on the economic status of country; richer countries tend to buy more EEE and then discard more WEEE. The Swiss Federal Laboratories for Material Testing and Research (EMPA, 2009) have collected data on the different unit generation rates for WEEE in 15 different countries. It is however difficult to compare the different countries as there are variations as to which waste types are included in the category 'WEEE'. Table 11.2.2 presents an overview of the unit generation rates for WEEE in 15 different countries.

**Table 11.2.2**  *Unit generation rates for 15 different countries. The countries are not directly comparable as there might be difference between the countries as to what is included in the 'WEEE category' (EMPA, 2009). EMPA 2009, Swiss Federal Laboratories for Materials Testing and Research (EMPA) 2009 Lerchenfeldstrasse 5, 9014 St. Gallen, Switzerland.*

| Country | Tonnes | Kilograms/capita | Year estimated | Source |
|---------|--------|------------------|----------------|--------|
| Bulgaria | 9600 | 5.7 | 2006 | Huisman *et al.*, 2008 |
| China | 2 425 000 | 1.9 | 2005 | Liu, 2006 |
| Denmark | 34 600 | 23 | 2000 | Huisman *et al.*, 2008 |
| Estonia | 16 700 | 8.2 | 2005 | Huisman *et al.*, 2008 |
| EU | 8 700 000 | 18 | 2006 | Huisman *et al.*, 2008 |
| Finland | 30 900 | 23 | 2003 | Huisman *et al.*, 2008 |
| France | 29 900 | 24 | 2005 | Huisman *et al.*, 2008 |
| Germany | 30 400 | 15 | 2005 | Huisman *et al.*, 2008 |
| Hungary | 16 300 | 11 | 2005 | Huisman *et al.*, 2008 |
| India | 800 000 | 0.79 | 2006 | Estimation by EMPA |
| Lithuania | 13 700 | 6.3 | 2003 | Huisman *et al.*, 2008 |
| Poland | 13 300 | 8.4 | 2008 | Huisman *et al.*, 2008 |
| Sweden | 29 800 | 24 | 1999 | Huisman *et al.*, 2008 |
| Thailand | 850 000 | 1.4 | 2003 | Cobbing, 2008 |
| UK | 30 300 | 29 | 2005 | Huisman *et al.*, 2008 |

**Table 11.2.3**   *Main compounds found in WEEE (Hagelüken, 2006). Reproduced with permission from the International Symposium of Electronics and Environment © (2006) IEEE.*

| Category | Compounds |
|---|---|
| Precious metals | Gold (Au), silver (Ag), palladium (Pd) and platinum (Pt) |
| Base metals | Copper (Cu), aluminium (Al), nickel (Ni), tin (Sn), zinc (Zn) and iron (Fe) |
| Compounds of concern | Mercury (Hg), beryllium (Be), indium (In), lead (Pd), cadmium (Cd), arsenic (As) and antimony (Sb) |
| Halogens | Bromine (Br), fluorine (F) and chlorine (Cl) |
| Combustibles | Plastics and organic fluids |

### 11.2.1.2   Composition

WEEE is a very heterogeneous waste type among the categories, but varies also significantly within each category. The main and most common substances found in WEEE can be seen in Table 11.2.3.

Materials such as iron, aluminium, plastics, and glass account for more than 80 % (by weight) but smaller amounts of valuable and toxic metals are also very important (EMPA, 2009). Table 11.2.4 provides an overview of the material compositions of the different WEEE categories.

## 11.2.2   Legislation

In the EU two key legislations regulate WEEE: namely the WEEE directive (2002/96/EC) and the restriction of hazardous substances (RoHS) directive (2002/95/EC). The two directives aim at having a significant impact on the way manufacturers design, produce and dispose of their products via the introduced concept of producer responsibility (PR), which makes the producers legally responsible for the collection and subsequent management of the WEEE. Also the energy-using products (EuP) directive (2005/32/EC) and the regulation on the registration, evaluation, authorisation, and restriction on chemical substances (REACH; EC 1907/2006) set demands on the EEE producers in the EU (Hester and Harrison, 2009).

**Table 11.2.4**   *Material composition (%) of the different WEEE categories. Note that WEEE is a heterogeneous waste type and that these data should be used with care (EMPA, 2009). EMPA 2009, Swiss Federal Laboratories for Materials Testing and Research (EMPA) 2009 Lerchenfeldstrasse 5, 9014 St. Gallen, Switzerland.*

| Material | Cooling and heating white goods | Small household appliances | IT and consumer equipment | Lighting equipment |
|---|---|---|---|---|
| WEEE directive categories | 1, 10 | 2, 6, 7, 8, 9 | 3, 4 | 5 |
| Ferrous metal | 43 | 29 | 36 | — |
| Aluminium | 14 | 9.3 | 5 | 14 |
| Copper | 12 | 17 | 4 | 0.22 |
| Lead | 1.6 | 0.57 | 0.29 | — |
| Cadmium | $1.4 \times 10^{-3}$ | $6.8 \times 10^{-3}$ | $1.8 \times 10^{-2}$ | — |
| Mercury | $3.8 \times 10^{-5}$ | $1.8 \times 10^{-5}$ | $7.0 \times 10^{-5}$ | $2 \times 10^{-2}$ |
| Gold | $6.7 \times 10^{-7}$ | $6.1 \times 10^{-7}$ | $2.4 \times 10^{-4}$ | — |
| Silver | $7.7 \times 10^{-6}$ | $7.0 \times 10^{-6}$ | $1.2 \times 10^{-3}$ | — |
| Palladium | $3.0 \times 10^{-7}$ | $2.4 \times 10^{-7}$ | $6.0 \times 10^{-5}$ | — |
| Indium | 0 | 0 | $5.0 \times 10^{-4}$ | $5.0 \times 10^{-4}$ |
| Brominated plastics | 0.29 | 0.75 | 18 | 3.7 |
| Plastics | 19 | 37 | 12 | 0 |
| Lead glass | 0 | 0 | 19 | 0 |
| Glass | $1.7 \times 10^{-2}$ | 0.16 | 0.3 | 77 |
| Other | 10 | 6.9 | 5.7 | 5 |
| Total | 100 | 100 | 100 | 100 |

Japan is one of the leading nations on management and recycling of WEEE; the Japanese approach differs from the approach in the EU. The home appliance recycling law (HARL) includes the four major types of household WEEE: televisions, refrigerators, washing machines and air conditioners. The HARL includes a takeback system where retailers when selling a new product are required to take an old one back. Although the HARL focuses on only four categories of WEEE, it is estimated that still 80 % (by weight) of WEEE is collected (Hester and Harrison, 2009).

In the United States the Environmental Protection Agency (EPA) is becoming aware of the problems related to unmanaged WEEE. The management of WEEE is however determined by the member states and only seven states out of 51 states have banned WEEE from landfills and only four states have established some sort of recovery system. The primary WEEE management in the United States therefore is landfilling and incineration (EPA, 2008).

### 11.2.2.1  EU – WEEE Directive

In 2003 the WEEE directive was implemented in the EU. This directive introduced 10 waste categories for WEEE, shown in Table 11.2.1. The 10 WEEE categories were already used in some countries but the directive was passed to raise the degree of recycling and recovery for WEEE in all member states. In each member state local declarations were passed to fit the specific conditions. The concept of producer responsibility was introduced by the WEEE directive. The producer responsibility means that the producers have to finance the handling of products that becomes waste. This is done as an attempt to ensure that all waste is properly treated. Producers outside the EU who market their product on the European market and the importers are also affected by the WEEE directive as they are obliged to label all EEE with a WEEE label. This label is used to inform the user to dispose of the WEEE correctly.

The extra costs for the industry due to the producer responsibility most likely affect the cost of the products, and were in 2003 estimated to constitute 0.2–3.0 % of the price of the product (MST, 2003).

### 11.2.2.2  EU – RoHS Directive

The restriction of the use of certain hazardous substances (RoHS) in electrical and electronic equipment banned the use of certain substances in EEE in the EU in 2006. The restriction was passed to protect human health and the environment. The banned substances can be found in Table 11.2.5. Even though the substances are banned today, they are still present in equipment produced before the implementation of the RoHS directive. The sorting facilities and the treatment plants should be aware of this kind of waste. There are some exceptions to where it is still possible to use the banned substances. These are, among others: lead in cathode ray tubes and mercury in fluorescent tubes.

## 11.2.3  Treatment Categories

In the EU the WEEE waste management industry divides the different types of WEEE into six treatment categories dependent on the necessary further waste management options. The six treatment categories are shown in Table 11.2.6, along with the corresponding WEEE categories.

**Table 11.2.5**  *The substances banned by the European RoHS directive (2002/95/EC). Reprinted with permission from Final report to the Directorate General Environment, Costs for municipal waste management in the EU by D. Hogg © (2002) European Commission.*

| Compounds | Symbol/abbreviation |
| --- | --- |
| Lead | Pb |
| Mercury | Hg |
| Cadmium | Cd |
| Hexavalent chromium | Cr |
| Polybrominated biphenyls | PBB |
| Polybrominated diphenyls ethers | PBDE |

**Table 11.2.6**   *The six WEEE treatment categories and the corresponding WEEE categories.*

| WEEE categories | Treatment categories |
| --- | --- |
| 1, 10 | Cooling white goods |
| 1,10 | Heating white goods |
| 2, 6, 7, 8, 9 | Small household appliances (Low grade WEEE) |
| 3, 4 | Small household appliances (High grade WEEE) |
| 4 | TV and monitors |
| 5 | Lighting equipment |

### 11.2.3.1   Cooling White Goods

The treatment category 'Cooling white goods' contains products like refrigerators and freezers. The main problem with old refrigerators is that the cooling system might contain chlorofluorocarbon (CFC) gasses which are highly ozone depleting. These CFC gasses were therefore abandoned by the Montreal Protocol in 1987 (UNEP, 2000). Today the cooling white goods contain pentane in the cooling system; it is however still possible to encounter CFC in old appliances.

When the white goods are received at the waste facility, the outer cables, the compressor, the capacitors (which might contain polychlorinated biphenyls; PCB), printed circuit boards, and the mercury containing switches are removed. PCB was banned in 1972 and has not been used in equipment since 1986. This means that it is very unlikely that products discarded contain PCB (Harding and Willis, 2004).

The oil is drained from the compressors and can be incinerated, which destroys the freon and thereby the ozone depleting potential. If the oil is pentane from a newer appliance, the oil is drained and the pentane is emitted without further treatment. The oil and the components removed are further treated at specialized facilities (Petersen, 2008). Modern electrolyte capacitors are normally found situated directly upon the printed circuit board and it is unlikely to contain hazardous compounds (Harding and Willis, 2004). The capacitors contain around 65 % aluminum and these are sent for aluminum recovery and controlled incineration. The rest of the cooling white goods are shredded under vacuum and up to 84 % is sent for recycling (Petersen, 2008).

### 11.2.3.2   Heating White Goods

The treatment category 'Heating white goods' contains stoves, ovens, microwave ovens and washing machines. These products consist of iron, plastic and other metals that can be recycled. The white goods are manually checked to see if they contain asbestos. Any asbestos, capacitors, printed circuit boards, components containing ceramic fibers and the outer cables are removed and treated separately as specified in the WEEE directive. The components containing asbestos and ceramic fibers are not treated but are sent to special landfills. Asbestos has been banned for 20 years but treatment facilities still may receive asbestos containing products (Harding and Willis, 2004). Similar to the cooling white goods, the capacitors are sent for aluminum recovery and controlled incineration. Washing machines contain concrete which is most often landfilled (Petersen, 2008).

### 11.2.3.3   Small Household Appliances

'Small household appliances' is often divided into 'low grade' and 'high grade' depending on the content of valuable metals. The low grade fraction contains products like vacuum cleaners, toasters, clocks, and sewing machines. The high grade fraction contains products like radios, digital cameras, computers, and mobile phones. The basic treatment of the high and low grade fractions are the same but managed separately because it is more economical for the industry.

Typically small household appliances are pretreated with the first step being manual decontamination where larger items of value as well as components that are specified by the WEEE directive (2002/95/EC) and the European battery directive (2006/66/EC) are removed (cables, central processing units (CPU), big metallic parts etc.). This is then followed

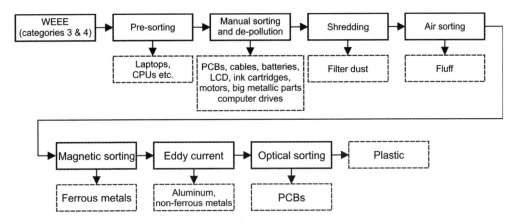

**Figure 11.2.1**  *An example of a WEEE pretreatment facility treating small household appliances. The dotted boxes correspond to the resulting output fractions (Bigum and Brogaard, 2009).*

by shredding of the fraction and subsequent removal of metals and valuable fractions, by means of different reefing technologies. However it is also possible to shred the waste first and then afterwards manually remove the desired compounds. An example of a pretreatment facility capable of treating small household appliances with the typical output fractions from high grade WEEE is shown in Figure 11.2.1.

### 11.2.3.4  TV and Monitors

The treatment category 'TV and monitors' contains primarily television and computer monitors. Monitors are basically treated similarly and two types of recovery technologies exist: 'Glass to glass' recycling and 'glass to lead' recycling. Glass to glass recycling is the process where cathode ray tube (CRT) cullet is used for the production of new CRT, so-called closed loop recycling (Herat, 2008). The CRT consists of two types of glass: Front glass and cone glass containing lead (Petersen, 2008).

Like most kinds of WEEE, the outer cables are removed as the first treatment. The appliances are then transferred to a conveyor belt for manually dismantling. First the screen is removed from the plastic or wood casing. Then the glass screen, demagnetizer, printed circuit board and electron gun are separated into different containers to be transported to the next treatment facility. Glass to glass recycling may result in 100 % recycling of the front glass and 96 % of the cone glass (Hansen, 2008).

Glass to lead recycling can be done by shredding the entire television and monitor, followed by a mechanical removal of the fluorescent coatings and recovery of the glass cullet. The cullet hereafter undergoes a melting process where all the recovered CRT (both the front and the cone glass) is melted and lead and copper are recovered. When lead and copper is removed, the remaining melted glass is still polluted with compounds such as lead, cadmium, barium, and strontium (Harding and Willis, 2004; Herat, 2008).

The main advantage of glass to glass recycling is the avoidance of having to landfill the glass. The disadvantage of recycling the CRT glass is that the composition of the glass cullet varies and many manufacturers are therefore reluctant to use CRT cullet. Another disadvantage is that this treatment requires manual labour, which means that the cost of the treatment is high. Glass to lead recycling is an automated process and is as such cheaper than glass to glass recycling. The disadvantage of this method is that it can be difficult to find uses for the recycled glass because of its level of contamination (Herat, 2008).

### 11.2.3.5  Lighting Equipment

The treatment category 'Lighting equipment' contains primarily fluorescent tubes. The tubes are manually sorted according to different lengths and are fed into a machine specially designed for the treatment. The metal ends (10 wt%) are cut off and these are later cleaned and recycled as metal scrap. The fluorescent powder (around 0.1 wt%) is blown out and collected. The collected powder is sent for further treatment, where it can be treated in a specialized treatment facility to be reused directly for new fluorescent tubes or be landfilled (Skårup *et al.*, 2003; Hansen, 2008). The glass (90 wt%) is crushed and sent for recycling for new fluorescent tubes. During the crushing process the air is polluted with mercury and therefore has to be cleaned before being emitted (Petersen, 2008).

## 11.2.4   Treatment Technologies

The treatment of WEEE depends not only on the treatment category, but also on the individual company and the metal prices. However, the overall treatment of WEEE can be described as a two step process:

1.   Manual and mechanical pretreatment
2.   Refining

In the EU, Article 6 of the WEEE directive requires that member states ensure that the treatment of WEEE is done using the best available technology for treatment, recovery and recycling (BATRRT). This can be done by the individual producers or collectively, but the member states are required to ensure that BATRRT is used also when WEEE is managed by third parties (DEFRA, 2006).

### 11.2.4.1   Manual and Mechanical Pretreatment

Pretreatment of WEEE has the overall purpose of removing larger valuable parts and/or hazardous parts as specified in the WEEE directive, as well as upgrading the fractions which are routed to recycling facilities. The pretreatment can be a mix of manual and mechanical dismantling.

Manual dismantling can be performed both before and after mechanical treatment (Hagelüken, 2006; Dalrympel *et al.*, 2007) and has the potential of a very high recovery rate of the outsorted fractions, but it also has some disadvantages. One of the disadvantages is the high cost as it is very labour intensive. Another problem with manual sorting is that although it has the possibility of being very effective, the people who actually conduct the sorting might for various reasons be less effective. In some countries, e.g. Germany, manual sorting is a protected job performed by people who, for some reasons, might have difficulties maintaining a job. This in practice means that the outsorted fractions sometimes have to be resorted several times (Sons, 2008).

Deciding on the optimal treatment technologies for the pretreatment of WEEE is a concentration dilemma, seeking a compromise between purity and recovery of an output stream. If a high purity (high concentration) of a given metal in an output is targeted it causes less recovery and a loss of the same metal to other streams. Likewise, a high recovery rate means that the output stream contains more impurities because of unintended coseparation (Hagelüken, 2006).

Mechanical processing has the advantage of being relatively cheap in investments and operating cost compared to manual sorting, but has some limitations that affect the sorting efficiency (Hagelüken, 2006). Some mechanical treatment technologies are shredders, magnetic separators, eddy current separators, optical sorters, and air classifiers (see also Chapter 7.1 for more information on the technologies):

- **Shredding:** The aim of shredding is to reduce the size of WEEE which makes the removal of certain fractions more effective. Cui and Forssberg (2003) suggested that WEEE should be shredded into fine particles below 5 or 10 mm in order to achieve the maximum separation of metals. However recent studies by Chancerel *et al.* (2008) suggest that the shredding might have a significant negative impact on the precious metal recovery, as the shredding of the WEEE increases the risk of precious metal-rich materials ending up in wrong fractions from where the metals are never recovered. Chancerel *et al.* (2008) suggests three ways that the loss of precious metal-rich materials could be minimized:
  - Manual sorting could be expanded to also remove smaller precious metal-rich materials.
  - High grade and complex materials such as mobile phones and digital cameras should not be shredded and only the battery should be removed prior to the metallurgical recycling processes.
  - The automated shredding and sorting technologies could be improved.

- *Air classifier:* Air classifiers remove less dense materials by blowing air through the waste fraction. The light fraction removed from the high grade WEEE by the air classifier is called fluff and is a waste fraction of very little value which is not recyclable (Chancerel, 2008).
- *Magnetic separation:* Magnetic separators remove the ferrous metals and are most efficiently used on shredded waste fractions. The ferrous iron can easily be recycled.
- *Eddy current separation:* Eddy current separators use alternating magnetic fields which create a force that repels nonmagnetic metals such as aluminum, copper, and stainless steel (Cui and Forssberg, 2003). The nonferrous metals can easily be recycled.
- *Optical sorting:* Optical sorters use color sensitive cameras to identify different types of plastics. For the pretreatment of high grade WEEE the optical sorters are used to sort out the high value shredded printed circuit boards (PCBs). The PCBs are rich on precious metals and are sent for metallurgical treatment for the recovery of copper, gold, silver, and platinum group metals.

The plastic fraction is the fraction remaining on the conveyor belt when all other fractions have been removed. The plastic fraction is a rather large part of the high grade WEEE (about 25 %) and is sent for recycling. A study performed by Chancerel *et al.* (2008) showed that significant amounts of the valuable precious metals from small household appliances end up in the plastic fraction from where it is not subsequently recovered.

### 11.2.4.2 Refining

The refining of the output fractions from the pretreatment is the process where the desired compound or metal is turned into a secondary raw material. The refining processes are therefore numerous, depending on whether or not the output fractions are metals, glass or plastic, and are not covered here. An example of refining: A copper fraction including precious metals from the pre-treatment will be sent to a metal smelter for metallurgical treatment. The traditional way to recover non-ferrous metals and precious metals from WEEE is by pyrometallurgical processing which includes extraction of metals via incineration, smelting, and other techniques at high temperature (Cui and Zhang, 2008). The three main metallurgical facilities in Europe are Boliden in Sweden, Aurubis (formerly known as Norddeutsche Affinerie) in Germany and Umicore in Belgium (Jalkanen, 2008).

## 11.2.5   Environmental Considerations/Aspects

Proper waste management of WEEE is necessary because WEEE contain many compounds that can be a threat to both human health and the environment if not treated and disposed off properly. The waste management of WEEE is however also important from another environmental perspective as it contains many metals that can be recycled and thus substitute virgin production and preserve resources.

### 11.2.5.1   Hazardous Compounds

Heavy metals are found in WEEE and are, among others: lead, cadmium, and mercury which are damaging to human health and to the environment, even in small doses. WEEE also contains heavy metals like nickel, copper, chromium, and zinc, which are damaging to humans and environment in higher amounts.

Heavy metals are emitted to the environment if WEEE is not collected and treated properly. If WEEE is wrongfully incinerated, the emission of heavy metals to air depends on the efficiency of the flue gas cleaning system of the incinerator. If the incinerator has a highly efficient gas cleaning system, the major part of the heavy metals ends up in the solid residues, which might limit the further utilization of the residues. Landfilled WEEE could lead to increased heavy metal concentrations in landfill leachate.

Flame retardants are used for the prevention of fire in electrical compounds by delaying the ignition and spread of fire. The European Commission has estimated that, in the period 1992–2002, the use of flame retardants reduced fire related deaths by 20 % (Landry and Dawson, 2002). Flame retardants like polybrominated diphenyl ethers (PBDEs) and triphenyl phosphate (TPP) are hazardous compounds which are also found in WEEE. The PBDEs are persistent and have been detected several places in the environment, in breast milk, and in human blood. PBDEs can cause long-term effects on the brain of animals and cause effects on the immune system. TPP is toxic and can be hormone disrupting (Brigden, 2008). When plastics containing brominated flame retardants are burned hydrogenbromide (HBr) is formed. HBr is known to contribute to the formation of polybrominated dibenzodioxins and dibenzofurans, which is very toxic and bioaccumulative in humans, animals, and plants (Thestrup, 2005). Both the PBDEs and some of the heavy metals have been banned through the RoHS directive for several years now and it is not very likely that they are found in high amounts in WEEE today. Damages from managing WEEE containing hazardous compounds have been reported for employees in the WEEE recycling industry (Brigden, 2008). It is therefore important for the recycling industry to be aware of both the compounds that have been banned and hazardous compounds which are currently in use.

### 11.2.5.2  Valuable Compounds

WEEE contains valuable compounds, in particular metals in IT and telecommunication equipment; the high value metals include: Gold, silver, palladium, platinum, and copper. Aluminium and iron are also valuable metals, because the lower market prices are compensated by larger amounts.

Copper is a metal with a high value because the known reserve is small. If the production and use of copper had continued as it did in 2001, but without any recycling of the metal, it would only be possible to produce virgin copper until 2027 (MST, 2003). The recovery of aluminium from WEEE has an additional environmental benefit to the resource aspect as it is much less energy consuming to extract and recycle aluminium from WEEE compared to the virgin production (See Chapter 5.4).

Scarce resources and lower environmental burdens from recycling compared to virgin metal production are the main arguments for the recycling of metals.

### 11.2.5.3  Environmental Assessments of WEEE Management

WEEE is a very heterogeneous waste type with varying treatment possibilities. The environmental costs of the energy, materials, and the chemicals used for the treatment should not be higher than the environmental benefits of the recycling of WEEE. There is therefore a need for a better understanding and control of the existing treatment technologies and also environmental assessments for the treatments possibilities in order to make the best environmental choice for the treatment of WEEE by for instance optimizing technologies and/or avoiding certain treatment options. The environmental assessment could be done by means of life cycle assessments (LCA), which is however considered to be a quite elaborate assessment method which needs much data as well as some allocation aspects that are considered problematic. Another method could be the determination of the environmental rucksack called the total mass requirement (TMR) method or the gross intrinsic value (CIV) methods, which are both considered simpler methods but which are still under development (Chancerel *et al.*, 2009). The quotes for environmentally weighted recyclability (QWERTY) concept is a fourth approach that determines environmentally based recyclability scores to give insight to the environmental aspects of WEEE management by also considering losses of metals during the treatment steps.

Chancerel *et al.* (2008) showed that there were significant losses of metals during pretreatment. This loss of metals was found to have a significant negative impact on the environmental benefits of the WEEE management as determined by Bigum and Brogaard (2009).

Barba-Gutiérrez *et al.* (2007) used LCA to determine the threshold beyond which the collection of WEEE would constitute a higher environmental burden compared to the environmental benefits gained from the proper management as determined by the WEEE directive. Different appliances were examined (washing machines, refrigerators, TV sets, personal computers) and the result showed that the travel distances (return trip) for the different appliances should not exceed 200–300 km, depending on the appliance type and the environmental categories. The transportation of WEEE should therefore be considered when selecting WEEE management systems.

# References

Barba-Gutiérrez, Y., Adenso-Diaz, B. and Hopp, M. (2008): An analysis of some environmental consequences of European electrical and electronic waste regulation, *Resources, Conservation and Recycling*, 52, 481–495.

Bigum, M. and Brogaard L. (2009): *LCA Modelling of Metal Recovery from WEEE*. Master thesis project. Technical University of Denmark, Copenhagen, Denmark.

Brigden, K., Labunska, I., Santillo D. and Johnston, P. (2008): *Chemical contamination at e-waste recycling and disposal sites in Accra and Korforidua*. Technical Note. Greenpeace Research Laboratories, Greenpeace International, London, UK.

Chancerel, P. (2008): Personal communication. Institute of Environmental Technology, Technische Universität Berlin, Germany, chancerel@ut.tuberlin.de.

Chancerel, P., Meskers, C., Hagelüken C. and Rotter, S. (2008): E-scrap, metals too precious to ignore. *E-Scrap Research*, 2008, 42–45.

Chancerel, P., Meskers, C., Hagelüken C. and Rotter, S. (2009): Setting priorities for the end-of-life management of complex products. In: CISA (ed.) *Proceedings of the 12th International Waste Management and Landfill Symposium*. CD-ROM. CISA – Environmental Sanitary Engineering Centre, Cagliari, Italy.

Cobbing, M., (2008), *Toxic tech: Not in our Backyard*, Greenpeace International, http://www.greenpeace.org/raw/content/international/press/reports/not-in-our-backyard.pdf, (Accessed the 23rd of March 2010).

Cui, J. and Forssberg, E. (2003): Mechanical recycling of waste electric and electronic equipment: a review. *Journal of Hazardous Materials*, B99, 243–263.

Cui J. and Zhang, L. (2008): Metallurgical recovery of metal from electronic waste: A review. *Journal of Hazardous Materials*, 158, 228–256.

Dalrymple I., Wright N., Kellner R., Bains N., Geraghty K., Goosey M. and Lightfoot L. (2007): An integrated approach to electronic waste (WEEE) recycling. *Circuit World*, 33, 52–58.

DEFRA (2006): *Guidance on best available treatment recovery and recycling techniques (BATRRT) and treatment of waste electrical and electronic equipment (WEEE) located*. Department for Environment, Food and Rural Affairs, London, UK.

EMPA (2009): Swiss Federal Laboratories for Materials Testing and Research, St.Gallen, Switzerland. Located on the internet: 3 March 2009. http://ewasteguide.info.

EPA (2008): *Fact sheet: Management of electronic waste in the United States*. United States Environmental Protection Agency, Washington, D.C., USA. Located on the internet: 27 February 2009. www.epa.gov/ecycling.

Hagelüken, C. (2006): Improving metal returns and eco-efficiency in electronic recycling. *Proceedings of the IEEE International Symposium on Electronics and Environment*, 2006, 218–223.

Hansen, H.J. (2008): The H.J. Hansen website. Located on the internet: 19 July 2009. www.elektro-miljoe.dk/dk/sider/f behand.html.

Harding, A. and Willis, K. (2004): *Consultation on guidance of treatment of waste electrical and electronic equipment (WEEE)*. Consultation draft. Harding and Willis, London, UK.

Herat, S. (2008): Recycling of cathode ray tubes (CRTs) in electronic waste, Griffith School of Engineering, Griffith University, Queensland, Australia. *Clean*, 36, 19–24.

Hester, R.E. and Harrison, R.M. (2009): Electronic waste management, design, analysis and application. *Issues in Environmental Science and Technology*, 27, RSC Publishing, London, UK.

Huisman, J., Ansems, T., Feenstra, L., & Stevels A., (2001) *The QWERTY concept, A Powerful Concept for Evaluating the Environmental Consequences of End-of-Life Processing for Consumer Electronic Products*, ecodesign, pp.929, 2nd International Symposium on Environmentally Conscious Design and Inverse Manufacturing, Tokyo, Japan.

Huisman, J., Magalini F., Kuehr R., Maurer C., Ogilvie S., Poll J., Delgado C., Artim E., Szlezak J., Stevels A., (2008), *2008 Review of Directive 2002/96 on Waste Electrical and Electronic Equipment (WEEE), Final Report,* United Nations University, Tokyo, Japan. http://www.ewasteguide.info/biblio/2008-review-d (Accessed 23rd of March 2010).

Jalkanen, H. (2006): *General view of WEEE and SR recycling and comments on pyrotreatments*. Technical University of Kosice, Faculty of Metallurgy. Located on the internet: 3 March 2009. http://web.tuke.sk/hfknkaso/slovak/konferencie/stofko%20jubileum/pdf/Jalkanen2.pdf.

Landry, S.D. and Dawson, R.B. (2002): *Life-cycle environmental impact of flame retarded electrical and electronic equipment*. Albemarle Corporation, London, UK.

Lui, X., Tanaka, M., Matsui, Y., (2006), Electrical and electronic waste management in China: progress and the barriers to overcome, *Waste Management & Research*, 24, 92–101.

MST (2003): *Affaldsstrategi 2005–2008* (Waste strategy 2005–2008; in Danish). Miljøstyrelsen, Copenhagen, Denmark.

Petersen, P.B. (2008): Personal communication with Poul Bengt Petersen from the company H.J. Hansen. www.elektro-miljoe.dk/dk/sider/f behand.html.

Schafer, T., van Looy, E., Weingart A. and Pretz T. (2003): Automatic separation devices in mechanical recycling processes. *Proceedings of the International Electronics Recycling Congress*, 2003.

Skårup, S., Christensen, C. L., Maag, J. and Jensen, Heilemann, S. (2003): *Massestrømsanalyse for kviksølv 2001*. Miljøprojekt Nr. 808. COWI A/S, Miljøstyrelsen, Copenhagen, Denmark.

Sons, C.T. (2008): personal communication. Der Steg, Berlin.

Thestrup, J. (2005): *Elektronikprodukter og miljøet* (Electronic products and the environment; in Danish). Chairman for Elektronikpanelet. In: *LMFK magazine*. Located on the internet: 27 February 2009. http://www.lmfk.dk/artikler/data/artikler/0502/0502 06.pdf.

UNEP (2000): *The Montreal protocol on substances that deplete the ozone layer as either adjusted and/or amended in London 1990, Copenhagen 1992, Vienna 1995, Montreal 1997, Beijing 1999*. Ozone Secretariat, United Nations Environment Programme, Nairobi, Kenya.

# 11.3

# Preservative Treated Wood

**Ina Körner**

*Hamburg University of Technology, Hamburg, Germany*

**Jenna Jambeck**

*University of Georgia, Athens, USA*

**Hans Leithoff**

*Johann Heinrich von Thünen-Institut, Hamburg, Germany*

**Volker Lenz**

*German Biomass Research Center, Leipzig, Germany*

Wood is an attractive renewable material for many different applications. At its end of life, it becomes waste wood. Waste wood includes various products from dimensional lumber to wood-based materials (e.g. particle boards, plywood). The wood products may have been in use for a short period (e.g. wood for casing of buildings: < 1 year) or a long period (wood from demolition of old buildings: > 100 years). The wood products may be untreated or treated with preservatives, laminates, paints or coatings. Products often composed of wood with some treatment include particle board, floor board, form board, interior doors, panels, and furniture. In general, wood is biologically degradable. Treatment of wood with biocides inhibits wood damage by fungi or insects. Wood must be protected for most exterior applications, especially in the residential and commercial building sectors. Preservatives are used for construction timber, window frames, claddings, transmission poles, posts, fences, and railway sleepers. Other types of preservatives can be used to reduce flammability (Marutzky *et al.*, 1993). This chapter focus on wood preserved by biocides only, since this type of wood constitutes a growing waste type and it is often considered as a special waste because of the biocide content.

*Solid Waste Technology & Management*   Edited by Thomas Christensen
© 2011 Blackwell Publishing Ltd

## 11.3.1   Wood Preservation

Wood treatment preservatives are typically liquid substances applied by coating, brushing, spraying, dipping, or impregnation. The preservation substances contain an active agent to protect the wood and may also contain other compounds such as corrosion inhibitors, emulsifiers, adjuvants, pigments, colours, binding agents, as well as organic solutions or water as a solvent (Marutzky *et al.*, 1993; Lay and Marutzky, 2003). The variety of active agents may generally be distinguished into three groups:

- *Group 1: Water soluble media* contain mostly inorganic salts with the elements As, B, Cr, F, Cu, Zn, and Hg in different combinations, but also organic agents are available (Marutzky *et al.*, 1993; Lay and Marutzky, 2003).
- *Group 2: Media with solvents* contain organic agents, of which a large range of active ingredients can be found (at least 100 exist). The solvents can be aliphatic or aromatic hydrocarbons as well as esters of carboxylic acid, alcohols, and glycols (Marutzky *et al.*, 1993; Voß, 1998; Lay and Marutzky, 2003).
- *Group 3: Creosotes* are byproducts of pyrolysis processes with an undefined composition and a multitude of compounds of primarily organic nature (probably more than 1000). PAHs and benzo(a)pyrene are often identified as indicators of creosote compounds. It is not possible to distinguish between dissolving and active agent (Lay and Marutzky, 2003; Schwarz, 2003).

The active ingredients differ strongly with respect to how long they have been in use. Some agents have been in use for 150 years (e.g. $CuSO_4$, coal creosote; Marutzky *et al.*, 1993; Lay and Marutzky, 2003). Others were on the market only several years because of laws and regulations. Cr-Cu-As salts like chromated copper arsenate (CCA) (Group 1, e.g., forbidden in Germany) dominate in most countries, but also PCB-containing preservatives (Group 2, e.g., forbidden in Germany) and creosotes (Group 3) are very common. CCA treated wood was a very popular wood product in the USA for more than 25 years, but as of 1 January 2004, became restricted from most residential uses of treated wood (EPA 2002) and now other non-arsenic-based (Group 1) products are in use. The numerous treated wood products that exist throughout the world result in a variety of possible waste wood contaminations. For example, in Germany more than 80 different active agents belonging to Groups 1 and 2 were detected in more than 1000 different wood preservatives (Brückner and Willeitner, 1992; Voß, 1998). In addition, while some treatment compounds are now banned/restricted, because wood products are in service for many years (typically 10–40 years or more), the waste wood also often contains agents which are no longer produced.

## 11.3.2   Hazardous Potential

The active agents of wood preservatives differ with respect to their wood protection efficiency and also to their human toxicology and ecotoxicology. Hazards by exposure to the preservative contained in the wood during its use are commonly less significant for most biocides. Harmful volatile compounds are contained only in some preservatives (e.g., creosotes) or their solving agents. Adverse health effects could mainly occur if the wood preservative or the preservative treated wood is used improperly (e.g., on wood inside a building). A continuous direct contact exposure to preservative treated wood (e.g., at children playgrounds) could result in health problems for some agents such as As. In the USA, the phase out of CCA treated wood production was initiated because of concerns associated with possible health effects from contact with treated wood and the availability of non-arsenical wood preservatives (EPA 2002, 2003).

During use of the wood product, preserving agents are released over time. Organic preservatives decrease approximately 50 % in 10 years and inorganic agents decrease up to 9 % each year (Thurmann, 1999; Klipp, 1994). These emissions contribute to the uncontrolled distribution of hazardous substances into the environment (mainly in soils). Although the end of life management is not often considered when a product is developed, the hazardous potential of preservative treated wood remains a significant concern during treatment or disposal.

Due to the hazardous potential the use of some substances is forbidden by law in some countries. The regulations and the limiting values vary worldwide. In Germany, Hg, As, and PCBs are forbidden in wood preservatives. PCP is limited under 5 mg/kg but only, if the waste wood will not be environmental friendly re-used or incinerated (ChemVerbotsV,

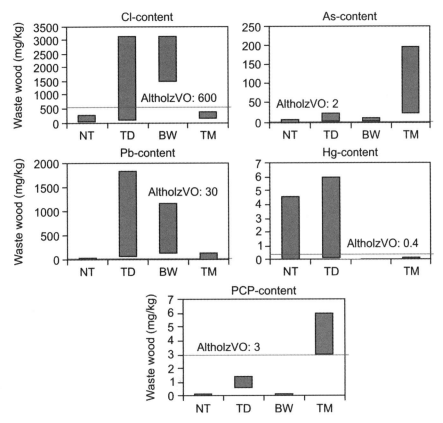

***Figure 11.3.1*** *Analytical results for selected hazardous substances occurring in different German waste wood fractions (data from Jasper, 1997; Scheithauer et al., 2009) compared to the limiting values (AltholzVO) for particle boards according to the German waste wood regulation (AltholzV, 2001). NT – Untreated wood; TD – timber and demolition wood; BW – wood from bulky refuse; TM – telephone mast as example of preservative treated wood.*

2003). PCBs have been prohibited in the European Union (EU) since 2004 by EU regulation 850/2004. CCA treated wood is banned for most residential applications in the USA, pentachlorophenol (PCP) treated wood is also similarly banned. Both can still be used in nonresidential products (e.g., telephone poles). PCBs have been banned since 1977 from production in the USA.

The content of hazardous compounds may vary in the different waste wood fractions (Figure 11.3.1). Traces are also found in untreated wood fractions due to a ubiquitous loading in the environment. Furthermore other wood treatments may add certain compounds as well (e.g., heavy metals from coatings).

### 11.3.3 Characterization of Preserved Wood

Since regulations vary from country to country on the use of various treated wood products, the quantities of treated wood entering the waste stream vary greatly throughout the world and even within countries. Furthermore, numbers on the amount of preservative treated wood depend very strongly on definitions and the statistical completeness within a country. Table 11.3.1 provides some estimates of amounts per person per year:

**Table 11.3.1**   *Generation of waste wood and preservative treated wood for different countries (kg/person/year).*

| Country | Waste wood | Preservative treated waste wood |
|---------|------------|--------------------------------|
| Denmark[a] | Not reported | Approximately 20 |
| Germany | 107 | 32 |
| USA | 172[b] | 16[c,d] |
| Canada | 52[e] | 12[d,f] |

[a]DEPA (2008)
[b]EPA (2003)
[c]Jambeck (2006)
[d]CCA treated wood only in 2010
[e]StatCan (2006)
[f]Cooper (1994)

- For Germany the total waste wood amounts to about 107 kg/person/year, including the import of waste wood of about 13 %. Hereof about 64 % is treated wood and about 30 % is preservative treated wood. So the amount of preservative treated wood waste is about 32 kg/person/year (IE, 2007).

- For the USA, one of the most common types of treated wood was CCA treated wood until CCA was restricted. Based upon disposal estimates the quantity entering the waste stream is forecasted to be 6–10 million m³/year in the years 2000–2030 (Jambeck *et al.*, 2007). Cooper (1994) estimated the amounts of CCA treated wood removed from service in the year 2010 at approximately 10 million m³ in the USA and 0.8 million m³ in Canada. Based upon 2009 projections of CCA treated wood waste and population, these estimates equal 16 kg/person/year for the USA and 12 kg/person/year in Canada. These quantities do not include any other type of treated wood. While this quantity will decrease since CCA treated wood is now restricted from residential uses in the USA, quantities of the alternative treated wood products (e.g., ammonical copper quaternary, ACQ), are expected to increase.

Different possibilities to distinguish between treated and non treated waste wood are discussed in the following. However, practical relevance today has only the characterization of the assortment and the origin. Various qualitative and quantitative methods for sorting and identification of wood which were primarily treated with water-soluble media have been evaluated regarding their effectiveness and costs by Solo-Gabriele *et al.* (2004), Block *et al.* (2007), Jacobi *et al.* (2007), and Omae *et al.* (2007).

- *Chemical analysis:* General chemical analytical methods can be used to identify individual substances in the waste wood. Most of the methods available require destruction of the wood matrix prior to analysis, and most methods determine only selected substances (Lay and Marutzky, 2003). These measurements are usually very costly and labour intensive. Some easy methods, commonly based on colour reactions, exist for qualitative determinations. At the moment no fast detection method covering the wide spectra of hazardous substances in waste wood is known (Meier et al., 1998). A further problem is a representative sampling. Waste wood fractions are often mixed and also the single piece of waste wood is not homogeneous.

- *Sensorical detections:* Colours may be used for characterization. Waste wood with a brown/black colour and a strong tarry smell may be treated with creosotes. Yellow/green waste wood may be impregnated with Cr or Cu salts. But the colour may not be visible very clearly after several years of usage, due to dirt, greying, and leaching. Furthermore very often wood may be coloured anyway for optical reasons (Marutzky et al., 1993; Lay and Marutzky, 2003). Some of the waste wood assortments may also have markings which determine the preservation (e.g., some telephone masts, particle boards). All these detections methods are very afflicted with failures (Meier *et al.*, 1998). But if the assortment and origin of the waste wood is known, they may be useful.

- *Characterization of the assortment and the origin:* The possibility of tracing a preservative treatment is easiest at the source of the waste wood. If the assortment of the waste wood is known, it may be possible to estimate what kind of preservation has been used. For example, load bearing parts of a roof are most likely impregnated, whereas interior casings are most likely untreated. It is also possible to consider regional differences. For example, packaging wood from Germany is most likely untreated, whereas packaging wood from overseas may be treated.

## 11.3.4   Treatment and Disposal

Generally, preservative treated wood can be reused, used for energy production, incinerated, or landfilled. Additional treatment methods such as decontamination are still experimental; see Box 11.3.1.

### 11.3.4.1   Reuse

Preservative treated wood can be reused in areas which are similar to the first application. The remaining preservation property can still be used and coatings and sealants can keep preservative from leaching out at a fast rate. Some preservatives (applied by hand) can be refreshed if desired and the wood is kept in use. In that case, use of the same preservative type is suggested, so that the later disposal does not become more difficult. Examples are the reuse of railway sleepers in landscaping for the fixation of slopes and coating of lanes as well as the reuse of old masts and external casings to produce fence slats (Marutzky *et al.*, 1993). However, reuse is quantitatively less important and the danger of incorrect applications and impacts to the environment exist.

---

**Box 11.3.1   Detoxification of Treated Wood.**

Methods for detoxification of preservative treated wood are being researched in order to make it possible to utilize treated wood. Some of the methods are described below. However, these methods have still no practical relevance. Often they are in an early research stage, the technical concepts are not fully developed, or the concepts are too complicated to be economical viable. In some cases, it may not be possible to reduce the content of preservative agents to a sufficiently low level.

- *Mechanical treatment:* Preservative treated surfaces of waste wood may be removed mechanically to use the inner, untreated wood parts only. Examples could be wire masts and timber with large dimensions and low penetration depths (Marutzky *et al.*, 1993). Investigation were carried out e.g. by Kluge (1991) with Hg-containing waste wood. However, preservatives may partly have penetrated also into the inner parts e.g. by cracks (Marutzky *et al.*, 1993).
- *Biological treatment:* Special kinds of microorganisms can transform certain inorganic preservatives into water soluble compounds which afterwards can be washed out. Leithoff (1997) used a treatment with the brown rot fungi *Antrodia vaillantii* for Cu-Cr-containing waste wood. In cases of organic wood preservatives microorganisms can be used to degrade the preservative into $CO_2$ and $H_2O$, but the preservatives were partly also enriched in the microorganisms and transformed into other products. The best investigated organic preservatives are PCP (e.g. Lamar and Dietrich, 1992; Venkatadri *et al.*, 1992) and creosotes (e.g. Wilson *et al.*, 1985, Thomas *et al.*, 1989). Mostly white rot fungi or bacteria were used (Voß, 1998). The decontaminated wood could be used only for composting, since the wood substance was partly degraded as well.
- *Chemical reatment:* The separation between waste wood and preservative can be done by chemical extraction, pressure and increased temperature. Stephan *et al.* (1996) and Borgards (1997) described decontamination possibilities during pulping using acidic or alkaline conditions common for delignification. The decontaminated pulp was supposed to be used for paper production. Schwarz (2003) used defibration and washing with tensides for removal of creosotes from railway sleepers. The treated material was intended for use in particle boards for outdoor applications.

### 11.3.4.2 Incineration

Depending on the characterization of the waste wood various options for incineration are available. National regulations have to be considered. In Germany, e.g., the incineration of preservative treated wood has to fulfil the demands of the 17th BImschV (2003) regarding total dust, CO, TC, NO$_x$, SO$_2$, HCl, HF, Hg, dioxin/furan, Cd/Tl, and other heavy metals in the flue gas. Preservative treated wood without PCB and PCP can, e.g., be used for thermal utilization in biomass combustion plants, for cofiring in power plants using coal, and for production processes. Additionally these wood fractions can be handled in thermal waste disposal plants. Waste wood with a high load of PCB and PCP is normally incinerated in special thermal waste disposal units or in cement kilns.

### *Thermal Utilization*

Due to the calorific value of treated wood (11–15 MJ/kg; water content 15–25 %; Marutzky, 2003) energy can be recovered by combustion processes. The calorific value is low in comparison to fossil fuels (natural gas: 33 MJ/kg; fuel oil: 42 MJ/kg) but high compared to fresh wood (8 MJ/kg; water content 50 %) or household waste (10–12 MJ/kg). Preservative wood without PCB and PCP can be burned in power plants for waste wood. During burning, organic wood preservatives may be destroyed to nonhazardous compounds such as CO$_2$ and H$_2$O, however, sometimes not completely. Aromatic organic compounds (e.g., creosotes) are more stable than nonaromatic compounds. Dependent on the active agents, pollutants such as hydrofluoric acid, sulfur dioxide and nitrous oxides may be generated (Marutzky *et al.*, 1993). Chlorine containing wood preservatives are a special problem since dioxins and furans may be produced (Strecker and Marutzky, 1994). Inorganic agents are not destroyed during combustion. To a large extent they are enriched in the solid residues. Of special importance are As, F, and Hg, which will be transferred into the flue gas (Marutzky *et al.*, 1993). The fate of important pollutants during incineration is summarized in Table 11.3.2. In the waste wood power plants, the flue gas is cleaned for particulate matter and gaseous emissions, so that stringent emission levels can be kept.

Wood treated with organic agents may also be coincinerated in cement kilns substituting for fossil fuel. There, no additional emission control and purification measures are necessary when preservative treated wood is used, since the very high incineration temperatures of ~2000 °C allow complete burnout without emissions (Meier *et al.*, 1998). Inorganic agent such as heavy metals will be integrated into the product clinker. Only Hg and Cr containing waste wood fraction should be avoided (Hg: volatile and is not detained by filters; Cr: limiting values in cement; Voß, 1989).

In Germany the installed capacity for waste wood combustion including preservative treated wood was in 2007 about 3 500 000 t (IE, 2007). This corresponds to about 42 kg/person/year. The installations were mainly initiated by financial incentives resulting from the legislation. Therefore even preservative treated wood is becoming a valuable feed for the power plants, offering to pay for the waste wood. In other countries the cocombustion of waste wood including preservative treated wood especially with coal fired power plants is important.

### *Thermal Disposal*

In addition to the use of waste wood for energetic purposes there are fractions of the preservative treated wood – especially wood with PCB and PCP – which must be incinerated in special facilities. Central waste wood incineration facilities

**Table 11.3.2**   *Fate of pollutants in preservative treated wood during incineration (based on Voß, 1998).*

| Pollutant | Deposition | Limitation and purification measures |
|---|---|---|
| As | Flue gas, fly ash | Filtration in cooled flue gas |
| Hg | Flue gas | Dry sorption |
| Cr, Cu | Mainly bottom ash | Fixation by reduction, precipitation, solidification, wet de-ashing |
| HCl, HF | Flue gas | Filtration after lime addition |
| Non-incinerated organic agents | Flue gas, fly ash | Good burnout, filtration |
| Dioxins and furans | Flue gas, fly ash | Good burnout, avoidance of formation due to regulation of filtration temperatures, filtration by tissue filters, destruction by high incineration temperatures |

which also can handle preservative treated wood fractions should not be too small due to the additional efforts regarding control, regulation and flue gas treatment (Voß, 1998).

Incineration facilities have special firing systems to make sure to have high burnout temperatures for enough time to secure a complete combustion of all organic components. Continuous measurements of the cleaned flue gas allow quick adaptations of the combustion behaviour (e.g., adding heat with an oil or gas burner) to keep legislative emission levels. Furthermore, some fractions of preservative treated wood waste may be disposed in waste incineration facilities (e.g., for mixed municipal waste). In this facilities the main intention is not energy recovery, but save disposal. Although these plants also commonly recover energy, the calorific value of waste wood is not as efficiently recovered as in waste wood power plants.

### Waste Wood Ashes

The ash content at complete burnout is of fresh wood usually less than 1%. The ash content of preserved wood is often higher: treated with organic preservatives: 0.5–2.0%; with inorganic preservatives: 2–5% (Pohlandt *et al.*, 1995). However, the burnout is not always complete. For example, the ignition losses from different waste wood ashes may range from 0.1 to 56.6% w/w (mean value 15.5% w/w; Anonymous, 1997). It demonstrates that residues of the preservatives stay in the ashes. Besides this, noncombusted carbon, mineral wood constituents, and residues of foreign materials are also retained. The main elements of wood ashes are Ca, Fe, K, Mg, Mn, and Na, and no significant differences are observed between treated and non treated wood ashes. However, the content of heavy metals, Cl, and F is higher in ashes from preservative treated wood than in ashes from non treated wood (Table 11.3.3). The ashes from incineration of treated wood are typically landfilled. Since the ashes of waste wood incineration may have a hazardous potential, e.g., in the USA, limits exist on the quantity of CCA treated wood allowed in fuel combustion plants. The limit in the state of Maine is < 1.5% by volume, with concentrations in the blended fuel limited to 50 mg/kg As, 375 mg/kg Pb, and 0.74 mg/kg PCBs (MDEP, 2006).

### 11.3.4.3 Landfilling

### *Landfilling of Treated Wood*

Preservative treated wood may affect, in particular, the leachate quality of the landfill, since preservative agents from the landfilled wood may become solubilized. Landfilled wood has been examined through laboratory batch leaching studies and leaching columns by various authors. Pohland (1998) examined the disposal of pentachlorophenol (PCP) treated wood. Jambeck (2004), Jambeck *et al.* (2006, 2008), and Khan *et al.* (2006) specifically examined CCA treated wood. It was found that the treated wood had an impact on the leachate quality. Landfilled in a monofill, with construction and demolition waste, and with MSW, the treated wood increase at least one of the elements As, Cr, or Cu in the leachate (Table 11.3.4). Other waterborne preservatives used on treated wood have also been evaluated for leachability in batch leaching studies conducted with landfill leachate (Table 11.3.5; Dubey *et al.*, 2004).

**Table 11.3.3** *Examples of composition of ashes (mg/kg dry matter) from different wood assortments (Pohlandt, 1995). Reprinted with pemrission from Pohlandt K. Zusammensetzung, Verwertung und Entsorgung von Holzaschen, in Holz-Zentralblatt, Jahrgang, Nr 79, pp 1305–1315 © (1995) DRW Verlag.*

| Element (mg/kg dry matter) | Ash | | |
| --- | --- | --- | --- |
| | Fresh wood | Residual wood | Treated wood[a] |
| Cl | 33–4510 | < 10–138 000 | 48–53 980 |
| F | 1–530 | < 2–564 | 1600–41 800 |
| As | < 1–35 | < 2–60 | 370–23 400 |
| Cr | < 10–592 | 16–810 | 4400–196 000 |
| Cu | 99–498 | 43–1450 | 4000–156 000 |
| Zn | 54–1900 | 16–6200 | 100–6550 |

[a] Treated with inorganic preservatives and including particle board

**Table 11.3.4**   *Leaching of As, Cr, and Cu from preservative-treated wood waste in a simulated landfill study. Data compiled from Dubey* et al. *(2004) and Jambeck (2004).*

| Parameter | L/S | As | Cr | Cu |
|---|---|---|---|---|
| Column leaching studies | | | | |
| Unit | l/kg | mg/l[a] | mg/l[a] | mg/l[a] |
| 100 % CCA treated wood | 0.6 | 42 | 9.4 | 2.4 |
| 10 % CCA treated wood with C&D debris | 0.8 | 2.2 | 1.3 | 0.007 |
| 2 % CCA treated wood with MSW | 1.1 | 0.5 | 0.20 | 0.13 |
| Batch leaching studies | | | | |
| ($L/S = 20$) | % leached | | | |
| CCA treated | 1.0–4.7 | 0.5–2.0 | 0.6–5.1 | |
| CCA treated[b] | 0.6–6.1 | 0.2–2.8 | 0.25–23.0 | |
| ACQ | BDL | BDL | 10–20 | |
| CBA | BDL | BDL | 33–55 | |
| CC | BDL | BDL | 15–38 | |
| CDDC | BDL | BDL | 5–15 | |

[a] An overall leachate concentration ($C_{overall}$) was calculated by dividing the total mass of each metal leached (mg) minus the total amount leached from the control (mg) by the total volume of leachate produced from the experimental column.
[b] Weathered wood. Notes: BDL = below detection limit; ACQ = ammonical copper quaternary; CBA = copper boron azole; CC = copper citrate; CDDC = copper dimethyldithiocarbamate.

## Landfilling of Ashes from Incineration of Treated Wood

Measurements on waste wood ashes showed that leaching – especially for Cr – often exceeded limit values. For example, when incinerated, CCA treated wood increases the As and Cr concentrations in the ash and the leachate produced from the ash (Solo-Gabriele *et al.*, 2002; Iida *et al.*, 2004). Table 11.3.6 contains leachability data of ash produced from mixed wood waste (construction and demolition wood). If only 5 % of the wood combusted is CCA treated wood, the ash can be characterized as a hazardous waste (Solo-Gabriele *et al.*, 2002) regarding USA guidelines. Investigations into minimizing the Cr load of waste wood ashes focus on washing, extraction, reduction by organic substances, fixation in building materials as well as on a thermal aftertreatment (Brill and Schlothmann, 2001).

**Table 11.3.5**   *Leaching data (minimum, mean and maximum values; eluate preparation regarding the guidelines in the AbfAblV, 2001) for different waste wood ashes as well as fresh and stored common waste incineration ashes in comparison to the German limiting values.*

| Parameter | Unit | Waste wood ashes[a] | Waste ashes[b] | | Limiting values[c] | |
|---|---|---|---|---|---|---|
| | | | Fresh | Stored | Class I | Class II |
| Pb | mg/l | 0.005–0.4–2.6 | ~0.1–3.2 | ~0.1–0.3 | 0.2 | 1.0 |
| Ca | mg/l | 0.0005–0.81–24.50 | — | — | 0.05 | 0.1 |
| Cr | mg/l | 0.01–5.48–300.0[d] | ~0.04–0.15 | ~0.03–0.13 | 0.05 | 0.1 |
| Cu | mg/l | 0.005–0.1–1.7 | ~0.04–0.54 | ~0.02–0.38 | 1 | 5 |
| Zn | mg/l | 0.005–1.06–27.0 | ~0.08–0.80 | ~0.02–0.21 | 2 | 5 |

[a] 33–64 analysed samples (Anonymous, 1997).
[b] 15 samples each (Pfrang-Stotz and Reichelt, 1999).
[c] German waste disposal regulation (AbfAblV, 2001): Class I: Landfills for waste containing a very reduced organic content (e.g. ashes); Class II: Landfills for waste containing a higher organic content (e.g. mechanically biologically pretreated waste).
[d] Data as Cr(VI).

**Table 11.3.6**  *Wood waste ash leachability (Solo-Gabriele et al., 2002). Reprinted with permission from Journal of Hazardous Materials, Characteristics of chromated copper arsenate-treated wood ash by H.M. Solo-Gabriele, T.G. Townsend, B. messick et al., 89, 2–3, 213–232 © (2002) Elsevier Ltd.*

| Parameter | Unit | Mixed wood ash 1 | | Mixed wood ash 1 | | Mixed wood ash 1 | |
|---|---|---|---|---|---|---|---|
| | | TCLP[a] | SPLP[b] | TCLP[a] | SPLP[b] | TCLP[a] | SPLP[b] |
| As | mg/l | 3.46 | 0.79 | 0.21 | 0.29 | 6.74 | 0.54 |
| Cr | mg/l | 24.1 | 55.0 | < 0.1 | 5.29 | 2.72 | 25.5 |
| Cu | mg/l | 0.45 | < 0.05 | 0.064 | < 0.05 | 0.093 | < 0.05 |

[a]USA regulatory test for leachability and hazardous waste characterization. Acetic acid based solution, pH = 4.98, $L/S = 20$.
[b]USA test for leachability. Dilute sulfuric/nitric acid solution, pH = 4.2, $L/S = 20$. Reproduced with permission from *Journal of Hazardous Materials.* © (2002) Elsevier Ltd.

### 11.3.4.4   Comparison of Incineration and Landfilling

Landfilling of CCA treated wood in a MSW landfill (landfill scenario) was compared with incineration including energy recovery and ash disposal (incineration scenario) for the same mass of waste wood by LCA studies (Jambeck *et al.*, 2007). The incineration scenario was more expensive, when operated in accordance with USA regulations, but it produces energy. From the ecological point of view, the incineration scenario releases As at a slower rate compared to the landfill scenario. Therefore, the incineration scenario releases less As on an annual basis, although As is more concentrated in the ash and in the leachate. But in contrast, in the incineration scenario Cr becomes concentrated and is often oxidized to Cr(VI), making it more toxic and mobile (Song *et al.*, 2006). If, in both scenarios, the landfills are lined and the leachate is collected, the groundwater may be protected and the pollutants released could then be managed through subsequent treatment. Higher concentrations of As and Cr in the ash leachate would increase management costs, however. If the wood is managed via incineration less landfill area is required (Jambeck *et al.*, 2007).

## References

AltholzV (2002): *Verordnung über Anforderungen an die Verwertung und Beseitigung von Altholz, Altholzverordnung vom 15. August 2002.* BGBl. I, Nr. 59, S. 3302.

Anonymous (1997): *Schadstoffströme bei der Gebrauchtholzverwertung für ausgewählte Abfallarten.* Fachbuchreihe des LUA NRW, LUA-Materialien Nr. 37. Landesumweltamt Nordreihn-Westfalen, Essen, Germany; ISSN 0947-5206.

BImSchV (2003): *17. BImSchV: Siebzehnte Verordnung zur Durchführung des Bundes-Immissionsschutzgesetzes.* Verordnung über die Verbrennung und die Mitverbrennung von Abfällen. BGBl. I, Nr. 41, p. 1633 et sqq.

Block, C.N., Shibata, T., Solo-Gabriele, H.M. and Townsend, T.G. (2007): Use of handheld X-ray fluorescence spectrometry units for identification of arsenic in treated wood. *Environmental Pollution*, 148, 627–633.

Borgards, A. (1997): *Untersuchungen zur Herstellung von gebleichtem ASAM-Zellstoff aus schutzsalzimprägnierten Hölzern und zu Wasserkreislaufschließung in der chlorfreien Bleiche.* Dissertation an der Universität Hamburg, Fachbereich Biologie, Hamburg, Germany.

Brill, S. and Schlothmann, V. (2001): Die Entsorgung von Holzaschen – ein Lösungsansatz. *Müll und Abfall*, 33, 668.

Brückner, G. and Willeitner, H. (1992): *Einsatz von Holzschutzmitteln und damit behandelter Produkte in der Bundesrepublik Deutschland.* Texte 48/92, Umweltbundesamt. Brückner and Willeitner, Berlin, Germany.

ChemVerbotsV (2003): *Chemikalien-Verbotsverordnung Verordnung über Verbote und Beschränkungen des Inverkehrbringens gefährlicher Stoffe, Zubereitungen und Erzeugnisse nach dem Chemikaliengesetz.* BGBl. I, Nr. S. 867, Gl.-Nr. 8053-6-20.

Cooper, P.A. (1994): Disposal of treated wood removed from service: the issue. in environmental considerations in the manufacture, use and disposal of preservative-treated wood. Forest Products Society, Madison, USA, pp. 85–90.

DEPA (2008): Nyttiggørelse af trykimprægneret træ – Nyttiggørelse af trykimprægneret træ ved metalekstraktion (Utilization of impregnated wood; in Danish). Miljøprojekt Nr. 1207. Danish EPA, Copenhagen, Denmark.

Dubey, B., Townsend, T. and Solo-Gabriele, H. (2004): *Metal leaching from pressure treated wood in sanitary landfill leachate*. IRG/WP 04-50220. IRG, Stockholm, Sweden.

EPA (2002): Notice of receipt of requests to cancel certain chromated copper arsenate (CCA) wood preservative products and amend to terminate certain uses of CCA products. United States Environmental Protection Agency, Washington, D.C., USA. *Federal Register*, 67, 36, 8244–8246.

EPA (2003): *A probabilistic risk assessment for children who contact CCA-treated playsets and decks*. Draft preliminary report. Office of Pesticide Programs, Antimicrobials Division, United States Environmental Protection Agency, Washington, D.C., USA.

IE (2007): *Monitoring zur Wirkung der Biomasseverordnung*. Endbericht 2007. Institut für Energetik und Umwelt gGmbH, Berlin, Germany.

Iida, K., Pierman, J., Tolaymat, T., Townsend, T. and Wu, C. (2004): Control of heavy metal emissions and leaching from incineration of CCA-treated wood using mineral sorbents. *Journal of Environmental Engineering*, 1302, 2, 184–192.

Jacobi, G., Solo-Gabriele, H., Townsend, T. and Dubey, B. (2007): Evaluation of methods for sorting CCA-treated wood. *Waste Management*, 27, 1617–1625.

Jambeck, J.R. (2004): *The disposal of CCA-treated wood in simulated landfills: Potential impacts*. PhD Dissertation, University of Florida, Gainesville, USA.

Jambeck, J., Townsend, T. and Solo-Gabriele, H. (2006): Leaching of chromated copper arsenate (CCA)-treated wood in a simulated monofill and potential impacts to landfill leachate. *Journal of Hazardous Materials*, A135, 21–31.

Jambeck, J., Weitz, K., Townsend, T. and Solo-Gabriele, H. (2007): CCA-treated wood disposed in landfills and life-cycle trade-offs with waste-to-energy and MSW landfill disposal in the U.S. *Waste Management*, 27, 8, S21–S28.

Jambeck, J., Townsend, T. and Solo-Gabriele, H. (2008): Landfill disposal of CCA-treated wood with construction and demolition (C&D) debris: arsenic, chromium and copper concentrations in leachate. *Environmental Science and Technology*, 42, 5740–5745.

Jasper (1997): *Altholzverwertung – die neuen Möglichkeiten: Markt- und Technologiestudie. UZD-Verlags- und -Beratungs-GmbH in Zusammenarbeit mit BVMW, Bundesverband Mittelständische Wirtschaft, Unternehmerverband e.V., Interessengemeinschaft Mittelständischer Verbände und Unternehmer–Umweltkommission*. UZD-Verlags- und Beratungs-GmbH, Dortmund, Germany; ISBN 3-9805292-1-5.

Khan, B., Jambeck, J., Solo-Gabriele, H., Townsend, T. and Cai, Y. (2006): Release of arsenic to the environment from CCA-treated wood: Part II – Leaching and speciation during disposal. *Environmental Science and Technology*, 40, 3, 994–999.

Klipp, H. (1994): *Auswaschung von Holzschutzmitteln aus behandelten Produkten und der Eintrag ihrer Wirkstoffe in die Umwelt*. Dissertation an der Universität Hamburg, Fachbereich Biologie, Hamburg, Germany.

Kluge, R. (1991): *Quecksilber (Hg)-Gehalt kyanisierter Hopfenstangen – Grundlage für ein Entsorgungskonzept*. In: Umweltaspekte der Tierproduktion, VDLUFA- Schriftenreihe 33, Kongressband 1991, VDLUFA-Verlag, Bonn, Germany, pp. 779–784; ISBN 3-922712-43-6.

Lamar, R.T. and Dietrich, D.M. (1992): Use of lignin-degrading fungi in the disposal of pentachlorphenol-treated wood. *Journal of Industrial Microbiology*, 9, 181–191.

Lay, J.P. and Marutzky, R. (2003): *Kompendium der Holzschutzmittelanalytik*. Springer Verlag, Berlin, Germany; ISBN 3-935065-11-6.

Leithoff, H. (1997): *Möglichkeiten und Grenzen der Überführung eines biologischen Reinigungsverfahrens für schutzsalzgetränkte Hölzer in den Technikumsmaßstab*. Dissertation an der Universität Hamburg, Fachbereich Biologie, Hamburg, Germany.

Marutzky, R., Peek, R.-D. and Willeitner, H. (1993): *Entsorgung von schutzmittelhaltigen Hölzern und Reststoffen*. Deutsche Gesellschaft für Holzforschung e. V., Munich, Germany; ISBN: 0466-2114.

MDEP (2006): *Solid waste management rules: Chapter 418*. Maine Department of Environmental Protection. Located on the internet: 30 October 2009. http://maine.gov/dep/rwm/solidwaste/index.htm#ru.

Meier, D., Wehlte, S., Simon, C. and Ollesch, T. (1998): *Stoffliche Verwertung von nicht naturbelassenen Holzresten durch Pyrolyse in der Wirbelschicht*. Abschlussbericht zum Projekt 03 631 financed by the Deutsche Bundesstiftung Umwelt. Bundesforschungsanstalt für Forst und Holzwirtschaft, Hamburg, Germany.

Omae, A., Solo-Gabriele, H. and Townsend, T. (2007): A chemical stain for identifying arsenic-treated wood products. *Journal of Wood Chemistry and Technology*, 27, 3/4, 201–217.

Pfrang-Stotz, G. and Reichelt, J. (1999): Charakterisierung und Bewertung von Müllverbrennungsschlacken aus 15 Müllverbrennungsanlagen unterschiedlicher Verfahrenstechnik. *Müll und Abfall*, 31, 5, 262–268.

Pohland, F.G., Karadagli, F., Kim, J.C. and Battaglia, F.P. (1998): Landfill codisposal of pentachlorophenol (PCP)-treated waste wood with municipal solid waste. *Water Science Technology*, 38, 2, 169–175.

Pohlandt, K. (1995): Zusammensetzung, Verwertung und Entsorgung von Holzaschen. *Holz-Zentralblatt, Jahrgang*, 79, 1305–1315.

Scheithauer, M.; Swaboda, C.; Aehlig, C. (2009): *Beitrag zur stofflichen Verwertung von nicht naturbelassenen Altholzsortimenten.* Undated report of the Institut für Holztechnologie Dresden GGMBH, Dresden, Germany.

Schwarz, U. (2003): *Entwicklung eines Verfahrens zur Dekontaminierung und stofflichen Nutzung von Altholz mit öligen Verunreinigungen, dargestellt am Beispiel der Eisenbahnschwelle.* Dissertation an der TU Dresden, Fakultät Maschinenwesen, Dresden, Germany.

Solo-Gabriele, H., Townsend, T., Messick, B. and Calitu, V. (2002): Characteristics of chromated copper arsenate-treated wood ash. *Journal of Hazardous Materials*, 89, 213–232.

Solo-Gabriele, H.M., Townsend, T.G., Hahn, D.W., Moskal, T.M., Hosein, N., Jambeck, J. and Jacobi, G. (2004): Evaluation of XRF and LIBS technologies for on-line sorting of CCA-treated wood waste. *Waste Management*, 24, 413–424.

Song, J., Dubey, B., Jang, Y., Townsend, T. and Solo-Gabriele, H. (2006): Implication of chromium speciation on disposal of discarded CCA-treated wood. *Journal of Hazardous Material*, B128, 280–288.

Stephan, I., Peek, R.-D. and Nimz, H. (1996): Detoxification of salt-impregnated wood by organic acids in a pulping process. *Holzforschung*, 50, 183–187.

Strecker, M. and Marutzky, R. (1994): Zur Dioxinbildung bei der Verbrennung von unbehandelten und behandeltem Holz und Spanplatten. *Holz Roh- Werkstoff*, 52, 33–38.

Thomas, J. M., Lee, M. D., Scott, M. J. and Ward, C. H. (1989): Microbial ecology of the subsurface at an abandoned creosote waste site. *Journal of Industrial Microbiology*, 4, 109–120.

Thurmann, U. (1999): Erfassung und umwelttoxikologische Bewertung von ausgewählten Holzschutzmittelwirkstoffen in Althölzern. Verlag Bremer Umweltinstitut, Bremen, Germany; ISBN 3-9803930-4-6.

Venkatadri, R., Tsai, S.-P., Vucanic, N. and Hein, L.B. (1992): Use of a biofilm membrane reactor for the production of lignin peroxidase and treatment of pentachlorophenol by *Phanerochaete chrysosporium*. *Hazardous Waste and Hazardous Materials*, 9, 3, 231–243.

Voß, A. (1998) : *Aufkommen und Zusammensetzung schutzmittelbehandelter Althölzer und ihre Entsorgung.* Mitteilungen der Bundesforschungsanstalt für Forst- und Holzwirtschaft Hamburg, Bd. 188. Kommissionsverlag Buchhandlung Max Wiedebusch, Hamburg, Germany; ISBN 0368-8798.

Wilson, J.T., McNabb, J.F., Cochran, J.T., Wang, T.H., Tomson, M.B. and Bedient, P.B. (1985): Influence of microbial adaption on the fate of organic pollutants in ground water. *Environmental Toxicology and Chemistry*, 4, 721–726.

# 11.4

# Hazardous Waste

**Joan Maj Nielsen**

*COWI, Lyngby, Denmark*

**Jørn Lauridsen**

*COWI, Svendborg, Denmark*

The USA introduced in 1976 in its 'Resource Conservation and Recovery Act' a general definition of hazardous waste as wastes that 'cause or significantly contribute to an increase in mortality or an increase in serious irreversible, or incapacitating reversible illness, or pose a substantial present or potential hazard to human health or the environment when improperly treated, stored, transported, disposed of, or otherwise managed'. This definition clearly signals the complexity in management of hazardous waste.

In most countries, the distinction between non-hazardous waste and hazardous waste is fundamental to waste management. Chapter 1.1 presents which characteristics make a waste hazardous and thereby demand it be treated as hazardous waste with special attention to all steps from collection, over transport to treatment and disposal.

This chapter briefly outlines some of the important issues in general hazardous waste management. Specific types of hazardous waste from individual large industries are beyond the scope of this chapter.

## 11.4.1   Characterization of Hazardous Waste

Hazardous waste stems from industry, manufacturing, maintenance and services. Everybody is today surrounded by products, which at least in the production phase have resulted in hazardous waste. From the cloth, spectacles and jewelry we wear, to the floor, wall, ceiling, kitchen elements, water tap, sink, light bulbs, lamps, windows, pots and pans in our homes. From the bicycles, cars, busses, trains and airplanes to the roads and rails for our transport.

The types of hazardous waste generated in each country depend strongly on the industry present. However, the common types of hazardous waste are: waste oil (lubricant waste oil), liquid waste containing sulfur and chlorine, high calorific waste, acid and bases, solvents, paint, cellulose paint mining waste, slag (waste from metal processing), shredder, waste, and flue gas cleaning products.

*Solid Waste Technology & Management*   Edited by Thomas Christensen
© 2011 Blackwell Publishing Ltd

**Table 11.4.1** *Hazardous waste generation in Europe by sector for 2007. Reprinted with permission from Final report to the Directorate General Environment, Costs for municipal waste management in the EU by D. Hogg © (2002) European Commission.*

| Sector | Hazardous waste contribution (%) |
| --- | --- |
| Manufacturing excluding recycling | 31 |
| Construction | 20 |
| Other economic activities (services) | 19 |
| Waste management | 16 |
| Electricity, gas and water supply | 9 |
| Mining and quarrying | 4 |
| Agriculture, hunting and forestry | 1 |
| Fishing | < 1 |

Two international conventions are highly relevant for hazardous waste management:

- The Stockholm Convention regulates one of the hazardous waste types, the persistent organic pollutants (POP).
- The Basel Convention regulates the transboundary movements of hazardous waste and their disposal. The Basel Convention has established lists of hazardous waste based on their hazardous characteristics (UN classification) and their origin, which must be used when hazardous waste is transported internationally.

### 11.4.1.1 Sources

The hazardous waste generation in 2007 in Europe and in the USA per industrial sector is presented in Tables 11.4.1 and 11.4.2, respectively. Chemical manufacturing industry is the largest producer of hazardous waste in the USA, producing more than 65 % of the total hazardous waste quantity. The second largest producer of hazardous waste in the USA is the petroleum and coal manufacturing industry producing 11 % of the total quantity. In the European Union (EU) the picture is significantly different, as the hazardous waste generation is divided between four large sectors: manufacturing, construction, service and waste management. The waste management sector is the fourth largest hazardous waste producer in Europe, due to the flue gas cleaning products from waste incineration which are classified as hazardous waste.

### 11.4.1.2 Quantity Per Capita

The average hazardous waste generation per capita in European countries, and the USA, is shown in Table 11.4.3. The quantity of hazardous waste generated per capita varies widely between the counties in Europe. This is mainly caused by the fact that:

- The statistical records are poor in many countries.
- Hazardous waste is treated at the waste producer and not registered in the national statistics.
- Hazardous waste is treated together with municipal waste and hence not registered as hazardous waste.

**Table 11.4.2** *Hazardous waste generation in the USA by sector for 2007. Reprinted with permission from US Environmental Protection Agency. © (2010) USEPA.*

| Sector | Hazardous waste contribution (%) |
| --- | --- |
| Basic chemical manufacturing | 68 |
| Petroleum and coal products manufacturing | 11 |
| Waste treatment and disposal | 4 |
| Iron and steel mills and manufacturing | 3 |
| Semiconductor and electronic components | 2 |
| Remediation of other waste sites | 1 |
| Coating, engraving, heat treatment, etc | 1 |
| Others | 10 |

**Table 11.4.3** *The hazardous waste generation in European countries, the USA in 2006.*

| Country | Quantity (×1000 t) | Quantity (kg/person) |
| --- | --- | --- |
| Austria | 962 | 116 |
| Belgium | 4039 | 384 |
| Bulgaria | 785 | 102 |
| Cyprus | 80 | 104 |
| Czech Republic | 1307 | 128 |
| Germany | 21 705 | 263 |
| Denmark | 372 | 69 |
| Estonia | 6619 | 4922 |
| Spain | 4028 | 92 |
| Finland | 2711 | 516 |
| France | 9622 | 153 |
| Greece | 275 | 25 |
| Hungary | 1300 | 129 |
| Ireland | 709 | 168 |
| Italy | 7465 | 127 |
| Lithuania | 127 | 37 |
| Luxembourg | 234 | 499 |
| Latvia | 65 | 28 |
| Malta | 51 | 125 |
| Netherlands | 4949 | 303 |
| Norway | 757 | 163 |
| Poland | 2381 | 62 |
| Portugal | 6063 | 574 |
| Romania | 1041 | 48 |
| Sweden | 2654 | 293 |
| Slovenia | 116 | 58 |
| Slovakia | 533 | 99 |
| United Kingdom | 8448 | 140 |
| United States | 46 271 | 155 |

The large average quantity of hazardous waste generated per capita in Estonia is the result of oil shale production.

## 11.4.2   Collection and Transport of Hazardous Waste

One of the important issues in relation to hazardous waste management is collection and transport. Empirical studies indicate that most incidents of improper disposal of hazardous waste occur during transport. These incidents are more often related to the hauler than to the hazardous waste generator or the hazardous waste disposal facility.

### 11.4.2.1   Collection

Different collection schemes for hazardous waste may be established and could be in the form:

- Free choice of hauler.
- Free choice of licensed hauler.
- Collection by a licensed hauler contracted by the authority responsible for establishment of the collection scheme.

Wherever there is a free choice of hauler, high demands should be placed on the qualifications and performance of such companies. If collection and transport of hazardous waste is licensed, the threat of loss of license for malpractice is a serious deterrent.

Responsibility for declaring the type of hazardous waste rests with the generator of the hazardous waste. The generator and the disposal facility should check the suitability of the waste for disposal or treatment prior to the shipment.

In many countries reception centers or transfer centers have been established for interim storage of hazardous waste. Their establishment may be recommended where disposal facilities are relative far away from the point of generation. The transfer centers may be equipped with dewatering equipment to minimize the unnecessary transport of large volumes of water, e.g. waste oil. Transfer centers may be used for reception of small quantities of waste, which are repacked prior to shipment to the disposal facility. Furthermore, the transfer centers may be used to minimize the disposal costs for waste producers generating small quantities of hazardous waste, as the transfer center may be able to negotiate lower disposal fees for larger quantities.

### 11.4.2.2 Packaging and Labeling

Prior to the shipment of hazardous waste, the waste should be packed in approved packaging and labeled in accordance with national and international regulation, e.g. ADR: European agreement concerning the international carriage of dangerous goods by road. Worldwide, UN approved packaging is used for the transport of hazardous waste (dangerous goods). The type and size of packaging depends on the type (liquid or solid), hazard and quantity of the hazardous waste.

Most common types of packaging are:

- Steel or plastic drums with removable head or nonremovable head.
- Intermediate bulk containers (IBC).
- Tank containers (tank vehicles) for large quantities of hazardous waste.

Special packaging has been developed for specific types of waste, e.g. fluorescent tubes and lead acid batteries.

The package should be labeled in order for the operator to know the hazards of the hazardous waste carried. The types of hazards specified by the UN classification system (ADR, RID, IMDG, IATA) are listed in Table 11.4.4. A label indicating the hazard class should be affixed to each package and the package should be marked with the UN number of the dangerous goods (hazardous waste) preceded by the letters UN.

**Table 11.4.4**   *The classification of dangerous goods according to the UN classification system.*

| UN hazard class | Type of hazard |
| --- | --- |
| 1 | Explosives substances and articles |
| 2.1 | Flammable gases |
| 2.2 | Nonflammable, nontoxic gases |
| 2.3 | Toxic gases |
| 3 | Flammable liquids |
| 4.1 | Flammable solids, self-reactive substances and solid desensitized explosives |
| 4.2 | Substances liable to spontaneous combustion |
| 4.3 | Substances which, in contact with water, emit flammable gases |
| 5.1 | Oxidizing substances other than organic peroxides |
| 5.2 | Organic peroxides |
| 6.1 | Toxic substances |
| 6.2 | Infectious substances |
| 7 | Radioactive material |
| 8 | Corrosive substances |
| 9 | Miscellaneous dangerous substances and articles |

Some hazardous wastes are not subject to the requirements of the ADR, e.g. used lead acid accumulators, if their cases are undamaged.

The transport unit carrying hazardous waste (dangerous goods) must also be placarded and marked in conformity with the hazardous waste (dangerous goods) carried. The placard includes the UN hazard class and should in general be affixed to both sides and the rear end of the vehicle. Furthermore, the vehicle should be marked with orange plates at both front and rear which include information on the UN number and the hazardous identification number of the dangerous goods (hazardous waste) carried.

### 11.4.2.3  Transport

The transport of hazardous waste can be carried out in different types of packaging and by different types of vehicles depending on the type and quantity of hazardous waste.

On the European continent the road transport of hazardous waste is regulated by the European agreement concerning the international carriage of dangerous good by road (ADR). Similar regulations exist for the transport of dangerous goods (hazardous waste) by rail (RID), by air (AIAT) and by sea (IMDG). The ADR sets up requirements to the packaging, labeling and transport of the dangerous goods (hazardous waste). Furthermore, the ADR sets up requirements to the appointment of a safety adviser in companies, which are involved with packing, loading, filling, unloading or carriage of dangerous goods. The safety adviser is *inter alia* responsible for helping to prevent the risks inherent in such activities with regard to persons, property and the environment.

The specific requirements to transport of dangerous goods by road are listed in Box 11.4.1.

---

**Box 11.4.1  Specific Requirements in Relation to Transport of Dangerous Goods (Hazardous Waste) According to ADR.**

A shipment of dangerous goods must be accompanied with different documents, including:

- Transport document describing:
  - √  the UN number and the proper shipping name of the goods carried
  - √  the class of goods
  - √  the number and description of the packages
  - √  the total quantity of dangerous goods covered
  - √  the name and address of the consignor and consignee.
- Instructions in writing. As a precaution against any accident or emergency that may occur or arise during carriage, the driver shall be given instructions in writing, specifying concisely:
  - √  the nature of danger related to the goods carried
  - √  the personal protection equipment to be used
  - √  general and special actions to be taken by the driver.
- Driver's training certificate, which describes what types of goods the driver is licensed to carry.

The transport unit and the driver should be equipped with the following equipment when carrying out a shipment (ADR):

- At least one portable fire extinguisher of minimum capacity 2 kg dry powder suitable for fighting a fire in the engine.
- At least one portable fire extinguisher of minimum capacity 6 kg dry powder suitable for fighting a tire/brake fire or one involving the load.
- One scotch of a size suited to the weight of the vehicle and the diameter of the wheels.
- Two self-standing warning signs (e.g. reflective cones or triangles or flashing lights).

---

- Suitable warning vest or warning clothing.
- Pocket lamp.
- Respiratory protective device (requirements depends on the type of goods carried).

All enterprises involved in the carriage, or the related packing, loading, filling or unloading of dangerous goods (hazardous waste) must appoint one or more safety advisers for the carriage of dangerous goods, responsible for helping to prevent the risks inherent in such activities with regard to persons, property and the environment.

In the USA the transport of hazardous waste is regulated by the federal hazardous materials transportation law, which is similar to the European regulation (ADR).

## 11.4.3   Treatment and Disposal of Hazardous Waste

The aim of all hazardous waste treatment is to treat the waste in such a way that the hazards of the waste are removed, neutralized and/or reduced.

The worst way to treat hazardous waste is to hide the hazardous characteristics of the waste. Furthermore, it must be clear that whatever treatment is given to hazardous waste, residues in terms of byproducts, emission gasses and wastewater are produced. Incineration of hazardous waste potentially releases POPs, $CO_2$, $NO_X$, dust, acids and heavy metals. Other treatment or disposal methods than incineration produce other residues, e.g. landfilling produces leachate. These issues have to be accounted for in a proper hazardous waste management system.

Treatment of hazardous waste in Europe includes recovery (48 %), landfilling and deep mine disposal (38 %) and incineration (14 %). The high percentage of recovery for hazardous waste in Europe includes the recovery of lead acid batteries, waste oil and the recovery of flue gas treatment products. The incineration includes incineration of hazardous waste in cement kilns and dedicated hazardous waste treatment plants with energy recovery. In the United States more than 40 % of the hazardous waste (liquids) is disposed off by deep well or underground injection, incineration is approximately 6 % and landfilling is 4 %. Table 11.4.5 provides some details about hazardous waste treatment and disposal in the USA.

The most commonly used hazardous waste treatment methods are:

- Landfilling.
- Incineration (thermal treatment).
- Deep well injection.
- Physical chemical treatment (including recovery).

**Table 11.4.5**   *Hazardous waste treatment and disposal in the USA for 2007. Reprinted with permission from US Environmental Protection Agency. © (2010) USEPA*

| Sector | Treatment and disposal (%) |
| --- | --- |
| Deep well or underground injection | 42 |
| Other disposal | 24 |
| Aqueous organic treatment | 6 |
| Incineration | 6 |
| Landfilling | 4 |
| Aqueous inorganic treatment | 4 |
| Recovery | 7 |
| Other treatment | 3 |
| Fuel blending | 1 |
| Stabilization | 1 |
| Sludge treatment | 1 |

### 11.4.3.1   Landfilling

Landfills for hazardous waste are designed in similar ways as landfills for nonhazardous waste. However, often hazardous waste landfills have stricter requirements regarding the number of bottom liners and/or the permeability of bottom liners. Landfills are described in Chapters 10.1 to 10.14.

The types of hazardous waste to be disposed in landfills depend on the national regulation. According to the EU regulation it is prohibited to dispose of the following hazardous waste types in landfills: liquid waste, waste which is explosive, corrosive, oxidizing, flammable or highly flammable and healthcare waste arising from medical or veterinary establishments. The EU also requires that hazardous waste has to be treated prior to landfilling.

### 11.4.3.2   Incineration

Typical hazardous waste incinerators include rotary kiln, cement kiln and fluidized bed.

According to the Basel guidelines on hazardous waste from the production and use of organic solvents, incineration provides a generally accepted disposal route for solvent wastes not being recovered. Other possibilities may sometimes exist, although these alternative options are not equally appropriate or equally satisfactory in all circumstances. Incineration is a very flexible technique in that, by judicious selection of incinerator design and of the various options for flue gas cleaning, together with operating conditions selected for the purpose, a plant can be able to handle many types or combinations of waste. Incinerators may sometimes be constructed as part of a manufacturing process, to deal with the hazardous waste streams from that process. In such cases, the nature and composition of the hazardous wastes will usually be reasonably well defined, and the plant can be designed and constructed on the basis of a narrow scope of operation.

#### *Rotary Kiln*

Rotary kilns were introduced about 40 years ago for incineration of hazardous waste and developed according to the 'all burn' principle. It consists of a rotating drum, supported by two massive rings carried in cradles at an angle of two degrees to the horizontal and rotating slowly (0.2–0.5 rpm).

Usually designed for capacities exceeding 2 t/h, they can accept large batches of hazardous waste, drums and containers, pastes, liquids and gaseous waste, which is fed at the front wall of the kiln via a chute.

Rotary kilns usually operate in the primary chamber at temperatures above 1000 °C in order to achieve an acceptable combustion of the waste. Complete combustion of the gases is achieved in the second combustion chamber at a temperature above 850 °C (up to 1200 °C) and with a residence time of at least 2 s. High energy and maintenance (seals, refractory) costs, fused slag formation, difficult operation and high investment cost make this technology often only interesting for very large plants (>5 t/h) and when the variety of waste is wide (typically for waste management companies). Rotary kilns are the most used technology in EU for hazardous waste incineration.

#### *Cement Kiln*

Cement kilns burning hazardous waste are normally equipped with either an electrostatic precipitator (ESP) or fabric filter (FF) to control emissions of particulate matter (PM). Typical wastes incinerated in cement kilns include paint, ink, spent halogenated and non-halogenated waste, still bottoms from solvent recovery operations, petroleum industry wastes and waste oils. Proper selected cement kilns can in many cases incinerate hazardous waste just as well as dedicated rotary kilns. Since 1991 cement kilns in USA have incinerated around 1 million tons of hazardous waste every year as fuel. The use of waste in European cement plants also saves fossil fuels, equivalent to about 3 million tons of coal per year.

#### *Fluidized Bed Incinerator*

There are two types of fluidized bed incinerator technologies for hazardous waste: bubbling bed and circulating bed. The differences are reflected in the relationship between air flow and bed material and have implications for the type of wastes that can be incinerated, as well as the heat transfer to the energy recovery system.

Unlike mass burn incinerators, fluidized bed incinerators require front-end pre processing, also called fuel preparation. Glass and metals are often removed at the source, because these materials are not suited for the fluidized bed incinerator.

Fluidized bed systems can successfully incinerate wastes of widely varying moisture and heat content. Its high contact surface waste/air, high heat transfer, turbulence and mixing properties provide the best combustion efficiency and lower air emissions. However, in order to take full advantage of a fluidized bed, waste should be fed preferably in small batches. Waste containing high salts content may cause formation of eutectics and risks of defluidization of the bed. This makes the fluidized bed more suitable for 'selected' waste streams and less attractive for a wide range of hazardous waste types. Cost comparisons with mass burn are inconclusive. In general, however, fluidized bed incinerators appear to operate efficiently on smaller scales than do mass burn incinerators, which may make them attractive in some situations. Fluidized bed incineration is very much used in Japan.

### 11.4.3.3  Deep Well Injection

In the USA the most used treatment method for hazardous waste is deep well injection. Deep well injection is a liquid waste disposal technology. This technology uses injection of treated or untreated liquid waste into geologic formations that have no potential to allow migration of contaminants into potential potable water aquifers.

A typical injection well consists of concentric pipes, which extend several thousand feet (hundreds of metres) down from the ground surface into highly saline, permeable injection zones that are confined vertically by impermeable strata. The outermost pipe or surface casing extends below the base of any underground sources of drinking water (USDW) and is cemented back to the surface to prevent contamination of the USDW.

Directly inside the surface casing is a long string casing that extends to and sometimes into the injection zone. This casing is filled in with cement all the way back to the surface in order to seal off the injected waste from the formations above the injection zone back to the surface. The casing provides a seal between the wastes in the injection zone and the upper formations. The waste is injected through the injection tubing inside the long string casing either through perforations in the long string or in the open hole below the bottom of the long string. The space between the string casing and the injection tube, called the annulus, is filled with an inert, pressurized fluid, and is sealed at the bottom by a removable packer preventing injected liquid waste from backing up into the annulus. A typical injection well is illustrated in Figure 11.4.1.

The target contaminant groups for deep well injection are volatile organic carbons (VOCs), semi VOCs, fuels, explosives and pesticides. However, existing permitted deep well injection facilities are limited to a narrow range of specific wastes.

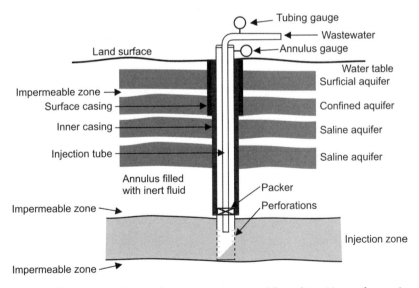

**Figure 11.4.1**  *Deep well injection of hazardous waste. Reprinted from http://www.frtr.gov/matrix2/section4/ DO1-4-54.html © (2010) FRTR.*

### 11.4.3.4   Physicochemical Treatment

A traditional plant for the physicochemical treatment of inorganic waste consists of:

- Detoxification of cyanide waste through oxidation with hypochlorite.
- Reduction of chrome (VI) and precipitation of chrome (III).
- Neutralization of acids and bases.
- Precipitation of heavy metals as low-solubility metal hydroxides.
- Separation of the precipitated substances in a filter press.
- Wastewater treatment and filter cake disposal at a hazardous waste landfill.

A modern physicochemical treatment plant is located in Iserlohn, Germany (ZEA). Here, in seven modules, liquid industrial waste is regenerated, recycled or pretreated in a way that the recovered materials can be recycled. Innovative process technologies are used to maximize the recovery from the waste and to minimize the amount of waste for disposal.

- The chromic acid is first cleaned with an ion exchanger and then concentrated in an evaporation tower. The process results in a cleaned and concentrated chromic acid, which can be used as a product in industrial processes.
- Emulsions such as cooling lubricants, oil/water mixtures and alkaline degreasing agents are recycled. The objective is to recover oil with low water content from these waste products.
- Cyanide containing wastes are recycled with an electrochemical method. Cyanide is detoxified without use of chemicals by an anodic oxidation. The previously dissolved metals are simultaneously collected at the cathodes. This allows the recovery of precious metals such as gold, silver and copper.
- Chemical nickel and alloy baths are destroyed by ozone. This makes the complex bonded metals accessible for a subsequent precipitation method. After the metal precipitation and subsequent dewatering, the metal rich filter cake sludge is suitable for metal recovery.
- Nitric acid and nitric/hydrofluoric acid pickling liquor contaminated with metals from the stainless steel industry are neutralized, dewatered and the metal rich filter cake sludge is suitable for metal recovery.
- Hydrochloric acid pickle liquor containing iron can be recycled diversely. Depending on the quality, these acids can be used as precipitant in the biological waste water treatment plant. The dissolved iron is useful in removal of phosphorous in wastewater treatment.

## References

Blackman, W.C. (2001): *Basic hazardous waste management*, third edition. Lewis Publishers, Boca Raton, USA.
Basel Convention (2002): *Technical guidelines on hazardous waste from the production and use of organic solvents.* Basel Convention series/SBC No. 02/06.ISBN: 92-1-158606-2. ISSN: 1020-8364.
Basel Convention and UNEP (2002): *Destruction and decontamination technologies for PCBs and other POP wastes.* A training manual for hazardous waste project managers, volume A. Located on the internet: 19 July 2009. http://www.basel.int/meetings/sbc/workdoc/TM-A.pdf.
Kloek W. and Blumenthal, K. (2009): *Generation and treatment of waste.* Eurostat 30/2009. Statistics in focus. Located on the internet: 19 July 2009. http://epp.eurostat.ec.europa.eu/cache/ity_offpub/ks-sf-09-030/en/ks-sf-09-030-en.pdf.

# 11.5

# Other Special Waste

### Line Brogaard and Thomas H. Christensen

*Technical University of Denmark, Denmark*

In addition to the main types of special waste related to municipal solid waste (MSW) mentioned in the previous chapters (health care risk waste, WEEE, impregnated wood, hazardous waste) a range of other fractions of waste have in some countries been defined as special waste that must be handled separately from MSW. Some of these other special wastes are briefly described in this chapter with respect to their definition, quantity and composition, and management options. The special wastes mentioned here are batteries, tires, polyvinylchloride (PVC) and food waste.

## 11.5.1   Batteries

Batteries are considered special waste because most batteries contain heavy metals, which are the main cause for environmental concern. Small batteries used in households are often disposed together with the household waste and are then incinerated or landfilled together with MSW. Metals are emitted together with the flue gas and end up in the slag and ashes when batteries are incinerated. In landfills, batteries may be a source of increased heavy metal concentrations in leachate.

### 11.5.1.1   Batteries: Definition, Quantity and Composition

Batteries are categorized as nonrechargeable (primary) or rechargeable (secondary) batteries. Both categories contain a number of different types of batteries for different purposes. Examples of nonrechargeable batteries are manganese, alkaline and lithium batteries. Rechargeable batteries are e.g. nickel–cadmium (NiCd), nickel–metal hydride (NiMH), and lithium ion (Li ion) batteries. The most common type of battery in households is the alkaline battery. The typical chemical composition of some of the most common batteries is presented in Table 11.5.1.

The quantities of waste batteries and accumulators in Europe are around $1.6 \times 10^6$ t/year. The unit generation rate in 2005 and 2006 in Europe was 2–3 kg/person/year (Eurostat, 2009).

The European Union (EU) demanded in 1991 (EEC, 1991) that batteries containing more than 25 mg of mercury (except alkaline–manganese batteries), 0.025 % of cadmium by weight and 0.4 % lead by weight to be collected separately from household waste for recycling or special disposal. The directive also defined permissible limits for the content of these

**Table 11.5.1** *Chemical composition of the most commonly used portable batteries. Depending on the size of the battery the composition will vary for most of the batteries. Data collected from EPBA (2007).*

| Type of battery | Composition |
| --- | --- |
| Nonrechargeable batteries | |
| Alkaline manganese | $MnO_2$: 37%, Fe: 23%, Zn: 16%, $H_2O$: 9%, KOH: 5%, C: 4%, brass: 2%, others: 4% |
| Zinc–carbon | $MnO_2$: 27%, Zn: 23%, $H_2O$: 18%, C: 10%, $ZnCl/NH_4Cl$: 5%, Fe: 4%, others: 13% |
| Lithium–manganese dioxide (major batteries) | Fe: 50%, $MnO_2$: 30%, plastic: 7%, dimethoxyethane: 6%, Li: 3%, C: 2%, Ni: 2% |
| Lithium–manganese dioxide (button cells) | Fe: 50%, $MnO_2$: 28%, Cr: 10%, plastic: 3%, Li: 3%, dimethoxyethane: 2%, C: 2%, Ni: 2% |
| Silver oxide (button cells) | Fe: 42%, $Ag_2O$: 33%, Zn: 9%, Cu: 4%, $MnO_2$: 3%, $H_2O$: 2%, plastic: 2%, Ni: 2%, KOH: 1%, C: 0.5%, Hg: 0.4%, others: 1.1% |
| Alkaline manganese dioxide (button cells) | Fe: 37%, $MnO_2$: 36%, Zn: 11%, $H_2O$: 6%, plastic: 3%, KOH: 2%, C: 2%, Ni: 1%, Hg: 0.6%, others: 1.4% |
| Zinc air (button cells) | Fe: 42%, Zn: 35%, $H_2O$: 10%, plastic: 4%, KOH: 4%, C: 1%, Hg: 1%, others: 3% |
| Rechargeable batteries | |
| Nickel cadmium | Fe: 40%, Ni: 22%, Cd: 15%, plastic: 5%, KOH: 2%, others: 16% |
| Nickel metal hydride | Ni: 33%, Fe: 30%, lanthanides: 10%, $H_2O$: 8%, Co: 3%, plastic: 5%, KOH: 2%, Mn: 1%, Zn: 1%, others: 7% |
| Lead–acid | Lead (incl. lead oxides): 72%, electrolyte ($H_2SO_4$): 17%, plastics: 9%, others: 2% |
| Lithium ion | Aluminum: 15–25%, carbon, amorphous, powder: 0.1–1.0%, copper foil: 5–15%, diethylcarbonate: 1–10%, ethylenecarbonate: 1–10%, methylethylcarbonate: 1–10%, lithiumhexafluorophosphate: 1–5%, graphite, powder: 10–30%, lithium cobalt oxide: 25–45%, polyvinylidene fluoride: 0.5–2.0%, and steel, nickel and inert polymer |

heavy metals in batteries. In 1998, the limits were reduced further prohibiting the marketing of batteries containing more than 0.0005% of mercury and button cells containing more than 2% of mercury by weight from 1 January 2000 (EEC, 1998).

Battery producers are today forced through the producer responsibility to pay for the treatment of batteries. This could help the development of more environmental friendly products in order to decrease the treatment costs.

### 11.5.1.2 Batteries: Management Options

Today only a small percentage of the waste batteries collected are recycled. Most waste batteries are disposed off in landfill sites. In the EU (2006/66/EC), collection targets are that 45% of all portable batteries should be collected by 2016. Car batteries are not included in the collection target since legislation already demands the collection of this kind of batteries and the collection efficiency is already almost 90% in EU. The recycling of car batteries is described in Box 11.5.1.

---

**Box 11.5.1 Recycling of Car Batteries. Reproduced with permission from Environmental Management.Salomone, R., Mondello, F., Lanuzza, F. and Micali, G. (2005): Environmental assessment, an eco-balance of a recycling plant for spent lead–acid batteries 35, 206–219 © (2005) NSRC.**

The amount of car battery waste in the European Union is estimated to be 800 000 t/year (European Commission, 2009).This corresponds to approximately 2 kg/person/year. The main contents of car batteries are lead, acid and plastic (Salomone *et al.*, 2005); a more detailed composition is presented in the table below. Not much waste is produced when car batteries are recycled, since almost 75% is recycled (Salomone *et al.*, 2005).

The first treatment step is to collect and clean the sulphuric acid from the accumulators. The acid is neutralized by addition of slaked lime [$Ca(OH)_2$]. Water and calcium sulfate is produced from the neutralization. The empty

---

accumulators are shredded and sorted out in plastic and lead containing material. The plastic is sent for plastic recycling. The battery paste and the grids are mixed with coke, iron scraps and sodium carbonate in the smelting process. The remains of the sulphuric acid are bound to the added iron and ends up in the slag. The outputs are unrefined lead and a smelting residue. The latter is not reused but landfilled. Lead is collected and refined for reuse in new accumulators (Salomone *et al.*, 2005). In a LCA study performed by Salomone *et al.* (2005) it was assessed that the smelting process contributes the most to the environmental impacts from the treatment and recycling of acid–lead batteries.

Typical composition of new and spent acid–lead 14-kg car battery (Salomone *et al.*, 2005).

| Component of battery | New battery (%) | Spent battery (%) |
|---|---|---|
| Lead metal (grid and connections) | 28–30 | 28–30 |
| Active paste (PbO) | 35–36 | — |
| Battery paste (PbO$_2$, PbSO$_4$, PbO·PbSO$_4$) | — | 48–50 |
| Plastic (PP, PE, PVC) | 7–8 | 7–8 |
| Acid solution (H$_2$O + H$_2$SO$_4$) | 26–27 | 12–13 |
| Other material (paper, ebonite) | 0.3–1.0 | 0.3–1.0 |

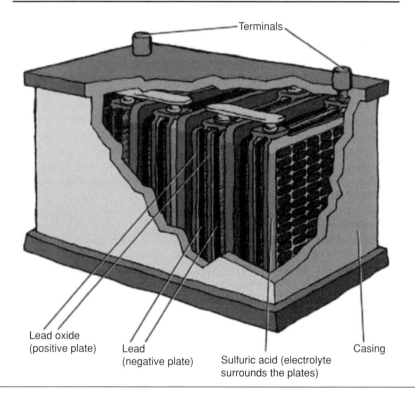

Terminals

Lead oxide (positive plate)　　Lead (negative plate)　　Sulfuric acid (electrolyte surrounds the plates)　　Casing

The metals in the batteries pose the pollution potential, but are also the valuable component. Several treatment methods are available for recovering metals from waste batteries. Some of treatment methods are briefly described here:

- Cadmium is distilled from the batteries in the treatment of NiCd batteries by heating. Small batteries are first treated by pyrolysis to remove plastics and organic material before distillation of cadmium. After the distillation the residue is a mixture of steel and nickel which can be used in the metal industry for alloys (Vestervang, 2005).

- Distillation is used to recover mercury in the treatment of button cell batteries. This is performed in special mercury distillation facilities (Vestervang, 2005).
- NiMH batteries are shredded under vacuum. During shredding the emitted hydrogen is collected from the vacuum system. The shredded material is sorted and a nickel fraction is further used in the steel industry (Vestervang, 2005).
- Alkaline batteries can be shredded to recover manganese and zinc. Another treatment option is the use in low grade steel as a feedstock with collection of zinc fumes for special recovery (batteryrecycling, 2009).
- Lithium batteries can be treated by pyrometallurgic processes (Fisher *et al.*, 2006). Lithium is recovered as lithium carbonate and also ferrous and non-ferrous metals are recovered (batteryrecycling, 2009).
- Lithium ion batteries can be treated by either pyrometallurgic processes or hydrometallurgic processes (Fisher *et al.*, 2006).

### 11.5.1.3   Batteries: Environmental Issues

Fisher *et al.* (2006) evaluated three different scenarios of collection and recycling of waste batteries against a baseline scenario where 89 % of the batteries were landfilled and 11 % of batteries were incinerated. The assessment included the collection, sorting, recycling and residual waste management of the waste batteries. The recycling scenarios provided the highest impacts on abiotic depletion, global warming and acidification due to the use of fuel and electricity. Nevertheless it was concluded from the final assessment that the metal recovery gave major savings of environmental impacts since the production of virgin metals was avoided. Emissions of heavy metals to air, soil and water were reduced compared to the baseline scenario through the metal recovery. With respect to global warming the assessment estimated $CO_2$ savings around 220 $CO_2$-equivalents/t of battery waste recycled.

The environmental assessment found that recycling is a better option than incineration and landfilling of batteries though it was added to the conclusion that the economical costs of recycling were very high compared to landfilling and incineration.

## 11.5.2   Tires

Waste tires are often considered a special waste fraction because it is so difficult to handle with traditional waste management technologies and because uncontrolled disposal of waste tires can be significant littering.

### 11.5.2.1   Tires: Definition, Quantity and Composition

Tires are used for all kinds of transport for various types of vehicles; passenger cars, busses, trucks, aircrafts and bikes. The major fraction of waste tires is tires from passenger cars and trucks.

The amount of collected waste tire in the EU is estimated to around 7 kg/person/year (ETRA, 2004). The total quantity of tire waste generated in the EU is around 2 660 000 t, corresponding to 250 million tires (WBCSD, 2008). In the USA more than 292 million scrap tires were disposed in 2003 (WBCSD, 2008), corresponding to approximately 10 kg/person.

Tires for different vehicles differ in composition due to the different use. The composition of waste tires from passenger cars and trucks are presented in Table 11.5.2. For example, tires for trucks contain more metal and no textile compared to the tires used for passenger cars.

### 11.5.2.2   Tires: Management Options

Waste tires are nondegradable and massive amounts have been landfilled for many years. In the EU, landfilling of waste tires has been banned since 2006 (ETRMA, 2008). Today waste tires are recycled or utilized by:

- Recycling into new tires.
- Utilization in different kinds of civil engineering projects. This could be in embankments, backfill for walls, road insulation, field drains, erosion control/rainwater runoff barriers and crash barriers (wcbsd, 2008).

**Table 11.5.2**  *Average composition of tires in the European Union used for passenger cars and for trucks/busses (ETRA, 2004). Reprinted with permission from http://www.etra-eu.org, ETRA 2004: Introduction to Tyre recycling: 2004 Annual Accumulation by EU State by Valerie L. Shulman © (2004) ETRA.*

| Material | Passenger cars (%) | Trucks/busses (%) |
|---|---|---|
| Rubber/elastomers | 48 | 43 |
| Carbon black | 22 | 21 |
| Metal | 15 | 27 |
| Textile | 5 | — |
| Zinc oxide | 1 | 2 |
| Sulfur | 1 | 1 |
| Additives | 8 | 6 |

- Remelting and use of the rubber for mixing with asphalt in order to reduce traffic noise (wcbsd 2008) or as crumb for carpet underlay and in new sports surfaces like soccer fields (GHK, 2006). Box 11.5. 2 describes such uses.
- Feeding as a fuel into cement kilns. Waste tires have a high energy content of around 32 GJ/t, which is similar to the value for coal of around 27 GJ/t (WBCSD, 2008).

---

**Box 11.5.2  Utilization of Tires in Sport Turfs and Flooring.**

Waste tires can be recycled and be a component in artificial turfs. It can be used both indoor as well as outdoor and is easy to maintain compared to conventional turfs. The artificial turf is used where grass is difficult to grow due to climate or intensive use and it is independent of the weather. Artificial turfs contain several types of plastic fibers such as polyethylene, polypropylene and polyamide (Nilsson *et al.*, 2008). These plastic fibers are coated and attached to a fabric of polypropylene or polyester (Nilsson *et al.*, 2008). Rubber from waste tires can be used as tire crumbs in an elastic sublayer in artificial turfs. Tire crumbs are produced by shredding and then grinding of waste rubber from tires. The tire crumbs has the size of coarse sand grains (Nilsson *et al.*, 2008).

Reprinted from http://www.waterlessgrass.com/waterless-grass-info.php © (2010) Rejuvalawn.com

Waste tires can be recycled as rubber in surfaces used for sports, e.g. athletics. Sports surfaces are produced from rubber crumbs mixed with urethane binders. This kind of sports surfaces are used many places even though the surface will be slightly harder than sports surfaces made from virgin rubber. (Western Rubber Group, 2009).

Reprinted from Http://sportsbuilders.org/court_of_the_year_05/tk_09_yellowmedicine.jpg © (2010) Greg Buchal.

Prior to recycling and utilization, the tires are sorted by size and composition and then shredded. The first shredding step is cutting of the waste tires into chips $50 \times 50$ mm (Kahl, 2009). The mixed fraction of mixed rubber, metal and textile are then granulated and ground into different degrees of fineness depending on the further recycling (SIMS, 2009). Then metal, textile and rubber fractions are sorted by gravity separators and different kinds of sieves (Kahl, 2009).

### 11.5.2.3   Tires: Environmental Issues

Illegal dumping or littering of waste tires may cause fires and enhance mosquito problems. Self ignition of waste tires is not common in waste tire dumps but accidental ignition has caused several tire fires. This type of fire is very difficult to put out and putting out a tire fire with water can cause an oily runoff that can contaminate nearby surface waters. Depending on region and climate, illegally dumped waste tires may provide breeding ground for mosquitoes by providing stagnant warm water because of the shape and heat-absorbing material properties of waste tires. Mosquitoes can be carriers of diseases (Environment Australia, 2001).

Several studies have been addressing the environmental impacts of tire waste treatment. Villanueva *et al.* (2008) compared material recycling of tires used in asphalt production as rubber granulates and co-incineration in cement kilns. Recycling was the best solution especially with respect to global warming potential, energy demand and acidification potential.

### 11.5.3   Polyvinylchloride

Polyvinylchloride (PVC) is considered as a special waste fraction due to the heavy metals used as stabilizers and the formation of hydrochloric acids when PVC is incinerated.

### 11.5.3.1   PVC: Definition, Quantity and Composition

Polyvinylchloride (PVC) is a strong plastic material often with a long lifetime. It is used for water pipes, cables and wire insulation, computers, medical devices, packaging, building materials and many other products. PVC is produced mainly from salt and fossil fuel such as natural gas or oil. Stabilizers, softeners and fire retardants are added to PVC in the production to optimize the technical properties of the product. Compounds containing barium, zinc, lead and cadmium are used as stabilizers (Rahbek and Rasmussen, 2006). Other problematic compounds in PVC are phthalates used as softeners. These compounds are considered to be hormone disrupting as well as carcinogenic (Ministry of Environment and Energy, 1999). The additives are problematic in different ways during disposal of PVC both for humans and the environment.

PVC is used for many purposes and PVC waste is generated both in households and in industries. Available data about the quantities of generated PVC waste are very uncertain because of the many purposes. The best estimates are obtained by combining available production data from the PVC producing industry with expected lifetimes of PVC products. In the EU, PVC waste collected is estimated at $4.7 \times 10^6$ t in 2010 and $6.2 \times 10^6$ t in 2020 (COM, 2000). This corresponds to 10–13 kg/capita. Two-thirds of the total amount of PVC is estimated to be hard PVC and one-third soft PVC (COM, 2000).

### 11.5.3.2   PVC: Management Options

In the EU only 3 % of PVC waste was recycled in year 2000 where 82 % was landfilled and 15 % incinerated (Baitz *et al.*, 2004). Recycling of PVC is on the rise as new technologies develop, but the amounts recycled at present are still small compared to the potential amounts. The recycling of PVC in cables has been going on for some time since the value and thereby the driving force for recycling is in the recovered copper and aluminum (Baitz *et al.*, 2004).

PVC can be treated mechanically by shredding, sieving and grinding. The powder produced is reused in new PVC products (COM, 2000). PVC can also be melted and recovered as new PVC since it is a thermoplastic which melts when heated and hardens when cooled. However, consumers may be skeptical to the quality of recycled PVC. Recycling of PVC for new pipes could be a good option since here the quality is less important compared to the use in new cars or electronics. Studies have shown that PVC can be recovered seven times for pipes without any loss of quality. There are two different ways to use recovered PVC in pipes, one is in between two layers of new PVC another is to mix the recovered PVC with the new PVC and produce pipes that are totally homogeneous (PVC informationsrådet, 2009).

### 11.5.3.3   PVC: Environmental Issues

The chloride content in PVC reacts with water vapor if PVC is incinerated. This causes the formation of hydrochloric acid. HCl causes acidification of the environment and may increase corrosion of stacks and pipes in incineration plants. Another issue is the heavy metals used as stabilizers in PVC. If PVC is incinerated the heavy metals end up in the slags.

Landfilling of PVC causes metals and softeners like phthalates in the leachate (Spillmann, 2000). Several studies have been made on degradation of PVC, but none have observed any significant degradation of PVC (Spillman, 2000).

Baitz *et al.* (2004) collected around 100 LCA studies on the production, use and disposal of PVC. The study concluded from the literature that recycling of PVC has less environmental impact than landfilling of PVC waste. This was due to the savings of fossil fuels for the material production.

## 11.5.4   Food Waste

Food waste from industries, institutions, restaurants and commerce may be considered a special waste because of the resource that can be recovered if used as feed stuff for animals. Food waste from households is not considered in this context.

---

**Box 11.5.3   Creutzfeldt–Jakob Disease.**

Creutzfeldt–Jakob disease (CJD) in humans has been related to bovine spongiform encephalopathy (BSE) through consumption of meat from BSE-infected animals (Speedy, 2001). The disease is caused by prions for example in feed. It is difficult to control the presence of prions since they are resistant to any kind of heat treatment. BSE was recognized the first time in the United Kingdom in 1986. The disease has a long incubation period and is ultimately fatal.

---

### 11.5.4.1   Food Waste: Definition, Quantity and Composition

Food waste is defined as waste from all kinds of food preparation and uneaten food. It is generated in the food producing industry and all kinds of kitchens both industrial and commercial. It contains leftovers from e.g. plates, inedible, unopened and opened food. It contains fruits, vegetables, meat and dairy products.

The amount of nonhousehold food waste is hard to estimate. It is believed to constitute a few kilograms per person per year.

### 11.5.4.2   Food Waste: Management Options

Food waste from the industry is used as feed for animals in Japan and the USA. The use of food waste for animal feed stuff, except for mink, was banned in the EU in 2003 due to concerns about the risk of transfer of Creutzfeldt–Jakob disease (see Box 11.5.3).

In Japan food waste from the industry is crushed before it is mixed and heated together with used cooking oil. The mixture is dried and the final product is crushed again and then separated in a wind sifter. The final product is used as formula feed in poultry and pork producing industries (TMG, 2006). Box 11.5.4 describes the Japanese Ecofeed technology.

---

**Box 11.5.4   Japanese Processing of Food Waste into Animal Feed (Sugiura, 2009).**

In Japan food waste can be used for animal feed. The Japanese company Ecofeed has specialized in processing food waste into animal feed. The input of waste is highly variable and the company uses only food waste from food processing industries to be able to control the composition of the food waste. The composition and nutrition content in the produced feed is highly dependent on the collected waste and therefore it is not possible for the producers to put a specific declaration on the feed. It is a challenge for food waste processing to avoid contamination from rodents as well as microbiological contamination. Thus the production is indoor and the feed is cooked both at the production and sometimes also after arrival at the farms.

The food waste is received at the treatment plant and foreign objects like cutlery, chopsticks and toothpicks are sorted out. Ecofeed has divided the production of animal feed for ruminants and nonruminants to avoid ruminants being fed with mammal proteins. There are three main processes used by Ecofeed: Silaging, liquefying of food waste and drying. Silaging is used as fermentation with lactic acid bacteria of mainly vegetable scraps. Mixing the food waste with water or milk produces a liquid that can be distributed via pipelines and be used by farmers. Several types of treatment are used to dry food waste:

- Low-pressure frying.
- Boiling for reducing water content.
- High-temperature fermentation.
- High-temperature drying.

When the animal feed arrives at the farm, it must be used right away to avoid rodents and contamination. If this is not possible, farmers need to cook the feed at 70 °C for 30 min or 80 °C for 3 min before it can be used. The cooking is especially needed if the animal feed contains uncooked meat.

Food waste is a cheap feedstock for animal feed, though the processing can be quite expensive due to the energy consumption.

### 11.5.4.3  Food Waste: Environmental Issues

No studies have been identified that quantify the benefit of reusing food waste for animal feeding. LCA approaches may not suffice in the context because also hygienic aspects must be carefully assessed in order to minimize the risk of transfer of diseases (Lundie and Peters, 2005).

# References

Amandus Kahl GmbH and Co. KG (2009): *KAHL recycling plants for waste tyres*. Located on the internet: 19 July 2009. http://ceeindustrial.com/public/data/companyCatalogue1214246176.pdf.

Baitz, M., Kreißig, J., Byrne, E., Makishi, C., Kupfer, T., Frees, N., Bey, N., Söes Hansen, M., Hansen, A., Bosch, T., Borghi, V., Watson, J. and Miranda, M. (2004): *Life cycle assessment of PVC and of principal competing materials*. Report for the European Commission, Strasbourg, France.

Batteries Directive (2006): *Directive 2006/66/EC of the European parliament and of the council, of 6 September 2006, on batteries and accumulators and waste batteries and accumulators and repealing Directive 91/157/EEC*. European Commission, Strasbourg, France.

batteryrecycling (2009): *Battery solutions, recycling information*. Located on the internet: 19 October 2009. http://www.batteryrecycling.com/.

COM (2000): *Green Paper, Environmental Issues of PVC*. Report COM (2000) 469 Final. Commission of the European Communities, Brussels, Belgium.

EEC (1991): *Council directive of 18 March 1991 on batteries and accumulators containing certain dangerous substances*. 91/157/EEC. European Council, Strasbourg, France.

EEC (1998): *Commission directive of 22 December 1998, adapting to technical progress Council Directive 91/157/EEC on batteries and accumulators containing certain dangerous substances*. 98/101/EC. European Commission, Strasbourg, France.

Environment Australia (2001): *A national approach to waste tyres*. Report prepared by Atech Group. Commonwealth Department of Environment, Canberra, Australia; ISBN 0 642 54749 1.

EPBA (2007): *Product information on primary and rechargeable batteries*. The European Portable Battery Association, London, UK.

ETRA (2004): *Introduction to tyre recycling: 2004*. Annual Accumulation by EU States. Located on the internet: 19 October 2009. http://www.etra-eu.org/.

ETRMA (2008): *Annual activity report 2007–2008*. European Tyre and Rubber Manufacturers Association, London, UK.

Eurostat (2009): *Waste statistics for the European Union*. Uerostat. Located on the internet: 19 September 2009. http://epp.eurostat.ec.europa.eu/portal/page/portal/waste/data/waste_statistics.

Fisher, K., Collins, M., Laenen, P., Wallén, E., Garrett, P. and Aumônier, S. (2006): *Battery waste management life cycle assessment*. Paper from the third international conference on life cycle management. University of Zurich at Irchel, Switzerland.

GHK (2006): *A study to examine the benefits of the End of Life Vehicles Directive and the costs and benefits of a revision of the 2015 targets for recycling, re-use and recovery under the ELV Directive*. GHK in association with Bio Intelligence Service. Located on the internet: 19 September 2009. http://ec.europa.eu/environment/waste/pdf/study/final_report.pdf.

Lundie, S. and Peters, G.M. (2005): Life cycle assessment of food waste management options. *Journal of Cleaner Production*, 13, 275–286.

Ministry of Environment and Energy (1999): *Strategi for PVC-området, Statusredegørelse og fremtidige initiativer*. Copenhagen, Denmark.

Nilsson, N.H., Malmgren-Hansen, B. and Thomsen, U.S. (2008): *Mapping, emissions and environmental and health assessment of chemical substances in artificial turf*. Survey of chemical substances in consumer products, No. 100. The Danish Technological Institute, Copenhagen, Denmark.

PVC informationsrådet (2009): The homepage of the PVC information council. Located on the internet: 19 September 2009. http://www.pvc.dk/t2w_171.asp.

Rahbek, U. and Rasmussen, E. (2006): *Nyttiggørelse af kommunalt indsamlet PVC-affald*. RGS90 Watech, Miljøprojekt Nr. 1137. Miljøstyrelsen, Copenhagen, Denmark.

Salomone, R., Mondello, F., Lanuzza, F. and Micali, G. (2005): Environmental assessment, an eco-balance of a recycling plant for spent lead–acid batteries, *Environmental Management*, 35, 206–219.

SIMS Tyrecycle (2009): *How tyres are recycled*. Located on the internet: 19 October, 2009. http://www.simstyrecycle. com.au/home.

Speedy, A.W. (2001): *FAO and pre-harvest food safety in the livestock and animal feed industry*. WHO consultation on pre-harvest food safety. WHO, Berlin, Germany.

Spillman (2000): *The behaviour of PVC in landfill*. European Commission DGXI.E.3. ARGUS, in association with University Rostock, Carl Bro a/s and Sigma Plan S.A. European Commission, Strasbourg, France.

Sugiura, K., Yamatani, S., Watahara M. and Onodera, T. (2009): Ecofeed, animal feed produced from recycled food. *Veterinaria Italiana*, 45, 397–404.

TMG (2006): *Tokyo super eco town project outline, presentation of facilities*. Alfo Co, Ltd. Tokyo Metropolitan Government, Tokyo, Japan.

Vestervang, S. (2005): *Behandlingsteknologier for batterier, Fase 1. Kortlægning af eksisterende behandlingsmetoder*. Miljøprojekt Nr. 1009. Kommunekemi a/s, Copenhagen, Denmark.

Villanueva, A., Hedal, N., Carlsen, R., Vogt, R., Giegrich, J. (2008): *Comparative life cycle assessment of two options for waste tyre treatment: recycling in asphalt and incineration in cement kilns*. Executive summary, by Danish Topic Centre of Waste and Institut für Energie- und Umweltforschung Heidelberg GmbH. Genan Business and Development A/S, Copenhagen, Denmark.

WBCSD (2008): *Managing end-of-life tires*. World Business Council for Sustainable Development, London, UK; ISBN: 978-3-940388-31-5.

Western Rubber Group (2009): *Background of tire recycling*. Located on the internet: 19 October 2009. http://www.western-rubber.com/background_recycling.php.

# Index